Man Xie, Feng Wu, Yongxin Huang
**Sodium-Ion Batteries**

# Also of interest

*Polymer-based Solid State Batteries*
Daniel Brandell, Jonas Mindemark, Guiomar Hernández, 2021
ISBN 978-1-5015-2113-3, e-ISBN (PDF) 978-1-5015-2114-0

*Applied Electrochemistry*
Krystyna Jackowska, Paweł Krysiński, 2020
ISBN 978-3-11-060077-3, e-ISBN (PDF) 978-3-11-060083-4

*Electrochemical Energy Storage.*
*Physics and Chemistry of Batteries*
Reinhart Job, 2020
ISBN 978-3-11-048437-3, e-ISBN (PDF) 978-3-11-048442-7

*Wearable Energy Storage Devices*
Allibai Mohanan Vinu Mohan, 2021
ISBN 978-1-5015-2127-0, e-ISBN (PDF) 978-1-5015-2128-7

*Chemical Energy Storage*
Edited by Robert Schlögl, 2022
ISBN 978-3-11-060843-4, e-ISBN (PDF) 978-3-11-060845-8

Man Xie, Feng Wu, Yongxin Huang

# Sodium-Ion Batteries

Advanced Technology and Applications

**DE GRUYTER**

電子工業出版社·
Publishing House of Electronics Industry
http://www.phei.com.cn

**Authors**
Dr. Man Xie
School of Material Science and Engineering
Beijing Institute of Technology
South Zhongguancun Street
Haidian District
100081 Beijing
China
xmxm@bit.edu.cn

Feng Wu
School of Material Science and Engineering
Beijing Institute of Technology
South Zhongguancun Street
Haidian District
100081 Beijing
China

Yonxing Huang
School of Material Science and Engineering
Beijing Institute of Technology
South Zhongguancun Street
Haidian District
100081 Beijing
China

Translated by Man Xie, Jiahui Zhou, Ying Jiang, Yixin Zhang

ISBN 978-3-11-074903-8
e-ISBN (PDF) 978-3-11-074906-9
e-ISBN (EPUB) 978-3-11-074918-2

**Library of Congress Control Number: 2022937146**

**Bibliographic information published by the Deutsche Nationalbibliothek**
The Deutsche Nationalbibliothek lists this publication in the Deutsche Nationalbibliografie;
detailed bibliographic data are available on the Internet at http://dnb.dnb.de.

# Contents

## Chapter 5
## Sodium-ion battery electrolyte — 218

**Chapter 6**
**The commercialization of sodium-ion batteries —— 306**

# Introduction

Sodium-ion batteries (SIBs) were discovered in almost the same era as lithium-ion batteries, and they even preceded lithium-ion batteries in some pioneering works. However, because of the limited theoretical capacity and sluggish reaction kinetics, SIBs have not made any breakthrough in theoretical research and practical applications for a long time. It was not until the late 1990s and early 2000s, with the development of layered transition metal oxide cathodes and hard carbon anodes, that SIBs came back into the limelight. For example, the Institute of Physics of Chinese Academy of Sciences, Wuhan University, Fudan University, University of Science and Technology of China, Beijing Institute of Technology, Wollongong University, Nanyang Technological University, Argonne National Laboratory, and foreign research institutions are devoted to the basic research of SIBs. In addition, domestic production enterprises represented by HINA BATTERY, NATRIUM, and STAR SODIUM have also made important contributions to the practical advancement of SIBs. In 2021, numerous investors made significant investments in SIBs, such as Hui Capital, Phoenix Tree Capital Group, Contemporary Amperex Technology Co., Limited, and Guangzhou Great Power Energy and Technology Co., Ltd are also building their commercial lines for SIBs.

Compared with the traditional energy storage system, SIB presents the advantages of low cost, easy availability for raw materials, green environment, safety and reliability, and combine the high energy density, power density, and long cycling life. For large-scale energy storage devices without site constraints and environmental restrictions, SIBs are ideal. In addition, SIBs enable broad application prospects in low-speed electric vehicles, home energy storage, electric boats, and other fields.

In recent years, breakthroughs have been made in research work on SIBs, covering cathode/anode materials, electrolytes, and full-cell assemble. Representative cathode materials include layered transition metal oxides, polyanionic compounds, and Prussian blue and its analogs; anode materials include carbon-based materials, titanium-based compounds, conversion compounds, and intermetallic compounds; electrolytes include organic electrolytes, ionic liquid electrolytes, high-concentration electrolytes, and solid-state electrolytes. The assembly and matching of full batteries

https://doi.org/10.1515/9783110749069-203

involve the study of cathode/anode compatibility, and the development of finished full batteries, flexible batteries, and new devices. Meanwhile, testing and characterization technology have also greatly shortened the development time of SIBs. A series of advanced techniques such as spherical differential electron microscopy, synchrotron radiation, neutron diffraction, and solid-state nuclear magnetism have been used to study the sodium storage mechanism, interfacial evolution, and mechanical properties of electrode materials in depth. This book was written during the booming period of basic research and industrial applications for SIBs. In the next step, SIBs need to be studied from an application perspective, focusing on the interfacial compatibility of electrode materials and electrolytes, further solving the problem of sluggish reaction kinetics in SIBs and improving the cycling stability of SIBs.

During the preparation of this book, PhD students including Ziheng Wang, Jiahui Zhou, and Ying Jiang made important contributions to the writing of this book by combining their research work. Ditong Chu, Guangling Wei, Cheng Li, Yixin Zhang, Zehua Li, and Yutong Hao are the master students who have done a lot of literature collection, data compilation, and graph compilation in the process of writing this book. In addition, we would like to express our gratitude to Yutong Hao, Yaozong Zhou, Anni Liu, Yan Chen, and Zekai Lv, who contributed to the translation of this book.

Meanwhile, we would like to especially thank the relevant editors of Electronic Industry Press for their help and support during the publication of this book.

At present, SIB is in the stage of rapid development, and its research scope involves materials, physics, chemistry, and other multidisciplinary. Due to the limited knowledge and ability of the authors, there are inevitably omissions and deficiencies in this book, and we would appreciate the criticism of colleagues and readers.

Editors
October 2021
At Beijing Institute of Technology

# Chapter 1
# Development history and present situation of sodium-ion batteries

## 1.1 The rise and development of new energy

In the 1760s, the First Industrial Revolution brought mankind into "the age of steam," and coal became the main source of energy. The Second Industrial Revolution, which began in the late 1860s, brought mankind into "the age of electricity." Oil and natural gas promoted the rapid development of science and technology and civilization. The Third Industrial Revolution began in the late 1940s, marked by the breakthrough and development of the electronic information industry. Mankind has officially entered "the age of information," and electric power has become indispensable in human society. Primary energy such as oil and natural gas also gradually occupies a monopoly position in global energy consumption. Each industrialization process not only updates people's mode of production but also changes the energy consumption structure of the whole society.

Because of the limited reserves, traditional resources such as coal, oil, and natural gas have been difficult to fully meet the needs of the rapid development of human beings. According to statistics, global oil demand increased steadily in 2018, with an increase of 1.4 million barrels per day over 2017, an increase of 1.4% in coal consumption, and 4.3% in production. Global consumption and production of natural gas have increased by more than 5% [1]. On this basis, the growth rate of global primary energy consumption increased to 2.9%, the highest growth rate since 2010. As shown in Figure 1.1, China occupies an important position in the current primary energy consumption. At present, China is not only the largest consumer of primary energy in the world but also the largest importer of oil and natural gas in the world. On the one hand, it shows the rapid development of China's modernization construction, but on the other hand, it also shows that China has a serious external dependence on energy. Therefore, the energy problem is not only related to the sustainable development of economy and society but also related to the energy security and independence of the country.

The rapid growth of energy consumption demand is mainly due to the rapid rise in electricity demand, but the fuel structure of the power generation system has not been significantly optimized, still dominated by fossil fuels. The share of fossil fuels in 2018 has little changed compared with 20 years ago, which has also caused a series of environmental problems (such as acid rain, haze, and greenhouse effect). Therefore, how to reduce the negative environmental effects of the power generation system while meeting the rapid growth of electricity demand will be one of the main challenges facing the international community in the next 20 years. In the face of this

https://doi.org/10.1515/9783110749069-001

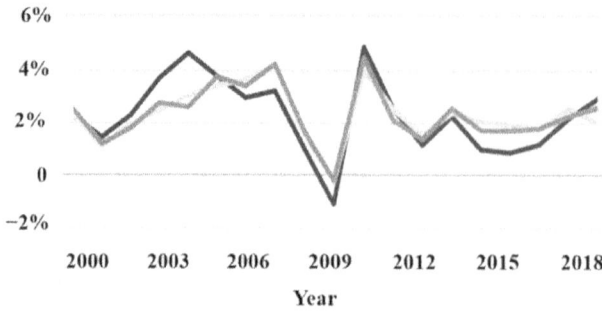

Primary energy consumption amplification

Primary energy consumption amplification (Without consideration of weather)

Primary energy consumption forecast amplification (With consideration of weather)

(a)

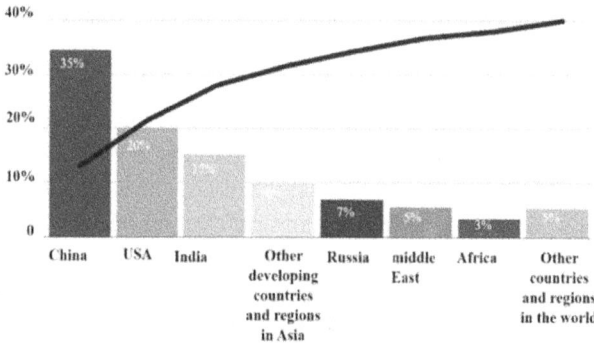

(b)

**Figure 1.1:** Status of primary energy in the world's energy structure [1]: (a) growth of global primary energy consumption and (b) contribution of countries to growth in primary energy consumption in 2018.

grim situation, in June 2018, the European Union formulated a "Horizon Europe" framework plan to explicitly support renewable energy storage technology and a competitive batteries industry chain, of which R&D funding in the fields of climate, energy, and transportation reached 15 billion euros. In July 2018, the Ministry of Economy, Trade and Industry of Japan issued the Fifth Energy Basic Plan, and NEDO, a new energy and industrial technology development organization under the Ministry of Economy, Trade, and Industry, approved the development project of "innovative battery-solid-state batteries." In September 2018, Germany announced the Seventh Energy Research Plan, which plans to invest 6.4 billion euros over the next 5 years to support multisectors to promote energy transformation through system innovation and explicitly support the research of electric energy storage materials. In view of the current situation of energy development and the international situation, China has successively issued documents such as the Energy Development Strategic Action Plan (2014–2020) and the 13th Five-Year Plan of Renewable Energy, in order to strengthen the support and guidance for the development of renewable energy. At present, new energy sources such as wind and solar energy have established a more mature industrial system (as shown in Figure 1.2), and the permeability of renewable energy generation in the power industry is also gradually increasing. In 2018, renewable energy generation reached $1.87 \times 10^{13}$ kW h, accounting for 26.7% of the total electricity generation. Although renewable energy power generation has made some development, they all have the disadvantages of discontinuity and instability; in addition, the lack of energy storage equipment that matched with the power generation system makes it difficult to be connected to the grid. These problems lead to the serious phenomenon of power abandonment at present, which reached $1{,}022.9 \times 10^{12}$ kW h in 2018 alone. Therefore, in order to improve the utilization rate of renewable energy, we must develop large-scale energy storage equipment. In other words, the construction and development of energy storage systems have become the key to solving the energy problem.

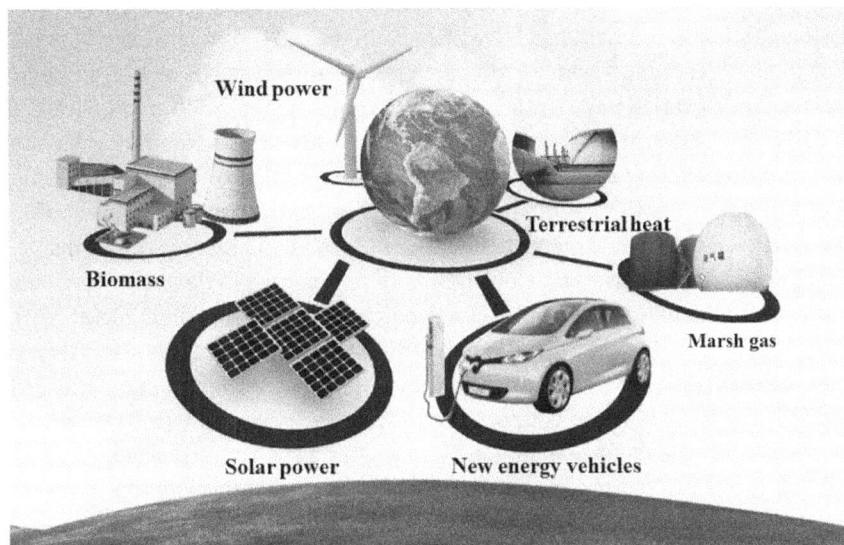

**Figure 1.2:** The industrial system of renewable energy.

At present, energy storage can be divided into physical energy storage and chemical energy storage [2]. Physical energy storage mainly includes pumped energy storage, compressed air energy storage, and flywheel energy storage. Among them, pumped storage is the most mature and widely used way of energy storage because of its high construction cost, long cycle, and many geographical constraints, which cannot meet the needs of large-scale energy storage and smart grid construction. Therefore, the chemical energy storage methods with high energy density, low environmental requirements, and strong environmental friendliness have attracted wide attention. Among them, lithium-ion batteries, high-temperature sodium–sulfur (Na–S) batteries, liquid flow batteries, and Ni-MH batteries all have practical application prospects [3].

The secondary batteries represented by lithium-ion batteries have many advantages, such as high energy density, high energy storage efficiency, no memory effect, small self-discharge, long cycle life, and wide application range. At present, lithium-ion batteries have been successfully applied in small electronic products, electric vehicles, aerospace, and other fields. At the same time, the research direction of lithium-ion batteries is gradually developing in the direction of ultra-high energy density, long service life, ultra-light portable form, and so on. As shown in Figure 1.3, although lithium-ion batteries have been developed for a long time, the cost of lithium-ion batteries is still high [1, 4]. And from a worldwide point of view, lithium resources are relatively concentrated in a few countries, the overall reserves are limited, and the mining conditions are relatively harsh, which will affect the promotion and development of lithium-ion batteries used in large-scale energy storage [5].

In order to solve these problems, recent studies have focused on finding new ways of energy storage to replace the current chemical energy storage devices. Among them, sodium-ion batteries have attracted great attention of researchers because of their good comprehensive properties [6]. Sodium salt is rich in storage, and sodium-ion batteries have similar chemical properties to lithium-ion batteries [7]. As early as 1970s and 1980s, people began to study lithium-ion batteries and sodium-ion batteries. However, due to the rapid development of commercial application of lithium-ion batteries, the research speed of sodium-ion batteries slowed down obviously. At that time, the material quality of sodium-ion batteries and the development level of electrolyte and glove box were not enough to deal with metal sodium, so it was difficult to observe the electrode performance of sodium metal. With the development of technology, the research of sodium-ion batteries has also ushered in the most exciting part of the new development. In recent years, the number of SCI papers published on sodium-ion batteries has increased exponentially (see Figure 1.4), and the number of patents for sodium-ion batteries has also increased synchronously. The solution to scientific problems will certainly promote the development of actual production.

Nowadays, the development of sodium-ion batteries has attracted the attention of countries all over the world. China has gradually grown into a big and powerful country in the research and development of sodium-ion batteries technology. The establishment of sodium-ion battery enterprises such as HiNa BATTERY, NATRIUM, and STAR SODIUM has opened up a new way for the commercialization of sodium-ion batteries. In 2021, ATL established its first commercial line of sodium-ion batteries (Prussian blue derivative for the cathode). It is believed that in the near future,

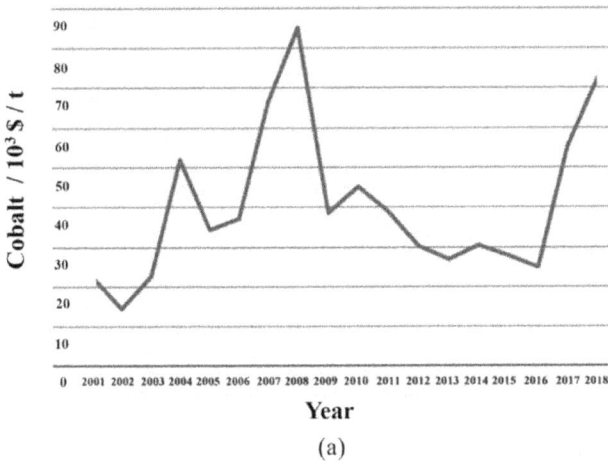

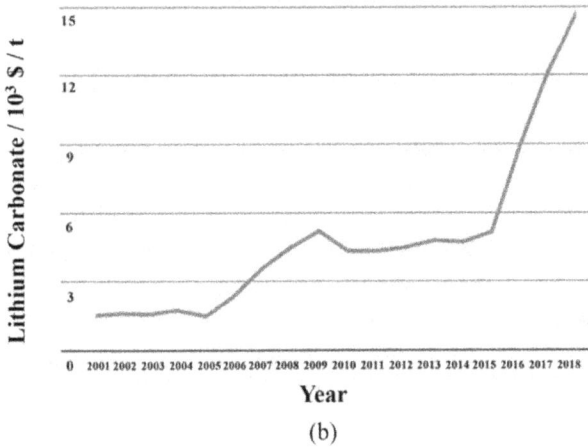

Figure 1.3: Raw material cost trend of lithium-ion battery from 2001 to 2018 [1, 4]: (a) changes in cobalt resource prices and (b) changes in lithium carbonate resource prices.

sodium-ion batteries are expected to take the lead in commercialization in China, contribute to the development of new energy, and provide a strong guarantee for national energy security.

## 1.2 The development of sodium-ion batteries

In the 1960s, Ford Company of the United States invented high-temperature Na–S batteries with sodium metal as anode and sulfur as cathode, which is the earliest sodium batteries in the world [8]. After that, sodium/nickel chloride batteries were invented, both of which possessed high energy density and have attracted wide

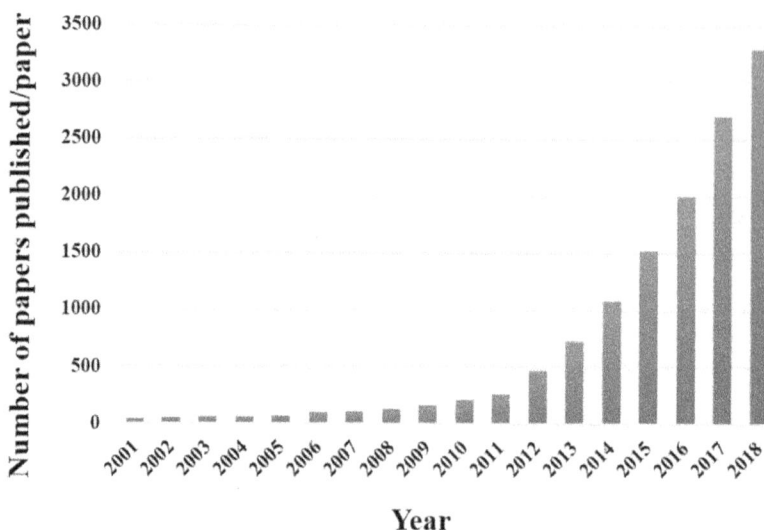

**Figure 1.4:** Number of SCI papers published on sodium-ion batteries in recent years.

attention since their inception [9]. Na–S batteries consist of molten sodium anode, elemental sulfur and sodium polysulfide molten salt cathode, and β-alumina ceramic electrolyte. During discharge, the anode loses electrons; at the same time, the sodium ions transfer to the ceramic electrolyte and move to the cathode, the electrons move from the anode to the cathode, and the sulfur cathode gets the sodium ions in the ceramic electrolyte and forms $Na_2S_x$. During charging, the positive electrons are reduced, the sodium ions are released, and the elemental sulfur is regenerated. At the same time, the sodium ions pass through the ceramic electrolyte to form metal sodium at the anode. The main advantages of Na–S batteries are as follows: ① high specific energy, the theoretical mass energy density up to 760 W h/kg; ② high power density, discharge current density up to 2–3 kA/m²; ③ high charge and discharge efficiencies, no self-discharge, long cycle life, small size, lightweight, and no pollution.

Sodium/nickel chloride batteries were first invented by Dr. Coetyer J of Zebra Power Systems Company in South Africa, referred to as Zebra batteries; its structure is similar to Na–S batteries, but the cathode uses molten transition metal chloride, such as $NiCl_2$, $FeCl_2$, and other materials [10]. So far, the sodium/nickel chloride batteries have been deeply studied. The anode uses liquid metal sodium, the electrolyte uses β-alumina tube, and the cathode is composed of solid Ni and $NiCl_2$ dispersed in $NaAlCl_4$ molten salt. The working principle of Zebra batteries is similar to that of Na–S batteries. It not only takes the advantages of high-power density and no self-discharge, but also possesses higher open-circuit voltage and more stable high-temperature performance than Na–S batteries.

The above two kinds of sodium metal batteries have excellent performance and show great potential in practical applications. The Na–S batteries produced by NGK

Company of Japan have entered the stage of commercialization in 2000, and now 8 MW energy storage Na–S battery device has been built, and more than 100 battery storage stations are in operation all over the world. Shanghai Silicate Research Institute of China has also successfully developed 650 A h Na–S battery monomers, and now has built the 2 MW Na–S battery monomer production line. China has become the second country in the world to master the core technology of large-capacity Na–S monomer batteries after Japan. At the same time, Zebra batteries are ideal batteries for electric vehicles. German Mercedes-Benz has tested the Zebra batteries for a long time as early as the 1990s, and its indicators have met the medium-term goals of USABC, and the battery performance is excellent. At present, electric vehicles using Zebra batteries have been tested for more than 3.2 million kilometers, and their feasibility has been recognized in many countries in the world. In 2003, British Rolls-Royce transferred the application of Zebra batteries from land to sea, which was used as the internal power of civil ships and military submarines. Although the development of sodium metal batteries has been on the rise, due to the high reducibility of sodium metal itself and the characteristic that it must work at high temperature, liquid sodium will react violently in the event of an accident, causing an explosion. At the same time, β-alumina tube is expensive and requires high technical requirements. These problems seriously limit the further industrial development of Na–S batteries and Zebra batteries.

In view of the above problems, scientists have proposed to use materials that can store sodium ions instead of metal sodium, and fully learn from the experience of lithium-ion batteries to develop sodium-ion batteries. Sodium-ion batteries and lithium-ion batteries are produced almost at the same time, but the performance and research conditions of sodium-ion batteries restrict their development. It was not until solid-state $Al_2O_3$ conductors were used for sodium-ion transport that the development and application of sodium-ion batteries became possible [11]. In the 1980s, before the commercialization of lithium-ion batteries, some companies in the United States and Japan developed sodium-ion batteries, in which sodium–lead alloy and P2-type $Na_xCoO_2$ were used as negative and cathodes, respectively. Although sodium-ion batteries can be stably cycled for more than 300 weeks, the average discharge voltage is less than 3.0 V, while the discharge voltage of $LiCoO_2$ batteries which also use carbon material can reach 3.7 V, so sodium-ion batteries have not attracted widespread attention. At the same time, due to the advantages of high specific energy, low cost, and no self-discharge, Na–S batteries have been developed to a certain extent [12]. Especially applied in the direction of energy storage, the maximum capacity of the single cell can reach 650 A h, with the power reaching more than 120 W, and the service life reaching 10–15 years. However, the safety problem of Na–S batteries has not been completely solved, in which the compactness of solid ceramic electrolyte is very critical, and high-temperature operation needs certain heat preservation conditions. Subsequently, the discovery of fast sodium-ion conductor provides a strong support for the development of all-solid-state sodium-ion batteries at room temperature. Until the end of 1990s, the application of layered transition metal oxides in room-temperature-reversible sodium storage cathode materials had sodium-ion batteries return to people's field of vision. In 2000, the first turning point occurred in the research of sodium-ion batteries. Stevens and Dahn used hard carbon as the anode of sodium-ion batteries, and the reversible cycle-specific capacity of the material reached 300 mA h/g, which was almost close to the reversible specific capacity of graphite anode in lithium-ion batteries at that time [13]. The second turning point came from the study of $NaFeO_2$ by the Okada team, which found that the $Fe^{3+}/Fe^{4+}$ redox pairs in $NaFeO_2$ have electrochemical activity in sodium-ion batteries [14]. The discovery of this material was of the same significance as the discovery of $LiCoO_2$ in lithium-ion batteries ($LiCoO_2$ has become a commercial cathode material for lithium-ion batteries). Based on these important findings, in the past few years, sodium-ion batteries have attracted the attention of researchers by virtue of their potential cost advantages. Finally, after 2010, the research of sodium-ion batteries came to a climax. From the improvement of the performance of certain electrodes to the development of sodium-ion battery pack, from finding new sodium storage electrode materials to explore the mechanism of sodium-ion storage, from studying the compatibility of electrode materials and electrolytes to the development of all-solid-state sodium batteries with safer and higher specific energy, the commercialization of sodium-ion batteries has unlimited potential. The development of sodium-ion batteries is shown in Figure 1.5.

From the changes in the number of patent applications for sodium-ion batteries, we can see that the research in the field of sodium-ion batteries in China started in the 1980s, moved on slowly before 2008, and began to rapidly develop after 2010. At present, under the background of the development of large-scale energy storage and smart grid in our country, the advantage of low cost of sodium-ion batteries is more and more obvious. As shown in Figure 1.6, more than 140 patents for key technologies of sodium-ion batteries were filed each year from 2014 to 2018. As shown in Table 1.1, patents are mainly focused on the selection of electrode materials, binders, and electrolytes. Layered compounds, phosphates, and Prussian blue in cathode materials are the current research hotspots. Toyota, Great Power, and the WanRun New Energy occupy an important position in electrode materials for sodium-ion batteries, among which Sumitomo's patent distribution and application volume are in the leading position [15]. Despite the rapid development of patent technologies related to sodium-ion batteries, the main applicants for patent technologies are still schools and scientific research institutes, which also reflect that the development of this technology is still in the stage of laboratory research and is still far from large-scale market application.

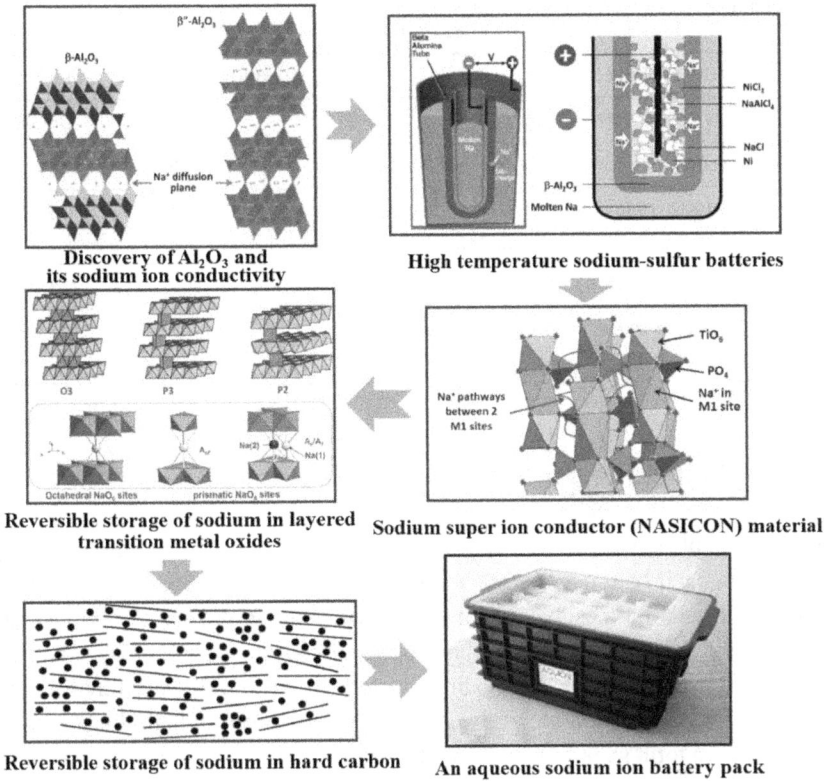

**Figure 1.5:** Development history of sodium-ion batteries [9].

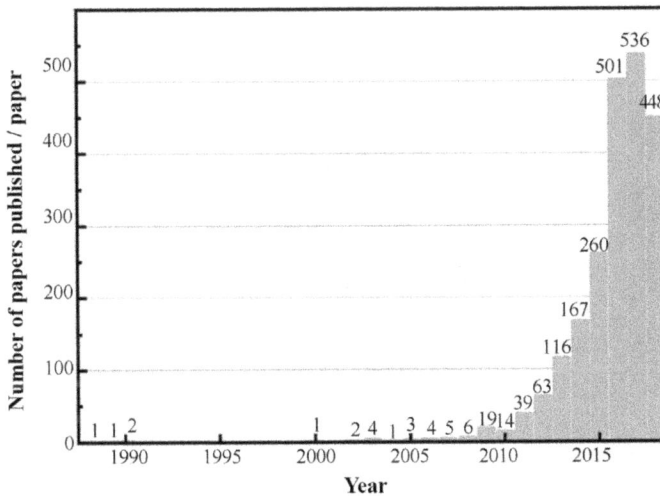

**Figure 1.6:** Changes in patent applications of key technologies of sodium-ion batteries in China from 1988 to 2018.

**Table 1.1:** Patent distribution of sodium-ion batteries.

| IPC code | Technical field | Quantity/item | Proportion |
| --- | --- | --- | --- |
| H01M4/00 | Electrode material | 1,990 | 36.38% |
| H01M10/00 | Secondary batteries and their manufacture | 1,786 | 32.65% |
| B82Y30/00 | Nanotechnology for materials and surface science | 313 | 5.72% |
| B82Y40/00 | Fabrication or treatment of nanostructures | 165 | 3.02% |
| H01M2/00 | Structural parts or manufacturing methods of inactive parts | 76 | 1.39% |

## 1.3 Basic principles of sodium-ion batteries

The working principle of sodium-ion batteries is similar to that of lithium-ion batteries, which is "rocking-chair batteries," as shown in Figure 1.7 [7]. In essence, it is a kind of concentration cell, and the positive and anode materials are composed of compounds with different sodium-ion contents. In the process of charging, sodium ions detach from the cathode materials with high sodium content, enter into the electrolyte, form solvated molecules with solvent molecules, migrate to the anode through the action of electric field, and re-form sodium ions through desolvation. Finally, it is transferred to

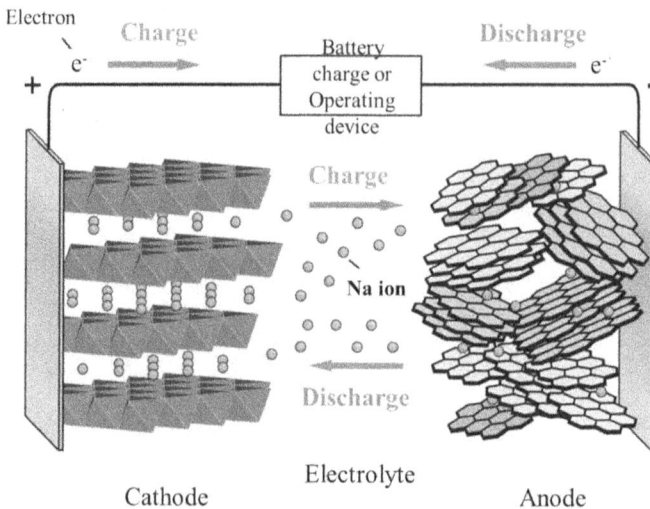

**Figure 1.7:** Schematic diagram of working principle of sodium-ion batteries [7].

the anode material with lower sodium content; at the same time, electrons are transported to the anode through the external circuit to maintain the charge balance between the cathode and anode. In the process of discharging, the migration path of sodium ions and electrons is opposite to that of charging. In terms of the overall reaction, the reversible storage and release of sodium ions in the cathode and anode materials, the transfer of sodium ions in the electrolyte, and the transfer of electrons outside the batteries were completed. Therefore, sodium-ion batteries are a kind of secondary batteries with charging and discharging abilities, which can be used in many fields, such as large-scale energy storage and low-speed electric vehicles. The research and modification methods can also draw lessons from the technology of lithium-ion batteries to some extent.

Although the working modes of sodium-ion batteries and lithium-ion batteries are similar, due to the differences in the properties of sodium and lithium, the two batteries also have different characteristics, as shown in Table 1.2 [16]. Sodium is located in the third cycle of the first main group of the periodic table, and its relative atomic mass is 23 g/mol. There are two kinds of electron clouds (spherical and dumbbell shaped) outside the nucleus, and the outermost electrons occupy 3S orbitals. Compared with lithium, the relative atomic mass and ion radius of sodium are larger, and the electrochemical equivalent is almost 3 times that of lithium. But the desolvation energy of sodium ion is lower and the diffusion ability is stronger. Therefore, compared with lithium-ion batteries, sodium-ion batteries also have unique system advantages: firstly, sodium-containing compounds are widely distributed and abundant on the Earth; secondly, sodium-containing chemicals are easy to mine and possess high quality; thirdly, the redox potential of sodium ion is about 300 mV higher than that of lithium ion, so it can be used to electrolyte solvents, electrolyte salts, and aluminum collectors with lower decomposition potential; fourthly, sodium-ion batteries are greener and in line with the principle of sustainable development. However, there are many technical difficulties in the development of sodium-ion batteries: first, the ion radius of sodium ion (1.06 Å) is larger than that of lithium ion (0.76 Å), and the resistance of migration in solid lattice is greater, showing a slow reaction kinetics; second, the larger radius of sodium ion will lead to larger lattice stress of the main material of sodium storage, which will lead to the collapse of the crystal structure and the deterioration of the cycling stability of the material during the electrochemical cycle; third, the relative atomic mass of sodium ion is larger than that of lithium ion, so the theoretical capacity of sodium storage electrode is usually smaller than that of lithium storage electrode. Among them, the slow reaction kinetics is one of the main reasons for the obvious gap between the actual capacity and the theoretical capacity of sodium-ion batteries.

**Table 1.2:** Comparison of the properties of sodium and lithium [16].

| Property | Sodium | Lithium |
|---|---|---|
| Ion radius (Å) | 1.06 | 0.76 |
| Relative atomic mass (g/mol) | 23 | 6.9 |
| Redox potential (vs. $Li^+/Li$) (V) | 0.3 | 0 |
| Precursor price (dollar/t) | 150 | 5,000 |
| Theoretical specific capacity (A h/g) | 1.165 | 3.829 |
| Melting point (℃) | 97.7 | 180.5 |
| Distribution | Global | 70% of the reserves are located in South America |

## 1.4 Overview of key materials for sodium-ion batteries

The key materials of sodium-ion batteries include cathode material, anode material, electrolyte, and separator. Because the theoretical specific capacity of the cathode material is relatively low, the cathode material determines the capacity of the batteries. The anode material affects the reaction kinetic performance of the batteries. And the electrolyte is directly related to the stability and safety of the batteries. The characteristic of the separators is designed to effectively improve the safety of battery operation while reducing the risk of explosion. Therefore, as the key to the construction of high-performance sodium-ion batteries, researchers need to conduct in-depth research on the above four materials to prepare battery materials with excellent electrochemical performance and good commercial application prospects.

The mechanism of sodium-ion batteries, which is similar to that of lithium-ion batteries, is the intercalation/deintercalation reaction. Even mildly structural reorganization occurring in the electrode during the process, the structure and composition of material remain almost intact, and the reaction is considered to be locally regular. Hence, the capacity and cycle life of the secondary batteries is mainly determined by the number of ions that can reversibly intercalate/deintercalate in electrodes and the stability of active material. According to formula (1.1), the theoretical capacity of the material can be calculated:

$$C_{\text{theory}} = \frac{nF}{3.6M} \tag{1.1}$$

where $C_{\text{theory}}$ is the theoretical specific capacity, $n$ is the number of moles embedded in lithium, $F$ is the Faraday constant, and $M$ is the molecular weight of the substance.

The ideal cathode material should have the following characteristics:

(1) The intercalation reaction should have a large Gibbs free energy, which can maintain a large potential difference between the cathode and the anode and provide a higher batteries voltage.

(2) In a certain range, the $\Delta G$ variation of sodium-ion intercalation reaction is smaller, that is, the amount of sodium-ion intercalation is larger, and the electrode potential is less dependent on the intercalation amount, in order to ensure that the batteries have higher electrochemical capacity and stable charging/discharging voltage.

(3) For crystals with layered or large aperture tunnel structure, sodium ion should have larger diffusion coefficient and migration coefficient in "interlayer" or "tunnel" to ensure higher diffusion rate and good electronic conductivity to ensure that the batteries have a good and fast charging–discharging performance.

(4) In the process of sodium-ion intercalation/deintercalation, the cathode material has a small volume change to ensure good cycle reversibility and improve the cycle performance of the batteries.

(5) In the required charging–discharging potential range, the electrode has good compatibility with the electrolyte solution, that is, the electrode/electrolyte interface has good thermal stability, chemical stability, and electrochemical stability.

(6) Low price, good storage in the air, no pollution to the environment, and light quality.

At present, the typical cathode materials for sodium-ion batteries can be roughly divided into two categories. The first kind is layered materials, which have anion dense or quasi-dense structure. The alternating layer between anion clusters is occupied by redox transition metal ions, while sodium ions are embedded in the remaining intercluster vacancies. The first kind of representative cathode materials are $Na_{1-x}FeO_2$, P2–$Na_{2/3}[Fe_{1/2}Mn_{1/2}]O_2$, $Na_{0.9}[Cu_{0.22}Fe_{0.30}Mn_{0.48}]O_2$, and so on [14]. The second kind of materials have a more open structure, including polyanionic compounds (such as $NaFePO_4$ and $Na_3V_2(PO_4)_3F_3$) and metal–organic framework compounds (such as Prussian blue and its analogs) [17, 18]. Due to the more compact crystal structure, the first kind of materials have an essential advantage in energy storage per unit volume. But some of the second kind of materials have more advantages in cost.

Due to the large ionic radius of sodium ion and the fact that sodium deposition takes precedence over the formation of sodium–graphite intercalation compounds, graphite anode materials that have been successfully applied in lithium-ion batteries are difficult to perform their performance in organic sodium-ion batteries [19, 20]. Therefore, one of the major bottlenecks in the development of sodium-ion batteries is still the lack of suitable commercially available anode materials. Currently, researches on sodium-storage anode materials have been extensive, and a series of high-performance anode materials (such as hard carbon, soft carbon, metal sulfide, titanium-based oxides, and alloy compounds) have been developed [21]. Generally, during the first charging process, electrolyte reduction reaction will occur on the surface of anode materials (especially carbon materials), resulting in insoluble sodium salt deposition on the surface of anode materials and forming a thin film, namely Solid elelctrolyte Interface (SEI) film. SEI film has the characteristics of allowing sodium ions to move in and preventing the passage of solvent molecules. When the SEI film reaches a certain thickness, the electrode is isolated from the electrolyte. At the same time, due to the electronic insulation of the SEI film, the reduction reaction of the electrolyte is prevented, and the irreversible reaction caused by the decomposition of the electrolyte stops. On the one hand, the formation of SEI film makes the subsequent cycle of the batteries stable and reversible; on the other hand, a large amount of irreversible capacity is generated in the first week, and the lost capacity cannot be recovered in the subsequent reversible cycle, resulting in the reduction of the actual capacity of the batteries. Therefore, the ideal sodium-ion anode material should not only have stable thermodynamic properties but also have a good match with the electrolyte, form a good SEI film with the electrolyte, and do not react with the electrolyte after the formation of the SEI film.

The electrolyte of sodium-ion batteries is composed of electrolyte salt, solvent and additive. The electrolyte salt is mainly sodium salt, and its solubility in the solvent directly affects the number of carriers in the electrolyte, and the redox potential plays an important role in the electrochemical window of the electrolyte system [22, 23]. The chemical inertia of anions/cations in sodium salt will affect the stability of separator, solvent, electrode, and fluid collector [24]. The thermal stability of sodium salt is directly related to the safety of the batteries. Sodium-ion electrolyte is mainly liquid, and the common systems in the study include organic electrolyte, aqueous electrolyte, and ionic electrolyte. The ideal electrolyte should have the following characteristics:

(1) The ion migration rate is high, and the conductivity is $3 \times 10^{-3}$–$2 \times 10^{-2}$ S/cm over a wide temperature range.
(2) Good thermal stability, that is, the electrolyte does not decompose in a wide temperature range.
(3) High chemical stability, that is, the electrode does not react with the electrolyte.
(4) Wide electrochemical window, that is, side reactions such as electrolyte decomposition do not occur in the process of charge and discharge.
(5) It has good film-forming (SEI) characteristics, and a stable passive film can be formed on the surface of the anode materials.
(6) Nontoxic, low steam pressure, safe to use, easy to prepare, low cost, and no environmental pollution.

At present, the study of organic electrolyte is the most extensive. This system has the characteristics of high dielectric constant, low viscosity, wide electrochemical window, and stable passivation film formed on the electrode surface during charge and discharge, so it is considered to be the most suitable commercial sodium-ion batteries electrolyte. In order to further improve the properties of materials, film-forming additives and flame-retardant additives are often used in electrolytes. The reduction mechanism of ester electrolytes and additives was studied by the first calculation principle (Density functional theory (DFT)) [25]. The results show that the SEI film formed by vinyl carbonate and propylene carbonate is unstable and may cause electrode failure, while the addition of fluoroethylene carbonate makes the electrolyte system possess priority to generate NaF and form a stable SEI film regardless of single electron reaction or double electron reaction. In addition, the addition of suitable additives can improve the performance of aqueous electrolytes, such as ethylene carbonate, can form a protective film on the electrode surface to prevent the entry of $O_2$, and inhibit the side reaction [26]. Therefore, the use of appropriate types and proportion of additives can improve the stability of the interface, so as to achieve better cycle stability and rate performance.

In order to solve the safety problems caused by sodium dendrites growth and poor thermal stability in organic electrolytes, the researchers proposed to use solid electrolytes with excellent thermal stability, high mechanical strength, and wide electrochemical window to construct high-safety sodium-ion batteries. At present, solid electrolytes have been able to achieve $10^{-2}$ S/cm ionic conductivity and 5 V electrochemical window. However, the problems such as interfacial compatibility, high sodium-ion diffusion barrier, and difficult processing still limit the further development of solid-state electrolytes.

Separator is an important part of sodium-ion batteries. Its main function is to isolate cathodes and anodes and prevent safety problems caused by short circuit. In addition, the separator also plays the role of transporting ions and isolating electrons. At present, the main commercial lithium-ion battery separators are polypropylene and polyethylene, but their thermal stability and mechanical properties are poor, and their wettability with sodium-ion battery electrolyte is very poor, so they are not suitable to be used as sodium-ion battery separators. Glass fiber is often used as sodium-ion battery separator in laboratory research. This material has good thermal stability and electrolyte compatibility, and is of low price. But its mechanical properties are poor, so it is still a challenge to use it as a commercial sodium-ion battery separator. The separators that can be used commercially in the future should at least meet the following requirements:

(1) Good chemical and electrochemical stability, can withstand the corrosion of the electrolyte, and be stable in the redox reaction at the same time.
(2) Thermodynamic stability, able to withstand a certain degree of temperature change.
(3) The pore diameter is smaller than any particle structure in the dielectric, and the pore diameter is uniform.
(4) It has good wettability with electrolyte.
(5) Good mechanical stability, able to withstand a certain degree of deformation.
(6) The production cost is low and the price is cheap.

## 1.5 Development trend of sodium-ion batteries

As a promising electrochemical energy storage system, sodium-ion batteries have been developed rapidly in recent years. The study of cathode materials, such as layered transition metal oxides, polyanion compounds, and Prussian blue and its analogs, has been gradually deepened. The application of modification methods such as material structure design, morphology control, and electrolyte optimization promotes the possibility of application. Especially, the open structure of Prussian blue and its analogs provides a channel for the rapid intercalation/intercalation of sodium ions to deliver promising commercial prospect. It can realize charge and discharge at ultra-high rate. In addition, the material can be used not only in organic electrolyte but also in aqueous electrolyte, showing the potential of industrial application. At present, it has been proved that what can be used as sodium storage anode materials is not limited to traditional carbon materials. Metal oxides, metal sulfides, and some metal and organic materials can be used as anode materials for sodium-ion batteries. It can achieve fast and reversible sodium-ion intercalation/intercalation reaction with excellent rate performance. However, cycling stability of this kind of materials is poor, so it is difficult to achieve industrial application.

Generally speaking, the advantages of high safety, low cost, and abundant sodium resources for sodium-ion batteries are fitter for industrial application in the field of large-scale energy storage in the future. At present, thousand number of materials have been found in laboratory research, but the research on sodium-ion full batteries is still on progress. In the future, the research of sodium-ion batteries should focus on problems of the electrode/electrolyte interface and improve the volume expansion and crystal structure collapse of materials in the process of charging and discharging by means of reasonable electrolyte design and material interface optimization to achieve better performance of cycle life, rapid charge–discharge capacity, and safety, so as to promote the competitiveness of products in the market, which really accelerate the commercialization of sodium-ion batteries.

## References

[1]    BP 中国. BP 世界能源统计年鉴 [EB/OL]. [2019-07-30]. https://www.bp.com/zh_cn/china/home/news/reports/statistical-review-2019.html.

[2]    袁亚琼, 司华清. 储能技术的发展前景与作用路线. 技术与市场, 2019, 26(2): 77–81.

[3]    李泓, 吕迎春. 电化学储能基本问题综述. 电化学, 2015, 21(5): 412–424.

[4]    Li M, Lu J, Chen Z. et al., 30 Years of Lithium-Ion Batteries. Advanced Materials, 2018, 30(33): 1800561.

[5]    钟财富, 刘坚, 吕斌, 等. 中国新能源汽车产业锂资源需求分析及政策建议. 中国能源, 2018, 40(10): 12–15.

[6]    Parant J P, Olazcuaga R, Devalette M. et al., Sur Quelques Nouvelles Phases De-Formule $Na_xMnO_2$ ($X \leq 1$). Journal of Solid State Chemistry, 1971, 3(1): 1–11.

[7]   Shacklette L W, Jow T R, Townsend L. Rechargeable Electrodes from Sodium Cobalt Bronzes. Journal of the Electrochemical Society, 1988, 135(11): 2669–2674.

[8]   孙文, 王培红. 钠硫电池的应用现状与发展. 上海节能, 2015, (2): 85–89.

[9]   郭永全, 闻俊锋, 赵晋峰, 等. 以氯化镍作正极材料的热电池研究. 电源技术, 2010(2): 85–87, 91.

[10]  曹佳弟. ZEBRA (钠-氯化镍)电池的研究新进展. 电源技术, 1999(3): 42–45.

[11]  Delmas C. Sodium and Sodium-Ion Batteries: 50 Years of Research. Advanced Energy Materials, 2018, 8(17): 1703137.

[12]  Ellis B L, Nazar L F. Sodium and Sodium-ion Energy Storage Batteries. Current Opinion in Solid State & Materials Science, 2012, 16(4): 168–177.

[13]  Stevens D A, Dahn J R. High Capacity Anode Materials for Rechargeable Sodium-Ion Batteries Articles. Journal of the Electrochemical Society, 2000, 147(4): 1271–1273.

[14]  Liu Q, Hu Z, Chen M. et al., Recent Progress of Layered Transition Metal Oxide Cathodes for Sodium-Ion Batteries. Small, 2019, 15(32): 1805381.

[15]  郑伟伟, 邓隽. 钠离子电池专利技术分析. 广东化工, 2017, 044(17): 124–126, 132.

[16]  Yabuuchi N, Kubota K, Dahbi M. et al., Research Development on Sodium-Ion Batteries. Chemical Review, 2014, 114(23): 11636–11682.

[17]  Liu R, Liang Z, Gong Z. et al., Research Progress in Multielectron Reactions in Polyanionic Materials for Sodium-Ion Batteries. Small Methods, 2019, 3(4): 1800221.

[18]  Qian J, Wu C, Cao Y. et al., Prussian Blue Cathode Materials for Sodium-Ion Batteries and Other Ion Batteries. Advanced Energy Materials, 2018, 8(17): 1702619.1-1702619.24.

[19]  Ge P, Fouletier M. Electrochemical Intercalation of Sodium in Graphite. Solid State Ionics, 1988, 28(2): 1172–1175.

[20]  Stevens D A, Dahn J R. The Mechanisms of Lithium and Sodium Insertion in Carbon Materials. Journal of the Electrochemical Society, 2001, 148(8): 87–91.

[21]  谢银斯, 孙丁武, 林维捐, 等. 钠离子电池负极材料研究进展. 电源技术, 2019, 43(02): 351–353.

[22]  Chen S, Ishii J, Horiuchi S. et al., Difference in Chemical Bonding between Lithium and Sodium Salts: Influence of Covalency on Their Solubility. Physical Chemistry Chemical Physics, 2017, 19(26): 17366–17372.

[23]  Ponrouch A, Marchante E, Courty M. et al., In Search of an Optimized Electrolyte for Na-ion Batteries. Energy & Environmental Science, 2012, 5(9): 8572–8583.

[24]  Plewa-Marczewska A, Trzeciak T, Bitner A. et al., New Tailored Sodium Salts for Battery Applications. Chemistry of Materials, 2014, 26(17): 4908–4914.

[25]  Cheng X, Jian P, Yang Z. et al., Gel Polymer Electrolytes for Electrochemical Energy Storage. Advanced Energy Materials, 2018, 8(7): 1702184.

[26]  Webb S A, Baggetto L, Bridges C. et al., The Electrochemical Reactions of Pure Indium with Li and Na: Anomalous Electrolyte Decomposition, Benefits of FEC Additive, Phase Transitions and Electrode Performance. Journal of Power Sources, 2014, 248(4): 1105–1117.

# Chapter 2
# Testing and research methods for sodium-ion batteries

In recent years, the researches on sodium-ion batteries and their related materials have developed rapidly and achieved the same results as other types of batteries in several decades, which is due to the continuous breakthroughs in testing and characterization techniques. Advanced testing and characterization techniques have clearly revealed the mechanisms of sodium intercalation/deintercalation, and material synthesis processes described the material's conformational relationships and guided people to conduct research on sodium-ion secondary batteries faster and better. For example, the latest development of electrochemical in situ characterization techniques has greatly advanced the research of the interface between electrode materials and electrolyte materials for sodium-ion batteries. Current test characterization techniques have been able to successfully test the performance of the full batteries and explore the changes in the battery during the charging and discharging process. The characterization of electrode materials can be divided into the characterization of basic physical parameters such as structure and composition and electrochemical properties.

## 2.1 Test analysis of electrode materials

### 2.1.1 Structure and composition test of electrode materials

The electrode materials of sodium-ion batteries are divided into cathode electrode materials and anode electrode materials. Among them, cathode materials mainly include layered transition metal oxide materials, polyanionic compound materials, metal–organic complex materials, and organic compound materials. Anode electrode materials mainly include carbon-based materials, titanium-based compound materials, conversion reaction-type compound materials, and intermetallic compound materials. The active electrode materials of sodium-ion battery spanning a wide range, types, and complex conformational relationship are a typical representative class of new battery system. Detecting the structural characteristics of active electrode materials is of great importance for the development of high-performance, long-life, and low-cost sodium-ion batteries. In general, the structural characteristics of materials include crystal structure, microscopic morphology, specific surface area and pore size distribution of particles, valence and elements distribution, thermal stability,

https://doi.org/10.1515/9783110749069-002

mechanical properties, and valence bonding coordination structure. Different characterization techniques of active electrode materials for sodium-ion batteries will be introduced one by one in the following.

### 2.1.1.1 X-ray diffraction

X-ray diffraction (XRD) is a technique to study the lattice parameters, atomic occupancy, symmetry, lattice defects, crystallinity, lattice stress, and heterogeneous structure of materials by using the diffraction phenomenon of X-rays with a short wavelength (200.06 Å) in the sample and is the most common and effective means to determine the composition and structure of materials. When X-rays irradiate a crystal, the incident X-rays can be scattered by each lattice point in the lattice, and the waves formed by each scattering will be coherently superimposed or canceled in space to produce the final diffraction result. When the Bragg equation is satisfied, a certain lattice surface produces mutually enhancing reflections of X-rays of a certain wavelength at a certain angle:

$$2d \sin \theta = n\lambda (n = 1, 2, 3, ...) \tag{2.1}$$

where $\lambda$ is the wavelength of X-rays, $d$ is the lattice plane layer spacing, and $\theta$ is the diffraction angle between X-rays and the lattice plane. The grain size of the material particles can be further estimated by combining Scherrer's formula:

$$D = K\lambda / (B \cos \theta) \tag{2.2}$$

where $D$ is the grain size; $K$ is the shape factor, a constant related to the crystal, generally taken as 1; $\lambda$ is the wavelength of X-rays; $B$ is the half-height width of the diffraction peak; and $\theta$ is the diffraction angle. When the crystal size is calculated, a diffraction curve with low angle is generally used. If the grain size is large, a diffraction curve with high angle can be used instead. Equation (2.2) is applicable to the calculation of materials with grain size of 1–100 nm.

The sharpness of XRD peaks can be used to discern the crystallinity of the material. The higher the intensity of the diffraction peak and the sharper the peak shape mean the higher the crystallinity of the material. The change of material lattice parameters can be discerned by the shift of diffraction peaks. When the diffraction peaks are shifted to a low angle, the material lattice parameters will become larger; when the diffraction peaks are shifted to a high angle, the material lattice parameters will become smaller.

For high-quality XRD spectra, more information about the crystal structure and phase composition can be obtained by refinement. The advantages of the refinement of XRD spectra include the following: ① determine the lattice parameters and structure of the crystal; ② identify the phase and quantify the components; ③ test the lattice size, crystallinity, and residual stress; ④ determine the atomic radial distribution function; and ⑤ obtain information such as bond length, bond angle, atomic occupancy, and occupancy rate. Three main refinement methods commonly used are Pawley, Lebail, and Rietveld.

In the Pawley method, the diffraction peaks are calculated from the lattice parameters. In the Lebail method, the diffraction peaks are calculated from the lattice parameters, and the lattice parameters and peak shape parameters are used as variables for least-squares fitting. A set of diffraction patterns is calculated based on the above theory and compared with the measured patterns, and the parameters are continuously adjusted using the Newton–Raphson mathematical principle to minimize the difference between the calculated and measured patterns. The results obtained from a modified structural model match with the actual one. A comparison of the advantages and disadvantages of these three refinement methods is shown in Table 2.1.

**Table 2.1:** Comparison of the advantages and disadvantages of Pawley, Lebail, and Rietveld refinement methods.

| Method | Pawley method | Lebail method | Rietveld method |
|---|---|---|---|
| Advantage | Crystal structure model of material is not required | Less refined parameters, convergence speed, less calculations, accurate | Wide range of applications, the lattice parameters may be determined, and the internal atomic mass, qualitative and quantitative analysis of the material phase composition, more exact calculations |
| Disadvantage | Refinement parameters are many, computationally intensive, and can only refine lattice parameters | The same or similar position fitting result diffraction peaks are equal, we need a final judgment | Needs a more accurate initial model, and the calculation is relatively complicated |

As shown in Figure 2.1, the composition and structural information of the layered transition metal oxide cathode material with P2 and O3 composite phases can be obtained by the Rietveld method refinement. The diffraction peaks corresponding to the P2 and O3 phases, respectively, and the lattice parameters and cell volume of the composite phase structure can be calculated.

### 2.1.1.2 X-ray absorption fine structure spectroscopy

X-ray absorption fine structure (XAFS) spectroscopy is used to characterize the local structure. The sample is probed by adjusting the X-ray energy to coincide with the electron layer within the element under study. And then monitors the number

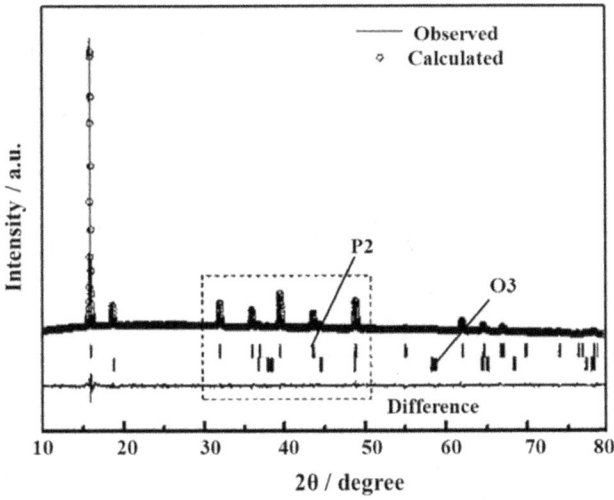

(a)

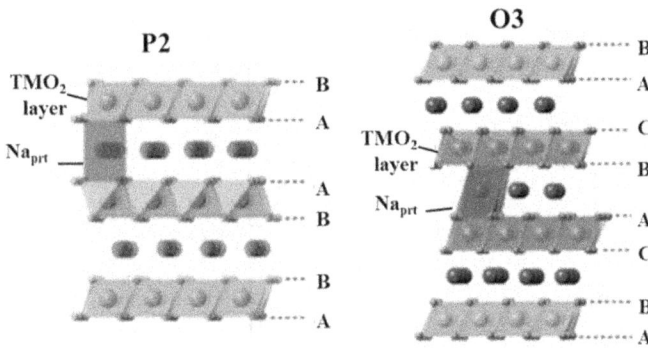

(b)

**Figure 2.1:** XDR refinement: (a) XRD Rietveld refined spectrum of $Na_{0.95}Li_{0.15}Ni_{0.15}Mn_{0.55}Co_{0.1}O_2$ cathode material with P2 + O3 phases; (b) schematic diagram of the crystal structure of P2 and O3 phases of the layered structure [1].

of absorbed X-rays as a function of energy. The spacing between the absorbing atoms and neighboring atoms, the number and type of atoms, and the oxidation state of the absorbing element are parameters that can be obtained from the XAFS spectrum to determine the local structure. By choosing different energies of X-rays, we can obtain information of all elements in the sample.

When the energy of X-rays resonates with the energy of one of the inner electron shell layers of an element in the sample, the absorption rate may suffer a sudden increase and the electrons are excited to form a continuous spectrum (Figure 2.2), which is also called the absorption edge. In most cases, the absorption edge is well separated and the target element is simply selected by scanning a suitable energy range. X-ray energy increases along the absorption edge, and the absorption rate decreases monotonically as the depth of X-ray penetration increases. Fine structures are observed when the spectrum is extended across a specific edge. The X-ray absorption near-edge structure (XANES) region appears when spectral peaks and spectral shoulders more than 20–30 electron volts (eV) just pass the beginning of the edge. The fine structure on the high-energy side of the edge at a few hundred electron volts is called the XAFS. The fine structure in XANES and XAFS has been studied more thoroughly, and it allows XAFS to determine the type and local structure of chemicals. Outside the edge region, the XAFS is superimposed in a series of undulating oscillations on a smoother absorption curve that would otherwise be characteristic of isolated atoms. These fine structures are formed by interference between the ionized photoelectron waves and the backscattered waves of neighboring atoms to some of these waves. As the X-ray energy changes, the interference conditions change accordingly, resulting in oscillatory fine structures in neighboring atoms.

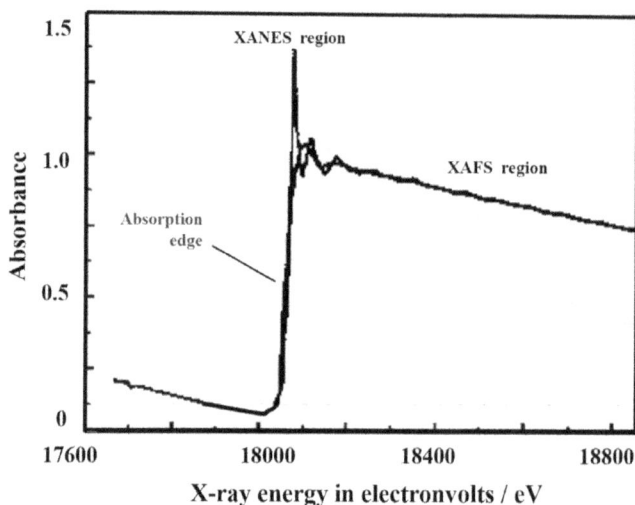

Figure 2.2: X-ray absorption spectrum of plutonium.

### 2.1.1.3 X-ray photoelectron spectroscopy

X-ray photoelectron spectroscopy (XPS) is a technique in which the inner electrons or valence electrons of an atom or molecule are excited and emitted through X-ray radiation of the sample, and the electrons excited by the photons are called photoelectrons. XPS technology can be used to study the nature of molecular structure and atomic valence states, and also qualitative and quantitative analyses of the surface of substances, providing information on the elemental composition and coordination structure of the surface of various compounds. In combination with techniques such as argon ion etching, it is also possible to study electronic, atomic, and elemental information on surfaces, microregions, and depth distributions. Current technological improvements allow for the acquisition of photoelectron images in a parallel fashion, from which XPS profiles can be obtained, and the elemental distribution on the surface of the material can then be analyzed.

The binding energy (BE) in XPS is the difference in energy of an atom before and after photoexcitation. Because the atomic structure of each element is different, the inner electronic BE of an atom is a specific reflection of the element's basic properties, which are special and identifiable. According to Einstein's law of the photoelectric effect, the electronic BE ($E_b$) in an atom can be calculated as follows:

$$E_b = h\upsilon - E_k \tag{2.3}$$

where $h$ is Planck's constant, $h\upsilon$ is the energy of the incident photon, and $E_k$ is the kinetic energy of the electron when it first escapes from the surface. Due to the known quantity of $h\upsilon$, the value of $E_b$ can be calculated by determining $E_k$.

In addition, the accuracy of the $E_b$ calculation can be further improved by taking into account the work function of $\Phi_s$ and the recoil energy of $E_r$ during the radiation process.

The same atom in a different chemical environment causes a change in the electronic BE of the inner shell layer, which is manifested as a displacement of the spectral lines on the XPS pattern, a phenomenon called chemical shift. In general, atoms are in different chemical environments: one is the different types and numbers of elements to form coordination structures, the other one is the different valence states. The electron BE of the inner shell layer of an atom increases with the oxidation state of the atom, and the higher the oxidation state, the larger the chemical shift, and conversely the electron BE of the inner shell layer will decrease, which is reflected in the magnitude of the shift of the peak position in the XPS spectrum. Therefore, it is possible to identify the element type based on the position of the peak, and identify the valence, coordination structure and content of the element based on the peak shift.

When applying the XPS method for material characterization, it should be noted that: ① XPS can detect all elements except H and He with the same sensitivity; ② chemical shifts can be used to discern atomic oxidation state and coordination structure information; ③ the concentration of different elements in the same compound can be quantitatively analyzed by calculating the peak area, and the concentration of different oxidation states of the same element can also be analyzed; and ④ it is sensitive to microsurface analysis, generally only 0.5–10 nm depth of the material surface can be detected.

There are five main types of characteristic curves in XPS spectra. The first is the photoelectron line, which is the spectral peak with the highest intensity, the smallest peak width, and the best symmetry in the spectrum. Each element corresponds to a photoelectron line at a specific position in the spectrum, so it can be used to qualitatively analyze the elemental species and valence states in the material. The second type is the O KLL. After the photoionization of the inner electron of an atom, a hole is left in the inner layer, and the atom is in an excited state. To transform to a stable state, the atom releases energy through a radiation leap, which excites another electron into a free electron and produces a spectral line called the O KLL. The common O KLL spectral peak is the Osher line, where the left letter "K" represents the electron layer of the starting hole, the middle letter "L" represents the electron layer to which the electron filling the starting hole belongs, and the right letter "L" represents the electron layer of the emitting hole. The third type is satellite lines, which are fluorescent X-rays produced when electrons on different energy levels jump into each other. The fourth type is the energy loss curve, which is the companion peak on the graph after the loss of energy when photoelectrons cross the sample surface in an inelastic collision. The magnitude of the characteristic energy loss is related to the nature of the sample. The fifth type is the vibration excitation line and vibration line of electrons, that is, due to the formation of vacancies in the inner shell layer, the atomic center potential changes abruptly caused by the valence shell layer electron leap in photoelectric emission. If the valence shell layer electrons leap to the higher energy level of the bound state, then that is called the vibration excitation of electrons. If the valence shell layer electrons leap to the unbound continuous state to become free electrons, then that is called the vibration separation of electrons.

As shown in Figure 2.3, the elemental composition in the material can be studied by XPS full spectrum in chemical power studies, and the three elements of cobalt (Co), selenium (Se), and carbon (C) may be present in the $Co_{0.85}Se@rGO$ composite [2]. The analysis of the electron ionization layers for the different elements allows obtaining the valence and structure of the different elements. The information on the coordination structure of different atoms can be obtained by performing peak splitting of the XPS pattern. Argon ion etching against the particles can probe the composition of the elements and valence states of the material at different depths. For example, the samples in Figure 2.3 show a gradual increase in the content of divalent cobalt ions ($Co^{2+}$) and a gradual decrease in the content of trivalent cobalt ions ($Co^{3+}$) in the material with increasing depth, which indicates that the material surface is oxidized. Similarly, the relationship between the components of the solid–electrolyte interphase (SEI) film formed on the surface of the electrode material and the depth change can be studied in the same way.

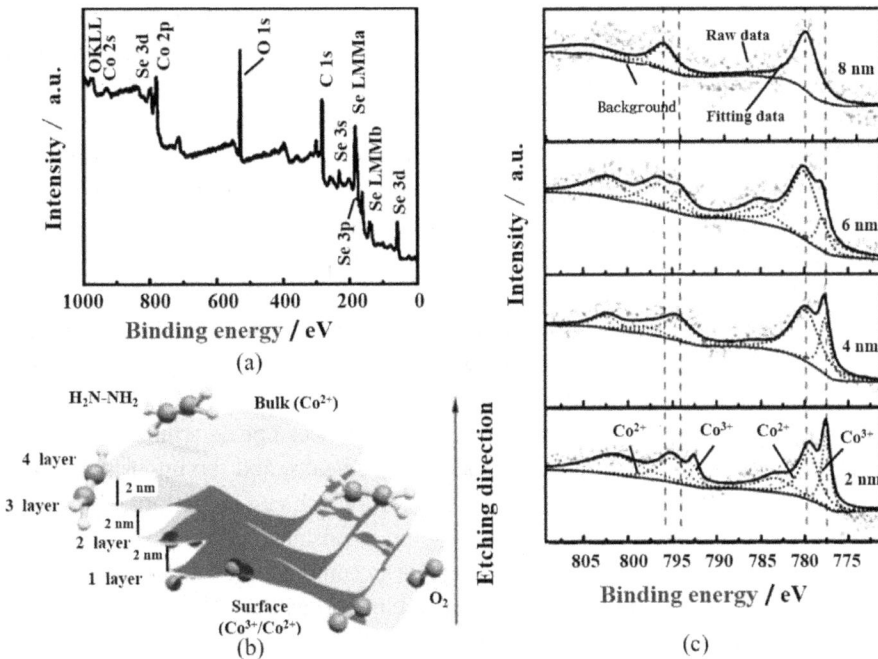

**Figure 2.3:** XPS spectrum: (a) XPS full spectrum of $Co_{0.85}Se@rGO$; (b) schematic diagram of components at different depths; and (c) XPS scan spectrum of different depths in $Co_{0.85}Se@rGO$ [2].

### 2.1.1.4 Spectroscopy analysis technique

Molecular spectra are spectra obtained by spectroscopy of light emitted by molecules or light absorbed by molecules, which reflect the internal motion of molecules. According to the classification of different wavelengths of the spectrum, it can be divided into ultraviolet and visible absorption spectra, fluorescence spectra, infrared (IR) spectra, Raman spectroscopy, and microwave spectra. The most common of them are IR spectroscopy and Raman spectroscopy, both of which are effective means to analyze the composition and structure of materials. Fourier-transform IR spectroscopy (FT-IR) mainly analyzes the asymmetric vibrations of polar groups. Raman spectroscopy mainly analyzes the symmetric vibrations of nonpolar groups and skeletons. The two methods can be used together to identify the structure and components of many types of cell materials, such as organic materials, inorganic materials, and metallic materials.

IR spectroscopy, also known as molecular vibration rotation spectroscopy, belongs to molecular absorption spectroscopy. When a sample is irradiated by IR light with continuously changing frequencies, the molecule absorbs some of the frequencies of radiation, and the vibration or rotation causes the change of dipole moment, so that the vibration–rotation energy-level jumps from the ground state to the excited state, the transmitted light intensity in the corresponding region is weakened, and the curve of the percentage transmittance $T\%$ against the wave number or wavelength is recorded, which is the IR spectrum. It is mainly used for the identification and molecular structure characterization of compounds and also for quantitative analysis. The production of IR spectrum needs to meet two conditions: ① the radiation energy to satisfy the vibrational leap of the material; ② the mutual coupling between the radiation and the material. According to the wavelength range of IR radiation, it can be classified into near-IR spectrum (0.5–2.5 μm), mid-IR spectrum (2.5–25 μm), and far-IR spectrum (25–1,000 μm), among which mid-IR spectrum is more commonly used.

The characteristics of IR spectroscopy include: ① the energy of IR light is low, only in the form of vibration–rotation leap; ② the application range is wide, except for single-atom molecules and mononuclear molecules, almost all organic substances have IR absorption spectrum; ③ the fine structure of molecules can be obtained, and the types of molecular groups and molecular structures can be determined according to the position of wave peaks, the number of wave peaks, and the intensity of wave peaks; ④ IR spectroscopy is a nondestructive test, which can be performed for samples in three states: solid, liquid, and gas; ⑤ the acquisition speed is fast, and the IR spectrum can be acquired in only a few seconds by combination with the liquid nitrogen cooling function; ⑥ it can be used in combination with other tests to determine the changes in the sample during the reaction, such as in combination with chromatography (GC-FT-IR) to achieve qualitative analysis.

IR spectra can be divided into two categories according to the form of vibration of molecules: telescopic vibration IR spectra and deformation vibration IR spectra. Telescopic vibrations are vibrations along the axis between the nuclei of the atoms, which only change in bond lengths and not in bond angles, and are represented by a letter $v$. Telescopic vibrations are divided into asymmetric telescopic vibrations ($v_{as}$) and symmetric telescopic vibrations ($v_s$). Deformational vibrations are vibrations in which the bond length remains constant while the bond angle changes, and are denoted by the letter $\delta$. Deformation vibration is divided into in-plane deformation vibration ($\delta$ in-plane) and out-of-plane deformation vibration ($\delta$ out-of-plane). Among them, in-plane deformation vibration can be further divided into in-plane wobble ($\rho$) and shear wobble ($\delta_s$), and out-of-plane deformation vibration can be divided into out-of-plane wobble $\omega$ and twist vibration $\tau$.

Usually, the horizontal coordinate of the IR spectrogram is the wavelength ($\lambda$, μm) or wave number ($1/\lambda$, cm$^{-1}$) of the absorbed light band, and the vertical coordinate is the percent transmittance $T\%$. The percentage transmittance is defined as the percentage of radiant light that passes through the sample substance:

$$T\% = I/I_0 \times 100\% \tag{2.4}$$

where $I$ is the transmitted intensity and $I_0$ is the incident intensity.

The information in the IR spectrum contains peak position, number of peaks, and peak intensity. For peak position, the larger the force constant of chemical bond, the smaller the atomic folding mass, the higher the vibration frequency of the bond, the higher the absorption peak will appear in the high wave number region (in short wavelength region); instead, the absorption peak appears in the low wave number region (in long wavelength region). While the peak number is related to the molecular degrees of freedom. There is no IR absorption when there is no instantaneous dipole moment change. For peak intensity, the greater the difference in the electronegativity of the atoms at the two ends of the bond (the greater the polarity), the intensity of the absorption peak may stronger. There are two main types of peaks in IR spectra, one is a jump from the ground state to the first excited state, producing a strong absorption peak called the fundamental peak; the other is a direct jump from the ground state to the second excited state, producing a weak absorption peak called the doubling peak.

The organic ligand components in the structure of Prussian blue and its analogs as viable cathode materials for sodium-ion batteries can be identified by FT-IR curves [3]. As shown in Figure 2.4, various groups appear in specific regions of the IR spectrum in the corresponding absorption bands whose approximate positions are fixed. In the first peak region from 4,000 to 2,500 $cm^{-1}$, the main absorption range is the X–H stretching vibration, where X represents O, N, C, and S, corresponding to the O–H bond, N–H bond, and C–H bond stretching vibrations. In the second peak region from 2,500 to 2,000 $cm^{-1}$, the absorption range of the stretching vibrations of triple bonds and cumulative double bonds, including $-C \equiv C-$, $-C \equiv N$, $-N = C = O$, and $-N = C = S$ and other functional group vibrations, which interfere little with each other and are easily identified. In the third peak region from 2,000 to 1,500 $cm^{-1}$, it is mainly the stretching vibrations of double bonds, including $C = O$ bond, $C = C$ bond, $C = N$ bond, $N = O$ bond, and N–H bond. In the fourth peak region from 1,500 to 600 $cm^{-1}$, it is mainly the stretching vibration of X–C bond and X–O bond.

When a transparent sample is irradiated with monochromatic light, most of the light is transmitted while a small portion of the light is scattered by the sample in all directions. Scattering can be divided into Rayleigh scattering and Raman scattering.

Rayleigh scattering is named after the British physicist Rayleigh. It is also known as molecular scattering. Rayleigh scattering is the scattering of incident light by particles whose radius is much smaller than the wavelength of light.

Rayleigh scattering has the following characteristics: ① the intensity of the scattered light is inversely proportional to the fourth power of the wavelength of the incident light; ② the scattered fluxes of the front and back halves of the particle are equal and distributed according to the relationship of $(1 + \cos \theta)$; ③ the scattered light in the forward direction $(\theta = 0°)$ and the backward direction $(\theta = 180°)$ is the strongest, which is one time stronger than that in the vertical direction $(\theta = 90°, 270°)$; ④ the scattered light in the forward and backward directions is in the same state of polarization as the incident light, while the scattered light in the vertical direction is fully polarized, that is, its parallel component (the component whose

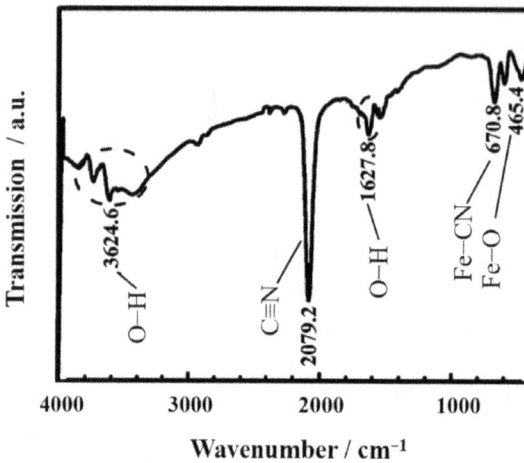

**Figure 2.4:** FT-IR curve of BR-FeHCF [3].

vibration direction is parallel to the plane of observation, which is composed of the incident and scattered light) is zero, and only the vertical component exists.

The intensity of Rayleigh scattered light is inversely proportional to the fourth power of the incident light wavelength $\lambda$:

$$I(\lambda) \text{ scattering} = I(\lambda) \text{ incident}/\lambda^4 \tag{2.5}$$

where $I(\lambda)$incident is the light intensity distribution function of the incident light.

According to the function relationship, the blue light with shorter wavelength is more easily scattered than red light with longer wavelength.

Indian physicist C. V. Raman first discovered the Raman scattering effect in 1928 and was awarded the Nobel Prize in Physics in 1930.

Raman scattering is the scattering that occurs when a photon collides inelastically with a molecule. There are two types of jump energy differences in Raman scattering. The first is that the molecule in the vibrational ground state ($v_0$) is excited to a higher unstable state under the action of photons and then returns to the lower energy level of the vibrational excited state ($v_1$), at which time the excitation light energy is greater than the scattered light energy, producing Stokes lines of Raman scattering, and the scattered light frequency is less than the incident light frequency. The second is the interaction of photons with molecules in the vibrational excited state ($v_1$), which excites the molecules to a higher unstable state and then returns to the vibrational ground state ($v_0$), where the energy of the scattered light is greater than the energy of the excited light, producing anti-Stokes scattering, and the scattered light frequency is greater than the incident light frequency. Stokes lines and anti-Stokes lines are collectively referred to as Raman spectral lines. In general, the vast majority of molecules are in the vibrational energy-level ground state, thus, the intensity of the Stokes lines is much greater than that of the anti-Stokes lines. But the Raman shift is the frequency difference ($\Delta v$) between the Raman scattered light and the incident light. For different substances, $\Delta v$ is different. For the same substance, $\Delta v$, independent of the frequency of the incident light, is a characteristic physical quantity characterizing the vibrational–rotational energy levels of molecules and is the basis for qualitative and structural analyses.

Raman spectroscopy mainly captures the information that a molecule has a change in polarization rate during the vibrational leap. IR spectroscopy captures the information of dipole moment change during the vibrational leap. Therefore, Raman spectroscopy is mainly for nonpolar groups and symmetric molecules, and IR spectroscopy is mainly for polar groups and nonsymmetric molecules. For vibrations with symmetry relationship, IR spectra are invisible and Raman spectroscopy are visible; for vibrations without symmetry relationship with the center of symmetry, IR spectra are visible and Raman spectroscopy are invisible, and they form a complementary relationship with each other.

Compared with IR spectroscopy, Raman spectroscopy has wider application scenarios and can measure and analyze more microstructural information. The specific advantages of Raman spectroscopy contain the following points:

(1) Low requirements for samples. Various samples such as solid powder, polymer, fiber material, single crystal material, and solution material can be analyzed by Raman spectroscopy. Raman spectroscopy also has no special requirements for the shape, size, and transparency of the sample.

(2) Test directly without preparation. However, for laser micro-Raman, the laser irradiating the sample for a long time may cause the sample with low crystallinity to be scorched.

(3) Some bands which are weakly absorbed in IR spectroscopy may be strongly absorbed in Raman spectroscopy, thus facilitating the detection of these groups, such as S–S bond, C–C bond, C = C bond, N = N bond, and other groups, which are weakly absorbed in IR spectroscopy have stronger signals in Raman spectroscopy.

(4) Wide measurement range ($25\ \mathrm{cm^{-1}}$) in the low wavelength direction of Raman spectroscopy is favorable for providing information on the vibrations of heavy atoms.

(5) Raman spectroscopy is especially suitable for the study of aqueous systems, where the Raman scattering of water is extremely weak, which is more advantageous for the study of biological macromolecular materials.

(6) Better resolution than IR spectroscopy.

The Raman spectroscopy of r-PBMN (PB materials containing $Mn^{2+}$ and $Ni^{2+}$ obtained by direct physical mixing) and s-PBMN (PB materials containing $Mn^{2+}$ and $Ni^{2+}$ obtained by adding chemical inhibitors) materials with different synthesis conditions is shown in Figure 2.5 [4]. The variation of Fe in the compounds shows intensity peaks at $2,132\ \mathrm{cm^{-1}}$ and $2,092\ \mathrm{cm^{-1}}$, indicating the presence of Fe and Mn ions with one divalent ion and nearly two Na ions in the materials, respectively ($F^{II}$–CN–$Mn^{III}$). Based on the increase of the peak intensity, it can be assumed that the amount of Na ions increases with the slowing down of the reaction rate. For some materials with poor crystallinity, Raman spectroscopy can effectively determine the structure and composition of the material and is also an effective complement to the XRD technique.

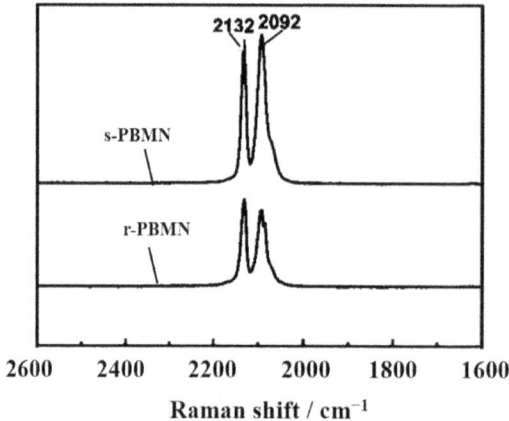

**Figure 2.5:** Raman spectra of PBMN under different synthesis conditions [4].

To determine the Raman spectroscopy of an unknown substance, it is only necessary to find out the characteristic Raman spectroscopy of the substance to identify the species of the substance.

### 2.1.1.5 Scanning electron microscope technique

Scanning electron microscope (SEM) technique is an electron microscopic technique that uses an electron beam to scan the surface of a sample to obtain information about the sample. The image signals include secondary electrons, backscattered electrons, and absorbed electrons, where secondary electrons are the main image signals.

A secondary electron is an electron outside the nucleus that is bombarded by the incident electron beam and leaves the surface of the sample. When an incident electron with high energy is shot into the sample, many free electrons can be produced, and 90% of which come from the valence electrons in the outer layers of the sample atoms. The secondary electrons have the following characteristics: ① the energy is low (not more than 50 eV) and can be easily confused with backscattered electrons; ② the secondary electron emission depth is 5–10 nm from the surface layer, which can effectively identify the surface morphology of the samples; ③ no obvious dependence exists between the yield of secondary electrons and the atomic number, therefore, the identification of components cannot be performed. Backscattered electrons are part of the incident electrons that are bounced back by the nuclei in the solid sample, including elastic backscattered electrons and inelastic backscattered electrons. Backscattered electrons have the following characteristics: ① they come from a depth range of several hundred nanometers from the sample surface, reflecting the three-dimensional shape of the materials; ② the yield increases as the atomic number of the sample increases, so they can be used to show the atomic number lining and qualitative component analysis. Absorbed electrons are the electrons that are finally absorbed after the incident electrons, after which the incident electrons are injected into the sample. In addition to the above three forms of electrons, there are also transmission electrons, characteristic X-rays, and oscillating electrons. Transmission electrons are produced when a high-energy electron beam with a diameter of less than 10 nm irradiates a thin sample, so the transmission electron signal is determined by the thickness, composition, and crystal structure of the microregion. When the inner electrons of a sample atom are excited or ionized by the incident electrons, the atom is in a higher energy excited state, and the outer electrons will leap to the inner layers to fill the vacancies of the inner electrons. Thus, the characteristic X-rays are produced. The electron is created by the energy released during the energy leap of an electron in the inner layer of an atom to excite another electron in the vacancy layer. This excited electron is called an oscillator.

The SEM works by emitting an electron beam from the uppermost electron gun, which is focused by the gate, and then, under the action of the accelerating voltage, passes through an electron optical system. This consists of electromagnetic lenses and converges into a fine electron beam focused on the sample surface. A scanning coil is equipped on the final lens, and the electron beam is scanned on the sample surface by its action. The generated signals, such as secondary electrons, backscattered electrons, absorbed electrons, and characteristic X-rays, are received by the corresponding receivers and sent to the gate of the picture tube after amplification. When the electron beam hits a point on the sample, a corresponding bright spot appears on the fluorescent screen of the Cathode ray tube (CRT). By scanning the image point by point, the overall morphology of the sample is obtained.

SEM can be used in conjunction with other instruments to perform many different functions. Energy-dispersive spectrometer (EDS) is an instrument used to analyze the composition of microregions of materials, including point analysis (electron beam is fixed in a certain point range for qualitative or quantitative analysis), line scan (electron beam is scanned along a line of analysis to obtain the distribution curve of elemental content changes), and surface distribution (when the electron beam is scanned on the surface of the sample, the distribution of the elements on the surface of the sample is shown by colored bright spots, and the number of which represents the elemental content). Electron backscattered diffraction is a technique used to analyze the orientation of crystal microregions and the crystal structure of materials.

SEM can be used in sodium-ion batteries to reveal the morphology/microstructure, size and homogeneity, composition, and crystallographic properties of the materials. For cathode materials, especially transition metal oxide materials, the main observations are the layered structure, the particle size and agglomeration, and so on. For the anode electrode materials, especially the transformation reaction anode materials, we mainly observe the special morphology, the spatial construction to relieve the volume expansion, the homogeneity of the composite materials, and the possible hierarchical structure.

### 2.1.1.6 Transmission electron microscopy technique

Transmission electron microscope (TEM) technique is an important method to characterize the microscopic morphology, crystal structure, and chemical composition of materials. TEM is mainly composed of the following parts: ① illumination system, mainly composed of electron gun and spotting mirror; ② imaging part, mainly composed of sample chamber, objective lens, intermediate mirror, and projection mirror; ③ image observation and recording system, mainly composed of fluorescent screen, camera, data display, and other components. The overall working principle is that the system, such as objective, intermediate mirror, and lens, is imaged in a building block manner. The image of the previous lens is the object of the next lens $i$, and the total magnification is the product of the magnification of each lens. The sample of TEM requires high thickness of the material. For bulk samples, the surface replication techniques (including plastic primary replication, carbon primary replication, plastic carbon secondary replication, and extractive replication) and sample thinning techniques (including cutting, metallographic sandpaper grinding, ion beam thinning, and electrolytic polishing) are usually used. For powder samples, samples can be prepared by ultrasonic dispersion.

Electron diffraction is an important application of TEM. The ability to combine morphological observation and structural analysis on the same specimen make electron diffraction of TEM pivotal in the application. In the diffraction pattern of TEM, for different specimens, different diffraction methods can be used to observe various forms of diffraction results, such as single-crystal electron diffraction pattern, polycrystalline electron diffraction pattern, and amorphous electron diffraction pattern. The electron diffraction pattern of a single crystal is like a magnified projection of a two-dimensional cross section of a crystal inverted easy dot matrix on a negative. From the electron diffraction pattern on the negative, some characteristics of the crystal structure and symmetry can be visually identified, which makes the study of crystal structure simpler than that by X-ray. That also has the advantages of strong scattering ability of the substance for electrons (about 10,000 times that of X-ray) and short exposure time. In addition, TEM can also be combined with energy spectroscopy to characterize the distribution of elements in nanomaterials, especially for special heterogeneous structures.

### 2.1.1.7 Nuclear magnetic resonance technique

A spin nucleus in an external magnetic field ($H_0$) receives electromagnetic radiation of a certain frequency ($v$). When the energy of the radiation is exactly equal to the energy difference between two different orientations of the spin nucleus, the spin nucleus in the low energy state absorbs the electromagnetic radiation and jumps to the high energy state. This phenomenon is called nuclear magnetic resonance (NMR). In 1951, Arnold discovered the NMR signal of ethanol and its relationship to structure. In 1953, Varian prototyped the first NMR instrument to study the effect of molecular structure on the magnetic field around hydrogen atoms and developed NMR spectra for resolving molecular structures. Over time, NMR techniques have evolved from the initial $^1H$ to advanced NMR techniques such as $^{13}C$ spectroscopy and two-dimensional NMR spectroscopy mapping. At the same time, the ability of NMR techniques to resolve molecular structures has become more and more powerful. After the 1990s, techniques relying on NMR information to determine the tertiary structure of protein molecules have been investigated, making the precise determination of the molecular structure of solution-phase proteins possible. Currently, the most studied and applied NMR spectra are $^1H$- and $^{13}C$-NMR. $^1H$-NMR (or PMR) of hydrogen can point out the position of hydrogen in the carbon skeleton. $^{13}C$-NMR (or CMR) of carbon can point out how the carbon skeleton composed.

The NMR instrument can be divided into continuous-wave NMR spectrometer and pulsed Fourier-transform NMR spectrometer according to the RF source and scanning method. According to the magnet, the NMR can be divided into permanent magnet, electromagnet, and superconducting magnet. Based on RF (resonance frequency of $^1H$ nucleus), it can be divided into 60, 80, 90, 100, and 200 MHz. The main components of the NMR instrument are magnet (provide a strong and uniform magnetic field), sample tube (a glass tube of uniform mass with a diameter of 4 mm and a length of 15 cm), RF oscillator (provide an RF wave irradiating the sample in the direction perpendicular to the main magnetic field), scanning generator (a Helmholtz coil mounted on the magnetic pole, providing an additional variable magnetic field for scanning measurements), and RF receiver (used to detect the NMR signal; this coil, the RF generator, and the scan generator are perpendicular to each other).

Characterization of the storage processes and coordination structures of sodium ions in both electrode and electrolyte materials requires the solid-state NMR (SS-NMR) technique. The principle of SS-NMR technique is to make the powder sample rotate rapidly to simulate the rapid motion of molecules in solution and eliminate the unfavorable factors that make the peaks broaden in order to obtain a spectrum with good resolution. This rotation technique is called magic angle rotation. At static or low rotation speed, the solid hydrogen spectrum is very broad. Increasing the rotation speed will make the SS-NMR become some continuous peaks with some distinction. By increasing the rotation speed again, independent peaks can be clearly distinguished. The SS-NMR technique also has the following characteristics:

① The SS-NMR technique can measure far more samples than the solution NMR technique, because the latter is limited by the solubility of the sample, and it is often difficult to analyze samples with poor solubility or easy deterioration after dissolution, which does not exist in the SS-NMR. ② the SS-NMR technique is better than the solution NMR technique, which can determine not only $^1$H, $^{19}$F, $^{13}$C, $^{15}$N, $^{29}$Si, $^{31}$P, $^{207}$Pb with a spin quantum number of 1/2 but also quadrupole nuclei, such as $^2$H and $^{17}$O. ③ The kinetics of the corresponding physical processes can be analyzed in situ technique, which helps to fully understand the relevant processes. ④ The pulse program can be set according to the information obtained, thus purposefully and selectively suppressing the unwanted information, retaining the desired information. As shown in Figure 2.6, at lower temperatures, the $^{23}$Na resonance broadens under second-order quadrupole interactions and dipole interactions; at 393 K, the line shape of the cubic phase Na$_3$PS$_4$ becomes isotropic and the central gravity moves toward higher ppm ($10^{-6}$) values, indicating that the mobility of the sodium ion at this temperature is able to balance the quadrupole interactions.

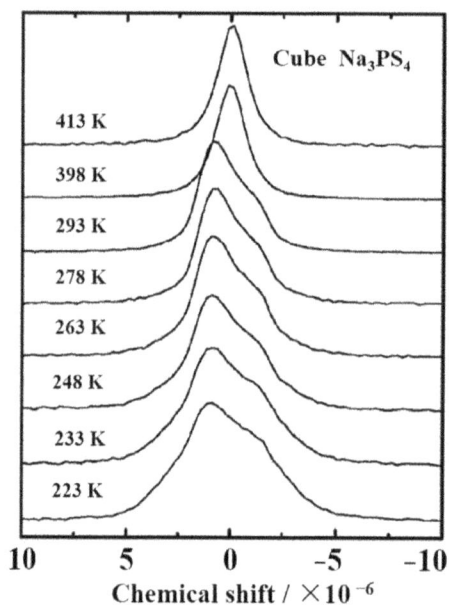

**Figure 2.6:** NMR spectrum of cubic structure Na$_3$PS$_4$ at different temperatures [5].

### 2.1.1.8 Thermogravimetric analysis technique

Thermogravimetric analysis (TGA) technique is a technique that measures the relationship between the mass of the sample and temperature change at a programmed controlled temperature. TGA is used to study the thermal stability and components of materials. TGA is the observation of the substantial mass change of a specimen during the heat exposure process. The basic principle of TGA is to convert the displacement of the balance caused by the change in the mass of the sample into electrical power, and this small amount of power is amplified by an amplifier and sent to a recorder for recording; and the magnitude of the power is proportional to the amount of change in the mass of the sample. When the measured substance has sublimation, vaporization, decomposition of gas, or loss of crystalline water in the heating process, the mass of the measured substance will change. This occurs when the thermogravimetric curve is not straight but decreases. By analyzing the thermogravimetric curve, it is possible to know at what temperature the substance under test starts to change, and calculate how much material lost based on the amount of mass lost.

TGA can usually be divided into two categories: static and dynamic methods. ① Static method includes isobaric mass change measurement and isothermal mass change measurement. Isobaric mass change measurement is a method to measure the equilibrium mass of a substance at a constant volatile partial pressure in relation to temperature under a programmed temperature control, and isothermal mass change measurement is a method to measure the mass of a substance in relation to pressure at a constant temperature, which is highly accurate but time-consuming. ② The dynamic method is what we often call TGA and microquotient TGA. Also known as derivative thermogravimetry (DTG), it is the first-order derivative of the TGA curve with respect to temperature (or time). The DTG curve is obtained by plotting the rate of mass change of a substance ($dm/dt$) against temperature $T$ (or time $t$).

In research for sodium-ion batteries, DTG is mainly used to study the thermal stability of electrode materials and electrolytes. By calcination in an inert gas, the decomposition temperature and crystallization temperature of the electrode material can be studied. Calcination in air allows the study of the composite components and content of the electrode material, especially for carbon composite.

Differential thermal analysis (DTA) is an indirect expression of the change of heat (heat absorption and exotherm) during the change in temperature difference ($\Delta T$). There are many factors affecting the DTA curve and it is difficult to analyze quantitatively, for which the differential scanning calorimetry (DSC) was developed. DSC is a thermal analysis method that uses a program to control the temperature and measure the relationship between the difference in power (heat flow) input to the sample and the reference material and the temperature. The recorded curves are called DSC curves. The main features of DSC are high resolution and sensitivity. DSC can not only cover the functions of DTA but also quantitatively determine various thermodynamic parameters such as enthalpy, entropy, and specific heat; hence, it is widely used in applied material science and theoretical research. It takes the rate of heat absorption or exotherm of the sample, that is, heat flow rate $dH/dt$ (in mJ s$^{-1}$) as the vertical coordinate and the temperature $T$ or time $t$ as the horizontal coordinate, and can determine a variety of thermodynamic and kinetic parameters, such as specific heat capacity, heat of reaction, heat of transformation, phase diagram, reaction rate, crystallization rate, crystallinity of polymers, and sample purity.

### 2.1.1.9 Specific surface area and pore size analysis techniques

The specific surface area is the total area per unit mass of the material. The main methods of specific surface area testing are adsorption, gas permeation, and other methods. Adsorption is the most commonly used method with high accuracy. The adsorption method is divided into iodine adsorption, mercury adsorption, and low-temperature nitrogen adsorption according to the different adsorption media. Among them, low-temperature nitrogen adsorption method is the most widely used. Based on the different methods of determining the adsorbed mass and adsorption amount, the low-temperature nitrogen adsorption method can be divided into dynamic chromatography, static volumetric method, and weight method. Here we mainly introduce the dynamic chromatographic method and the static volumetric method.

The dynamic chromatographic method is to first place the powder sample in a U-shaped tube, then pass in a certain proportion of a mixture of adsorbent gas ($N_2$) and carrier gas (He), and after the mixture flows completely through the sample, calculate the adsorption amount of the sample to be measured according to the change of gas concentration before and after adsorption. This method has high analytical speed, low lower analytical limit, high accuracy, resolution, and repeatability, which is suitable for testing small and medium adsorption amounts with small specific surface area of the sample.

The static capacity method is to first put the powder sample to be tested into a closed test tube of a certain volume, and then inject a certain pressure of adsorbent gas ($N_2$) into the sample tube, and finally determine the adsorption amount of the sample on the adsorbent molecule ($N_2$) according to the change of pressure or mass before and after adsorption. This method has high resolution and accuracy for medium to large specific surface area and developed pore samples, and is suitable for testing the specific surface area and pore size distribution of samples with large comparative surface area.

Methods for pore size analysis include gas adsorption, pressure pumping, bubble point, suspension filtration, and liquid–liquid exclusion. The gas adsorption method is based on low-temperature nitrogen adsorption to obtain the pore volume and thus the porosity. The pore size test range is from 0.35 to 500 nm, but this method cannot test micron-sized pores. The pressure pump method is to press the mercury into the dry porous sample with the help of external force, determine the variation of the volume of mercury penetrated into the sample with the pressure, and calculate the pore size distribution of the sample from it. This method includes the impermeable U-shaped pores, so the reference value of the measurement results is not very high. The testing principle of bubble point method is: when the pore is blocked by the liquid wetting agent, due to the effect of the surface tension of the wetting agent, at this time, if the pore is opened by gas, it is necessary to apply a certain pressure to the gas. And the smaller the pore, the greater the pressure required to open the pore. The pore size distribution of the sample can be obtained by comparing the relationship curves between pressure and gas flow rate of porous materials under dry and wet conditions and calculating according to a certain mathematical model. The pore size test range is from 20 nm to 500 μm, and this method is not applicable to the measurement of membrane materials with small pore size. The pore size distribution can be calculated by comparing the changes of particle size distribution in the original suspension and the permeate, and the diameter of the particles in the permeate is the pore size of the porous material. The principle of liquid–liquid exclusion method is similar to the bubble point method, that is, using another liquid that is insoluble with the wetting agent instead of gas as the pore-opening agent. Due to the small liquid–liquid interfacial tension, only a very small pressure is needed to measure the larger pore size; thus, the measurement error of pressure is larger, and the best measurement range is 10 nm to 200 μm.

### 2.1.1.10 Neutron diffraction analysis technique

Neutron diffraction (ND) analysis technique refers to the Bragg diffraction that occurs when neutrons with a Broglie wavelength of about 1 Å pass through crystalline matter. Currently, the ND method is one of the most important tools to study the structure of electrode materials and solid-state electrolytes. Neutrons, like other microscopic particles, have wave-particle duality. When a neutron wave is directed to a crystal surface at a grazing angle $\theta$, the difference in the neutron wave range reflected on two adjacent crystal surfaces is $2d \sin \theta$. Like X-rays, when $2d \sin \theta$ is equal to an integer multiple of the neutron wavelength, these two reflected waves are coherent and strengthened, and a clear diffraction peak appears due to the coherence effect of many layers. Compared to XRD, ND is characterized as follows: ① neutrons are more sensitive to scattering of lighter C, H, O, and their isotopes than XRD, and electrons are less sensitive to scattering of lighter nuclei and their isotopes than neutrons due to the Coulomb shielding effect of electrons in the nucleus. ② For magnetic materials, the ND spectrum contains magnetic ions and magnetic structure information, and the specific magnetic periodic structure can be obtained by ND-specific peak positions and certain model fitting.

### 2.1.2 Electrochemical performance testing of electrode materials

Electrochemical performance testing of electrode materials is important. The tests described in this chapter are based on a two-electrode system assembled with active electrode materials and metallic sodium electrodes. Because of the differences in carriers and electrode systems, sodium-ion batteries also have many differences in electrochemical performance and electrochemical mechanisms compared to conventional lithium-ion batteries. The electrochemical performance of reactive materials is mainly controlled by both thermodynamic and kinetic factors. Based on the Nernst equation, thermodynamic information such as the electric potential, theoretical specific capacity, and electrode potential of the electrode active material can be calculated. These properties are inherent to the active material, are related to the type and structure of the material, and are theoretical upper limits. While in the process of electrochemical reaction, the kinetic factors such as the process and mechanism of redox reaction, the speed of electron conduction and exchange, and the diffusion rate of ions in the active material directly determine the actual electrochemical performance of the electrode material. The kinetic performance of electrode materials is often influenced by the microstructure of the materials, ion doping in the materials, and compounding of the materials, which is one of the key elements for sodium-ion batteries. This section introduces the electrochemical performance test characterization means to study the important issues of specific capacity, cycling stability, multiplicity performance, ion mobility, and electrochemical reaction mechanism.

### 2.1.2.1 Galvanostatic charge/discharge

Galvanostatic charge/discharge is to apply a constant current to the battery and record the change of electrode potential (for one electrode) or cell voltage (for the whole battery) with time. Since charging and discharging are done with a constant current, the time axis is easily converted into a power (capacity) axis.

During charging or discharging, the charging curve or discharging curve has two parameters: time (capacity) and potential (or voltage). The former indicates the charging (discharging) extent, and the latter indicates the state of the electrodes. In other words, the charge/discharge curve is a reflection of the state of the electrode (or battery). The current indicates that an electrochemical reaction is taking place, but it does not directly reflect which electrochemical reaction is taking place. The magnitude of the current indicates the rate of the electrochemical reaction. For potential, the electrochemical reaction can only occur if the corresponding electrode potential is reached.

From the charge/discharge curve, we can get the capacity and specific capacity, the relationship between electrode potential and charge state (for electrode), the working voltage and voltage plateau (for battery), working voltage and charge/discharge current (rate charge/discharge), and the information of electrode process (formation and decomposition of surface film layer, change of crystal shape of electrode material, change of structure, etc.).

### 2.1.2.2 Cyclic voltammetry

Cyclic voltammetry (CV) is to apply a linear scanning voltage to the electrode, scan at a constant rate, and then return to a set starting potential when a set turnaround potential is reached. In the process of positive scanning from the starting potential to the turning potential, the potential changes from low to high, and the active material in the electrode is oxidized, generating oxidation current. In the process of negative scanning from the turning potential to the starting potential, the potential changes from high to low, and the active material in the electrode is reduced, generating reduction current. According to the change of scanning potential, it can be determined whether the peak of CV curves is an oxidation peak or a reduction peak.

When setting the parameters, for a system with good reversibility, the initial setting is open-circuit voltage (OCV), and the cut-off voltage is set to be the same as the initial voltage in order to get a closed loop. The scanning direction is related to the material, and the oxidation reaction occurs in the first step, which should be scanned in the forward direction, and vice versa in the reverse direction. This measurement is commonly used for cathode materials. For systems with poor reversibility, if the system is set according to the previous system, it is not always possible to get a closed-loop voltammetry curve, so the initial setting should look at the first step. Depending on whether the first step is a reduction reaction or an oxidation reaction, the setting is either high or low potential, and the scanning direction is determined at this point. This setting mode is often used for anode materials, but there will be a period of rapid discharge at the beginning of the measurement.

CV is a commonly used method for transient electrochemical measurements of dynamic potentials and is one of the most important tools for electrode reaction kinetics, reaction mechanism, and reversibility studies. As shown in Figure 2.7, the redox reaction processes of HCS-PBMN and PBM electrodes were characterized by CV curves [6]. Among them, two pairs of redox peaks exist at (3.69/3.51) V and (3.46/3.16) V for the HCS-PBMN electrode, corresponding to the high-spin $Mn^{III}/Mn^{II}$ (Mn-HS) electric pair and the low-spin $[Fe(CN)_6]^{3-}/[Fe(CN)_6]^{4-}$ (FeCN-LS) electric pair. It is also shown that he reversible intercalation/deintercalation of Na ion in the HCS-PBMN electrode, especially the Mn-HS electric pair at high pressure, is highly reversible and active. The HCS-PBMN electrode has fast reaction kinetics and voltage stability because the voltage polarization between the two redox pairs is only 180 and 300 mV. The PBM electrode also shows two redox pairs at (3.06/3.52) V and (3.49/3.88) V, respectively. However, the polarization voltage between these two redox pairs increased significantly to 460 and 390 mV, which indicates a relatively sluggish kinetics of Na ion transport. The irreversible peaks located at 3.25 and 3.7 V correspond to side reactions between the electrode and electrolyte interfaces, which may be caused by the Jahn–Teller effect and the dissolution of $Mn^{2+}$.

Different CV curves can be obtained at different scan rates, and different CV curves are usually used for pseudocapacitive contribution calculations. B. E. Conway defined the pseudocapacitive process as a Faraday process occurring on the surface of the electrode material. In contrast to conventional capacitance or bilayer capacitance, although both store charge at the electrode surface, pseudocapacitive behavior is a Faraday process based on ion adsorption/desorption, which is the primary characteristic of pseudocapacitive processes. Whether a material exhibits pseudocapacitive or cell behavior generally depends on the electrode material structure and carrier type [7]. For example, when a material achieves nanoscale dimensions, it usually exhibits pseudo-capacitive properties.

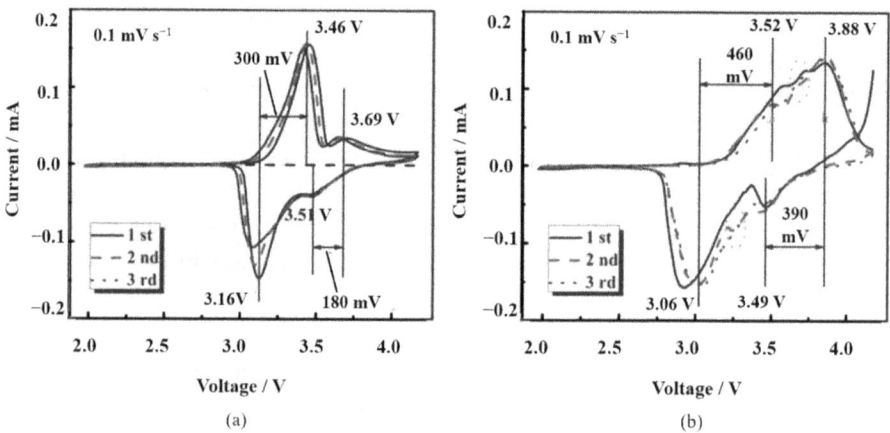

**Figure 2.7:** Cyclic voltammetric curves: (a) HCS-PBMN; (b) PBM (scan rate: 0.1 mV/s, voltage range: 2.0–4.2 V [6]).

As shown in Figure 2.8, bilayer capacitors generally exhibit a capacitive behavior that is not dependent on voltage changes, with the voltage decreasing linearly with time during discharge. On the opposite, batteries usually exhibit independent redox peaks at a specific potential, with the voltage remaining constant for a certain period of time during discharge. The pseudocapacitive behavior, on the other hand, is generally somewhere in between, mediated by both mechanisms. Pseudocapacitive materials usually have one or more of the following electrochemical properties: ① surface redox materials (e.g., $MnO_2$ in neutral, aqueous electrolytes); ② embedded materials (e.g., $Li^+$ in organic systems embedded in $Nb_2O_5$); and ③ embedded materials that exhibit broad and electrochemically reversible redox peaks (e.g., $Ti_3C_2$ in acidic, aqueous electrolytes).

The pseudocapacitive behavior can not only qualitatively explain the process and kinetic mechanism of charge storage in electrode materials but also the contribution of pseudocapacitance to charge storage can be precisely obtained by tests and calculations. In CV tests, different peak current values ($i$, in mA) are obtained

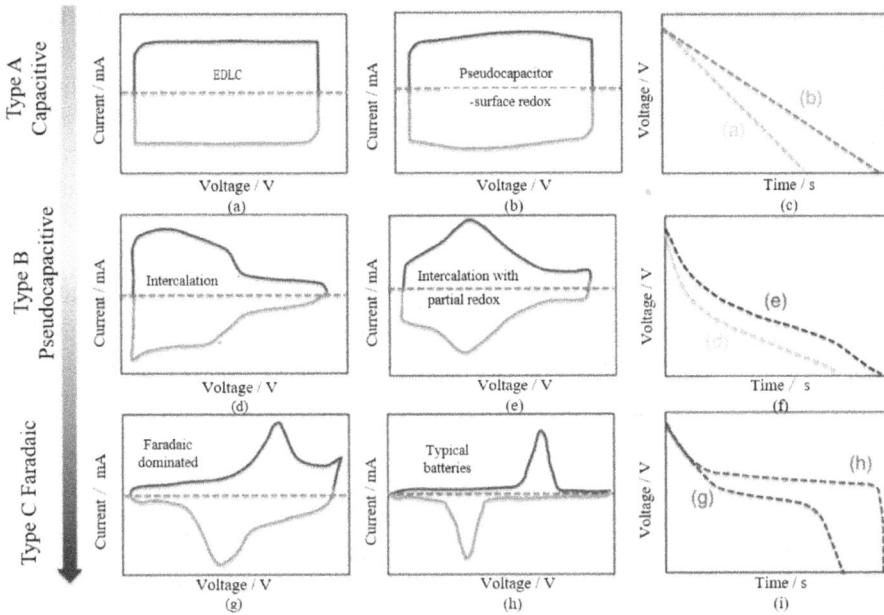

**Figure 2.8:** Cyclic voltammetry (a, b, d, e, g, h) of various energy storage materials and corresponding schematic diagram of constant current discharge curves [8] (c, f, i).

for different voltage scan rates ($v$, in $mV \cdot s^{-1}$) [9]. To distinguish whether the batteries behave diffusely or pseudocapacitively during charging and discharging by the corresponding scan rate to the resulting peak current response, the following equation is calculated:

$$i = av^b \tag{2.6}$$

$$\lg_{10} i = b \lg_{10} v + \lg_{10} a \tag{2.7}$$

where $i$ is the current value of the test, $v$ is the voltage scan rate, and $a$ and $b$ are constants.

If the value of $b$ is 0.5, the electrode material exhibits a diffusion-dominated behavior. If the value of $b$ is in the range of 0.5–1, the electrode material exhibits the behavior of diffusion and pseudocapacitance. If the value of $b$ is greater than or equal to 1, the electrode material exhibits a pseudocapacitive behavior. In addition, the contribution of the pseudocapacitive behavior to the charge storage in the electrode material can be further calculated by

$$i = k_1 v + k_2 v^{1/2} \tag{2.8}$$

$$i/v^{1/2} = k_1 v^{1/2} + k_2 \tag{2.9}$$

where $k_1$ and $k_2$ are adjustable constants. $k_1v$ and $k_2v^{1/2}$ describe the contribution of capacitive and diffusive storage behaviors to the total charge storage, respectively.

A linear fit for $i/v^{1/2}$ and $v^{1/2}$ yields the value of $k_1$ for specific voltage conditions. The corresponding $k_1$ is fitted separately at each specific voltage, and then $k_1v$ is calculated for each specific voltage, which is the contribution of the pseudocapacitor to the current. The $k_1v$ at different specific voltages is connected by a smooth curve for nonlinear fitting. Then the fitted closed curve is integrated to find the area, and the CV curve at a specific scan rate is integrated to find the area. The area of the fitted curve is divided by the area of the CV curve, and the resulting value is the pseudocapacitive contribution at a specific scan rate. In general, the pseudocapacitive contribution of the electrode material increases as the scan rate increases. This indicates that the kinetics of the pseudocapacitance storage reaction is faster than that of the diffusion reaction.

### 2.1.2.3 Electrochemical impedance spectroscopy

Electrochemical impedance spectroscopy (EIS) is one of the important methods for electrochemical measurements. AC impedance method is a method to measure the change of system potential (or current) with time by controlling the current (or potential of the system) through the electrochemical system under the condition of small amplitude and changes with time according to the sinusoidal law, or directly measures the AC impedance of the system, analyzing the reaction mechanism of the electrochemical system and calculating the relevant parameters of the system.

When conducting experimental measurements, parameters such as initial potential, high frequency, low frequency, amplitude, and resting time need to be set. The initial potential, that is, the set potential condition, is usually set to the OCV, then the impedance is measured under the open-circuit condition. If it is lower than the OCV, it is the impedance under the discharge condition, and vice versa for charging. High frequency refers to the highest frequency given to the AC disturbance, and low frequency refers to the lowest frequency given to the AC disturbance. Amplitude refers to the size of the amplitude of the AC disturbance given, the smaller the result obtained the more accurate, while the noise signal will also be larger. Resting time refers to the time the system rests before measurement. The resting time is set to ensure that the electrode material is in full contact with the electrolyte to ensure the accuracy.

For AC impedance controlled by a mixture of charge transfer and diffusion processes, the Nyquist curve consists of a semicircle in the high-frequency region and a slope line in the low-frequency region when the electrode process is controlled by both charge transfer and diffusion processes. The high-frequency region is controlled by the electrode reaction kinetics (charge transfer process) and the low-frequency region is controlled by the diffusion of reactants or products of the electrode reaction. The slope line of diffusion impedance may deviate from 45°, either because the electrode surface is so rough that the diffusion process is partially equivalent to spherical diffusion, or another state variable other than the electrode potential. This variable causes impedance during the measurement. Impedance can be better understood through equivalent circuit diagrams, which are simple series and parallel circuits that are drawn by adapting a complex circuit in an appropriate way. Think of an electrochemical system as an equivalent circuit, which is made up of basic components such as resistance ($R$), capacitance ($C$), and inductance ($L$) combined in different ways (series or parallel). In this case, the AC equivalent circuit diagram draws only the circuit associated with the AC signal in the original circuit, omitting the DC circuit. In the AC equivalent circuit, the coupling capacitor in the original circuit is to be viewed as a path and the coil as an open circuit. The equivalent circuit with an impedance of $R-jX$ is a circuit with a resistor $R$ and an inductor $L$ in series. The equivalent circuit with an impedance of $R-jX$ is a circuit with a resistor $R$ and a capacitor $C$ in series. With the EIS, the composition of the equivalent circuit and the size of each element can be determined, and the electrochemical meaning of these elements can be used to analyze the electrochemical system and the nature of the electrode process.

Figure 2.9 shows the Nyquist curve for PBM and HCS-PBMN electrodes, which consists of two parts, a depressed semicircle in the high-frequency region and a sloping line in the low-frequency region [6]. Usually, the distance from the zero point to the origin of the curve indicates the solution resistance ($R_0$), which is determined by the electrolyte. The diameter of the depressed semicircle indicates the charge transfer impedance ($R_{ct}$) at the interface. The sloping line reflects the Warburg diffusion process ($W_s$) in the electrode. It is clear that the $R_{ct}$ value of the PBM electrode (707.7 $\Omega$) is larger than that of the HCS-PBMN electrode (340.5 $\Omega$). The low interfacial impedance value of the HCS-PBMN electrode illustrates the fast reaction kinetics of Na ion diffusion between the electrode and the electrolyte interface.

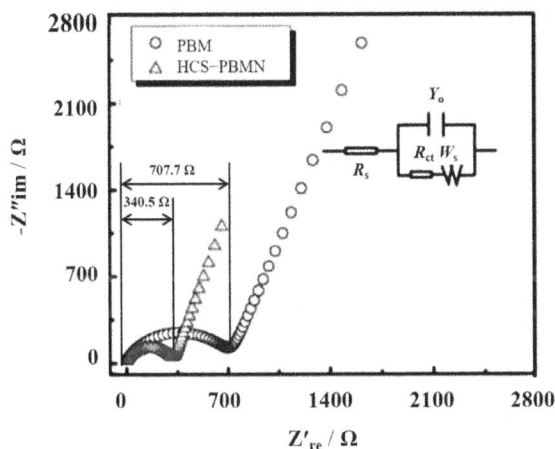

**Figure 2.9:** Nyquist plots of PBM and HCS-PBMN [6].

### 2.1.2.4 Galvanostatic intermittent titration technique

The galvanostatic intermittent titration technique, proposed by the German scientist W. Weppner, is an effective electrochemical tool for determining the diffusion rate of ions in electrode materials. The basic principle is to apply a constant current to the measurement system in a specific environment and cut off the current after a period of time, observe the changes of the applied current and the potential of the system with time, as well as the equilibrium voltage after relaxation, analyzing the changes of the potential with time. The relaxation information of the overpotential of the electrode process can be derived, and then infer and calculate the reaction kinetic information.

The lithium-ion diffusion coefficient can be calculated when the system meets the following conditions: ① the electrode system is isothermal adiabatic; ② the electrode system has no significant volume deformation when current is applied; ③ the electrode response is completely controlled by the diffusion of ions inside the electrode; and ④ the electronic conductivity of the electrode material is much larger than the ionic conductivity. The specific calculation equation is as follows:

$$D_{Na^+} = 4(mV_M)^2 \Delta E_s / \pi (MA)^2 \Delta E_\tau \tag{2.10}$$

where $D_{Na^+}$ is the chemical diffusion coefficient, $V_M$ is the molar volume of the active material, $M$ is the molar mass of the electrode material, $A$ is the real electrode area in contact with the electrolyte, $\Delta E_s$ is the voltage drop between the initial and steady states, and $\Delta E_\tau$ is the total change in cell voltage over a constant pulse time.

During a single voltage response, $\Delta E_s$ and $\Delta E_\tau$ correspond to the positions shown in Figure 2.10(a), respectively. According to equation (2.10), the diffusion coefficient of the electrode material in the cell can be calculated, which will vary with the charging and discharging process (Figure 2.10(b)). In general, the minimum value of the diffusion coefficient generally occurs during the phase transition reaction, mainly because Na ion needs to overcome a large impedance when passing through the two-phase interface.

### 2.1.2.5 Potentiostatic intermittent titration technique

The potentiostatic intermittent titration technique (PITT) is a technique for quantifying the diffusion coefficient based on the current response under bias, assuming that the mobile ions are uniformly distributed in the electrolyte and that Fick's law of diffusion applies. When a small voltage bias (typically a few millivolts open circuit) is applied to a planar electrode, diffusion of the mobile ions (supplied from or into the electrolyte) will occur before equilibrium is reached and the current response can be recorded throughout the process until the current response drops to zero. In the long-time approximation ($t \gg L^2/D$), the relationship between the diffusion coefficient $D$ and the current response $I$ can be approximated by the following equation:

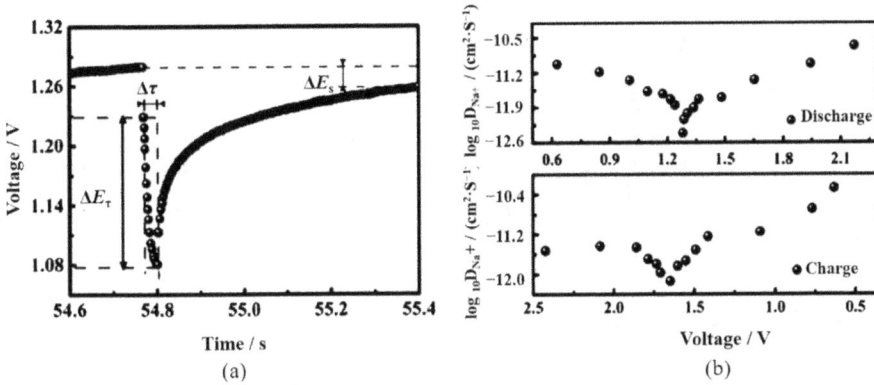

**Figure 2.10:** Calculation of the diffusion coefficient based on the voltage response: (a) at a pulse current duration of 10 min followed by disconnection at 60 min, the voltage response process of Co$_{0.85}$Se NSs@rGO electrode; (b) the charge/discharge process of chemical diffusion coefficient of Na$^+$ in the Co$_{0.85}$Se NSs@rGO electrode [2].

$$D = -(\mathrm{d}\ln i/\mathrm{d}t)(4L^2/\pi^2) \tag{2.11}$$

where $L$ is the thickness of the active material layer and $t$ is the recording time.

Equation (2.11) holds only if the diffusion coefficient $D$ remains constant throughout the time period, and the diffusion coefficient $d$ usually depends on the charge state, so the voltage deviation should be small. Another advantage of small voltage deviation is that the system can be kept in the single-phase range. When the system reaches a new equilibrium state, an additional voltage bias can be applied to study the diffusion characteristics, and thus over the entire voltage range. When using PITT, the transfer of mobile ions is the only influence that must be taken carefully and other effects can be ignored or eliminated [10].

## 2.2 Test analysis of electrolytes

In the study of the physical/chemical properties and electrochemical performance of electrolytes and interfaces, the ionic conductivity, viscosity, density, thermal stability, and electrochemical windows need to be characterized by various characterizations. In the characterization of electrolytes, some traditional characterizations, such as CV, SEM, and XRD, have been very common. And with the progress and development of technology, new characterizations are constantly used for the characterization of electrolytes, such as in situ ND, atomic force microscope (AFM), and SS-NMR. Based on these characterization results, chemical, electrochemical, thermal stability, and some other basic information of the electrolyte and interface can be obtained to support theoretical studies. Figure 2.11 summarizes the commonly used methods for the electrolyte and interface characterization of sodium-ion batteries.

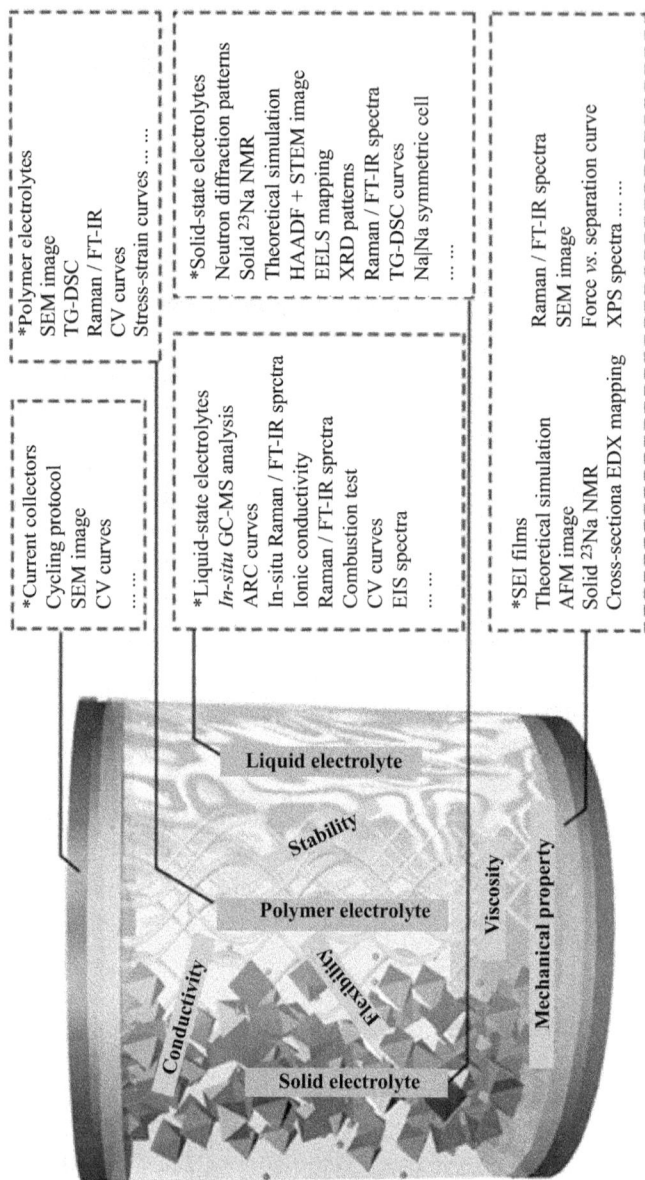

**Figure 2.11:** Summary of research methods on electrolytes (liquid, polymer, and solid) and interfaces of sodium-ion batteries, including electrode and electrolyte interface (SEI film), and the interface between electrode and current collector (Al corrosion).

### 2.2.1 Liquid electrolyte analysis of sodium-ion batteries

The characterization and analysis methods of liquid electrolytes are mainly focused on the components and electrochemical properties of the electrolyte. Among them, FT-IR spectroscopy can be used to study the components of liquid electrolytes [11]. In FT-IR spectroscopy, the bands located at different absorption wavelengths correspond to different organic functional groups in the electrolyte. Different polar functional groups, such as carbonyl ($C = O$), nitrile ($C \equiv N$), sulfonyl ($S = O$), and ether bonds (-O-), are usually observed in the FT-IR spectra of electrolytes. Raman spectroscopy can be used to determine the solubility of solvents for sodium salts and the coordination relationships between cations and solvent molecules [12].

CV and linear sweep voltammetry (LSV) curves can also be used to study the electrochemical stable windows (ESW) of liquid electrolytes. EIS profiles of liquid electrolytes consist of internal components, passivation layers, charge transfer, diffusion, and Warburg resistance [13]. Based on the EIS results, the internal resistance caused by the interfacial compatibility between the separators/electrolyte and the electrolyte/electrode can be obtained and used to evaluate the electrochemical performance of the electrolyte. The ionic conductivity of the liquid electrolyte can be determined by conductivity meter with a DC voltage. A typical EIS profile of sodium-ion detachment and embedding processes in an inlay electrode consists of five components:

(1) Ultrahigh-frequency region: the ohmic resistance associated with the transport of sodium ions and electrons through the electrolyte, porous septum, wires, and active material particles. Represented as a point on the EIS map, this process can be represented by $R_s$.

(2) High-frequency region: a semicircle related to the diffusive migration of sodium ions through the insulating layer on the surface of the active material particles. This process can be represented by $R_{SEI}/C_{SEI}$, where $R_{SEI}$ is the resistance of sodium-ion diffusion migration through the SEI films.

(3) Mid-frequency region: a semicircle associated with the charge transfer process can be represented by $R_{ct}/C_{edl}$, where $R_{ct}$ is the charge transfer resistance or the electrochemical reaction resistance and $C_{edl}$ is the double layer capacitance.

(4) Low-frequency region: a diagonal line associated with the solid diffusion process of sodium ions inside the active material particles, which can be represented by a Warburg impedance $Z_W$ describing the diffusion.

(5) Ultralow-frequency region: consisting of a semicircle associated with the change of the crystal structure of the active material particles or the generation of new phases, a vertical line associated with the accumulation and consumption of sodium ions in the active material is generally rarely measured in the frequency range below 0.01 Hz.

To study the chemical and electrochemical reactions occurring during the discharge process, gas chromatography and mass spectrometry are commonly used to analyze the decomposition gas products of the electrolyte during the cycle [14]. Organic electrolytes may produce decomposition products, such as carbon dioxide, oxygen, and sulfur dioxide, and the gases generated by the structural evolution of the active material should not be neglected. The evolution of the components in the electrolyte during battery operation can also be detected by in situ FT-IR, Raman spectroscopy, and imaging techniques [15, 16].

In addition to the traditional TGA and DSC tests, accelerating rate calorimetry can be used to further explore thermally stable liquid electrolytes by studying the thermal stability of the active material in the electrolyte and the slow pressure change process in the initial thermal decomposition. For this new characterization technique, the onset temperature represents the onset of self-heating, while the self-heating rate shows the decomposition rate and flammability [17].

### 2.2.2 Solid electrolyte analysis for sodium-ion batteries

Conventional XRD has been widely used for structural characterization of electrolytes, especially for polymer electrolytes and inorganic solid electrolytes. For polymer electrolytes, the effect of inorganic fillers and polymer blends on the crystallinity of solid polymer electrolytes is investigated by XRD to guide the direction of improving ionic conductivity. For inorganic solid electrolytes, XRD refinement can further understand the crystal structure parameters of electrolytes.

For the thermal stability of solid electrolytes, TGA and DSC tests are commonly used. The initial value of mass loss in the TGA curve or the initial value of exothermic peak in DSC mode corresponds to the decomposition temperature of the electrolyte. At the same time, the upper and lower operating temperatures of the solid electrolytes can be obtained.

CV and LSV characterization are the most commonly used methods to study the electrochemical stability of electrolytes. For example, CV tests and LSV tests can be used to evaluate the ESW of electrolytes, as shown in Figure 2.12(a) and (b) [18, 19]. In the inert electrode system, the initial potential of electrolyte decomposition, which is the upper limit of ESW, can be directly observed from the LSV curve. And the CV test can be used to study the sodium storage process and to analyze the interfacial pseudocapacitance and the sodium-ion embedding/detachment reaction on the electrode material. Also, the CV test can be used to understand the formation of SEI and other irreversible processes of side reactions, and thus determine the lower limit of ESW.

In addition to CV test, EIS is also an effective method to study the electrochemical properties of electrolytes [20].The EIS curve consists of two parts, high-frequency semi-circular and low-frequency sloping line, corresponding to charge transfer impedance and Warburg impedance, respectively. Generally, at low voltages, the increase of charge transfer impedance indicates the formation of SEI film; at high voltages and OCV, the polarization state of EIS indicates the electrochemical stability of the electrolyte at high voltages. In addition, the ionic diffusion coefficients of Na ion in the electrolyte and interface can be obtained from the EIS results according to eqs. (2.12) and (2.13), as shown in Figure 2.12(c) [20]:

$$Z' = R_e + R_{ct} + \sigma\omega^{-1/2} \tag{2.12}$$

$$D_{Na} = R^2 T^2 / 2A^2 n^4 F^4 C_{Na}^2 \sigma^2 \tag{2.13}$$

where $Z'$ is the impedance (real part), $R_e$ is the SEI membrane resistance, $R_{ct}$ is the charge transfer resistance, $\omega$ is the frequency, $\sigma$ is the Warburg coefficient, $D_{Na}$ is the sodium-ion diffusion coefficient, $R$ is the gas constant, $T$ is the absolute temperature, $A$ is the surface area, $n$ is the number of electrons transferred per unit molecule of oxidation, $F$ is the Faraday constant, and $C_{Na}$ is the concentration of sodium ions.

The steady-state polarization curve shows the relationship between current density and overpotential, revealing the kinetic parameters of the rate control step in the electrode reaction. Combining the EIS results, the Na ion mobility can be calculated according to the following equation:

$$t_{Na+} = I_{ss}(\Delta V - I_0 R^0_{ei}) / I_0(\Delta V - I_{ss} R^{ss}_{ei}) \tag{2.14}$$

$$I_0 = \Delta V / (R^0_b + R^0_{ei}) \tag{2.15}$$

where $I_0$ is the initial state current (A), $I_{ss}$ is the steady-state current (A), $\Delta V$ is the polarization voltage (V), $R^0_{ei}$ is the electrode/electrolyte interfacial impedance in the initial state ($\Omega$), $R^{ss}_{ei}$ is the electrode/electrolyte interfacial impedance in the steady state ($\Omega$), $R^0_b$ is the ac impedance in the initial state ($\Omega$), and $t_{Na+}$ is the Na ion mobility.

The Na ion mobility can be calculated from the linewidth of $^{23}$Na SS-NMR [22]. The conductivity of typical anions in electrolytes (e.g., $^{19}$F) can also be detected by SS-NMR methods. In addition, the error in electrochemical measurements can be effectively reduced and the measurement accuracy can be improved by using a three-electrode setup with the introduction of a reference electrode, as shown in Figure 2.12(d) [21].

In addition to these basic characterizations, researchers can use more advanced methods in their investigations to study the evolution mechanism of solid electrolytes during charge and discharge. For example, in situ ND studies can monitor the structural evolution due to Na ion migration during charge and discharge, which in turn can be inferred for Na ion migration in solid-state electrolytes [23]. Based on the results of in situ ND simulations, information on the local Na ion coordination environment and the pathways of Na ion migration in the electrolyte can be provided.

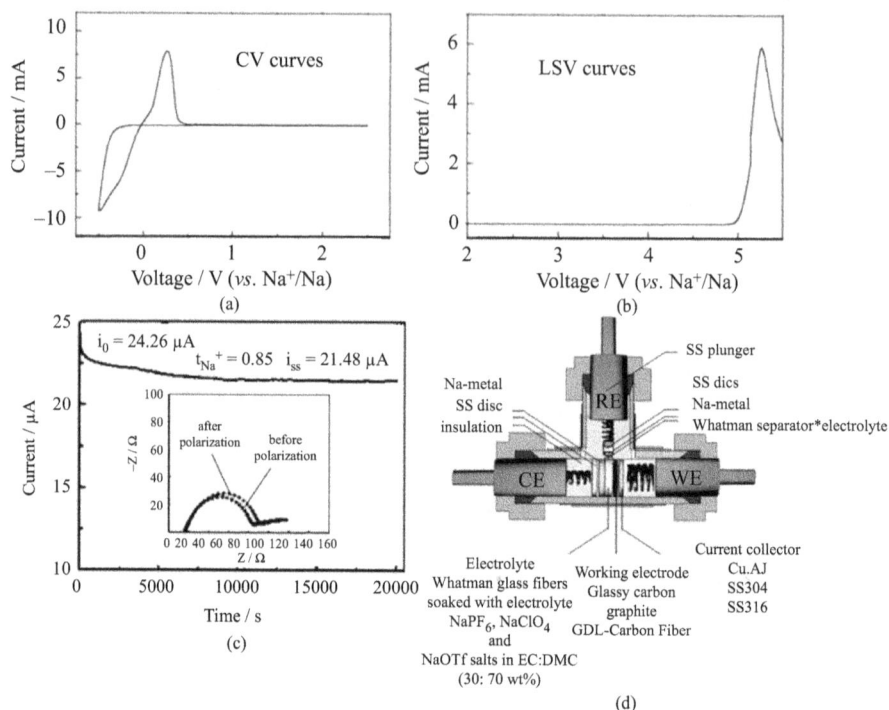

**Figure 2.12:** Electrochemical performance characterization of solid electrolyte of sodium-ion batteries: (a) CV curve of sodium intercalation/deintercalation in solid batteries; (b) ESW of solid electrolyte measured by LSV curve, with stainless steel as working electrode and sodium foil as the counter electrode; (c) current–time curve of a symmetrical sodium metal batteries with 5 mV DC voltage. The inset is the Nyquist impedance diagram of the batteries before and after polarization; and (d) schematic diagram of the three-electrode system.

## 2.2.3 Analysis of sodium-ion battery interface

The nature of the electrode/electrolyte interface plays a decisive role in sodium-ion batteries, so it is essential to study the composition and structure of SEI films in different cell systems containing solid and aqueous electrolytes. SEM images are usually used to observe the thickness and surface morphology of SEI. The random aggregation of electrode active materials, conductive additives, and binders leads to the formation of a rough surface on the electrode. After the initial cycle, the electrode surface appears slightly wrinkled and becomes gray, which indicates the formation of an SEI film. The thickness of the SEI film can be obtained from the SEM image of the cross section. And the EDS mapping can reveal the distribution of elements on the surface and inside the SEI film. In addition, AFM techniques are often used to capture the mechanical properties and roughness of SEI films.

### 2.2.3.1 Atomic force microscope techniques

Atomic force microscope (AFM) is a new instrument with atomic-level high resolution invented after scanning tunneling microscope, which can test the physical properties (including morphology) of various materials and samples in the nanoregion under atmospheric and liquid environments, or directly manipulate the nano. It is now widely used in the research experimental fields of semiconductors, nanofunctional materials, biological, chemical, food, and pharmaceutical research and research institutes of various nano-related disciplines. It has become a basic tool for nanoscience research. The working principle is that by approaching the sample surface up to a certain distance from the probe, the signal related to the distance between the probe and the sample surface is generated on the sample surface. By moving the scanner in both $X$- and $Y$-directions, the signal of the whole surface of the sample is obtained, and then according to the formula of signal and distance, the undulation degree of the sample surface is inverse solved, and the morphology of the sample surface is obtained. In this case, when the sample is scanned in contact mode, the tip of the needle is in contact with the sample at a distance of less than a few nanometers, and the intermolecular forces are mainly expressed as intermolecular repulsive forces. As the scan proceeds, the sample surface undulates and the intermolecular repulsive force changes, resulting in a change in the amount of cantilever bending, which is detected by the detector, and the feedback system sends a signal to the piezoelectric actuator to adjust the height of the sample stage so that the scan continues. In the noncontact mode, the distance between the needle tip and the sample surface is a few nanometers to tens of nanometers, and the interaction between the two is mainly characterized by intermolecular Van der Waals gravitational forces. At this time, the cantilever maintains its intrinsic frequency vibration. When the distance to the sample surface changes, the cantilever amplitude also changes, which is detected by the detection system. Tap mode is a mode of operation between contact mode and noncontact mode, which generates a morphological image by tapping the sample surface with a probe tip vibrating at a certain resonant frequency. During scanning, the cantilever vibrates at a higher amplitude than the noncontact mode, with an amplitude greater than 20 nm, and in this vibration mode, the probe tip is able to make intermittent contact with the sample surface. The force between the sample surface and the tip is kept constant by adjusting the distance between the tip and the sample surface.

Three main AFM imaging modes, and the advantages and disadvantages of these three modes are shown in Table 2.2. Based on the results of the AFM testing of the elastic modulus, the components and stability of the SEI film can be further analyzed [24]. In the electrochemical reaction, the development of in situ SEM and AFM techniques can avoid the oxidation and stripping of SEI, and the relevant in situ testing means will be described in detail in the in situ testing section.

**Table 2.2:** The advantages and disadvantages of the three imaging modes for AFM.

| Mode | Contact mode | Noncontact mode | Tap mode |
|---|---|---|---|
| Advantage | Direct contact with the sample surface and the tip, a stable high-resolution image can be obtained | Tip and sample surface is not in contact with almost no damage to the sample surface | Resolution is almost comparable to the accuracy of the contact pattern, little damage on the surface of the sample; it can also be used for liquid scanning |
| Disadvantage | Might damage the surface of the sample and tip | Low resolution, slow scanning rate, and only for hydrophobic surface | Scan rate may be slower than the contact mode |

### 2.2.3.2 Other SEI testing techniques

XPS is an important technique to study the composition and evolution of SEI layers, and its basic principle is to determine the change of chemical coordination based on the change of BE of different elements. For example, the XPS of SEI films on $Na_2Ti_3O_7$ indicates that SEI films are formed during discharge and are not always stable during cycling [25]. In combination with argon ion etching technique, XPS spectra of SEI layers can be collected at different depths to reveal the inhomogeneous components distribution. In addition, SS-NMR techniques are widely used for interfacial composition studies, mainly to investigate the different coordination structures of elements in SEI films. For example, in the SS-NMR spectra of discharge electrodes, the typical $^{19}F$ signals of PVDF ($-78 \times 10^{-6}$, $-88 \times 10^{-6}$, and $-93 \times 10^{-6}$) disappearance and the appearance of new double peaks ($-68 \times 10^{-6}$ and $-70 \times 10^{-6}$) have demonstrated the degradation of PVDF binder. FT-IR spectroscopy and Raman spectroscopy are also effective techniques to characterize the composition of electrolyte and electrode surface species.

## 2.3 In situ techniques for sodium-ion batteries

The development of electrochemical in situ characterization techniques has promoted the research work on electrode materials and electrolyte materials for sodium-ion batteries. The average research speed of sodium-ion batteries mechanism exploration and refinement is two times faster than that of lithium-ion batteries, which is mainly attributed to the rapid development and application of in situ technology. Currently, in situ characterization techniques can reveal the structural transformation, redox processes, solid–liquid interface formation, occurrence of side reactions, and sodium-ion transport properties, which are mainly used to study the structural stability of electrode materials, cell reaction kinetic processes, chemical reaction mechanisms, and morphological changes of electrode materials in sodium-ion batteries. In situ characterization techniques specifically include in situ XRD, in situ ND, in situ spectroscopy, in situ scanning probe technique, and in situ electron microscopy. We need an in-depth understanding of the working principle, device design, and experimental methods of in situ techniques in order to apply in situ techniques to the study of sodium-ion batteries in a mature way.

### 2.3.1 In situ structural and component evolution studies

Redox reactions in sodium-ion batteries are usually accompanied by structural evolution of the electrode materials. A wide variety of structural changes, including phase transitions, transformation reactions, and solid-solution reactions, are usually studied to investigate the electrochemical behavior during charge/discharge cycles and to develop new electrode materials. This section introduces the related test characterization of in situ XRD, in situ NMR, and in situ ND. XRD mainly studies the crystal structure and crystallinity of materials. NMR mainly probes the local structure and kinetic properties of materials, and ND mainly detects the coordination structure of light elements (Na).

#### 2.3.1.1 In situ XRD technique

The in situ XRD technique has been widely used in the study of crystalline sodium-ion battery cathode and anode materials, mainly to monitor the phase transition during the sodium storage process of the electrode material. In situ XRD tests usually require materials that can pass X-rays as the optical window of the in situ device, such as Be metal or Kapton film. During the electrochemical reaction, the structural evolution, phase transition, mechanical properties, and kinetic performance of the electrode material can be studied by in situ XRD. In sodium-ion batteries, layered transition metal oxide materials have complex crystal structure change processes, and further data can be calculated for properties such as expansion and contraction of lattice parameters of the materials. As shown in Figure 2.13(a), the manganese-rich P2-type $Na_{2/3}Mn_{0.8}Fe_{0.1}Ti_{0.1}O_2$ electrode material exhibits hexagonal structure with P63/mmc symmetry in "in situ" XRD [26]. There is no obvious new peak or peak shift in the first cycle or peak splitting, which indicates that the sodium storage mechanism is a solid-solution reaction. And it can be seen by in situ XRD that the $Ni^{2+}/Ni^{3+}/Ni^{4+}$ redox reaction process in the $Na_{2/3}Ni_{1/3}Mn_{2/3}O_2$ electrode is shown in Figure 2.13(b). The phase change of P2–O2 and the ordering of Na ion vacancies during this process lead to the capacity decay of the electrode material [27]. When the charging potential of the composite phase layered transition metal oxide is increased, it may exhibit a more complex phase transition process.

For example, the P2/O3/O1 composite phase of $Na_xNi_{1/3}Co_{1/3}Mn_{1/3}O_2$ exhibited better reversibility. This P-/O-coexistence structure can effectively suppress the irreversible phase transition between P2 and O2, while the presence of O1 and O3 phases is beneficial to improve the stability of the structure (Figure 2.13(c)). In addition to oxide cathode studies, complex transformation reactions usually need to be studied by in situ XRD [28]. As shown in Figure 2.13(d), the NiS-transformed anode undergoes the intercalation sodium storage reaction at potentials above 1.2 V, the conversion reaction between $Na_xNiS$ and $Ni_3S_2$ in the range of 1.2–0.95 V, and the conversion reaction between $Ni_3S_2$ and Ni in the range of 0.95–0.01 V, while the reverse charging generates $Ni_3S_2$ instead of NiS [29]. The XRD results proved that some conversion anode electrodes undergo irreversible redox reactions in the first cycling, influenced by the reaction kinetics. Therefore, in situ XRD analysis provides important reference information for the study of sodium storage mechanism of sodium-ion batteries, but in situ XRD cannot determine the internal phase transition process of amorphous materials.

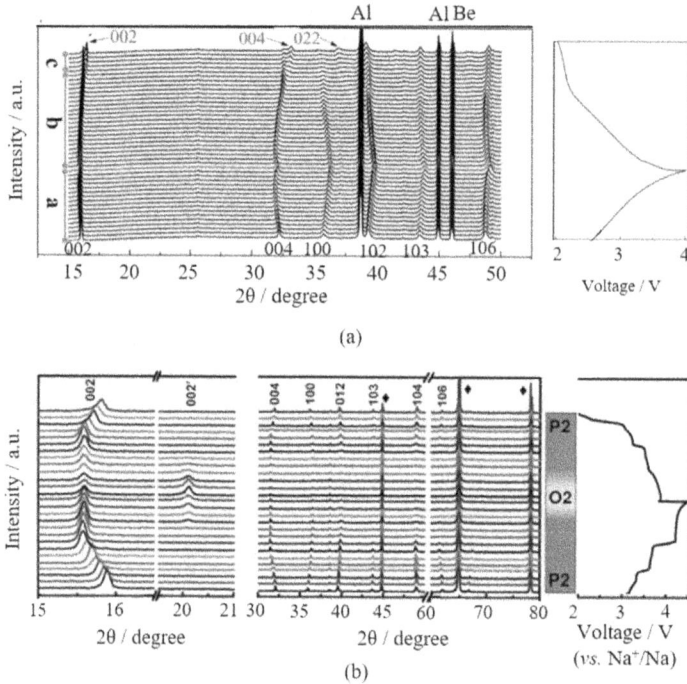

**Figure 2.13:** In situ XRD test: (a) in situ XRD and charge/discharge curve of $Na_{2/3}Mn_{0.8}Fe_{0.1}Ti_{0.1}O_2$; (b) in situ XRD and charge/discharge curves of $Na_{2/3}Ni_{1/3}Mn_{2/3}O_2$; (c) in situ XRD and charge/discharge curve of $Na_xNi_{1/3}Co_{1/3}Mn_{1/3}O_2$ with P2/O3/O1 phase; and (d) in situ XRD and charge/discharge curve of NiS electrode material.

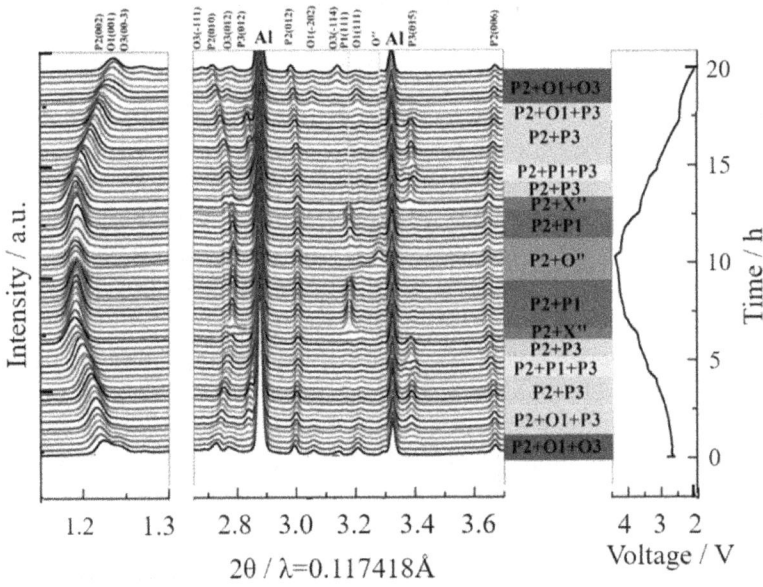

(c)

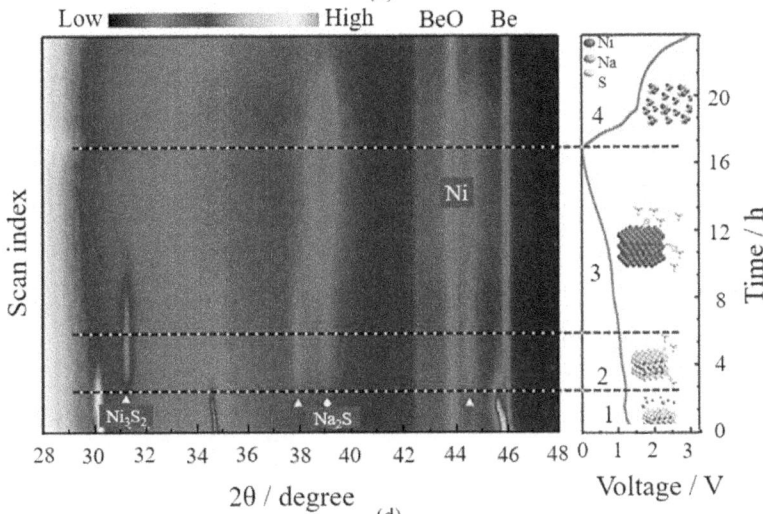

(d)

**Figure 2.13** (continued)

### 2.3.1.2 In situ NMR technique

The structural information of electrode materials can be obtained based on the response of quadrupole moments and nuclear spins, as well as their resonance at characteristic frequencies by in situ NMR technique. The local structure and sodium storage dynamics can be obtained based on the interaction of nuclear spins with the local environment, including chemical shifts and dipole and quadrupole interactions. The combination of SS-NMR spectroscopy and electrochemical techniques is expected to track the dynamic reactions during the charging and discharging of crystalline and amorphous electrode materials in real time on a large scale. A primary real-time NMR cell was prepared as a ring cell using a carbon anode, and the in situ NMR cell was assembled and mounted on an NMR probe for testing. In addition, there is a modified buckle-type in situ NMR cell that uses a copper sheet as the collector and NMR detector. The electrode material is coated on the copper sheet and assembled with the septum and sodium metal to form an in situ cell. The response signal of $^{23}$Na is usually used to study the structural evolution of the active material, electrolyte, and SEI films. Yang Yong's group has made important progress in the sodium storage process by in situ NMR technique. The group investigated the sodium storage sites of $Na_2FePO_4F$ electrode material using in situ NMR (Figure 2.14), revealing that it contains two-phase reactions with three different structures during sodium storage, including $Na_2FePO_4F$ with Pbcn space group, $Na_{1.5}FePO_4F$ with P21/c space group, and $Na_2FePO_4F$ with Pbcn space group, which is in agreement with the results of in situ XRD [30].

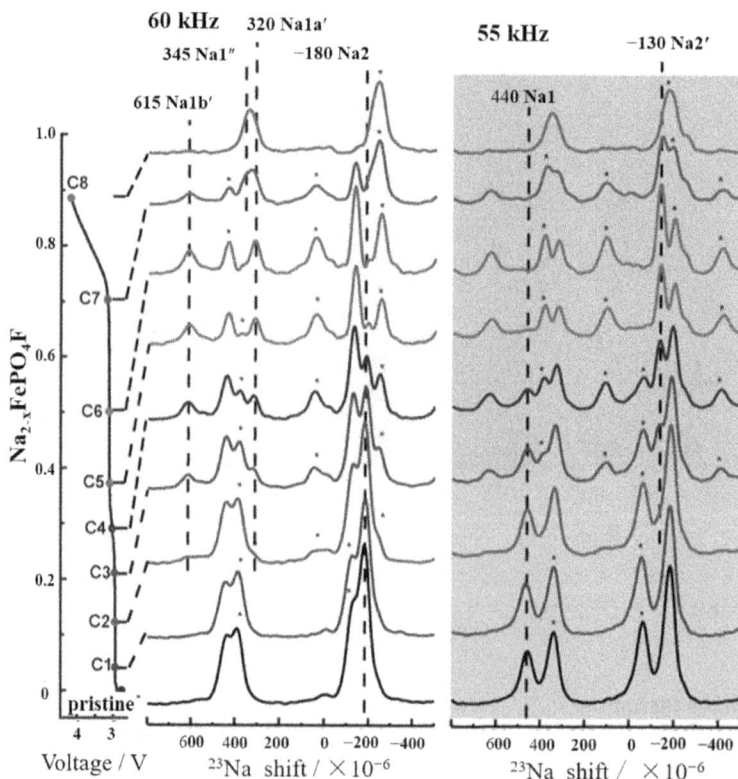

**Figure 2.14:** In situ Na MAS NMR spectrum of $Na_2FePO_4F$ during the first cycle and the distribution of $Na^+$ at different sodium storage sites during the cycle [30].

### 2.3.1.3 In situ spectroscopy

In situ spectroscopy techniques can also be used to analyze the structural evolution process of amorphous and organic materials, and even the formation and properties of the electrode/electrolyte interface. Both Raman spectroscopy and FT-IR spectroscopies study bond vibrations in molecular structures, and they are nondestructive detection modes that can be used to monitor in situ the production and disappearance of unstable intermediates in cells. The difference between Raman and FT-IR spectroscopies is that Raman spectroscopy is the inelastic scattering of photons by structural or molecular vibrations that lead to changes in polarizability (the degree of susceptibility of molecular electron clouds to deformation), whereas FT-IR spectroscopy is the direct absorption of photons by vibrations that lead to changes in molecular dipole moments. Thus, Raman and IR spectroscopies are highly complementary, and their combination represents an advanced characterization method. As shown in Figure 2.15(a), the setup of in situ Raman is similar to that of a button cell, and the Raman laser can be directed to the electrode materials or separators through the optical window on the cell surface to collect information on the changes of the electrode material or interface. The material of the optical window is usually quartz or calcium fluoride with high purity [31]. As for FT-IR, there are mainly transmission mode, total reflection mode, and attenuated total reflectance (ATR) mode. Because the energy of IR light is low, more information is lost when passing through the optical window. The current use of in situ IR spectroscopy to detect structural changes in electrode materials is mainly through the ATR mode (Figure 2.15(b)). Usually in situ Raman spectroscopy can be applied to carbon-based electrode materials to investigate the degree of order, sodium storage content, and binding sites of carbon-based materials by the relative intensity changes of G and D peaks in the spectrum. In addition, as introduced earlier, the sensitivity of Raman mapping to cyano is used to characterize the ligand metal valence states in Prussian blue and its analog materials to determine the variation of sodium content in the materials. FT-IR mapping, on the other hand, is mainly used for organic electrode materials or changes at the electrode/electrolyte interface to determine the components and coordination structures in the materials.

### 2.3.1.4 In situ synchrotron radiation technique

In situ synchrotron radiation technology can also be used to study the structural changes of electrode materials and ion/electron transport during the charging and discharging process. Compared with conventional X-ray sources, synchrotron radiation sources have the advantages of high intensity, continuously adjustable wavelength, good monochromaticity, polarizability, and high resolution. Current synchrotron radiation techniques widely used in sodium-ion batteries include high-resolution XRD, XAFS, and three-dimensional reconstruction of electrode material particles. Due to the high intensity of the synchrotron light source, the structural changes of the electrode material during sodium storage can be captured more rapidly, accurately, and finely.

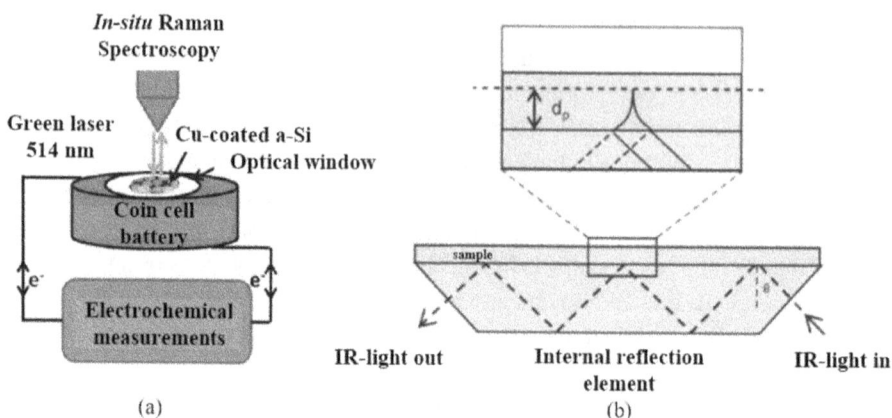

**Figure 2.15:** Schematic diagram of the working principle of in situ optical equipment: (a) schematic diagram of the structure and working principle of the Raman device; (b) schematic diagram of the working principle of the FT-IR device in ATR mode.

Refinement of the obtained patterns can also analyze the changes of lattice parameters and the process of intermediate phase transition, revealing new structural information and intermediate transition state products that are not captured by ordinary XRD. The synchrotron XAFS technique can adjust the X-ray energy to study the surrounding chemical environment, redox state, local electronic structure, and neighboring coordination atoms of each element in the electrode material separately. The three-dimensional reconstruction of the electrode material particles can be used to observe more intuitively important issues such as the stresses on the material during cycling and the relationship between electrode pulverization and cycle life. The design of in situ synchrotron electrochemical cell should ensure the conventional electrochemical performance in addition to reducing the X-ray absorption by the optical window material. The structure of the in situ cell can be either a button cell or a soft pack cell. The optical window material is generally beryllium, polyimide Kapton film, and Mylar film. As shown in Figure 2.16(a), the in situ synchrotron cell is assembled using Mylar film as the window material, two sheets of aluminum with open holes as the cell shell, aluminum and copper as the cathode and anode collectors, respectively, with a rubber ring implanted in the middle [32].

In situ synchrotron XRD has a more accurate pattern than general XRD, finer diffraction peaks can be observed, and more accurate information on lattice parameters, atomic occupancies, is obtained by refinement. The structural evolution of the $Na_3V_2(PO_4)_2F_3$ material during the charging and discharging process was characterized by in situ synchrotron XRD, and a clearer diffraction peak pattern was obtained (Figure 2.16(b)). The material follows the solid solution intercalation/deintercalation reaction mechanism during the whole charge/discharge process, and thus the material has good structural stability and reversibility. The rate of change of the material cell volume during charge/discharge is only 1.89% [33].

Electrochemical in situ synchrotron XAFS has been widely used to reveal the charge compensation mechanism, electronic structure, and local structure evolution during the charge/discharge process of sodium-ion battery electrode materials. As shown in Figure 2.16(c), the subtle structural changes and valence changes that occur during sodium storage in $Ni_3S_2$ electrode materials can be captured by XANES mapping, especially for composite components and multiple valence states [34]. Therefore, XANES mapping can be used to determine whether the electrode material reacts adequately, whether the short-range structure changes, and whether the reaction process is reversible.

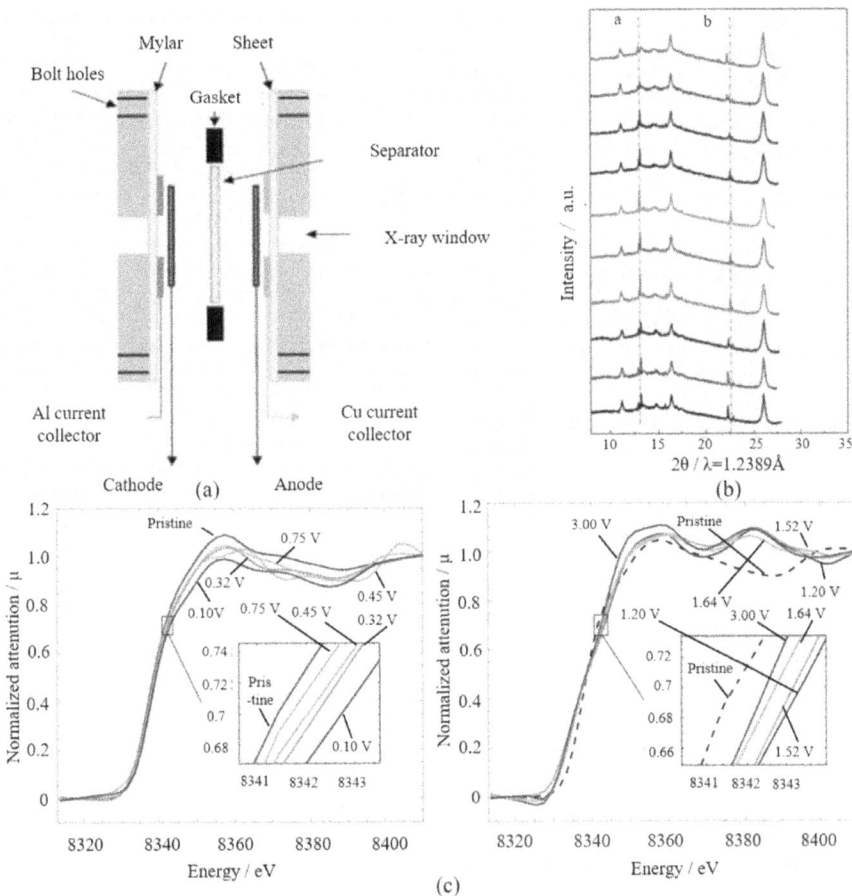

**Figure 2.16:** In situ synchrotron radiation characterization: (a) electrochemical in situ synchrotron radiation spectroscopy battery; (b) synchrotron in situ XRD patterns of $Na_3V_2(PO_4)_2F_3$ during the first cycle; and (c) in situ XANES mapping of $Ni_3S_2$ electrode material during the first cycle.

### 2.3.2 In situ morphology evolution

During the electrochemical reactions of sodium-ion batteries, observation of the morphology evolution processes of electrode materials and electrode/electrolyte interfaces is important for understanding the underlying electrochemical reactions, the sodium storage, and degradation mechanisms. The interfacial processes include electrolyte decomposition, SEI growth, redox reactions, and sodium-ion intercalation/de-intercalation. However, due to the dynamic nature of the interfacial processes, their real-time monitoring is a challenging task. The emergence of in situ research methods has brought new directions to interface studies. So far, many techniques have employed model cells, which consist of a working electrode, counter electrode, and electrolyte. In situ interface analysis techniques including TEM, AFM, XAS, and Raman imaging have been successfully used to reveal the morphology evolution, electrode expansion/contraction, local geometric and electronic structural changes, and the formation of new chemical bonds/species. In addition, in situ optical microscopy has observed the formation process of sodium dendrites. This section focuses on the in situ techniques to characterize the material morphology and interfacial structure changes during electrochemical reactions.

### 2.3.2.1 In situ TEM technique

The in situ TEM technique has proven to be an effective technique for directly observing microstructural changes and sodium-ion diffusion at the atomic scale, which is a powerful tool for studying electrochemical reactions during charging and discharging. Two main types of nanocells are used in "in situ" TEM technique studies. One type uses a nonvolatile ionic liquid electrolyte and the other type uses a solid-state electrolyte. Both types of cells generally use sodium metal as the counter electrode. In general, a nanocell consists of a working electrode coated with an ionic liquid electrolyte or a solid-state electrolyte and a sodium metal electrode. The structural and morphological changes of the electrode material are identified using real-time TEM images under an applied bias.

The actual lack of energy density and poor cycling stability of sodium-ion batteries is due to the large ionic radius and atomic mass of sodium ions, which leads to sluggish kinetics in the battery reaction. Therefore, the construction of volume expansion buffer space can effectively release the lattice stress during sodium storage, which is one of the important means to improve the cycling stability of sodium-ion batteries. As shown in Figure 2.17(a), the in situ lens device uses a gold wire dipped in active electrode material for the cathode electrode and a tungsten wire dipped in metallic sodium for the anode electrode [35]. To facilitate dipping, the front ends of the gold and tungsten wires need to be processed into a flat shape. During in situ device transfer, the surface of the sodium metal is oxidized to form sodium oxide, which can be used as solid electrolytes to transport sodium ions. By

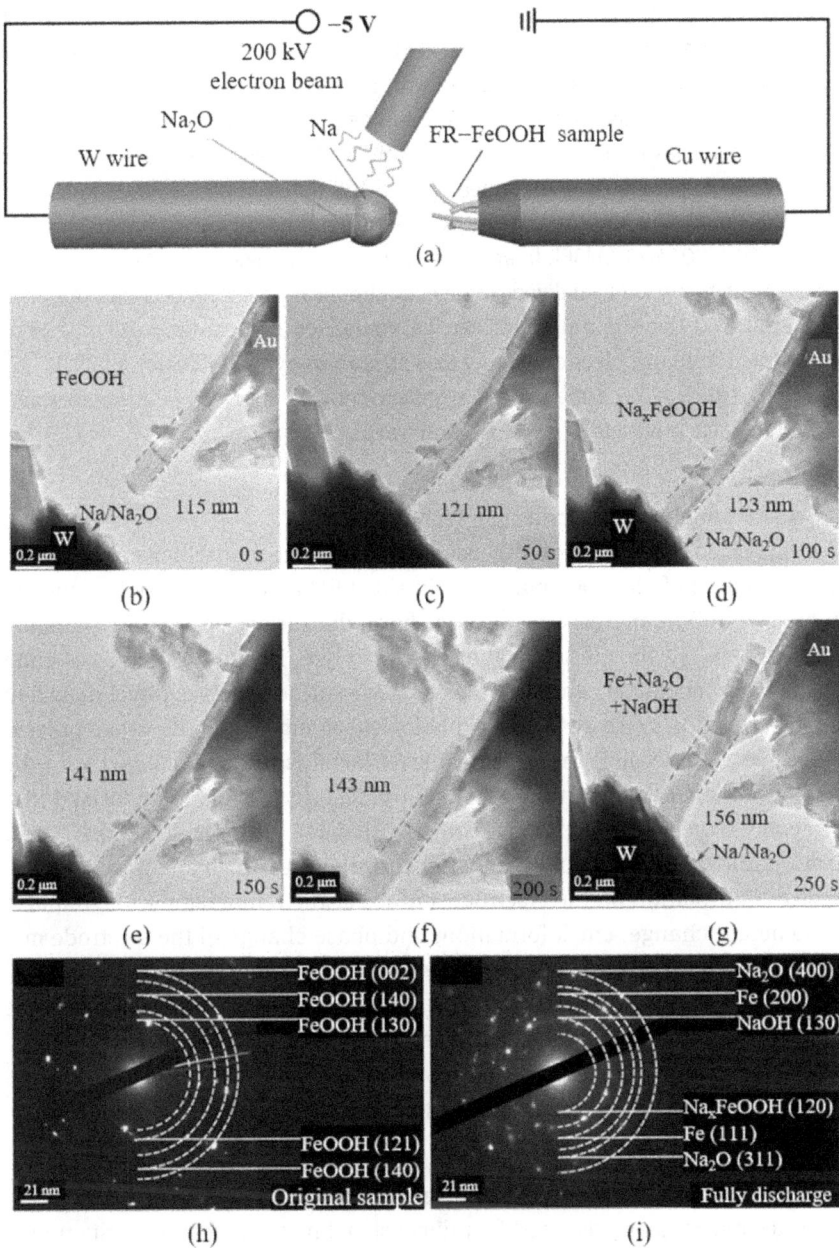

**Figure 2.17:** In situ TEM characterization: (a) schematic diagram of the in situ TEM device; (b)–(g) in situ TEM images of FR-FeOOH observed at 0, 50, 100, 150, 200, and 250 s; and (h, i) selected area electron diffraction images of FR-FeOOH before and after sodiation.

putting the electrode material in contact with the sodium metal, the sodium ions will be embedded in the electrode material under the action of external voltage, and the volume of the electrode material will change with time. As shown in Figure 2.17(b)–(g), after 250 s, the electrode material reaches the fully sodiated state and the diameter of the nanorods increases from 115 to 156 nm. According to previous studies reported in the literature, 36% increase in the diameter of the rod material corresponds to its volume expansion of 81%, a result significantly lower than that of typical conversion-type anode materials (200–400%) [36]. In addition to the morphological changes of the electrode material that can be visualized, the electrochemical reactions between the electrode material and the sodium metal can be visualized by real-time selected area electron diffraction maps, electron energy loss spectroscopy, and EDS for analysis [37] (Figure 2.17(h), (i)). The electrode materials underwent conversion reactions before and after sodiation, with products of $Na_2O$, Fe, and NaOH.

### 2.3.2.2 In situ atomic force microscope technique
While in situ TEM technique is mainly used to observe the morphological evolution and phase changes of electrode materials, in situ AFM analysis is based on the detection of needle-like cantilever beams and shows the surface changes by scanning the interface between the electrode and the electrolyte. In situ AFM analysis combined with electrochemical control of the electrode surface enables direct detection of interfacial reaction processes and morphological changes, and the whole process is noninvasive [38]. Because the AFM cantilever beam is in direct contact with the interface between electrode and electrolyte, the testing process is performed in a water-free and oxygen-free argon atmosphere. AFM can not only detect the microscopic morphology of the electrode material surface in real time and provide physicochemical information on the electrode surface at the nanometer scale, but also detect the height change, crack formation, and phase change of the electrode material in real time under electrochemical reaction conditions through the interaction between its needle tip atoms and the atoms on the electrode surface. Three main types, open buckle cells, metal-coil cells, and metal-foil cells, are concluded for in situ AFM. They are suitable for cell systems with different properties and meet the needs of interfacial composition studies [39].

### 2.3.2.3 In situ neutron diffraction technique
In situ ND technique is an advanced four-dimensional material characterization tool, which can be used to test the macroscopic crystal structure, crystal dot structure, electromagnetic properties, volume fraction of constituent phases, elastic stress–strain, grain and dislocation size, and dislocation density. of solid materials under external conditions such as temperature, atmosphere, external forces, and electromagnetic fields. It has significant technical advantages and application prospects, especially in

the analysis of material microstructure and tissue transformation mechanism, and has been successfully applied in the field of secondary batteries. As shown in Figure 2.18, the changes of $MoS_2$ during the first-cycle discharge were studied by in situ ND technique, and the changes of (002) and (103) crystal surfaces were observed. The lithium embedding into $MoS_2$ to form $Li_xMoS_2$ can be inferred based on the change of the

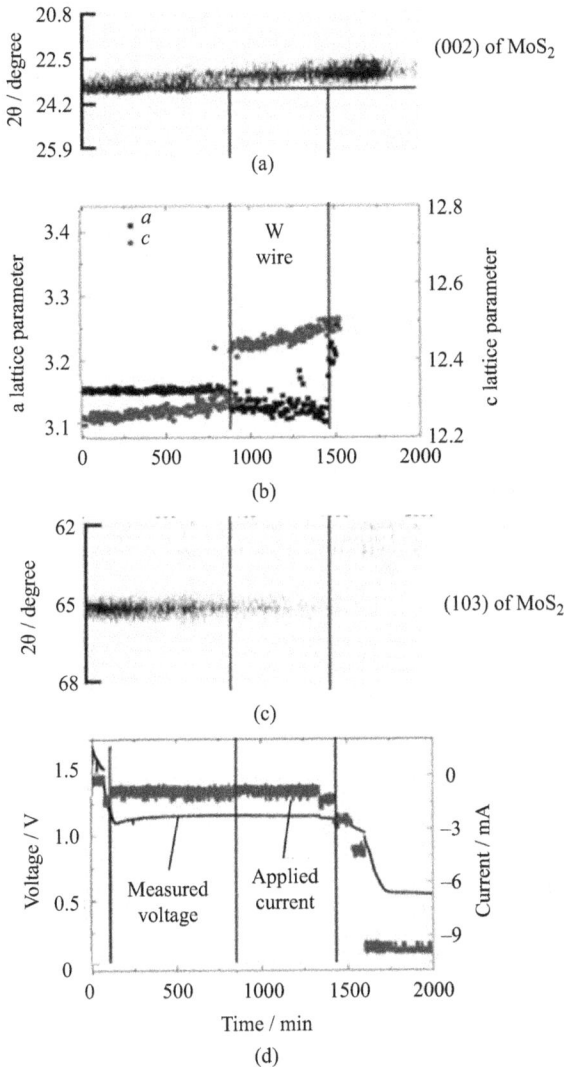

**Figure 2.18:** In situ ND of the selected area during the first cycle of the discharge process of the $MoS_2$: (a) the shift of peak of $2\theta$ based on (002) plane; (b) refined lattice parameters of $a$ and $c$ based on $MoS_2$; (c) (103) crystal plane peak of $MoS_2$; and (d) the measured voltage and the reflected intensity of the applied current related to in situ ND with scale [40].

diffraction peak $2\theta$ of (002) crystal plane; and the lithium embedding into $MoS_2$ is inferred based on the linear increase of the lattice constant $c$. However, in partial lithium embedding into $MoS_2$, the increase of lattice parameters is nonlinear, which indicates the appearance of an intermediate phase. The structure of this phase has similar characteristics to that of the final fully lithiated product (short-range ordered $LiMoS_2$), so this method is not yet very accurate. The in situ ND technique uses customary methods, such as cell testing using paired distribution functions, which make it difficult to determine the short-range ordered structure. Therefore, cells for in situ ND testing need to be redesigned to ensure the accuracy of diffraction data. The in situ ND test technique is able to test the changes of lattice parameters and the emergence of new phases in real time, which will be used on a large scale in sodium-ion batteries.

# References

[1]   Guo S, Liu P, Yu H. et al. A Layered P2- and O3-Type Composite as a High-Energy Cathode for Rechargeable Sodium-Ion Batteries. Angewandte Chemie, 2015, 54(20): 5894–5899.

[2]   Huang Y, Wang Z, Jiang Y. et al., Hierarchical Porous $Co_{0.85}Se$@Reduced Graphene Oxide Ultrathin Nanosheets with Vacancy-Enhanced Kinetics as Superior Anodes for Sodium-Ion Batteries. Nano Energy, 2018, 53: 524–535.

[3]   Huang Y, Xie M, Zhang J. et al., A Novel Border-Rich Prussian Blue Synthetized by Inhibitor Control as Cathode for Sodium Ion Batteries. Nano Energy, 2017, 39: 273–283.

[4]   Chen R, Huang Y, Xie M. et al. Chemical Inhibition Method to Synthesize Highly Crystalline Prussian Blue Analogs for Sodium-Ion Battery Cathodes. ACS Applied Materials & Interfaces, 2016, 8(46): 31669–31676.

[5]   Yu C, Ganapathy S, De Klerk N J. et al. Na-Ion Dynamics in Tetragonal and Cubic $Na_3PS_4$, a Na-Ion Conductor for Solid State Na-Ion Batteries. Journal of Materials Chemistry, 2016, 4(39): 15095–15105.

[6]   Huang Y, Xie M, Wang Z. et al. A Chemical Precipitation Method Preparing Hollow–Core–Shell Heterostructures Based on the Prussian Blue Analogs as Cathode for Sodium-Ion Batteries. Small, 2018, 14(28): 1801246.

[7]   Brousse T, Belanger D, Long J. et al., To Be or Not to Be Pseudocapacitive. Journal of the Electrochemical Society, 2015, 162: A5185–A5189.

[8]   Gogotsi Y, Penner R M. et al. Energy Storage in Nanomaterials – Capacitive Pseudocapacitive, or Battery-Like. ACS Nano, 2018, 12(3): 2081–2083.

[9]   Simon P, Gogotsi Y, Dunn B. et al. Where Do Batteries End and Supercapacitors Begin. Science, 2014, 343(6176): 1210–1211.

[10]  Yang X, Rogach A L. Electrochemical Techniques in Battery Research: A Tutorial for Nonelectrochemists. Advanced Energy Materials, 2019, 9(25): 1900747.

[11]  Zhang H, Xuan X, Wang J. et al. FT-IR Investigations of Ion Association in PEO–MSCN (M=Na, K) Polymer Electrolytes. Solid State Ionics, 2003, 164(1): 73–79.

[12]  Monti D, Ponrouch A, Palacin M R. et al., Towards Safer Sodium-Ion Batteries via Organic Solvent/Ionic Liquid Based Hybrid Electrolytes. Journal of Power Sources, 2016, 324: 712–721.

[13]  Xu K. Nonaqueous Liquid Electrolytes for Lithium-Based Rechargeable Batteries. Chemical Reviews, 2004, 104(10): 4303–4417.

[14] Wu L, Bresser D, Buchholz D. et al. Unfolding the Mechanism of Sodium Insertion in Anatase $TiO_2$ Nanoparticles. Advanced Energy Materials, 2015, 5(2): 140–142.

[15] Matsyshita T, Dokko K, Kanamura K. *In-Situ* FT-IR Measurement for Electrochemical Oxidation of Electrolyte with Ethylene Carbonate and Diethyl Carbonate on Cathode Active Material Used in Rechargeable Lithium Batteries. Journal of Power Sources, 2005, 146(1-2): 360–364.

[16] Hy S, Felix C Y. et al., In Situ Surface Enhanced Raman Spectroscopic Studies of Solid Electrolytes Interphase Formation in Lithium Ion Battery Electrodes. Journal of Power Sources, 2014, 256: 324–328.

[17] Fan J, Chen J, Zhang Q. et al. An Amorphous Carbon Nitride Composite Derived from ZIF-8 as Anode Material for Sodium-Ion Batteries. Chemsuschem, 2015, 8(11): 1856–1861.

[18] Gao H, Zhou W, Park K. et al. A Sodium-Ion Battery with a Low-Cost Cross-Linked Gel-Polymer Electrolyte. Advanced Energy Materials, 2016, 6(18): 1600467.

[19] Yu Z, Shang S, Seo J. et al. Exceptionally High Ionic Conductivity in $Na_3P_{0.62}As_{0.38}S_4$ with Improved Moisture Stability for Solid-State Sodium-Ion Batteries. Advanced Materials, 2017, 29(16): 1605561.

[20] Hou H, Xu Q, Pang Y. et al. Efficient Storing Energy Harvested by Triboelectric Nanogenerators Using a Safe and Durable All-Solid-State Sodium-Ion Battery. Advanced Science, 2017, 4(8): 1700072.

[21] Bhide A, Hofmann J D, Durr A K. et al. Electrochemical Stability of Non-aqueous Electrolytes for Sodium-Ion Batteries and Their Compatibility with $Na_{0.7}CoO_2$. Physical Chemistry Chemical Physics, 2014, 16(5): 1987–1998.

[22] Komoroski R A, Mauritz K A. A Sodium-23 Nuclear Magnetic Resonance Study of Ionic Mobility and Contact Ion Pairing in a Perfluorosulfonate Ionomer. Journal of the American Chemical Society, 1978, 100(24): 7487–7489.

[23] Zhu J, Wang Y, Li S. et al. Sodium Ion Transport Mechanisms in Antiperovskite Electrolytes $Na_3OBr$ and $Na_4OI_2$: An In Situ Neutron Diffraction Study. Inorganic Chemistry, 2016, 55(12): 5993–5998.

[24] 赵春花. 原子力显微镜的基本原理及应用. 化学教育(中英文), 2019, 40(4): 10–15.

[25] Munozmarquez M A, Zarrabeitia M, Castillomartinez E. et al. Composition and Evolution of the Solid-Electrolyte Interphase in $Na_2Ti_3O_7$ Electrodes for Na-Ion Batteries: XPS and Auger Parameter Analysis. ACS Applied Materials & Interfaces, 2015, 7(14): 7801–7808.

[26] Han M H, Gonzalo E, Sharma N. et al. High-Performance P2-Phase $Na_{2/3}Mn_{0.8}Fe_{0.1}Ti_{0.1}O_2$ Cathode Material for Ambient-Temperature Sodium-Ion Batteries. Chemistry of Materials, 2016, 28(1): 106–116.

[27] Wang L, Sun Y, Hu L. et al. Copper-Substituted $Na_{0.67}Ni_{0.3-x}Cu_xMn_{0.7}O_2$ Cathode Materials for Sodium-Ion Batteries with Suppressed P2–O2 Phase Transition. Journal of Materials Chemistry A, 2017, 5(18): 8752–8761.

[28] Xu G, Amine R, Xu Y. et al. Insights into the Structural Effects of Layered Cathode Materials for High Voltage Sodium-Ion Batteries. Energy & Environmental Science, 2017, 10(7): 1677–1693.

[29] Fan S, Huang S, Chen Y. et al., Construction of Complex NiS Multi-Shelled Hollow Structures with Enhanced Sodium Storage. Energy Storage Materials, 2019, 23: 17–24.

[30] Li Q, Liu Z, Zheng F. et al. Identifying the Structural Evolution of the Sodium Ion Battery $Na_2FePO_4F$ Cathode. Angewandte Chemie, 2018, 57(37): 11918–11923.

[31] Deng Y, Dong S, Li Z. et al. Applications of Conventional Vibrational Spectroscopic Methods for Batteries beyond Li-Ion. Small Methods, 2018, 2(8): 1700332.

[32] Balasubramanian M, Sun X, Yang X. et al. In Situ X-ray Diffraction and X-Ray Absorption Studies of High-Rate Lithium-Ion Batteries. Journal of Power Sources, 2001, 92(1): 1–8.

[33]  Hao X G, Liu Z G, Gong Z L. et al.. In Situ XRD and Solid State NMR Characterization of $Na_3V_2$ $(PO_4)_2F_3$ as Cathode Material for Lithium-Ion Batteries. Scientia Sinica Chimica, 2012, 42(1): 38–46.

[34]  Wang L, Wang J, Guo F. et al., Understanding the Initial Irreversibility of Metal Sulfides for Sodium-Ion Batteries via Operando Techniques. Nano Energy, 2018, 43: 184–191.

[35]  Huang Y, Xie M, Wang Z. et al., All-Iron Sodium-Ion Full-Cells Assembled via Stable Porous Goethite Nanorods with Low Strain and Fast Kinetics. Nano Energy, 2019, 60: 294–304.

[36]  Huang J Y, Zhong L, Wang C. et al. In Situ Observation of the Electrochemical Lithiation of a Single $SnO_2$ Nanowire Electrode. Science, 2010, 330(6010): 1515–1520.

[37]  Klein F, Jache B, Bhide A. et al. Conversion Reactions for Sodium-Ion Batteries. Physical Chemistry Chemical Physics, 2013, 15(38): 15876–15887.

[38]  Alessandrini A, Facci P. AFM: A Versatile Tool in Biophysics. Measurement Science and Technology, 2005, 16(6): 65–92.

[39]  Yang Y, Liu X, Dai Z. et al. In Situ Electrochemistry of Rechargeable Battery Materials: Status Report and Perspectives. Advanced Materials, 2017, 29(31): 1606922.

[40]  Sharma N, Du G, Studer A J. et al., In-Situ Neutron Diffraction Study of the $MoS_2$ Anode Using a Custom-Built Li-Ion Battery. Solid State Ion, 2011, 199–200: 37–43.

# Chapter 3
# Cathode materials for sodium-ion batteries

The diameter of Na ion is about 1.5 times larger than that of Li ion, and hence, the research on electrode materials of lithium-ion batteries cannot be directly applied to the system of sodium-ion batteries, especially for cathode materials. Due to the large radius of Na ion, the cathode materials with chemical formulas similar to lithium may not have suitable channels for Na-ion migration and sufficient sodium storage sites. At the same time, it may also show different crystal structures, and different crystal phase transitions may occur in the process of Na-ion intercalation/deintercalation, which leads to the difference in the number of embeddable ions and the reversibility of the electrochemical cycle. For example, in the development of lithium-ion batteries, the first generation of cathode materials with mature technology is $LiCoO_2$ layered oxide. However, when $NaCoO_2$ is directly used as the cathode material of sodium-ion batteries, complex phase transition reaction will occur in the process of Na-ion intercalation/deintercalation process, and multiple charge-discharge platforms will appear in the charge-discharge curve (Figure 3.1) [1]. When $NaCoO_2$ is used as the cathode material of sodium-ion batteries, the structural irreversibility and poor cycling performance of the material are more likely to appear.

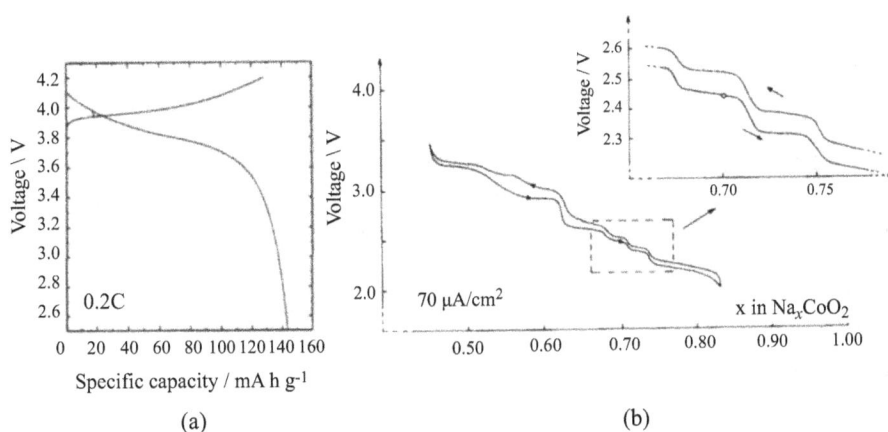

**Figure 3.1:** Galvanostatic charge and discharge curves: (a) cathode of $LiCoO_2$ lithium-ion batteries [2]; (b) cathode of $NaCoO_2$ sodium-ion batteries [1].

In order to improve the energy density, rate performance, cycle stability, and other electrochemical properties of sodium-ion batteries, the cathode materials must have the ability of highly reversible Na-ion intercalation/deintercalation, good electronic conductivity, ion diffusion rate, and other dynamic properties. Therefore, since the 1980s, cathode materials have been widely considered by researchers as the key to

https://doi.org/10.1515/9783110749069-003

the study of sodium-ion batteries. Up to now, many research reports have proposed cathode materials for different applications, for example, layered transition metal oxides with high energy density, polyanions with rapid Na-ion intercalation/deintercalation ability, and economical and efficient Prussian blue and its analogues. Typical materials of each type are shown in Figure 3.2. Researchers try to modify these materials to solve their own defects, strive to establish a perfect theoretical system, and improve other parts of the battery to promote the related performance of cathode materials. In this chapter, we will introduce the development history, basic properties, research problems, and development direction of all kinds of cathode materials in sodium-ion batteries, one by one.

## 3.1 Transition metal oxides

There are many transition metal oxides, and the same oxide may have different crystal structure. According to the different arrangements of Na ions after intercalation in these materials, transition metal oxides are divided into layered oxides and tunnel structures. Since 2010, a lot of research reports on the application of layered oxides in the cathode of sodium-ion batteries have appeared. Since the transition metal elements from Ti to Ni have high activity, the layered oxides of these metal elements are more suited for use as cathode materials for sodium-ion batteries. The general formula can be expressed as $Na_xMeO_2$ (Me: Mn, Co, Ni, Fe, Cu, and V).

The common layered structure of transition metal oxides is composed of a common-edge $MeO_6$ octahedron, which produces a variety of stacking forms with different $c$-axis orientation. $Na_xCoO_2$ is the first cathode material to be developed. The transition metal Co atoms are distributed in the octahedral space of O to form a stable $CoO_2$ layer, while Na ions are embedded in a separate layer, parallel to the $CoO_2$ layer. According to the $x$ value and the location of sodium-ion storage, Delmas et al. divided $Na_xCoO_2$ into four stable crystallographic structures: O3, O'3, P3 and P2. In the octahedral layered oxide, Na ion is located at the center of the octahedral site composed of oxygen atom. However, in the triangular prismatic layered oxide, Na ion is located in the prismatic site composed of oxygen atom, O. The numbers 2 and 3 in crystallographic structures indicate that each unit cell is composed of several stacked layered structures [3]. For example, when three different stacking forms of layered structure in $Na_xMeO_2$, and Na ion is stored in the octahedral O position between different $MeO_2$ layers, it is named O3 type. Similarly, P2-type $Na_xMeO_2$ is composed of two stacked layered structures, in which the sodium ion is located in the P position of the prismatic structure, and hence the name, P2 type. The crystal structures of several kinds of layered oxides are shown in Figure 3.3. In addition, when there is lattice distortion in the crystal structure, a ""symbol is added, which is named O'3 type and P'2 type [4].

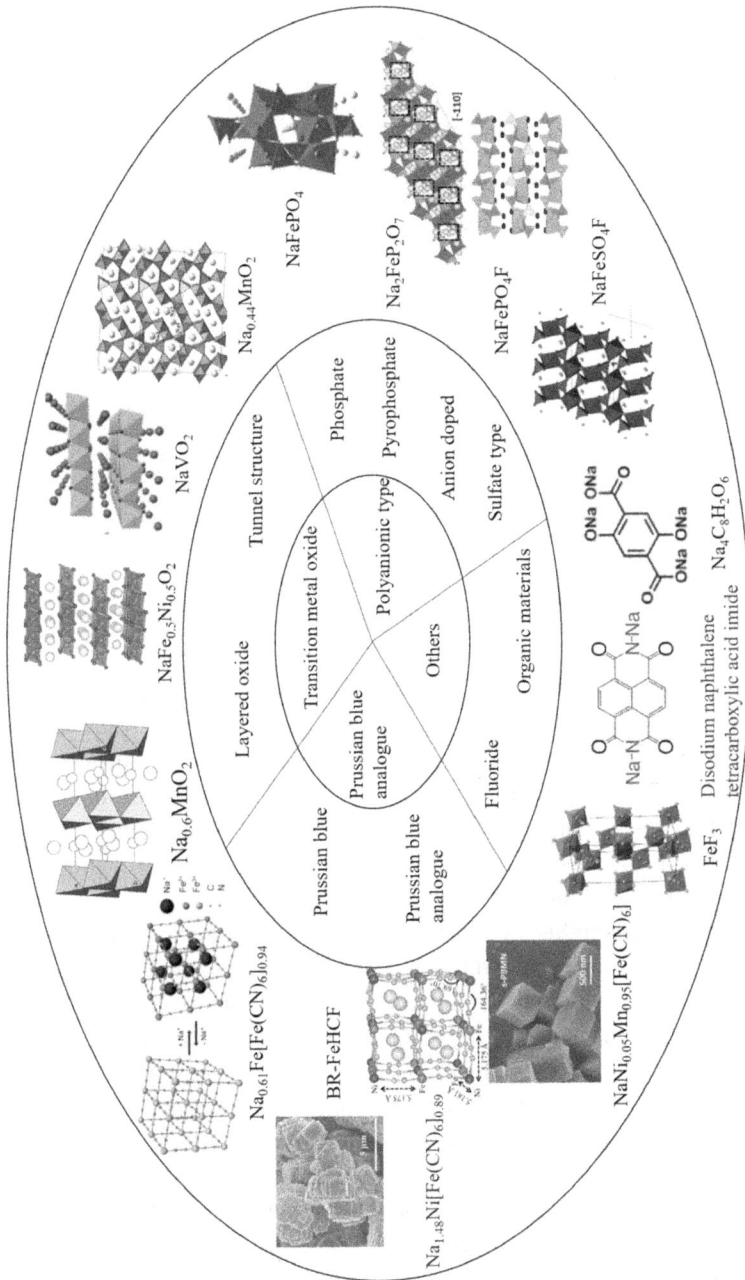

**Figure 3.2:** Typical cathode materials.

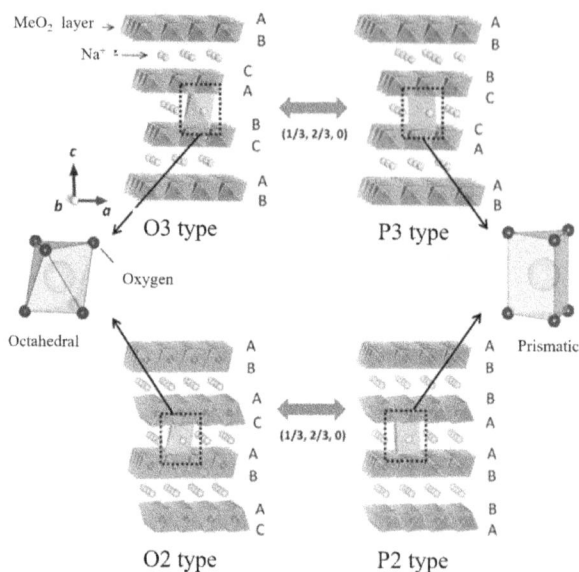

**Figure 3.3:** Crystal structure of several kinds of layered oxides [4].

Different layered transition metal oxides have different characteristics, such as iron-based oxides with low-cost and large-scale application prospects, cobalt-based materials with high energy density, and copper-based materials with certain stability. For the same components, different ratios may have O3 or P2 type. For example, $Na_xVO_2$ can be divided into O3-type $Na_xVO_2$ and P2-type $Na_xVO_2$. Furthermore, more possibilities can be achieved by mixing binary or ternary oxides. Fe-/Mn-based and Ni-/Co-based binary metal oxides improve the deficiency of a single metal-based oxide and improve the comprehensive electrochemical performance. In addition to the multiple metal oxides composed of different transition metal elements, composite transition metal oxides with different configurations can be designed to consider O3-type materials with high capacity and P2-type materials with high stability.

The cathode materials of sodium-ion batteries with tunnel structure generally refer to the one-dimensional tunnels connected in only one direction in the crystal structure, and these tunnels can be used for reversible intercalation/deintercalation of Na ions. Compared with layered materials, these materials are more stable in air, and the way of storing sodium is simpler, so there is no more phase transition. However, unlike layered materials, which have a wider Na-ion intercalation/deintercalation path, the ion transfer rate is limited to a certain extent, and the effect on the electrochemical performance is mainly reflected in the poor rate performance. If the material collapses after a long period of cycling, the tunnel may be blocked, which, further, has an adverse effect on the intercalation/deintercalation of Na ions.

The metal elements that can form transition metal oxides with tunnel structure mainly include Fe, Mn, V, and so on. Due to the different ratios of transition metal ions to oxygen atoms, the samples may show different crystal forms with the same element. The materials with the same molecular formula and different crystal forms can also be produced by using different synthesis processes and conditions. The difference of crystal form has an essential effect on the electrochemical performance. Therefore, the research of this kind of materials is mainly to explore the synthesis process, especially to grasp the influence of different element proportions and doping modification on the crystal structure and to produce high-performance transition metal oxide materials.

In this section, the sodium storage mechanism, application shortcomings and modification process of P2-type, O3-type, and tunnel-type transition metal oxides will be introduced in detail, so that it is convenient for readers to understand the properties of transition metal oxides and their application prospects in sodium-ion batteries.

### 3.1.1 O3-type transition metal oxides

Owing to the low energy density of sodium-ion batteries, the application prospect of sodium-ion batteries is mainly in the field of large-scale energy storage. Reducing the production cost is the best way to compete with the rest of the energy storage devices in the market. Due to the low price of iron-based compounds, the early research on O3-type transition metal oxides focused on iron-based materials. The $\alpha$-NaFeO$_2$ is a typical example of O3-type layered transition metal oxides. This kind of material can be prepared by simple solid-state reaction of Na$_2$O$_2$ and iron oxides (Fe$_2$O$_3$ or Fe$_3$O$_4$) calcined in air. The low price of raw materials and the low requirements of production environment make its cost advantage obvious. Okada et al. first introduced the sodium storage properties of O3-type $\alpha$-NaFeO$_2$, and the XRD pattern is well confirmed in the synthesis of the $\alpha$-NaFeO$_2$ (Figure 3.4) [5].

The electrochemical performance of O3 type $\alpha$-NaFeO$_2$ as the cathode material of sodium-ion batteries was tested by experiments in the early research. Under the current density of 0.2 mA g$^{-1}$, the discharge specific capacity is close to 85 mA h g$^{-1}$, and the average discharge voltage is about 3.3 V (vs. Na$^+$/Na). The electrochemical activity comes from Fe$^{3+/4+}$ redox couple, and the reaction equation is:

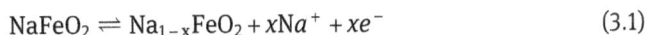

$$\text{NaFeO}_2 \rightleftharpoons \text{Na}_{1-x}\text{FeO}_2 + x\text{Na}^+ + xe^- \tag{3.1}$$

The reversibility of this electrode during the charge/discharge process is greatly affected by the cut-off voltage, as shown in Figure 3.5(a). Although the expansion of the voltage range is conducive to the improvement of the capacity, after 3.5 V, with the increase of the voltage, the reversible capacity during the cycle decreases significantly, and the cycle stability is also affected (Figure 3.5(b)) [5]. When $\alpha$-NaFeO$_2$ is used as the

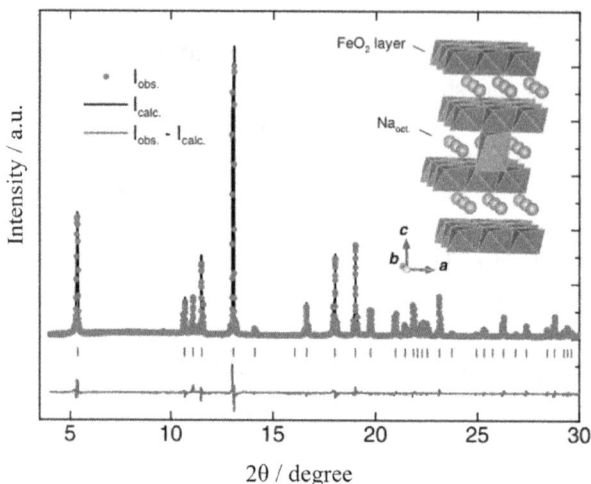

**Figure 3.4:** XRD spectrum and crystal structure of α-NaFeO$_2$ [5].

electrode, the capacity of the battery decreases due to the irreversible phase change. In order to prevent the change of matrix structure, the sodium content in Na$_{1-x}$FeO$_2$ should be maintained at $x > 0.5$ [6]. Okada et al. further revealed the attenuation mechanism of the cathode materials by Mossbauer spectroscopy. It is found that Fe$^{4+}$ increases under high voltage and a part of Fe$^{4+}$ will spontaneously reduce. Combined with the study of crystal transformation, it has been found that this part of spontaneously reduced iron ions will migrate to the adjacent sodium reservoir, which hinders the migration of Na ions and also affects the discharge capacity, which is manifested in the partial loss of capacity in electrochemical performance [7].

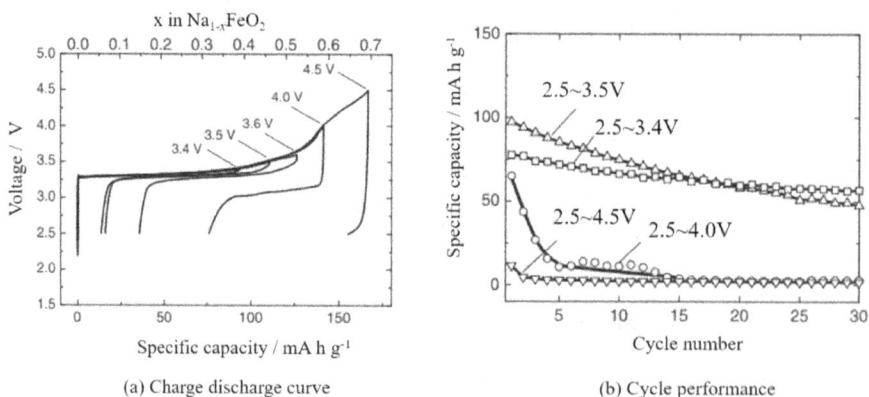

(a) Charge discharge curve

(b) Cycle performance

**Figure 3.5:** Electrochemical performance of O3-type NaFeO$_2$ cathode in different voltage ranges: (a) galvanostatic charge and discharge curves; (b) cycle performance [5].

Therefore, when the cut-off voltage is set in a suitable range, the O3-type $Na_xFeO_2$ material can maintain cycle stability. At the same time, this kind of material also has excellent rate performance in the application of sodium-ion batteries. As shown in Figure 3.6, this material as the cathode can provide more than 50% of the theoretical specific capacity at 1 C (C is the current rate). However, when $NaFeO_2$ comes into contact with water, $Na^+/H^+$-ion exchange reaction will occur, and $NaFeO_2$ will change into $FeOOH$ and $NaOH$, absorbing $CO_2$ to generate $Na_2CO_3$ or $NaHCO_3$. The side reaction of this electrode is very unfavorable to the overall battery performance [8].

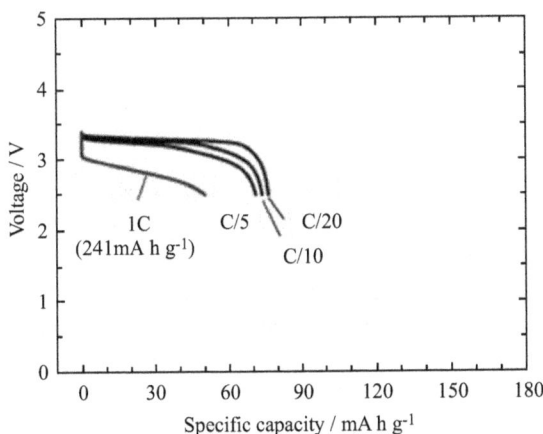

**Figure 3.6:** Rate performance of O3 type $NaFeO_2$ [8].

In addition to Fe based materials, O'3-type $NaMnO_2$ (α-$NaMnO_2$) with monoclinic crystal can also be used as thermodynamically stable layered Na–Mn-based oxide. The crystal structure of $NaMnO_2$ cathode material is shown in Figure 3.7 [9]. The reason of lattice distortion is related to the Ginger-Taylor effect of $Mn^{3+}$. In 1984, it was first reported that the $NaMnO_2$ was used as the electrochemical reversibility material for sodium-ion intercalation/deintercalation. However, in this work, researchers thought that only when x in $Na_{1-x}MnO_2$ was less than 0.2, this kind of material had reversible sodium storage properties, which seemed to be also limited by the voltage window [10]. Subsequently, in 2011, Cedar et al. studied this type of materials. They used $Na_2CO_3$ and $Mn_2O_3$ as raw materials to prepare $NaMnO_2$ materials, and found that when x < 0.8, $Na_{1-x}MnO_2$ materials have reversibility of sodium-ion intercalation/deintercalation. Researchers believe that the difference of x value is due to the difference of synthesis method and electrolyte system, and the specific reason has not been determined yet.

Compared to $Na_{1-x}FeO_2$, the voltage range of structural reversibility of $NaMnO_2$ is wider, but their cycling performance still cannot meet the requirements of sodium-ion batteries application. With the increase in number of cycles, the specific capacity of

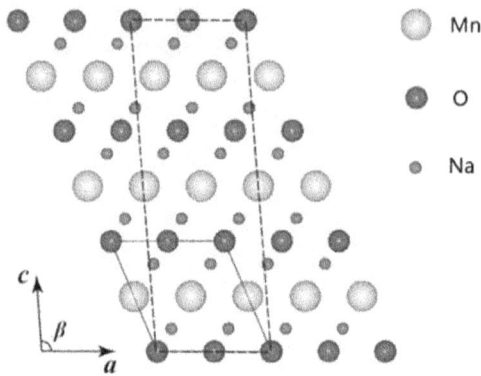

**Figure 3.7:** Crystal structure of α-NaMnO$_2$ cathode material [9].

NaMnO$_2$ electrode materials decreases obviously. As shown in Figure 3.8, the working voltage of Na/NaMnO$_2$ is 2.0–3.8 V. Under the condition of low current rate of C/10, the specific capacity of the first cycle is close to 185 mA h g$^{-1}$, but only about 150 mA h g$^{-1}$ is left in the 10th cycle. With the increase in cycle number, the specific capacity attenuation becomes more serious. It can be seen from the charge/discharge curve that when O3-type NaMnO$_2$ material is used as the cathode of sodium-ion batteries, there are many voltage platforms in the discharge process, which indicates that the phase transition is very serious, which is the main reason for its poor cycle performance.

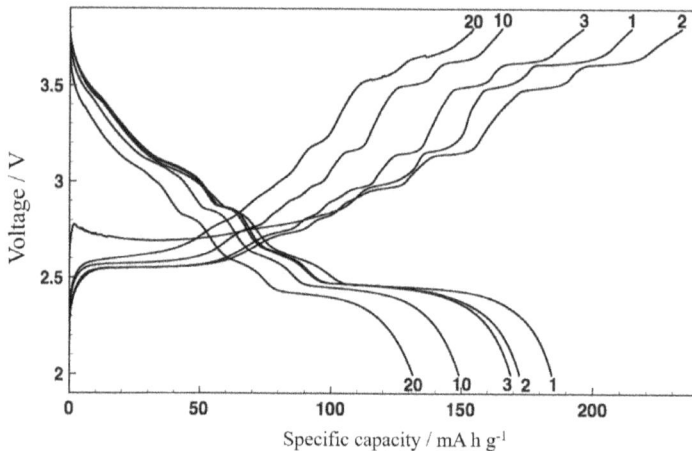

**Figure 3.8:** Charge and discharge curves and capacity attenuation of Na/NaMnO$_2$ material at C/10 [9].

In addition to poor cycling performance, sodium ions need to jump from one octahedron through the tetrahedral gap between octahedrons to another octahedron in

O3-type layered oxide structure, especially in O3-type $NaMnO_2$ materials. The whole Na-ion transport process needs to achieve high activation energy, in order for it to have a certain impact on the ion transport.

In addition to Fe and Mn, other transition metal elements with 3d orbital in Ti to Ni can also be used as high active sites for sodium storage matrix. When these layered oxides are used as cathode materials for sodium-ion batteries, their electrochemical properties have their own advantages. For example, layered $Na_xCoO_2$ materials are also the first to be studied and applied as the main object of sodium storage materials. $Na_x CoO_2$ can obtain O3 and P2 layered structures through a simple preparation method, and both have electrochemical reversibility of sodium storage [11]. A common synthetic reaction equation is [12]

$$2Co_3O_4 + 3xNa_2O_2 \rightarrow 6Na_xCoO_2 + (3x-2)O_2 \tag{3.2}$$

As early as 1981, O3-type $NaCoO_2$ was reported as the cathode of sodium-ion batteries. As shown in Figure 3.9(a), $Na_xCoO_2$ undergoes reversible phase transition evolution of O3$\leftrightarrow$O'3 $\leftrightarrow$ P'3 in the range of $x = 0.8-1$. The schematic diagram of layered structure is shown in Figure 3.9 (b). The reversible phase transition of O3$\leftrightarrow$P3 does not require the fracture and recombination of Co – O bond, but only the slip of $(CoO_2)_n$ lamellae [11].

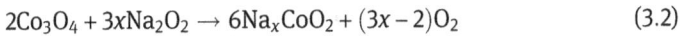

(a)

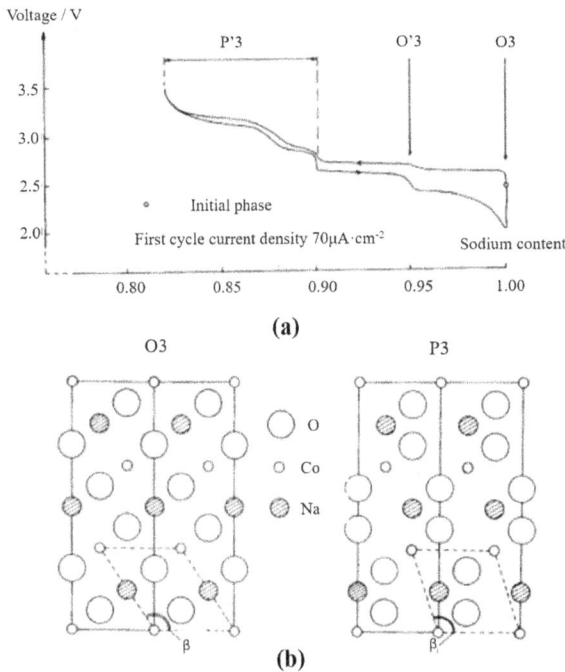

(b)

**Figure 3.9:** $Na_xCoO_2$: (a) charge and discharge curves; (b) diagram of crystal structure [11].

The value of $x$ in $Na_xCoO_2$ can be controlled by controlling the content of oxygen vacancies (Figure 3.10). By controlling oxygen vacancies and adjusting the value of $x$, $O3$-$NaCoO_2$, $O'3$-$Na_{0.77}CoO_{1.96}$, and $P'3$-$Na_{0.60}CoO_{1.92}$ can be synthesized initially. It can be seen that the voltage plateau of the three materials also increases in turn, because a small amount of oxygen vacancies reduce the resistance between the oxide layers when the phase transition occurs, improving the stability of the structure and working voltage. Melinda et al. also studied a small amount of anoxic phase ($y = 0.004$) and significant anoxic phase ($y = 0.073$) of $Na_{0.7}CoO_{2-y}$, proving that $Na_{0.7}CoO_{2-y}$ ($y = 0.004$) material can reversibly intercalate/deintercalate more Na ions and also has higher working voltage, and inferred that oxygen deficiency caused stoichiometric imbalance, resulting in a small number of electron holes and unstable $Co^{4+}$, which would lead to the loss of oxygen, and low conductivity [13].

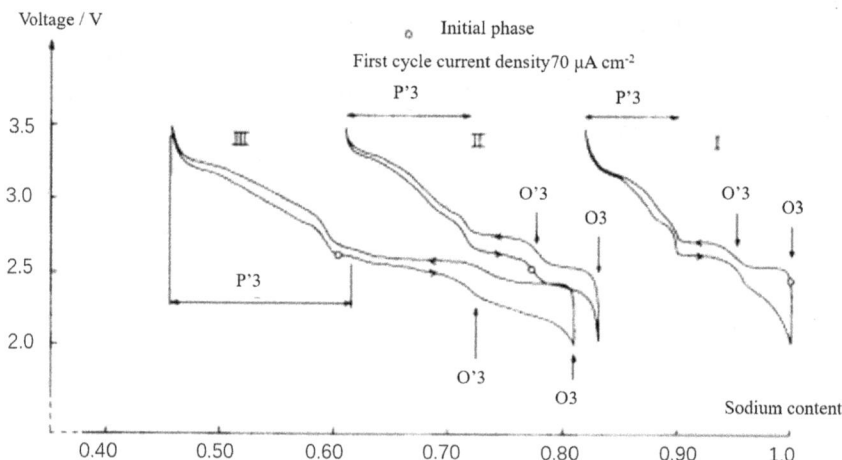

**Figure 3.10:** Charge and discharge curves of P'3, O'3, and O3 and potential range of phase transition [11].

The research on O3-$NaCrO_2$ electrode materials appeared in 1982 [14]. It was not until 2009 that there was a major breakthrough in research on the mechanism of Na-ion intercalation/deintercalation in $NaCrO_2$. Yabuuchi et al. have shown that about 50% of sodium ions can be reversibly intercalated/deintercalated from $NaCrO_2$ materials [15]. In recent research, researchers used $Cr_2O_3$ and $Na_2CO_3$ as raw materials to prepare O3-$NaCrO_2$ materials. It can be seen from Figure 3.11 that the voltage plateau of O3-$NaCrO_2$ material as the cathode of sodium-ion batteries is about 3.0 V (vs. $Na^+/Na$). When the sodium content becomes higher, the voltage plateau will rise to 3.3 V. In $Na_{1-x}CrO_2$ material, when $x \approx 0.5$, another voltage plateau will appear at 3.6 V [16].

Similar to $NaFeO_2$ materials, $NaCrO_2$ materials will undergo irreversible phase transition when charging to high voltage. However, the reversible range of $NaCrO_2$ is slightly wider than that of $NaFeO_2$ (the content of sodium ion is less than 40%).

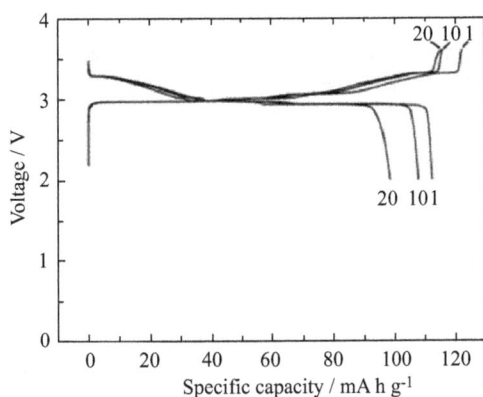

**Figure 3.11:** Charge and discharge curves of $NaCrO_2$ material for different cycles at 25 mA g$^{-1}$ [16].

Although the theoretical specific capacity of O3-$NaCrO_2$ can reach 250 mA h g$^{-1}$, the actual reversible specific capacity is only about 110 mA h g$^{-1}$ due to the large ion radius of Na ion and slow ion migration kinetics.

An important advantage of O3-$NaCrO_2$ as electrode material is its high thermal stability. Researchers studied the thermal stability of $Na_{0.5}CrO_2$ by accelerated calorimetry, and found that $Na_{0.5}CrO_2$ is more stable than $Li_xFePO_4$, which has the best thermal stability in lithium-ion batteries [17]. When tested at 80 °C, O3-$NaCrO_2$ shows excellent cycle performance and rate performance, and there is no structural damage and life decay at high temperature, indicating that the material has excellent thermal stability [18, 19].

In addition to the low capacity of O3-$NaCrO_2$, the main problem is that the material is easy to absorb water in the air, and generate electrical insulating NaOH and $Na_2CO_3$ on the surface, so that sodium diffuses to the surface of the material with the generation of particles inside, resulting in the deactivation of the whole electrode material. Carbon coating on the surface of the material can solve this problem to a certain extent, but improving the cycling performance of O3-$NaCrO_2$ cathode material and making it have a certain value still needs further exploration by researchers.

However, the Ni-based O3-$NaNiO_2$ has a monoclinic rhombohedral lattice, and its structure is similar to α-$NaFeO_2$. The Ni-O layer shares the same edge with each $NiO_6$ octahedral structure. In $NaNiO_2$ materials, $Ni^{2+}$ cannot be in the position of Na ions. Therefore, O'3-type $NaNiO_2$ as electrode materials has similar electrochemical properties as $NaCrO_2$.

$NaNiO_2$ can be synthesized by using $Na_2O$ and NiO in oxygen, and its electrochemical performance is obviously related to the cut-off voltage. $Ni^{3+/4+}$ has three phase transitions in the voltage range of 1.7–3.5 V and can provide high specific capacity of 145 mA h g$^{-1}$ in the voltage range of 2.2–4.5 V, but the cycle stability is extremely poor (Figure 3.12). When the cut-off voltage is set to 3.75 V, the cycle stability and coulomb efficiency of $NaNiO_2$ cathode material are greatly improved, and capacity retention of 94% can be obtained after 20 cycles. At this time, about 50% of Na ions can be reversibly intercalated/deintercalated [20].

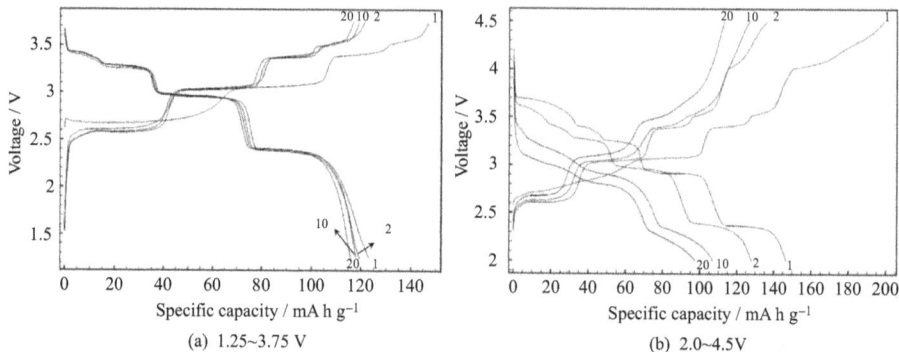

Figure 3.12: Charge and discharge curves of $NaNiO_2$ material in different voltage ranges at C/10 [20].

O3-type $Na_{1-x}VO_2$ materials are also used as cathode in sodium-ion batteries. Generally, when $x = 0$, the material structure is O3 type (Figure 3.13). $NaVO_2$ materials show the ability of reversible intercalation/deintercalation in the voltage range of 1.2–2.4 V with three voltage platforms. When the Na content in the O3-type $Na_{1-x}VO_2$ materials changes to $Na_{2/3}VO_2$, the electrode material has a specific capacity of about 120 mA h $g^{-1}$, and the O3-type structure will change to O'3-type structure during the Na ions intercalation/deintercalation process. When $x > 0.5$ in $Na_{1-x}VO_2$ materials, vanadium ions will migrate to the vacancies in the lattice, leading to a sharp decline in the performance of electrode materials [21]. $Na_{1-x}VO_2$ materials can react rapidly with air in a few seconds and can be coated with carbon in a reducing atmosphere. The coated carbon layer can not only reduce the contact between $Na_{1-x}VO_2$ materials and air but also improve the conductivity of the materials, thus improving the cycle stability and rate performance of $Na_{1-x}VO_2$.

The $NaTiO_2$ materials can also be used as the host materials for Na-ion intercalation/deintercalation. In the process of Na-ion intercalation/deintercalation, this material will also undergo a lattice transition from O3 type to O'3 type, and the structure of O type is shown in Figure 3.14 (a). Its chemical composition is similar to that of $Na_{0.7}TiO_2$, and it has a specific capacity of about 75 mA h $g^{-1}$ at an average working voltage of 1 V. Therefore, it was initially considered to be an electrode material that can only be used as anodes [22]. However, Yu et al. prepared $NaNi_{0.5}Ti_{0.5}O_2$ by using $NaNiO_2$ and O3-$NaTiO_2$ (Figure 3.14(a)). Ti can significantly improve the cycle stability of the material, while $Ni^{2+/4+}$ can provide relatively high specific capacity. The electrochemical performance is shown in Figure 3.14(b). When $NaNi_{0.5}Ti_{0.5}O_2$ material is used as the cathode of sodium-ion batteries, it has a capacity retention rate of 75% after 300 cycles

a

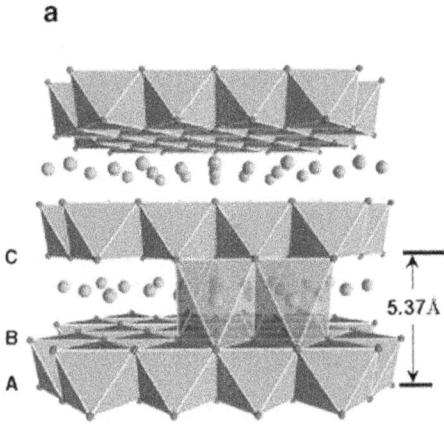

**Figure 3.13:** Crystal structure of O3-type $NaVO_2$ [21].

at 1 C, the coulomb efficiency is always close to 100%, and the cycle stability is significantly improved [23]. This kind of mixed element material can improve the cycle stability of $Ti^{3+/4+}$ and the capacity of $Ni^{2+/4+}$. The doping of multielement ions has always been an important method in materials science research. Through the doping of multielement ions, the competitiveness of layered oxide cathode materials can be improved in terms of electrochemical performance or cost.

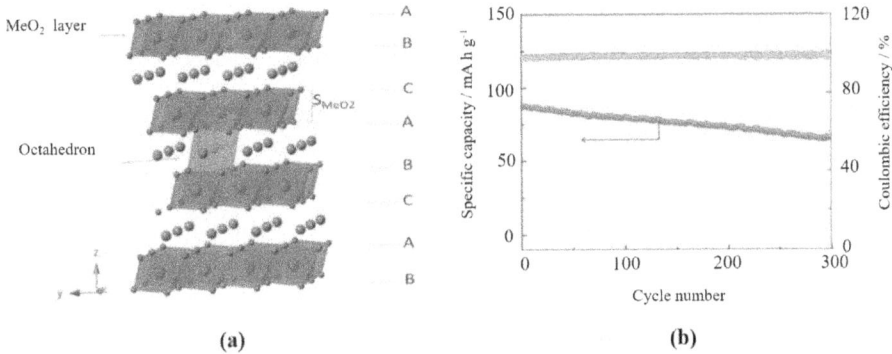

**Figure 3.14:** $NaNi_{0.5}Ti_{0.5}O_2$ material: (a) schematic diagram of crystal structure; (b) cycle performance at 1 C [23].

Layered oxide cathode materials with $Mn^{3+/4+}$ redox pairs can provide higher specific capacity, but the working voltage is relatively low, which cannot meet the demand of high working voltage of sodium-ion batteries. O3-type $Na_xFeO_2$ materials have higher working voltage of $Fe^{3+/4+}$ redox pairs, but the amount of Na ions reversibly intercalated/deintercalated within the working voltage range is less [24].

Yabuuchi et al. prepared O3-type $Na_x[Fe_yMn_{1-y}]O_2$ material by adjusting the raw material ratio of Na and Fe/Mn. The working voltage of the material is 1.5–4.2 V, and the initial specific capacity is about 125 mA h $g^{-1}$. It is found that the change of chemical composition will have a significant impact on the electrochemical performance of the electrode material [25].

The partial substitution of Fe by Co can also improve the performance of the electrode materials, because Co can improve the conductivity of the materials and inhibit the lattice structure changes caused by $Fe^{4+}$ migration [26]. Compared to $NaFeO_2$ and $NaCoO_2$, typical $NaFe_{0.5}Co_{0.5}O_2$ materials can provide a reversible specific capacity of about 165 mA $\cdot$ h $\cdot$ $g^{-1}$ under the working voltage of 2.5–4.0 V, and the cycle performance is significantly improved (Figure 3.15) [8].

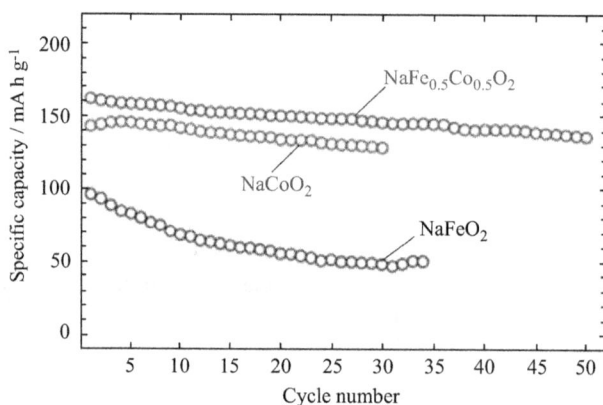

**Figure 3.15:** Comparison of cycle performance between $NaFe_{0.5}Co_{0.5}O_2$ and O3- type $NaCoO_2$ and $NaFeO_2$ [8].

$Ni^{2+/4+}$ can provide high specific capacity by ion doping. $NaFe_{0.3}Ni_{0.7}O_2$ material was prepared by ion doping. This material can increase the specific capacity to 135 mA h $g^{-1}$ with capacity retention rate of 74% after 30 cycling [27]. However, O3-type $Na[Ni_{1/2}Mn_{1/2}]O_2$ can be prepared from $Li[Ni_{1/2}Mn_{1/2}]O_2$ by $Na^+/Li^+$ exchange. The material has a high theoretical specific capacity of 185 mA h $g^{-1}$, and the O3-P3 phase transition occurs during the process of Na-ion intercalation/deintercalation. The cyclic stability of this material is relatively good and shows no irreversible crystal structure change under the working voltage of 4.5 V. When the working voltage is controlled at 3.8 V, it has a good capacity retention rate, achieving a stable specific capacity of 128 mA h $g^{-1}$ [28].

On the basis of binary metal oxides, a third element can be introduced to synthesize ternary metal oxides. For example, the ternary $Na[Ni_{1/2}Mn_{1/2}]O_2$ was synthesized by introducing Co into $NaNi_{1/3}Mn_{1/3}Co_{1/3}O_2$. This material has a specific capacity of about 120 mA h $g^{-1}$ at 2–3.75 V (Figure 3.16) [29]. Meanwhile, it undergoes a reversible crystal structure change of O3-O1-P3-P1 during the process of sodium-ion intercalation/ deintercalation. In order to reduce the use of high-cost Ni and Co elements, Hu et al. also introduced $Cu^{2+}$ to synthesize ternary O3-type $Na_{0.9}[Cu_{0.22}Fe_{0.30}Mn_{0.48}]O_2$, with a specific discharge capacity of 100 mA h $g^{-1}$ and a working voltage of 3.2 V [30].

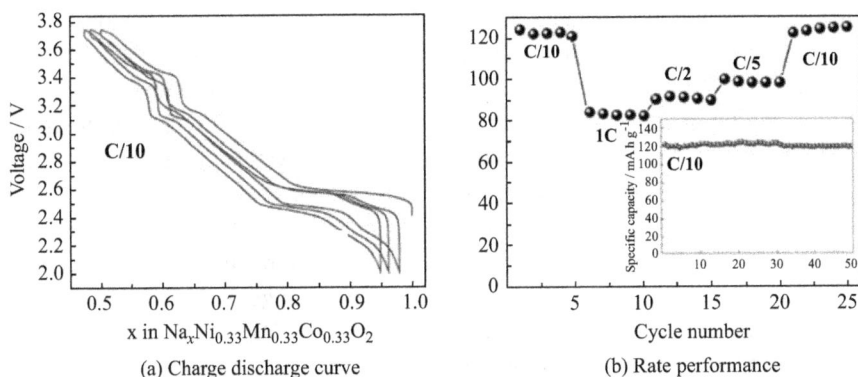

(a) Charge discharge curve

(b) Rate performance

**Figure 3.16:** Electrochemical performance of $NaNi_{1/3}Mn_{1/3}Co_{1/3}O_2$ material:(a) charge and discharge curves at C/10; (b) rate performance [29].

$Na_2MO_3$ (M = 4d) materials also have layered structure that can accommodate lithium-ion or sodium-ion reversible intercalation/deintercalation, so it can be used in the cathode of sodium-ion batteries. For example, O3-type $Na_2RuO_3$ is the most typical 4d-group metal oxide cathode material for sodium-ion batteries. This material shows similar characteristics to metal conduction and crystallizes between Na and $Na_{1/3}Ru_{2/3}$ layers. Researchers used $NaHCO_3$ and $RuO_2$ to prepare this material. This cathode delivered the first cycle discharge specific capacity of 150 mA h $g^{-1}$ at C/5, and achieved good rate performance (Figure 3.17) [31].

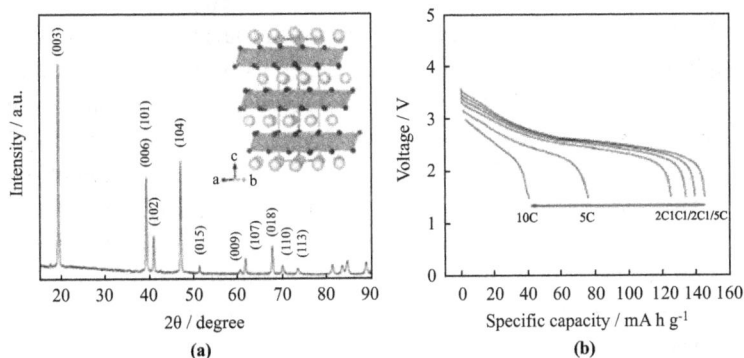

(a)

(b)

**Figure 3.17:** $Na_2RuO_3$ material: (a) XRD spectrum and crystal structure; (b) rate performance [31].

### 3.1.2 P2-type transition metal oxides

As opposed to O3-type layered oxides, for P2-type $Na_{1-x}MO_2$, when the value of $x$ is between 0.3 and 0.7, the structure of the material is relatively stable, and the average valence state of the active metal is also higher than +3.3. The oxygen in the adjacent A-A layers will have a significant electrostatic repulsion phenomenon, so the crystal plane spacing is larger, and Na ions with larger ion radius can stably exist in the triangular prism between the two $MO_6$ octahedrons [32]. Furthermore, the sites of Na ions can be divided into two types – $Na_f$ and $Na_e$. However, due to the large size of Na ions, the distance between the two sites is very close, and the intercalation/deintercalation of Na ions in the two sites is almost simultaneous. When $x$ in $Na_xMO_2$ is less than 0.46, if Na ions continue to be deintercalated, $MO_6$ will slip and deform, resulting in the shrinkage of crystal structure and the decrease of crystal plane spacing, thus becoming O2 type material. The change of P2$\leftrightarrow$O2 is also the difference between P2 type materials and O3-type (O3 $\leftrightarrow$ O'3 $\leftrightarrow$ P'3) materials).

Considering that the ability of Na-ion intercalation/deintercalation at the sites of $Na_f$ and $Na_e$ is similar, the whole Na layer can be regarded as the channel of Na-ion fast transport. Therefore, it can be judged that P2-type materials have lower Na-ion diffusion barrier than O3-type materials. Among the two materials with similar chemical formula, P2-type layered materials tend to show higher ionic conductivity than O3-type layered materials [33]. Using $Na_x[Ni_{1/3}Mn_{2/3}]O_2$ as the research object, Meng et al. verified that the ion diffusion rate of Na ions in P2-type materials is about one order of magnitude higher than that in O3-type materials through theoretical calculation and GITT data analysis [34].

As mentioned above, the crystal structure of O3-type $Na_xCoO_2$ is different based on the $x$ range. When the temperature of P3-type material is raised to more than 700 °C, P2 structure will replace P3 structure, which involves a process of $CoO_6$ octahedral spin and Co-O bond breaking [35]. Delmas et al. also studied the sodium storage process of P2-type materials with O3 and P3 (Figure 3.18(c)). The differential capacity curve of $Na_xCoO_2$ can be used to analyze the phase transition behavior. It is found that although the voltage platform of P2-type materials is similar to the other two materials in the range of 2.7–3.5 V, the reaction of P2-type materials becomes very complex in the range of less than 2.7 V. It is believed that more Na ions can be stored in the triangular prism structure, resulting in the continuous phase transition of the material structure [36]. Delmas et al. combined with in situ XRD to analyze these complex phase structures during the charge/discharge process, and clarified the corresponding relationship between the number of Na ions and voltage ($Na^+/V_{Na^+}$). Some single phases also show different thermal stability at different temperatures, and adjusting the stoichiometric ratio is an important factor affecting the electrochemical performance of materials [37].

Xia et al. studied the thermal stability of P2-type $Na_xCoO_2$ by accelerated calorimetry, and found that a side reaction between $Na_xCoO_2$ and $NaPF_6$ electrolyte salt to

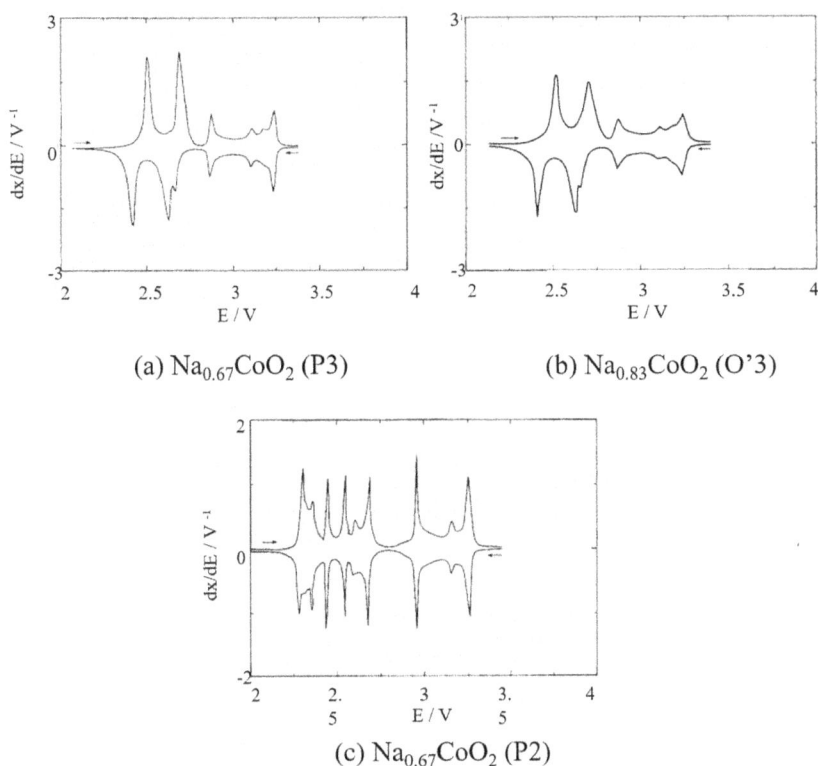

(a) $Na_{0.67}CoO_2$ (P3)

(b) $Na_{0.83}CoO_2$ (O'3)

(c) $Na_{0.67}CoO_2$ (P2)

**Figure 3.18:** $dx/dE$ curves of $NaCoO_2$ ($dx/dE$ is the differential capacity curve of $Na_xCoO_2$) [12].

form $NaCoF_3$. When using this type as cathode, the matching of electrolyte should be considered to avoid affecting the performance and safety of the whole batteries [38]. Ma et al. used $P(EO)_8NaCF_3SO_3$ electrolyte to test $P2\text{-}Na_{0.7}CoO_2$ at 90 °C and obtained a relatively stable long cycle performance [39].

The synthesis temperature is an important factor affecting the types of $Na_xMnO_2$ analogues. $Na_xMnO_2$ with hexagonal P2-type can be synthesized above 700 °C. However, when the material is used at room temperature, it is easy to cause irreversible collapse of the structure with the rapid simultaneous storage of Na ion at $Na_e$ and $Na_f$ sites and strong crystal plane repulsion, resulting in the transformation to amorphous material [40]. When Caballero studied $P2\text{-}Na_{0.6}MnO_2$, it was found that the structural distortion caused by the Ginger-Taylor effect caused by $Mn^{3+}$ was not obvious. The electrochemical test results of half-cell assembled with this electrode material are shown in Figure 3.19(a). The specific capacity of the material reaches 140 mA h g$^{-1}$ at 0.1 mA cm$^{-2}$ and voltage range of 2.0–3.8 V. However, due to intercalation/deintercalation of Na ions, it is easy to cause structural collapse. After 8 cycles, the crystal form changes to amorphous, resulting in a significant decrease in capacity [41]. Yabuuchi et al. could synthesize high purity $P2\text{-}Na_xMnO_2$ by raising the synthesis temperature to 900 °Cwith the water content of this material less than $20 \times 10^{-6}$. As shown in Figure 3.19(b), the specific capacity of the material can reach 190 mA · h · g$^{-1}$ in the charge-discharge range of 1.5–4.3 V [42].

(a)

(b)

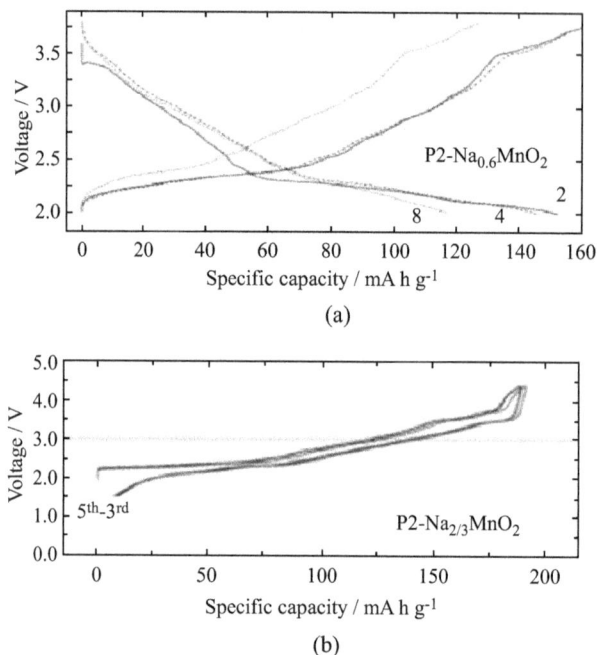

**Figure 3.19:** Charge and discharge curves of $NaMnO_2$: (a) $Na/Na_{0.6}MnO_2$ at 2.0–3.8 V [41]; (b) $Na/Na_{2/3}MnO_2$ at 1.5–4.3 V [42].

P2-type $Na_x[Fe_{0.5}Mn_{0.5}]O_2$ is not stable due to the instability of $Fe^{4+}$ in the oxide, which is different from Co and Mn. Yabuuchi et al. synthesized layered $Na_x[Fe_{0.5}Mn_{0.5}]O_2$ material by replacing part of Fe with Mn. The partial substitution of Mn makes $Fe^{3+/4+}$ have electrochemical activity in the range of 3.8–4.2 V. Before 3.8 V, the oxidation activity of the material is provided by $Mn^{3+/4+}$. P2-$Na_x[Fe_{0.5}Mn_{0.5}]O_2$ can provide a high specific capacity close to 190 mA h g$^{-1}$ (Figure 3.20(a) and (b)), which is much higher than the initial specific capacity of 120 mA h g$^{-1}$ [43] of O3-$Na[Fe_{0.5}Mn_{0.5}]O_2$. Rojo et al. used P2-$Na_{2/3}[Fe_{0.5}Mn_{0.5}]O_2$ to analyze the material structure (Figure 3.20(c)). It is found that the P2 structure is maintained in the range of Mn activity. When the oxidation process of $Fe^{3+/4+}$ begins (3.8–4.2 V), the structure of the material changes from P2 type to OP4 type. When the electrode is fully charged to 4.2 V, the chemical composition of the electrode material is $Na_{0.13}[Fe_{0.5}Mn_{0.5}]O$ [44].

Binary P2-type layered oxides composed of two transition metal elements of Co and Mn are also widely concerned. Carlier et al. synthesized P2-$Na_{2/3}Co_{2/3}Mn_{1/3}O_2$ materials, and confirmed that the valence states of Co and Mn are stable at +3 and +4,

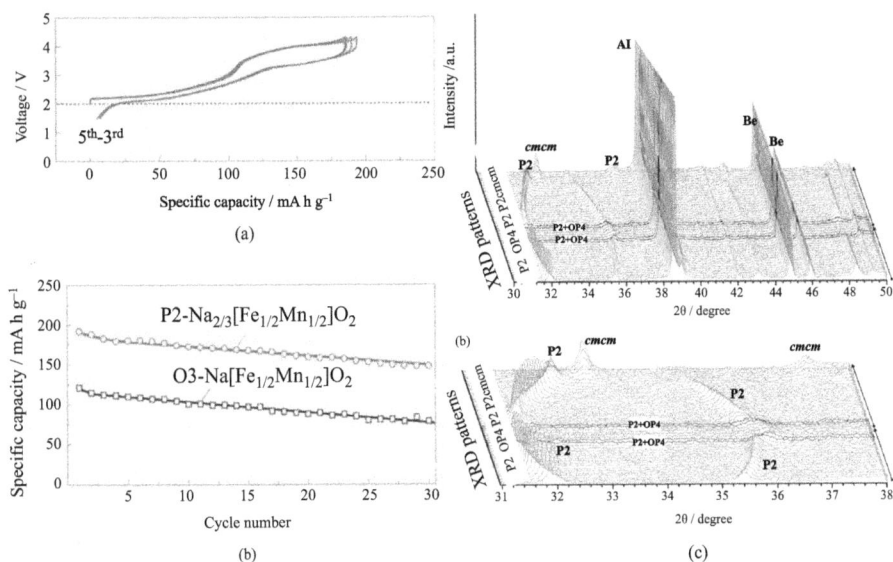

**Figure 3.20:** Electrochemical performance of $Na_{2/3}$ [$Fe_{1/2}Mn_{1/2}$] $O_2$ and in situ XRD pattern for the first cycle: (a) charge and discharge curves of Na / $Na_{2/3}$ [$Fe_{1/2}Mn_{1/2}$] $O_2$ at 1.5–4.3 V and 12 mA $g^{-1}$; (b) comparison of electrochemical performance between P2-Na/$Na_{2/3}$ [$Fe_{1/2}Mn_{1/2}$] $O_2$ and O3-Na/$Na_{2/3}$ [$Fe_{1/2}Mn_{1/2}$] $O_2$ [43]; (c) in situ XRD patterns of P2-$Na_{2/3}$ [$Fe_{1/2}Mn_{1/2}$] $O_2$ for the first cycle [44].

respectively. The half-cell test of this material was carried out and compared with P2-$Na_{0.74}CoO_2$. As shown in Figure 3.21(a), the stepped voltage platform of $Na_xCo_{2/3}$ $Mn_{1/3}O_2$ disappeared in the range of $0.5 \leq x \leq 0.83$ (compared with P2-$Na_{0.74}CoO_2$), but the voltage dropped sharply in the range of $0.65 \leq x \leq 0.83$, which was unfavorable to the energy density of the battery [45]. In order to stabilize the structure and expand the range of Na content, the P2-type $Na_{2/3}[Co_{1-x}Mn_x]O_2$ was synthesized by Yang et al. with low $x$ value of 0.5. Although only the initial specific capacity of 124.3 mA h $g^{-1}$ and reversible specific capacity of 85 mA h $g^{-1}$ were achieved, which happens mainly in the range of 1.5–2.1 V, the capacity retention rate can reach 99% after 100 cycles at 5 C (Figure 3.21(b)) [46]. In order to improve the specific capacity of P2-type $Na_{2/3}Co_{2/3}Mn_{1/3}O_2$ and reduce the synthesis temperature below 700 °C, the P2/P3-type $Na_{0.66}Co_{0.5}Mn_{0.5}O_2$ composite was synthesized by Zhou et al. Compared to the P2 type $Na_{2/3}Co_{2/3}$ $Mn_{1/3}O_2$, the first cycle discharge capacity of the material reaches 180 mA h $g^{-1}$ at 0.1 C in the voltage range of 1.5–4.3 V, and the cycling performance is good at high current density. Electrochemical performance test and in situ XRD test verified the stable existence of P2/P3 phase (Figure 3.21(c) and (d)) [47]

P2-$Na_{2/3}[Ni_{1/3}Mn_{2/3}]O_2$ is stable in the external environment because it does not hydrate with water in the air. Although the theoretical specific capacity of this material is only 173 mA h $g^{-1}$, the effect of $Ni^{2+/4+}$ in the high voltage range can give the

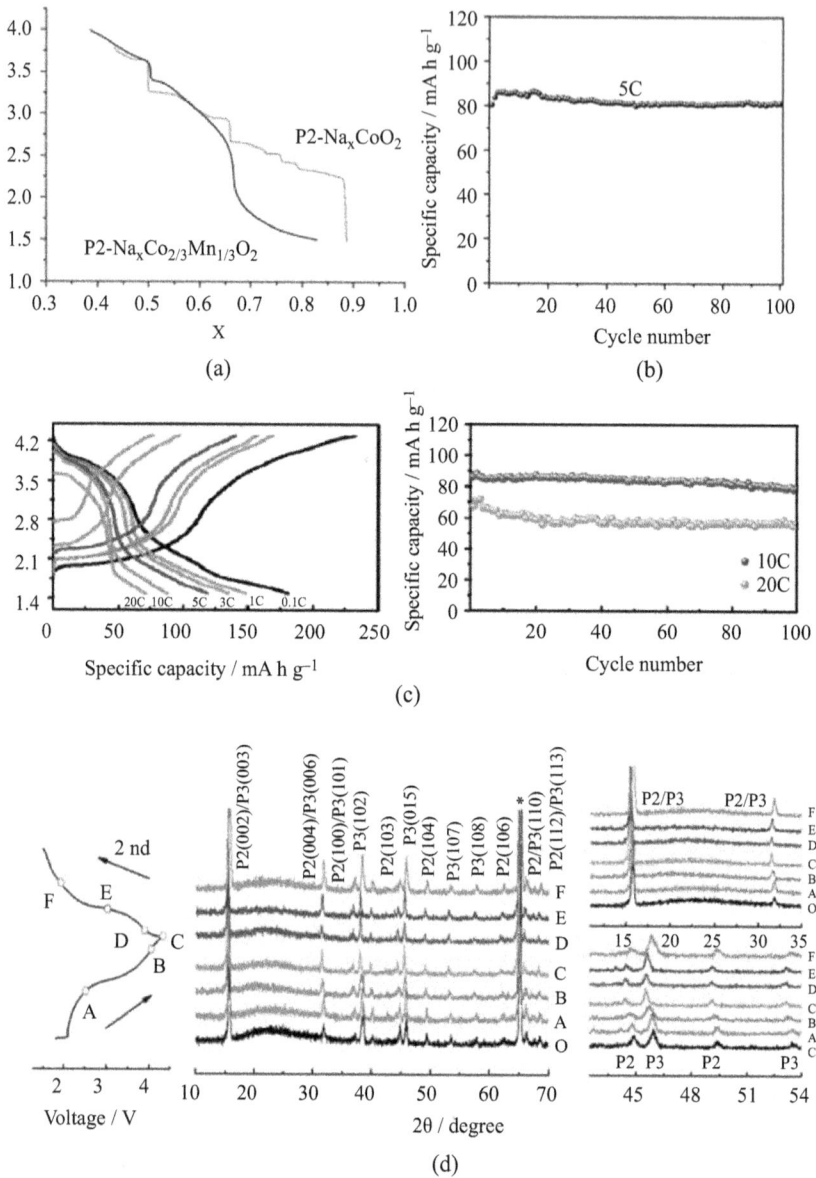

**Figure 3.21:** Electrochemical performance of $Na_xCo_{2/3}Mn_{1/3}O_2$ and in situ XRD pattern for the first cycle: (a) discharge curve of $Na/Na_xCo_{2/3}Mn_{1/3}O_2$ and $Na/Na_xCoO_2$ [45]; (b) capacity performance of P2-$NaCo_{0.5}Mn_{0.5}O_2$ at 5 C (0.6 A $g^{-1}$) for 100 cycles [46]; (c) rate curve of P2/P3-$Na_{0.66}Co_{0.5}Mn_{0.5}O_2$ (1 C = 170 mA $g^{-1}$) and cycle performance at different rates; (d) in situ XRD of P2/P3-$Na_{0.66}$ $Co_{0.5}Mn_{0.5}O_2$ (O is the initial position before charge and discharge) [47].

material a higher working voltage. As shown in Figure 3.22(a), P2-$Na_{2/3}[Ni_{1/3}Mn_{2/3}]O_2$ material can release the specific capacity of 160 mA h $g^{-1}$ in the voltage range of 2–4.5 V, and the average working voltage reaches 3.5 V under the action of $Ni^{2+/4+}$. In situ XRD technology revealed that the structural phase transition from P2 to O2 occurred at the high voltage platform [48]. As mentioned above, due to the small migration barrier of adjacent sites in P2 structure, the diffusion ability of Na ions in P2 structure is greater than that in O2 structure. The diffusion of Na ions in O2 structure shuttles through the tetrahedron between the two octahedral sites and needs to cross a larger energy barrier (Figure 3.22(b)). This phenomenon has a great influence on the electrochemical properties of materials, before and after phase transformation. When the cyclic voltage range is set before the P2- to O2-phase transition, the electrode material has extremely high cyclic stability (Figure 3.22(c)) [49]. Therefore, many studies on the modification of these materials are aimed at the phase transition from P2 to O2. The $Na_{0.67}[Ni_{0.2}Mg_{0.1}Mn_{0.7}]O_2$ material was synthesized by introducing divalent metal ions $Mg^{2+}$ (Figure 3.22(d)). It was observed by in situ XRD that OP4-structure phase was formed at about 4.2 V. This phase transformation is due to the fact that the inactive $Mg^{2+}$ occupies part of the Na ions sites, resulting in non-transformation to O2 in the structure. However, the reversibility of P2 ↔ OP4 was significantly higher than that of P2 ↔ O2. Therefore, $Na_{0.80}[Li_{0.12}Ni_{0.22}Mn_{0.66}]O_2$ has excellent cycle stability [50]. If $Li^+$ is added into the transition metal-ion layer, the reaction potential from P2 to O2 can be increased to 4.4 V. P2-type $Na_{0.80}[Li_{0.12}Ni_{0.22}Mn_{0.66}]O_2$ has a simple solid solution reaction process compared to P2-type $Na_{2/3}[Ni_{1/3}Mn_{2/3}]O_2$, and the formation of O2 phase can be delayed. The capacity retention rate can reach 91% after 50 cycles, which also significantly improves the cycle stability of P2-type $Na_{2/3}[Ni_{1/3}Mn_{2/3}]O_2$ [51].

There are many modification processes for P2 materials. Singh et al. added $NaN_3$ as the sodium salt to make up for the large irreversible capacity in the first cycle. Comparing the electrochemical performance of $Na_{2/3}[Fe_{1/2}Mn_{1/2}]O_2$ doped with 5% $NaN_3$ to that of the original electrode material, it is found that the specific capacity loss of the first cycle is reduced by 50%, only 27 mA h $g^{-1}$ [52]. Doping Ni into P2-$Na_{0.67}[Mn_{0.5}Fe_{0.5}]O_2$ to construct ternary P2-layered oxide materials can not only inhibit the production of electrochemically inert $Mn^{4+}$ on the material surface but also inhibit the structural irreversibility and polarization problems caused by the migration of $Fe^{3+}$ to tetrahedral sodium storage layer, under high voltage [53]. Yabuuchi and Komaba proposed a new process and principle to improve the specific capacity of P2-type materials. P2-$Na_{2/3}[Mg_{0.28}Mn_{0.72}]O_2$ [54] and P2-$Na_{5/6}[Li_{1/4}Mn_{3/4}]O_2$ [55] were synthesized by solid-state method, and the initial specific capacities of both materials were more than 200 mA h $g^{-1}$. Through structural analysis, it is found that these capacities exceeding the theoretical value of P2-type materials come from the oxidation of $O^{2-}$ under high pressure and the release of oxygen. In this way, in order to balance the charge at high voltage, $Mn^{3+/4+}$ firstly oxidizes and then reduces in the same charging process, resulting in the rearrangement of cations in the crystal plane.

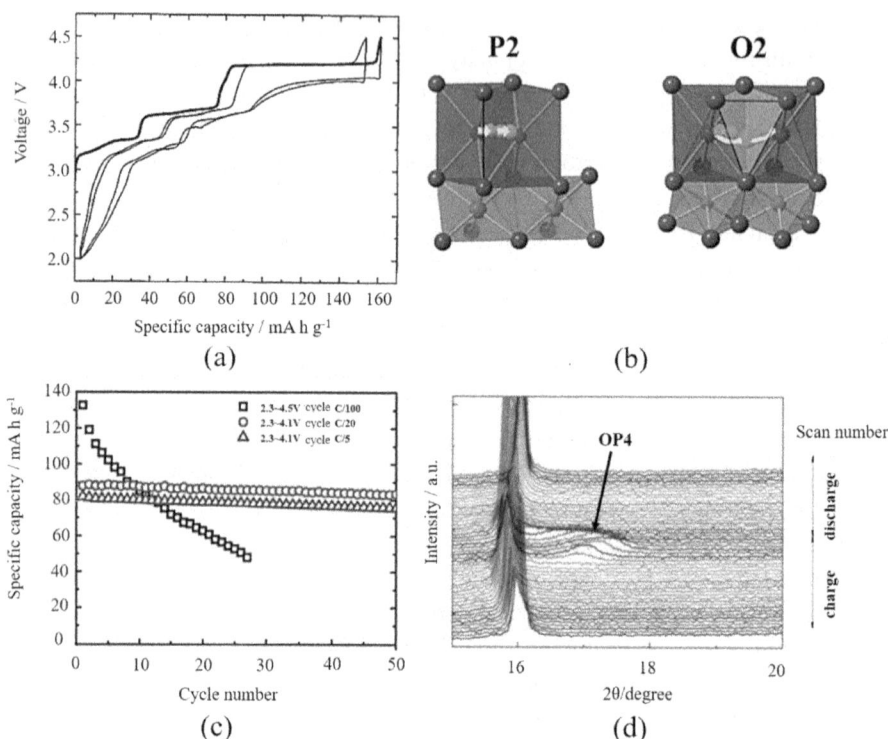

**Figure 3.22:** Electrochemical performance and structural characteristics of $Na_{2/3}Ni_{1/3}Mn_{2/3}O_2$:
(a) discharge curve of $Na_{2/3}Ni_{1/3}Mn_{2/3}O_2$ [48]; (b) $Na^+$ migration channels in P2 structure and O2 structure; (c) cycle performance of $Na/Na_{2/3}Ni_{1/3}Mn_{2/3}O_2$ half-cell in different voltage range and rate performance [49]; (d) in situ XRD of $Na_{0.67}Ni_{0.2}Mg_{0.1}Mn_{0.7}O_2$ in the phase transition process of P2 ⟷ OP4 [50].

Zhou et al. synthesized $Na_{0.66}Li_{0.18}Mn_{0.71}Ni_{0.21}Co_{0.08}O_{2+\delta}$ (NaLiMNC) by coprecipitation and solid-state reaction. This special structure with two-phase mixing can be observed by high-resolution TEM, as shown in Figure 3.23(a). This material not only suppresses the loss of a large number of irreversible capacities caused by the transformation from P2-type material to O2 at high potential, but also improves the structural distortion caused by the complex reaction process of O3-type material itself. The charge/discharge curve is shown in Figure 3.23(b), and the average working voltage of this material reaches to 3.2 V. Its electrochemical performance is shown in Figure 3.23(c). The specific capacity of the assembled half cell reaches 200 mA h $g^{-1}$ at 0.1 C. Even if the current rate reaches 0.5 C, the specific capacity of the electrode is greater than 105 mA h $g^{-1}$, and the retention rate can reach 75% after 150 cycles. Through rough calculation, the energy density of this material as the cathode material of sodium-ion batteries is about 650 W h $kg^{-1}$, which is of great significance for future research and large-scale energy storage applications [56].

(a)                                    (b)

(c)

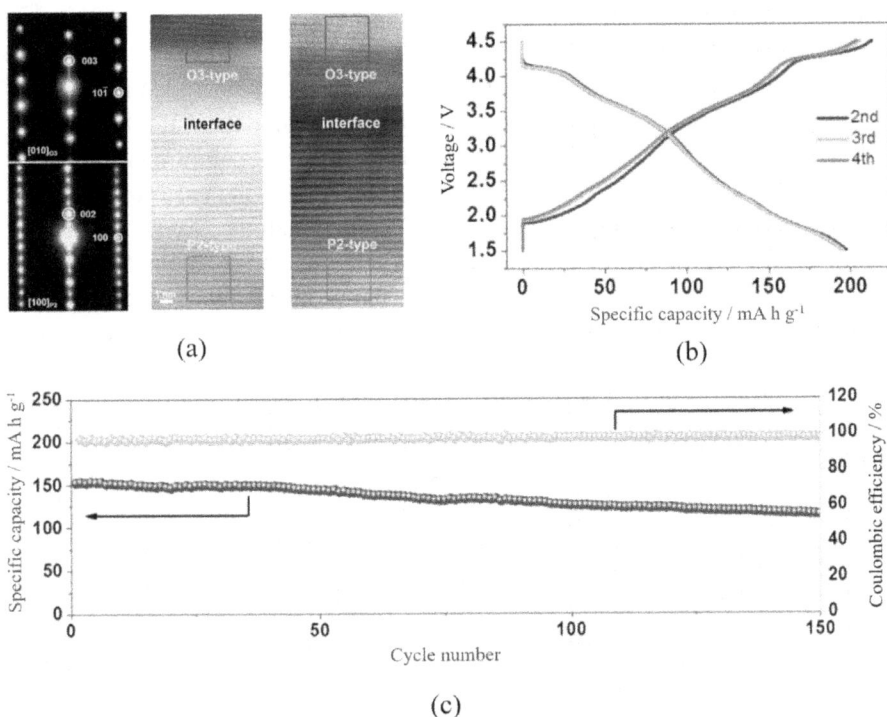

**Figure 3.23:** The structure and electrochemical performance of NaLiMNC: (a) the structure of O3 and P2 of NaLiMNC in selected area electron diffraction; (b) galvanostatic charge and discharge curves of NaLiMNC at 0.1 C in 1.5–4.5 V (1 C = 100 mA g$^{-1}$); and (c) long cycle performance and coulombic efficiency of NaLiMNC at 0.5 C [56].

## 3.1.3 Tunnel structure transition metal oxides

As opposed to the layered transition metal oxides that are rich in sodium before cell cycle, the tunnel type transition metal oxides that are poor in sodium or even lack sodium have good Na-on intercalation/deintercalation ability. Most of these materials have three-dimensional structure and two-dimensional structure. The synthesis of this kind of material is basically carried out at low temperature, so it is easy to produce particles with small particle size and large specific surface area, which is convenient for rapid Na-ion intercalation. The reason why this kind of material is called tunnel type material is that there are channels for Na ion to be intercalated/deintercalated in x, y, or z directions. Zero stress or low stress materials were prepared to reduce the deformation of the porous structure during the Na-ion intercalation/deintercalation process; otherwise, the structure would collapse with the battery cycle. This kind of cathode often has good performance when it is matched with sodium metal anode, but the problem is that in the actual assembly process of the whole battery, the anode is basically sodium-free material, which requires some initial sodium pretreatment or excitation of metal anode.

Among the transition metal oxides of tunnel type, Fe-based and Mn-based materials are the most common. As for Fe-based oxide materials, $NaFeO_2$ with $Fe^{3+}$ will crystallize into two different crystals under different conditions, one is O3 type $NaFeO_2$ calcined at 500 °C, the other is $\beta$-$NaFeO_2$ calcined at 760 °C, similar to wurtzite. As shown in Figure 3.24(a), in $\beta$-$NaFeO_2$ material, oxygen atoms are arranged in hexagonal close packing, and sodium ions and iron ions are in tetrahedral positions. The charge discharge performance of this material can be seen in Figure 3.24(b). In the voltage range of 2.0–4.0 V, the specific capacity of this material is only close to 0, which is close to the electrochemical inert electrode material. This is because it is difficult for iron ions in tetrahedral position to undergo redox reaction [4].

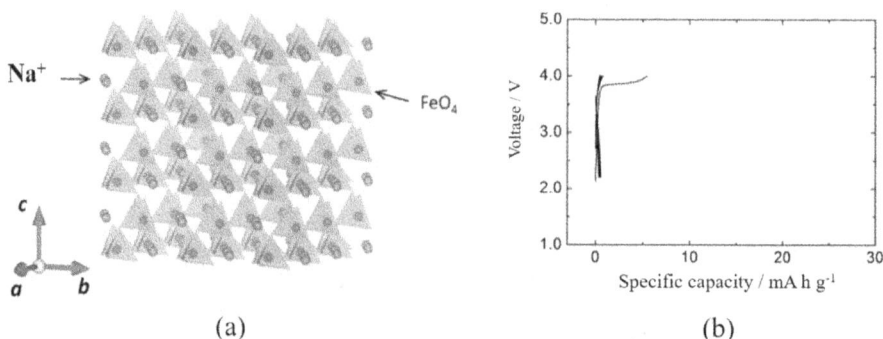

**Figure 3.24:** Crystal structure and electrochemical performance of $NaFeO_2$: (a) schematic diagram of crystal structure; and (b) charge and discharge curves [4].

The researchers also studied the electrochemical performance of anti-spinel $Fe_3O_4$, $\gamma$-$Fe_2O_3$ and $\alpha$-$Fe_2O_3$ as the cathode of sodium-ion batteries. The specific capacity of micron scale iron-based oxides is usually less than 10 mA h $g^{-1}$, which can be considered as electrochemical inert when used as electrode materials for sodium-ion intercalation/ deintercalation. However, nanoscale $Fe_2O_3$ and $Fe_3O_4$ show good electrochemical performance, which is affected by kinetic factors [57]. For example, $Fe_3O_4$ with different particle sizes was selected to match the electrolyte of EC:DEC = 1:1 in 1 M $NaClO_4$, and sodium metal was used as the reference electrode to assemble a three-electrode system to test the Na-ion intercalation/deintercalation performance of $Fe_3O_4$ with different particle sizes. As shown in Figure 3.25, it was found that even if the particle size was as small as 400 nm, it still indicated electrochemical inertia for Na-ion storage. However, if the particle size is about 10 nm, the reversible specific capacity can reach 170 mA h $g^{-1}$, and the capacity can be maintained well under 30 cycles, which indicates that the particle size control is necessary for this kind of material. Due to the

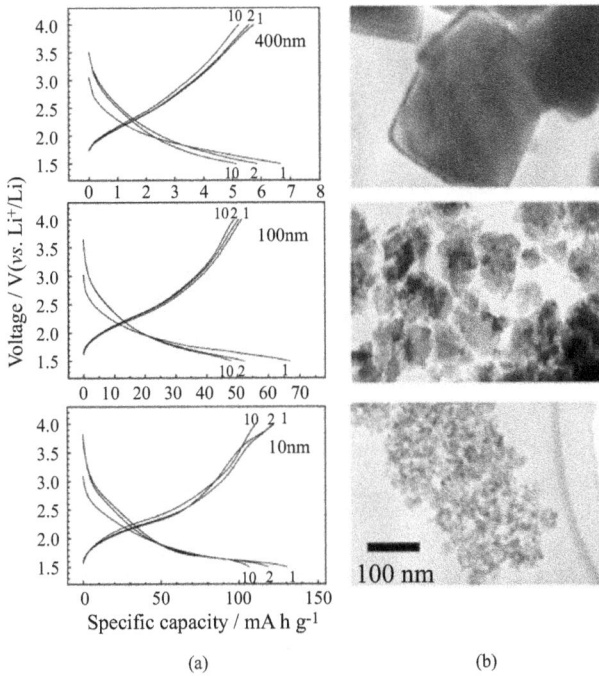

**Figure 3.25:** Charge and discharge curves of $Fe_3O_4$ with particle size of 400, 100, and 10 nm at 20 mA g$^{-1}$: (a) charge and discharge curves; and (b) comparison of particle morphology and size [58].

low working potential of these materials, they are mostly used as anode materials, so the further modification of materials is mainly introduced in detail in the following anode part [58].

Manganese-based oxide materials as the main materials of sodium-ion intercalation/deintercalation were first proposed in the 1980. Manganese is abundant in nature, mainly in the form of pyrolusite, with low price, which is conducive to the promotion of secondary batteries. Mn in manganese oxides mainly exists in +2, +3, and +4 valences, such as MnO, $Mn_2O_3$, $MnO_2$, and various manganate materials. Manganese-based oxide materials show various electrochemical activities due to their rich chemical composition, morphology, and crystal structure (Figure 3.26) [59]. Materials with different crystal structure parameters can directly affect the electrochemical performance of these materials as electrodes for sodium-ion batteries. The optimization of morphology and the change of the type and distribution of cationic active sites (doping other transition metal ions) can also indirectly change the arrangement of crystal structure, thus improving the capacity and stability of materials.

Among them, $\alpha$-$MnO_2$ and $\beta$-$MnO_2$ are the common research objects in the research of lithium-ion batteries. These two materials have also been widely studied as the main materials of sodium-ion intercalation/deintercalation. The framework

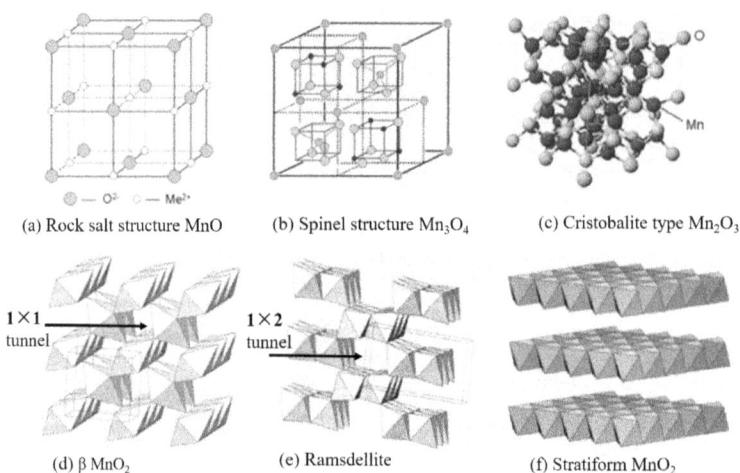

(a) Rock salt structure MnO     (b) Spinel structure $Mn_3O_4$     (c) Cristobalite type $Mn_2O_3$

(d) β $MnO_2$       (e) Ramsdellite       (f) Stratiform $MnO_2$

**Figure 3.26:** Schematic diagram of crystal structure of some manganese oxides [59].

structure of $\alpha$-$MnO_2$ and $\beta$-$MnO_2$ materials is an octahedral structure, and each unit cell is connected in the form of common angle. Among them, $\beta$-$MnO_2$ has a smaller $1 \times 1$ tunnel along the $c$-axis (Figure 3.26(d)), and the edge of the tunnel is composed of $MnO_6$. Even the smallest alkali metal ion, lithium ion, cannot be intercalated/deintercalated in such a tunnel, but the nano-sized $\beta$-$MnO_2$ has regular and ordered micropores, which can stimulate the electrochemical activity of the material and make it possible to accommodate the intercalation/deintercalation of sodium ions. Some researchers have prepared $\beta$-$MnO_2$ nanorods with a diameter of about 100 nm, which can provide a large specific capacity of more than 200 mA h $g^{-1}$ in the first cycle as the cathode of sodium-ion batteries. However, due to the large diameter of sodium ion, this material will undergo structural phase transition during the process of sodium-ion intercalation/deintercalation, which affects the cycle stability of the material. After 100 cycles, it only has a capacity of 145 mA h $g^{-1}$ [60].

The structure of $\alpha$-$MnO_2$ is almost similar to that of $\beta$-$MnO_2$. The difference is that $\alpha$-$MnO_2$ has a large tunnel structure with $2 \times 2$ size (Figure 3.27(c)). Such an open channel can accommodate relatively large alkali metal ions. Therefore, $\alpha$-$MnO_2$ nanorods have a high reversible specific capacity of about 280 mA h $g^{-1}$. However, during the charge and discharge process, the electrode materials also have crystals similar to those of $\beta$-$MnO_2$ materials. After 100 cycles, only about 75 mA h $g^{-1}$ is left. $\lambda$-$MnO_2$ type materials can also be used in the cathode of sodium-ion batteries. In the electrochemical cycle, part of them will be transformed into O'3 phase. These materials have $2 \times 3$ tunnels composed of two MnO and three $MnO_6$, so they have more open channels. On the basis of $2 \times 3$ channels, $Na_{0.4}MnO_2$ can be prepared by a simple solid-state method (Figure 3.27(d)).

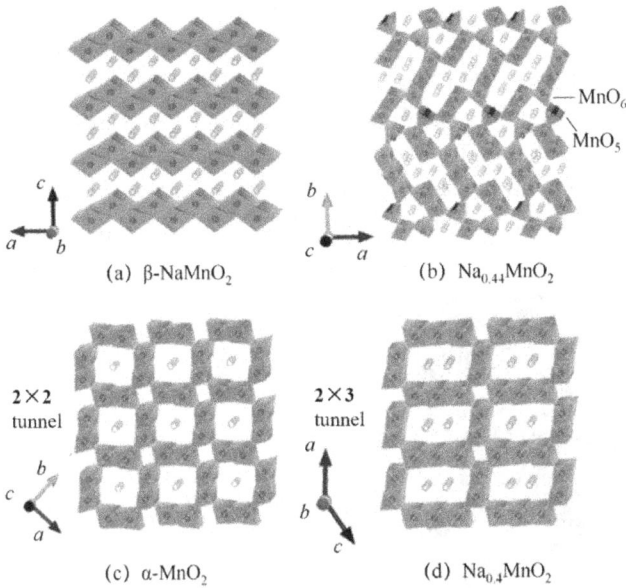

Figure 3.27: Schematic diagram of crystal structure of some manganese oxides [4].

The structure of $Na_xMnO$ is complex. When $x < 0.5$, a three-dimensional pore structure will be formed, such as $Na_{0.4}MnO_2$ and $Na_{0.44}MnO_2$; when $x > 0.67$, a two-dimensional layered structure will be formed. The tunnel structure crystal has a large transport channel of Na ions, but its capacity is relatively low due to less active sites. Similar to iron-based oxides, manganese-based oxides can prepare $\beta$-$NaMnO_2$ at higher temperatures above 900 °C. As shown in Figure 3.27(a), when $\beta$-$NaMnO_2$ materials are used as cathode materials for sodium-ion batteries, the charge-discharge range can only be 1.5–3.5 V, and the average discharge voltage is low, which is a disadvantage in practical battery applications.

$Na_{0.44}MnO_2$ has the best application prospects because of its good structural stability and high specific capacity (~ 120 mA h $g^{-1}$). Doeff et al. tested the mechanism of Na-ion intercalation in solid electrolyte at 85 °C for the first time. The crystal structure of $Na_{0.44}MnO_2$ consists of four $MnO_6$ octahedrons with $Mn^{4+}$ sites and one $MnO_5$ pyramid. As shown in Figure 3.28, these structures form two kinds of tunnels through the common angle structure. Each cell contains one S-shaped tunnel that can intercalate four Na ions and two identical Pentagon tunnels. When the Na3 site of the small tunnel is fully filled, the Na1 and Na2 sites of the large tunnel are only partially filled. Each Na ion can migrate rapidly in the $c$-axis direction along the Na ion, and is finally stored in the tunnel structure, providing capacity for the material [61]. Cao et al. proposed that the structure of electrode materials is prone to collapse under high-rate charge/discharge, resulting in the gradual increase of irreversible capacity. Kim proposed that the electrostatic repulsion in the structure at 2.6–2.8 V and the inherent

Ginger-Taylor effect of Mn-based materials will accelerate the capacity decay [62]. The nano-sized $Na_{0.44}MnO_2$ material prepared by Cao et al. through the thermal decomposition of polymer shows good rate performance and excellent capacity retention. It also has a reversible specific capacity of about 100 mA h g$^{-1}$ after 100 cycles as the cathode of sodium-ion batteries. Therefore, to improve the capacity attenuation of $Na_{0.44}MnO_2$, we should mainly consider two aspects: increasing the crystallinity of the composite and reducing the tunnel length in the crystal structure [63].

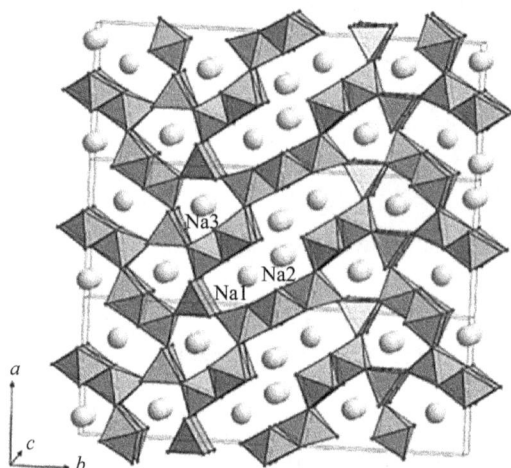

**Figure 3.28:** Schematic diagram of crystal structure of $Na_{0.44}MnO_2$ on *ab* plane [61].

The specific capacity of tunnel type materials is lower than that of layered oxide type materials, and the voltage is not dominant. Therefore, these materials are more suitable for the cathode materials of aqueous sodium-ion batteries. Whitacre et al. tested the voltage range of $Na_{0.44}MnO_2$ for reversible sodium storage and found that it has multistep Na-ion intercalation/deintercalation process at 0.3–3 V (Hg/HgSO$_4$), as shown in Figure 3.29 (a). And when it is assembled with the activated carbon anode as a water system full cell for testing, the full cell can be reversibly cycled for 1,000 cycles in the voltage range of 0.4–1.8 V, 4 C current density, and the capacity remains basically unchanged, as shown in Figure 3.29(b) [64]. Kim et al. found that the diffusion coefficient of Na ions in $Na_{0.44}MnO_2$ ($1.08 \times 10^{-13}$–$9.15 \times 10^{-12}$ cm$^2$ s$^{-1}$) in water system is close to three orders of magnitude higher than that in organic system ($1.08 \times 10^{-13}$–$9.15 \times 10^{-12}$ cm$^2$ s$^{-1}$) (Figure 3.29(c) and (d)). As a result, the rate performance of $Na_{0.44}MnO_2$ electrode material in water system is better than that in organic system. This may be caused by the strong interfacial resistance in organic system and the resistance of SEI film [65].

Vanadium-based oxides have also been widely studied as matrix materials for sodium-ion intercalation/deintercalation. As early as the 1980s, West et al. studied the mechanism of sodium storage of layered α-$V_2O_5$ and $Na_{1+x}V_3O_8$. Among them, α-$V_2O_5$

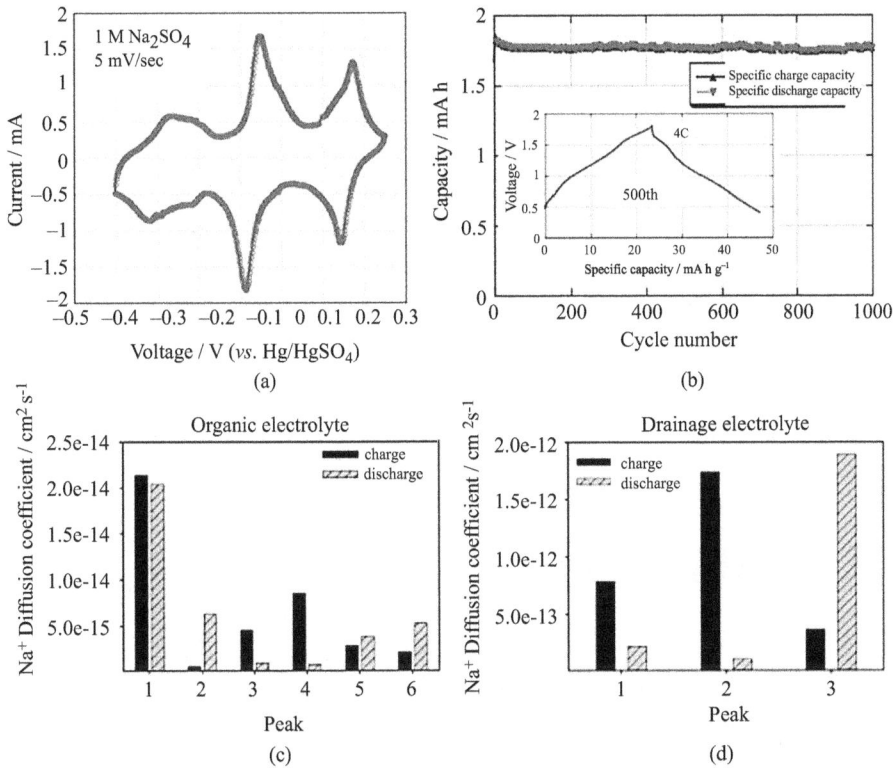

**Figure 3.29:** Electrochemical performance of $Na_{0.44}MnO_2$: (a) CV curves of $Na_{0.44}MnO_2$ at 0.5 mV s$^{-1}$ in aqueous electrolyte; (b) Cycle performance of the full battery at 4 C in 0.8–4 V [64]; (c) and (d) the Na$^+$ diffusion coefficient based on the peaks of CV curves of $Na_{0.44}MnO_2$ in organic system and aqueous system [65].

layered structure is composed of $VO_5$ with common edge and common angle, in which alkali metal ions of different sizes can achieve relatively high specific capacity [66]. Tapavcevic et al. proposed the concept of double-layer nanostructure and synthesized a special structural material with adaptive Na-ion intercalation/deintercalation adjustment (Figure 3.30). The α-$V_2O_5$ layer spacing of bilayer nanostructures is 13.5 Å, which is different from the general $V_2O_5$ layer spacing of 4.4 Å. The α-$V_2O_5$ layer spacing of bilayer nanostructures is 13.5 Å. The α-$V_2O_5$ material prepared by electrodeposition too has abundant nanoporous structure. Through the redox pair of $V^{4+/5+}$, the working voltage range of 1.5–3.8 V can approach 250 mA h g$^{-1}$ of the theoretical specific capacity, with cycle stability of more than 300 cycles. The energy density and power density of the battery with sodium as anode can exceed 700 W h kg$^{-1}$ and 1,200 W kg$^{-1}$, respectively [67].

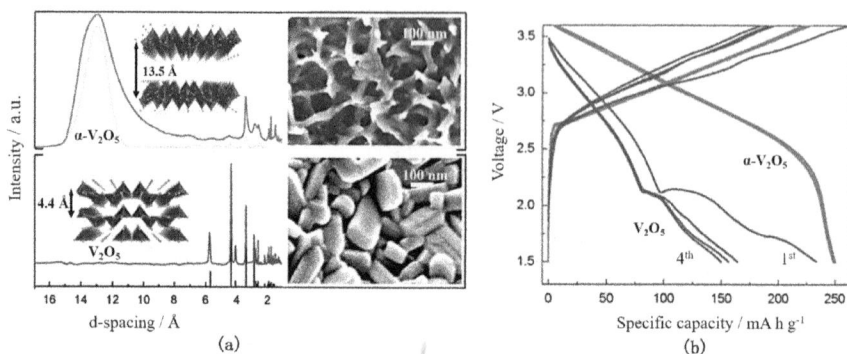

**Figure 3.30:** Synthesis of bilayer nanostructures of α-V$_2$O$_5$ and traditional V$_2$O$_5$ by deposition: (a) Synchrotron radiation XRD and SEM images of double-layer nanostructures of α-V$_2$O$_5$ synthesized by electrodeposition (above) of and traditional V$_2$O$_5$ (below) [65]; (b) Charge and discharge curves of double-layered nanostructures α-V$_2$O$_5$ and traditional V$_2$O$_5$.

## 3.2 Polyanionic cathode

Polyanionic compounds are also a key material for a class of sodium-ion batteries cathode materials. The rich variety and unique sodium storage properties make this material always attract researchers' attention. Polyanionic materials with refer to a series of tetrahedral $(XO_4)^{n-}$ anion units and their derivative units $(X_mO_{3m+1})^{n-}$ (X = S, P, Si, As, Mo, or W) and polyhedral units in the compound structure $MO_x$ (M stands for transition metal) is a class of compounds composed of structure. In most polyanionic compounds, the $(XO_4)^{n-}$ anion unit not only allows ions to be rapidly conducted in the open structural framework, but also stabilizes the redox couple of transition metals. Compared with the layered compound, the strong X–O covalent bond in the polyanionic compound can induce the M–O covalent bond to produce a stronger degree of ionization, thereby generating a higher transition metal redox couple. This is the "induction effect" in polyanionic compounds, so polyanionic electrode materials often have a higher working voltage. Moreover, the strong covalent bond between X and O stabilizes the O in the crystal lattice, so that polyanionic materials tend to have higher structural stability and safety. This is one of the reasons why polyanionic materials are more suitable for rechargeable secondary batteries [68].

Compared with oxides or other types of cathode materials, the significant advantages of polyanionic compounds are stable three-dimensional structure, wide voltage platform, and high safety. For example, phosphate materials have a P–O covalent bond, which gives them high thermal stability. At temperatures above 200 °C, it is very common for layered oxides to decompose to release oxygen, and the covalent bond of polyanionic compounds can effectively suppress this problem. However, compared with layered oxides and Prussian blue, the conductivity of polyanionic compounds is generally poor. Carbon coating is more frequently applied to the surface modification of polyanionic compounds than other materials, and improves the conductivity of this type of material to a certain extent. Another major problem of polyanionic compounds is that they have strong water absorption. When the surface contacts water, NaOH will be generated. The uneven surface of the material may adversely affect the electrochemical performance of the electrode.

Taking phosphate-based NaFePO$_4$ as an example, the current modification methods for improving the conductivity and energy density of polyanionic compound cathode materials for sodium-ion batteries mainly include the following four aspects [60]:
(1) Controlling the sodium storage of polyanionic compounds in the site.
(2) Partial or full replacement of Fe elements with transition metal elements.
(3) Preparation of mixed polyanion systems with F, OH, and CO$_2$.
(4) Combined phosphate (PO$_4$)$^{3-}$ and pyrophosphate (P$_2$O$_7$)$^{4-}$ to stabilize the crystal structure.

The following section describes the application and characteristics of various polyanionic compounds in the cathode of sodium-ion batteries. They have their own characteristics, but they can also discover commonalities. It is believed that as some of the defects of polyanionic materials are significantly improved, they will have more application value as cathode materials for sodium-ion batteries.

### 3.2.1 Phosphate-type polyanionic compounds

Phosphate type is the most representative type of polyanionic materials, mainly including olivine-type NaMPO$_4$ (M: Fe, Mn) and sodium fast ion conductor Na$_x$M$_2$(PO$_4$)$_3$ (M = V, Ti).

NaMPO$_4$ with olivine structure has the simplest structure among polyanionic materials. In Section 3.1, NaFeO$_2$ layered oxide materials are introduced. This material is based on the redox properties of Fe$^{2+/3+}$ to make the material electrochemically active. However, the low capacity and low stability of this kind of material make it lag the practical cathode material. Based on similar working principles, NaFePO$_4$ has also been widely studied and used in the cathode materials of sodium-ion batteries. The performance of the electrode depends on the crystal structure. For example, the amorphous NaFePO$_4$ has a high specific discharge capacity of 150 mA h g$^{-1}$, but the operating voltage (2.4 V) is only a very low [69], while the olivine-type NaFePO$_4$ can operate at a low above 2.8 V. In the olivine-type NaFePO$_4$ crystal structure, iron ions are located at the position of the octahedrons, and the FeO$_6$ octahedrons are coedged with each other (Figure 3.31(a)), with Na ion in the b-axis direction of the crystal structure one-dimensional tunnel of transmission. The other sodium phosphite-type NaFePO$_4$ does not have a storage channel (Figure 3.31(b)) [70]. However, it should be pointed out that olivine-type NaFePO$_4$ does not exist in nature. Olivine-type NaFePO$_4$ is mainly obtained by ion exchange with $LiFePO_4$, and the synthesis process is more complicated. Olivine-type $NaFePO_4$ has two discharge platforms, which are caused by the formation of an ordered phase of sodium ions during the discharge process. Early research tested that olivine-type $NaFePO_4$ can contribute discharge specific capacity of 120 $mA \cdot h \cdot g^{-1}$ at a working voltage of 2.8 V. At the same time, it is cycled for 100 cycles at a current density of 0.1 C, and the capacity retention reaches 90%. Such materials can show better electrochemical performance in batteries with aqueous electrolytes [71]. As shown in Figure 3.31(c) and (d), setting the charge and discharge range in the organic system to the same voltage range as the stable charge and discharge in water, the charge and discharge polarization and rate of $NaFePO_4$ in water-based batteries is even better. This may be due to the fact that the ionic conductivity and interface impedance of water are better than those of organic systems. Therefore, it can be considered that olivine-type $NaFePO_4$ is more suitable as a cathode material for aqueous sodium-ion batteries [72].

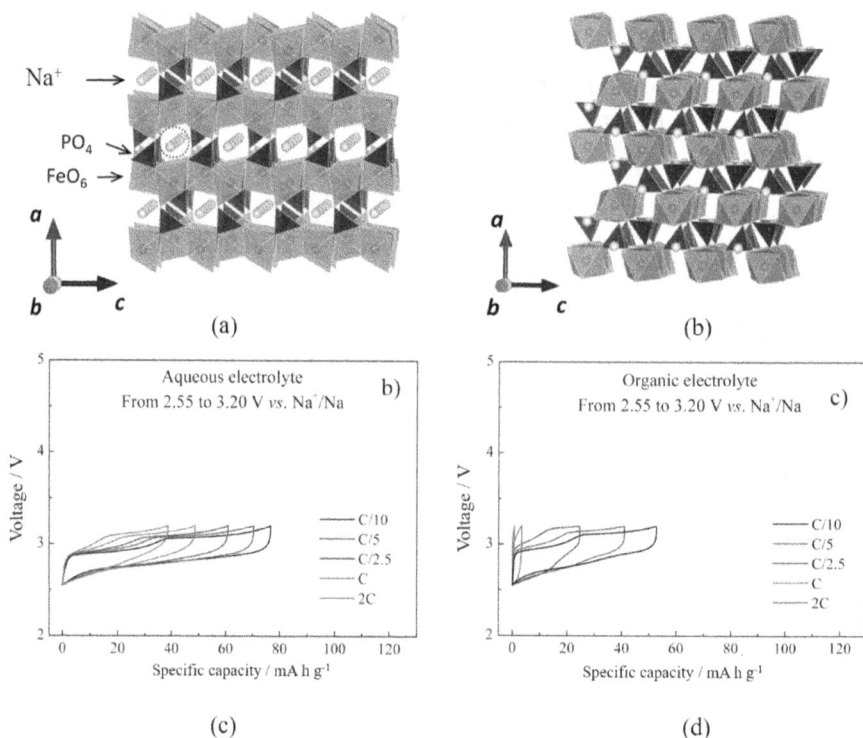

**Figure 3.31:** Structure and electrochemical performance of $NaFePO_4$: (a), (b) structural diagram of olivine-type $NaFePO_4$ and sodium phosphate type $NaFePO_4$ [70]; (c), (d) electrochemical properties of olivine-type $NaFePO_4$ in different electrolyte systems [72].

Olivine-type NaMnPO$_4$ also has electrochemical activity. The early synthesis of such materials was obtained by high-temperature solid-phase transformation of phosphite-type NaMnPO$_4$. In order to simplify the synthesis process, Nazar et al. used NH$_4$MnPO$_4 \cdot$ H$_2$O as the precursor, and prepared olivine-type NaMnPO$_4$ by surface chemical reaction at a temperature of 100 °C. Although Mn$^{2+}$ has a higher valence state than Fe$^{2+}$ and may have a higher capacity, the electrochemical performance is poor, with a reversible specific capacity of only about 80 mA h g$^{-1}$. For the modification of olivine-type polyanionic materials, a similar research idea as used for the layered oxide in the previous article can be considered, combining Fe and Mn to form a binary transition metal compound. The researchers synthesized Na[Fe$_{0.5}$Mn$_{0.5}$]PO$_4$, a dibasic phosphate material, which can reversibly intercalate/extract sodium ions up to 0.6 at a working voltage of 2.0–4.4 V (Figure 3.32) [73].

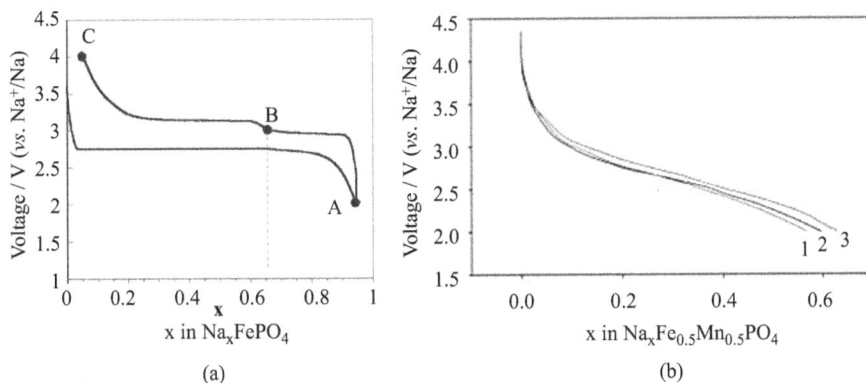

**Figure 3.32:** Comparison of electrochemical performance between NaFePO$_4$ and NaFe$_{0.5}$Mn$_{0.5}$PO$_4$: (a) charge and discharge curves of NaFePO$_4$; and (b) charge and discharge curves of NaFe$_{0.5}$Mn$_{0.5}$PO$_4$ [73].

Another type of sodium fast ion conductor (NASICON type framework structure) Na$_x$M$_2$(PO$_4$)$_3$ (M = V, Ti; x = 1, 2, 3) is a type of electrode material with a stable three-dimensional framework and high-ion transmission rate. Among them, NaTi$_2$(PO$_4$)$_3$ is the anode material, and the space structure of the Na$_3$V$_2$(PO$_4$)$_3$ cathode is composed of VO$_4$ octahedrons and PO$_4$ tetrahedrons that share angles and has good Na-ion conductivity. The ordered arrangement of sodium ions can be observed in the NASICON structure (Figure 3.33(a)). After simple carbon coating, this material can be used as cathode of non-aqueous sodium-ion batteries. It can be found that there is a relatively flat discharge platform at 3.4 V (Na$^+$/Na), and a discharge specific capacity of about 93 mA h g$^{-1}$ can be released [74]. In an in-depth study of the reaction mechanism of Na-ion insertion/extraction Na$_3$V$_2$(PO$_4$)$_3$, it can be found that there is a typical two-phase reaction at 3.4 V. With the extraction of Na ions, the second phase appears from Na$_3$V$_2$(PO$_4$)$_3$ to NaV$_2$(PO$_4$)$_3$, which is shown in Figure 3.33(b). In addition, from the analysis of the migration path of Na ions, it is combined with the experimental results and first-principles calculations, and it is found that there are two obvious migration paths of Na ions, which are distributed along the $x$ and $y$ directions of the three-dimensional crystal structure. At the same time, a 3D-based curved path of the structure is shown in Figure 3.33(c). These paths provide abundant channels for the migration of Na ions and bring a higher theoretical specific capacity of 117 mA $\cdot$ h $\cdot$ g$^{-1}$ [68].

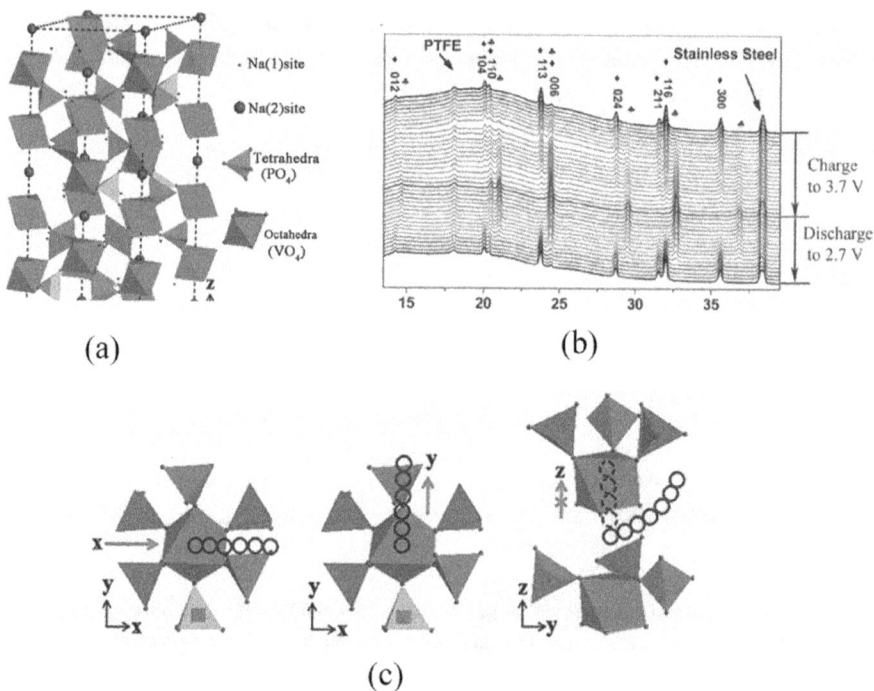

(a)

(b)

(c)

**Figure 3.33:** Structure and electrochemical process of $Na_3V_2(PO_4)_3$: (a) schematic diagram of $Na_3V_2$ $(PO_4)_3$; (b) in situ XRD patterns of $Na_3V_2(PO_4)_3$; (c) possible migration paths of $Na^+$ in $x$, $y$, and $z$ directions in $Na_3V_2(PO_4)_3$ structure [68].

Although $Na_3V_2(PO_4)_3$ has good structural stability, good ionic conductivity, and high discharge platform, the low electronic conductivity still limits the commercial application. This problem is also a key problem to be solved for other polyanionic materials. The surface modification of carbon materials is the most important modification strategy, and the optimization of element doping and synthesis process also has certain effects. While continuing to develop carbon coatings, researchers have also discovered that carbon materials can also connect $Na_3V_2(PO_4)_3$ particles to form a community of carbon and $Na_3V_2(PO_4)_3$, with a highly conductive and compact system. For example, by chemical vapor deposition method, a graphene structure coating layer is formed on the surface of $Na_3V_2(PO_4)_3$ particles, and the synthetic carbon nanofibers form a good connection channel between the particles (Figure 3.34). The synthesized crimped and layered carbon-coated material has good rate performance, and still has a discharge specific capacity of 38 mA h $g^{-1}$ at 500 C, and at 30 C, the capacity retention is still 54% after 20,000 cycles [75]. In addition, the modification method by ion doping is also beneficial to the improvement of the performance of $Na_3V_2(PO_4)_3$. For example, the $Na_3V_{2-x}Mg_x(PO_4)_3/C$ composite material is synthesized by doping $Mg^{2+}$ into $Na_3V_2(PO_4)_3$ and coating with carbon. Due to the properties of $Mg^{2+}$ and its light ion and the increase in potential, the electrochemical performance of overall material can be significantly improved [76].

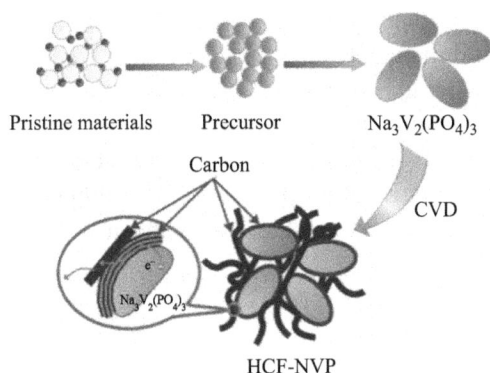

Figure 3.34: Morphology and structure of mesoporous carbon-coated $Na_3V_2(PO_4)_3$ (HCF-NVP) [75].

## 3.2.2 Pyrophosphate-type polyanionic compound

The pyrophosphate polyanionic compound obtained by replacing phosphate with pyrophosphate is widely used in lithium-ion batteries, and the same research ideas can also be applied to sodium-ion batteries. Sodium-based pyrophosphate materials are mainly divided into two categories: $NaMP_2O_7$ (M = Ti, V, Fe) and $Na_2MP_2O_7$ (M = Fe, Mn, Co), and each material may have multiple crystal types at the same time, such as $NaMP_2O_7$. It contains triclinic, monoclinic, or tetragonal structures, and can provide a channel for the migration of Na ions.

There are two types of crystal structure of $NaFeP_2O_7$ in sodium-based pyrophosphate materials, I-$NaFeP_2O_7$ and II-$NaFeP_2O_7$. At low temperature, $NaFeP_2O_7$ exists in the form of I-$NaFeP_2O_7$, and will transform into II-$NaFeP_2O_7$ at high temperature of 750 °C. II-$NaFeP_2O_7$ is composed of octahedral units of $P_2O_4$ and $FeO_3$ that share the same angle. The two crystal types are shown in [77] Figure 3.35(a) and (J). Based on the $Fe^{2+/3+}$ redox couple, $Na_2FeP_2O_7$ has a theoretical specific capacity of 97 mA h $g^{-1}$. But the charging capacity of $Na_2FeP_2O_7$ can reach 130 mA h $g^{-1}$, which is much higher than the theoretical specific capacity. This is because the $Fe^{3+}/Fe^{4+}$ redox couple can be converted at a potential as high as 5 V, but it is not suitable for charging to such a high voltage, due to the maximum voltage window of about 4.5 V in general electrolytes. Researchers synthesized this material by conventional solid-state synthesis, and its reversible specific capacity reached 90 mA h $g^{-1}$, and the charge/discharge curve is shown in Figure 3.35(c) [78]. Although in the structure of $Na_2FeP_2O_7$, the framework composed of $FeO_6$ octahedron and $FeO_5$ square pyramid structure is completely separated by pyrophosphate, the pyrophosphate anion forms a good open channel, improving the overall rate performance of the material largely compared to phosphoric acid, and the magnification performance is good (Figure 3.35(d)). $NaVP_2O_7$ in the sodium-based pyrophosphate material also has two crystal forms. It is used as a high-voltage cathode material in sodium-ion batteries. The discharge voltage is about 3.4 V, and the theoretical specific capacity is about 108 mA h $g^{-1}$, but this material has a higher impedance and will inhibit the transition from $NaVP_2O_7$ to $Na_{1-x}VP_2O_7$. The $Na_2MnP_2O_7$ material has a structure that is similar to $Na_2FeP_2O_7$. When used as cathode of sodium-ion batteries, excess raw materials can produce a sodium-rich structure, so even without carbon coating and nano-size structure, the material can show good electrochemistry, compared to the performance of other phosphate materials.

### 3.2.3 Sulfate-type polyanionic compounds

Sulfate has stronger electronegativity than phosphate, so the use of sulfate to replace phosphate can increase the working voltage as cathodes of sodium-ion batteries. As the thermodynamic decomposition temperature of $SO_4^{2-}$ is as high as 400°C, Yamada et al. synthesized $Na_2Fe_2(SO_4)_3$ of the sodium phosphite type through a solid-state reaction at about 350 °C (Figure 3.36(a)). The structure is a co-edged $FeO_6$ octahedron, which is connected together by $SO_4$ units to form a 3D structure, forming an Na-ion transmission channel in the $c$-axis direction. The highest working voltage of $Na_2Fe_2(SO_4)_3$ in the current $Fe^{2+/3+}$ redox pair is about 3.8 V (to $Na^+/Na$) (Figure 3.36(b)); reversible specific capacity of 102 mA h $g^{-1}$ can be obtained when the material is used as cathode of sodium-ion batteries (the theoretical specific capacity is the specific capacity of 12 mA h $g^{-1}$ contributed by the conversion of a single electron). Even after 30 cycles, under the condition of high current rate at 20 C, the specific capacity of the material can still maintain 60 mA h $g^{-1}$, showing good rate performance. At the same time, Fe in $Na_2Fe_2(SO_4)_3$ materials can also be expanded and replaced with transition metal elements, such as Ni, Co, V, or Mn [79].

When $NaFeSO_4F$ is used as the cathode material of sodium-ion batteries, the voltage range with reversible charge and discharge is very small (Figure 3.37), but the average working voltage has a larger increase than $Na_2FePO_4F$, which proves that sulfate can increase the working voltage [80]. In $NaMSO_4F$ materials, M can

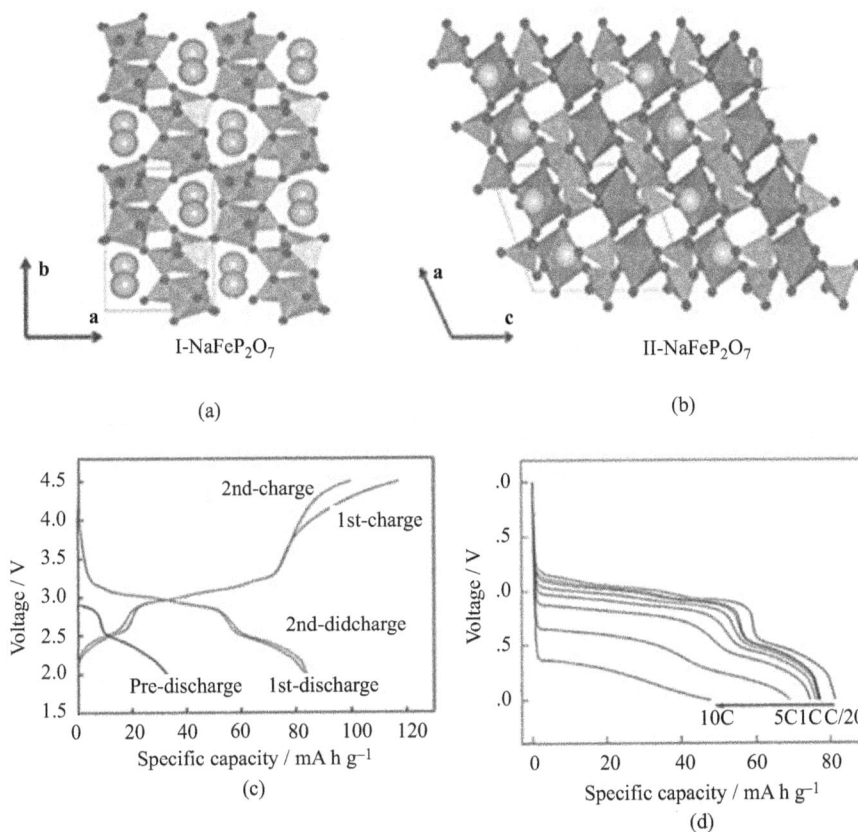

Figure 3.35: Structure and electrochemical performance of polycrystalline $NaFeP_2O_7$ with monoclinic structure: (a) the crystal structure of I -$NaFeP_2O_7$; (b) crystal structure of II -$NaFeP_2O_7$ [77]; (c) charge and discharge curves of $NaFeP_2O_7$; and (d) rate performance of $NaFeP_2O_7$ [78].

also be Ni, Co, and other transition metal compounds. The common features of this type of materials help improve the low working voltage and low conductivity of fluorophosphate to a certain extent. Various transition metal-based polyanion compounds with sulfate will play an important role, in future.

## 3.2.4 Doped phosphate-type polyanionic compounds

In Section 3.1, the structural diversity of transition metal oxides, such as $O_3$ type structure and $P_2$-type structure, are introduced. There are also various structures in polyanionic compounds. The combination of phosphate and pyrophosphate can also prepare a large class of structurally stable polyanionic compounds, which can

(a)

(b)

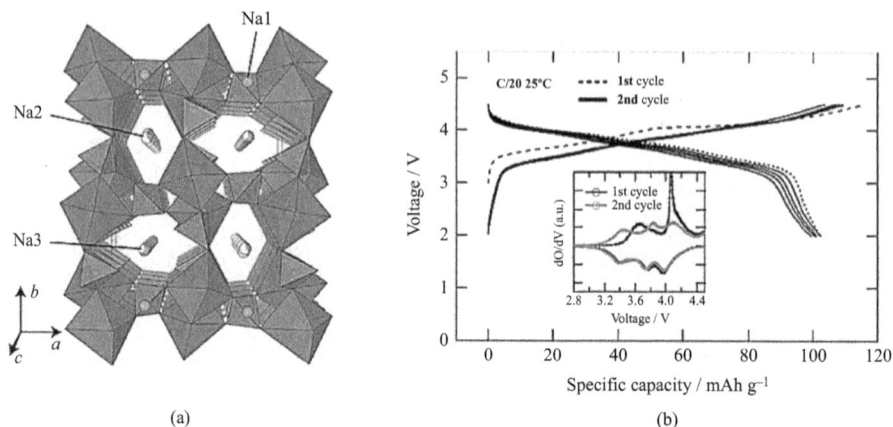

**Figure 3.36:** Structure and electrochemical performance of $Na_2Fe_2(SO_4)_3$: (a) Schematic structure; (b) Charge and discharge curves [79].

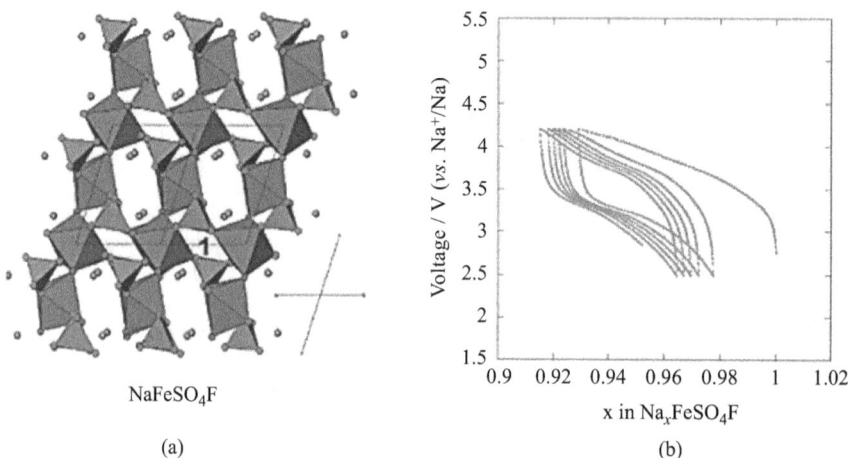

NaFeSO$_4$F

(a)

(b)

**Figure 3.37:** Structure and electrochemical performance of $NaMSO_4F$ materials: (a) crystal structure; and (b) charge and discharge curves [80].

be used in cathodes of sodium-ion batteries. The chemical formula of the mixed phosphate polyanion material is $Na_4M_3(PO_4)_2(P_2O_7)$ (M = Fe, Co, Ni).

The most representative of this type of material is $Na_4Fe_3(PO_4)_2(P_2O_7)$, which has an orthorhombic structure and belongs to the Pn21a space group. In the crystal structure of this material, $FeO_6$ octahedron and $PO_4$ tetrahedron are connected together by means of common edges and common bi-angles to form a layered unit. These layered units are further connected by pyrophosphate (Figure 3.38) [81]. When $Na_4Fe_3(PO_4)_2$ $(P_2O_7)$ is used as the cathode material of sodium-ion batteries, the redox couple based on $Fe^{2+/3+}$ can have a reversible specific capacity of 100 mA h g$^{-1}$, and the voltage platform is 3.2 V (vs. $Na^+/Na$) and has a good capacity retention [82]. The lattice parameters of $Na_xFe_3 (PO_4)_2(P_2O_7)$ $(1 < x < 4)$ are calculated. As the number of Na ions increases, the volume expansion rate of the material will not exceed 4%, so the cycle performance of the material is good. In addition, Fe-based materials have the weakest atomic binding energy and are prone to cause a large number of defects. This is an effective way to improve its electrochemical performance. Therefore, the electrochemical performance of the mixed phosphate $Na_4Fe_3(PO_4)_2(P_2O_7)$ polyanion compound is slightly better than that of the single phosphate $NaFePO_4$ material [83].

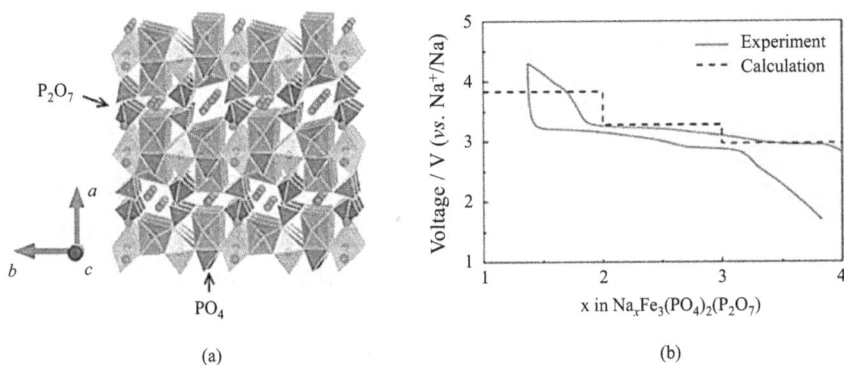

**Figure 3.38:** Crystal structure and electrochemical performance of $Na_4Fe_3(PO_4)_2(P_2O_7)$: (a) Crystal structure; (b) Charge and discharge curves [81].

The $Na_4Co_3(PO_4)_2(P_2O_7)$ material structure can be regarded as the element replacement of the iron element in the $Na_4Fe_3(PO_4)_2(P_2O_7)$ material. Based on the $Co^{2+/3+}$ redox couple, $Na_4Co_3(PO_4)_2(P_2O_7)$ can provide a reversible specific capacity of 95 mA h $g^{-1}$, and the working voltage is 4.0 V [84]. In the process, Na ions are intercalated/deintercalated and multiple voltage platforms appear in $Na_4CO_3 (PO_4)_2(P_2O_7)$, which is similar to the $O_3$ or $P_2$ $Na_xCoO_2$ materials mentioned above, indicating that complex phase transition occurs in the material itself, during this process. In addition, Mn element or Ni element can also be used to partially replace Co element, such as the $Na_4Co_{2.4}Mn_{0.3}Ni_{0.3}(PO_4)_2(P_2O_7)$ [85]. When $Na_4Co_{2.4}Mn_{0.3}Ni_{0.3}(PO_4)_2(P_2O_7)$ is used as the cathode material of sodium-ion batteries, it has a certain discharge capacity in the voltage range of 5.0–4.5 V, and shows a reversible specific capacity of more than 100 mA h $g^{-1}$. The ultra-high working voltage of the material comes from the effect of the highly electronegative polyanionic compound framework on the oxidation–reduction of nickel ions.

As highly electronegative F has strong induction effect when combined with $PO_4^{3-}$ and $SO_4^{3-}$, fluorophosphate and Fluor sulfate materials are suitable for high voltage cathode materials of sodium-ion batteries, including $Na_2FePO_4F$, $Na_3V_2$

$(PO_4)_3F_3$, $Na_2Fe_{0.5}Mn_{0.5}PO_4F$, $NaVPO_4F$, $Na_3V_2O_{2x}$ $(PO_4)$ $_2F_{3-x}$ and so on. The multi-channel properties of NASICON structure have attracted great attention, and NaV $(PO_4)F$, $Na_3V_2(PO_4)_2F_3$, and $Na_3V_2(PO_4)_2O_2F$ materials can provide higher reversible specific capacity, based on the redox properties of $V^{3+/4+}$ [86]. As shown in Figure 3.39, the frame structure of $Na_3V_2(PO_4)_2F_3$ is composed of $VO_4F_2$ octahedron and $PO_4$ tetrahedron. $Na_3V_2(PO_4)_2O_{1.6}F_{1.4}$ obtained through solid solution phase can provide a reversible specific capacity of 130 mA h $g^{-1}$, and it has good cycling stability due to the small volume change (3%) during charge and discharge [87].

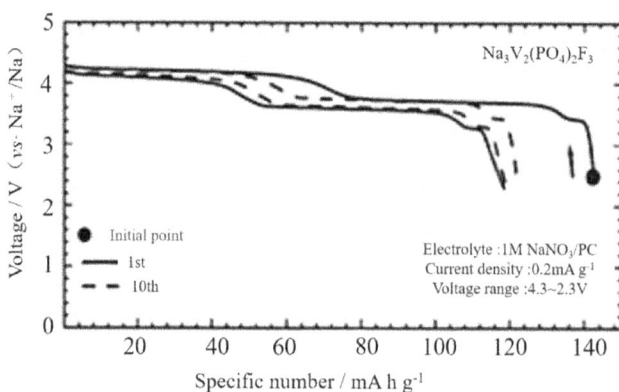

**Figure 3.39:** Charge and discharge curves of $Na_3V_2(PO_4)_2F_3$ [87].

$Na_2FePO_4F$ is also a typical fluoride ion-doped material, with a two-dimensional layered structure. Sodium ions are stored between the $FePO_4F$ layered structures. $FeO_4F_2$ octahedrons are co-edged and co-angular, and sodium-ion shuttle between the $Na_2$ $FePO_4F$ layers. When carbon-coated $Na_2FePO_4F$ is used as the cathode, the average working voltage is about 3.0 V, and it has a reversible specific capacity of 110 mA h $g^{-1}$, which is equivalent to 90% of the theory specific capacity of $Fe^{2+/3+}$ redox pair. As shown in Figure 3.40, it can be seen in the charge/discharge curve that there are two voltage platforms at 3.06 and 2.91 V, indicating that there is only a small amount of material polarization in the process of sodium-ion intercalation/deintercalation. The structure of $Na_2MnPO_4F$ material is completely different from that of $Na_2FePO_4F$.In $Na_2MnPO_4F$, all $MnPO_4F_2$ octahedral co-angles form a one-dimensional $Mn_2F_2O_8$ chain, which forms a 3D frame structure with low density through $PO_4$ tetrahedron, and this three-dimensional frame structure has good stability [88]. Based on the stability of $Na_2MnPO_4F$ material, researchers synthesized $Na_2[Fe_xMn_{1-x}]PO_4F$ material, which has thermodynamic stability under the condition of $x < 0.75$. Carbon-coated $Na_2[Fe_{1/2}Mn_{1/2}]PO_4F$ material has a reversible specific capacity of 110 mA h $g^{-1}$ when used as the cathode of sodium-ion batteries. Three voltage platforms can be seen during the charge and discharge process. Compared with $Na_2FePO_4F$, the carbon-coated nanostructured $Na_2[Fe_{1/2}Mn_{1/2}]PO_4F$ material has better reversibility. The structure and charge/discharge curve of $Na_2FePO_4F$ and $Na_2[Fe_{1/2}Mn_{1/2}]PO_4F$ are shown in Figure 3.40.

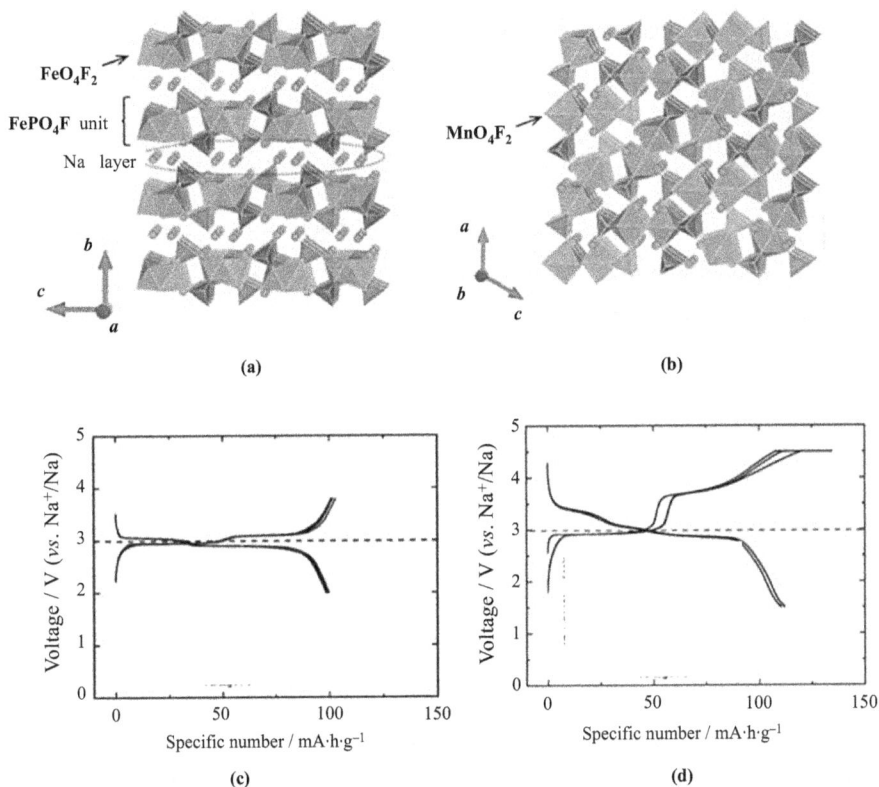

(a)

(b)

(c)

(d)

**Figure 3.40:** Structure diagram and charge/ discharge curves comparison of $Na_2FePO_4F$ and $Na_2$ $[Fe_{1/2}Mn_{1/2}]PO_4F$: (a) and (c) crystal structure and charge and discharge curves of $Na_2FePO_4F$; (b) and (d) crystal structure and charge and discharge curves of $Na_2[Fe_{1/2}Mn_{1/2}]PO_4F$ [88].

In recent years, the synthesis of $Na_3MePO_4CO_3$ materials and their application in the field of cathode materials have also been research hotspots. This type of material contains two different anions, phosphate and carbonate. $Na_3MnPO_4CO_3$ exists in natural minerals. Its crystal structure is composed of $MnO_6$ octahedron, $PO_4$ sharing an angle, and $CO_3$ triangle and $MnO_6$ octahedron sharing the same side, so that the structure of $FeO_6$ octahedron is deformed to a certain extent. In the laboratory, $Na_3$ $MnPO_4CO_3$ material can be prepared by hydrothermal reaction at 120°C. When used as a cathode material for sodium-ion batteries, the initial charging specific capacity is as high as 200 mA h $g^{-1}$, but the specific capacity decays rapidly in the following cycles, and the reversible specific capacity remains approximately 100 mA h $g^{-1}$ after 10 cycles (Figure 3.41) [89].

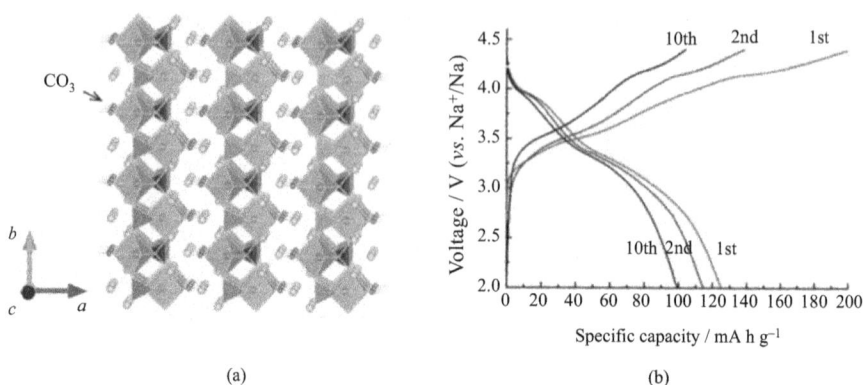

(a)                                                          (b)

**Figure 3.41:** Crystal structure and electrochemical performance of $Na_3MnPO_4CO_3$: (a) schematic structure; and (b) charge and discharge curves [89].

At present, only a few transition metal elements of 4d-group have been introduced into polyanionic compounds as cathode materials for sodium-ion batteries. The $Fe_2(MoO_4)_3$ material with monoclinic structure can be synthesized into sheet-like materials by magnetron sputtering. The reversible capacity of this electrode material is 91 mA h $g^{-1}$, and the redox activity is determined by $Fe^{2+/3+}$. With the insertion of Na ions, the molecular formula becomes $Na_2Fe_2(MoO_4)_3$ [90]. Earlier studies also proved that Na ions can carry out the intercalation/deintercalation reaction in the $Mo_2O_4P_2O_7$ material in the voltage range of 2.0–3.0 V, and the reversible specific capacity can reach 250 mA $\cdot$ h $\cdot$ $g^{-1}$ [91]. However, this type of material has poor reversibility, and more research is needed to obtain materials with application.

Polyanionic compounds are rich in types and forms. Compared with layered oxides, polyanionic compounds can not only introduce a variety of transition metal elements, but also combine a variety of anionic groups. This variety of combinations provides researchers with more research ideas. The anion electronegativity of the polyanionic compound is relatively large, so this type of material generally has a higher working voltage when used as the cathode material of sodium-ion batteries. However, research results show that the reversibility of polyanionic compounds is hard to compare with metal oxide materials.

## 3.3 Prussian blue and Prussian blue analogues

Prussian blue (PB, $Na_xFe[Fe(CN)_6]$) and Prussian blue analogues (PBAs, $Na_xM[Fe(CN)_6]$, M = Co, Mn, Ni, Cu, etc.) are a large class of transition metal hexacyano ferrites with open framework structure, abundant redox active sites and strong structural stability [92]. As PBAs have large ion channels and lattice gaps, they are easy to carry out reversible-ion intercalation/deintercalation reactions, so this type of material is one of the few cathode matrix materials that can accommodate larger alkaline cations such as (Na$^+$ and K$^+$). It is precisely because of this excellent structural feature that PB compounds are widely used in the research of new, low-cost sodium storage cathode materials. Each molecular formula of PB contains two redox centers, M$^{2+/3+}$ and Fe$^{2+/3+}$, so PB has the ability for two-electron redox reactions, and each molecular unit can reversibly store two Na ions. Studies have shown that Na$_2$FeFe-PB has a high specific capacity of 160 mA h g$^{-1}$, with an average potential as high as 3.1 V, while Na$_2$MnMn-PBAs can reach 209 mA h g$^{-1}$ at a high potential of 3.50 V. The above two materials have high energy densities of 496 W h kg$^{-1}$ and 730 W h kg$^{-1}$ [93]. The high energy density achieved by the PB cathode even exceeds those of spinel LiMn$_2$O$_4$ (430 W h kg$^{-1}$) and olivine LiFePO$_4$ (530 W h kg$^{-1}$). In addition to a variety of "M" metals, the Fe element in PB can also be replaced by other transition metals with variable valence properties, such as Co, Ni, Mn, Cu, and Zn, to form a series of cathode matrix material with similar structures but different electrochemical properties. Due to the low cost, high structural stability, and electrochemical stability, PBAs, especially Fe-based and Mn-based PBAs, have broad development prospects when used as high-performance cathode materials for commercial sodium-ion batteries [94, 95].

PBAs stand for hexacyanoferrate containing a variety of transition metals; these substances are usually face-centered cubic structure (Fm3-m) [96](Figure 3.42). The chemical formula of PBAs can be expressed as A$_x$M[Fe(CN)$_6$]$_y$ · zH$_2$O, such as Li$^+$, Na$^+$, or K$^+$. M represents a transition metal ion, such as Fe, Mn, Co, Ni, Mn, Cu, etc. In a cubic lattice ($0 < x < 2$, $0 < y < 1$), the M$^{2+}$ ion is coordinated with the nitrogen atom of the CN ligand. The Fe$^{2+}$ ion is coordinated with the carbon atom adjacent to the octahedron of the CN ligand. They form a 3D rigid frame, which contains open-ion channels and spacious interstitial spaces. The structural characteristics of the PBAs and tunability of the chemical composition are beneficial for the intercalation reaction of Na ions. As opposed to the conventional intercalation of transition metal oxides and phosphate compounds, the PBAs have large gap "A" sites and spacious channels, so they have a high ion diffusion coefficient ($10^{-9}$–$10^{-8}$ cm$^2$ s$^{-1}$) [97, 98]. This means that the ion conduction in the lattice of PBAs is much higher than that in traditional oxide and phosphate cathode materials. Obviously, the large channel structure of PBAs makes it easier to insert a larger volume of Na ions in the electrochemical intercalation reaction, which is also conducive to the reversibility of the reaction.

In addition, PBAs contain two different redox centers, M$^{2+/3+}$ and Fe$^{2+/3+}$, both of which can undergo a complete electrochemical redox reaction (when M = Fe, Co, Mn, etc.), providing double electron transfer and two intercalation reactions. The reaction equation is

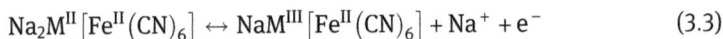

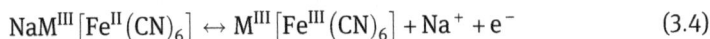

$$\text{Na}_2\text{M}^{\text{II}}\left[\text{Fe}^{\text{II}}(\text{CN})_6\right] \leftrightarrow \text{NaM}^{\text{III}}\left[\text{Fe}^{\text{II}}(\text{CN})_6\right] + \text{Na}^+ + \text{e}^- \tag{3.3}$$

$$\text{NaM}^{\text{III}}\left[\text{Fe}^{\text{II}}(\text{CN})_6\right] \leftrightarrow \text{M}^{\text{III}}\left[\text{Fe}^{\text{III}}(\text{CN})_6\right] + \text{Na}^+ + \text{e}^- \tag{3.4}$$

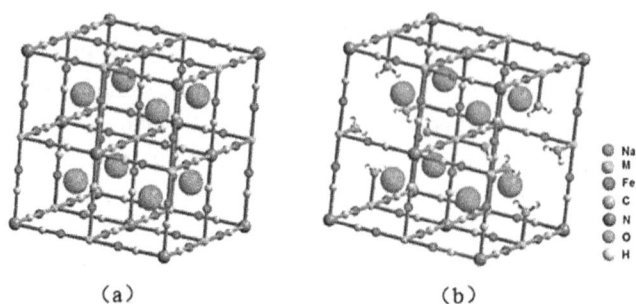

**Figure 3.42:** Schematic diagram of PBA structure: (a) Schematic diagram of anhydrous cube PBA structure; (b) Schematic diagram of PBA with crystal water cube structure [99].

According to the molecular weight of $Na_2FeFe(CN)_6$, which corresponds to the double Na-ion intercalation in each PBAs unit, the theoretical specific capacity of the PB compound is calculated to be 170 mA h g$^{-1}$, which is much higher than most transition metal oxides (100–150 mA h g$^{-1}$) and phosphates (120 mA h g$^{-1}$).

Another structural advantage of PB is that Fe can be partially or completely replaced by many transition metals with redox activity (such as Co, Ni and Mn) without destroying the crystal structure [100]. This structural stability allows people to conveniently adjust the electrochemical properties of PB. For example, $Na_2FeFe$-PB shows a charge–discharge platform of about 3.2 V, while $Na_2MnFe$-PBAs and $Na_2CoFe$-PBAs can increase the working potential to 3.6 V and 3.8 V. The cathode shift of the redox potential is due to the redox transition of $Fe^{2+}/Fe^{3+}$ coordinated with carbon atoms and the connection of $M^{2+}/M^{3+}$ with nitrogen atoms in the PBAs framework. Because of the strong charge-spin-lattice coupling effect in the PBAs, when calibrating the redox potential of the PBAs, the redox potentials of $Fe^{2+}/Fe^{3+}$ in the $Na_2CoFe$-PBAs, $Na_2NiFe$-PBAs, and $Na_2CuFe$-PBAs are + 0.90 V, + 0.45 V, and + 0.58 V, respectively, which are much higher than the standard redox potential of soluble $Fe(CN)_6^{-4}/Fe(CN)_6^{-3}$ in aqueous solution ($E_0 = 0.16$ V). Obviously, this coupling is very helpful to increase the energy density of the PB cathodes. In addition, element substitution can be used to develop high-capacity PB cathodes. When Fe element is completely replaced by Mn [101], the obtained $Na_2MnMn$-PBAs material can greatly enhance the specific capacity of sodium storage (>200 mA h g$^{-1}$), which is one of the strongest sodium storage capacity materials so far. Due to the advantage of adjustability of the structural chemical composition, the electrochemical properties of the PBAs framework can be customized to meet the various requirements of applications.

Due to the strong and wide three-dimensional channel framework, the PBAs are structurally very stable, with almost zero lattice strain [102], which can satisfy the intercalation/deintercalation process of larger ions. In recent years, many PBAs crystal lattices with high crystallinity, such as $Na_2NiFe(CN)_6$ [103], $Na_2CuFe(CN)_6$

[104], and $Na_2CoFe(CN)_6$, have been cycled more than 1,000 times; even in aqueous electrolytes, they have a capacity retention rate of not less than 85%.

Although PB and PBAs materials have various potential advantages, the lithium-ion or sodium-ion intercalation/deintercalation tests of almost all cathode materials of PBAs fail to show satisfactory performance; even the PBAs materials predicted in the early studies have not shown good electrochemical intercalation ability [105–107]. Through research, it was found that the structure of the PBAs crystal lattice is irregular, causing many vacancies and crystal water in the crystal structure, which greatly affects the electrochemical properties of PB materials, such as capacity, rate capability, and cycle stability.

In the early development of sodium storage materials, PBAs materials are usually prepared by a simple chemical precipitation method, which leads to a large amount of $Fe(CN)_6$ vacancies and coordination water in the PBAs lattice [108]. Such PBAs compounds should be expressed in a more accurate form, that is, $Na_{2-x}M[Fe(CN)_6]_y$ $\square_{1-y} \cdot mH_2O$, where $\square$ stands for $Fe(CN)_6$ vacancy, $0 < x < 2$, $0 < y < 1$; the oxygen atoms in water molecules, suspended M ions, and interstitial water molecules are located around the A site [109]. The presence of $Fe(CN)_6$ vacancies and coordination water in the PB compound lattice will have various adverse effects, which will seriously reduce the electrochemical performance of the PBAs framework material. First, the increase of $Fe(CN)_6$ vacancies will inevitably introduce more water molecules into the PB framework to coordinate with the suspended M ions, thereby reducing the available sites of Na ions and leading to the loss of sodium storage. Second, water molecules in the crystal have a strong tendency to stay in the interstitial space and compete with Na ions to occupy the interstitial space, hinder the migration of Na ions, and reduce the capacity utilization rate of the PB framework material. Moreover, the water molecules remaining in the crystal lattice may enter the electrolyte and undergo electrochemical decomposition, resulting in the deterioration of the organic electrolyte and even a safety hazard. In addition, randomly distributed $Fe(CN)_6$ vacancies will break the bridging of the Fe-CN-M framework and form a distorted and defective lattice. This lattice will cause structural instability during the intercalation/deintercalation process of Na ion. The presence of $Fe(CN)_6$ vacancies can also block the electronic conduction of the Fe-CN-M framework, block the Na-ion intercalation reaction, and increase the ohmic polarization of the material.

In order to solve the various problems existing in PB and PBAs, researchers have done a lot of work. The main research ideas are as follows:
(1) Use transition metal elements to partially or completely replace iron elements.
(2) Combine PB or PBAs materials with conductive materials.
(3) Reduce lattice defects and coordination water in PB or PBAs.
(4) Improve the interface between PB or PBAs and electrolyte.

### 3.3.1 Prussian blue analogues based on single-point reaction

It is known that the volume change generated in the process of lithium-ion intercalation/deintercalation is the main reason for the collapse of the structure of the cathode matrix material of the lithium-ion batteries. The intercalation reaction of larger sodium ions may cause more serious structural instability. In order to alleviate the volume change of the PBAs, people usually adopt the method of introducing electrochemically inert metal elements into the lattice to reduce the mechanical stress in the lattice and maintain the structural stability of the PBAs framework.

Guo et al. replaced the equimolar $Fe^{2+}$ with electrochemically inert $Ni^{2+}$, and synthesized the "zero-strain" NaNiFe-PBAs material ($Na_{0.84}Ni[Fe(CN)_6]_{0.71}$). The change is less than 1%, which is almost considered as "zero strain." Benefiting from its "zero-strain" characteristics, NaNiFe-PBAs material exhibits good cycle stability, with capacity retention of 99.7% after 200 cycles, and a coulombic efficiency close to 100%. A special advantage of the zero-strain framework is its stable electrode/electrolyte interface, which can prevent the rupture of the solid electrolyte mesophase and the unstable reconstruction, thereby significantly enhancing the charge and discharge efficiency [110]. Choi et al. replaced Fe with Zn and synthesized Na₂ZnFe-PBAs ($Na_2Zn_3[Fe(CN)_6]_2 \cdot xH_2O$) material. Compared with typical PBAs materials, the framework of Na₂ZnFe-PBAs has larger ion channels, which can adapt to the volume changes caused by intercalation/deintercalation and prevent the capacity degradation caused by lattice fracture [111]. However, due to the large amount of water in the crystal lattice, the movement of water during the battery cycle will cause structural instability, making the cycle life of the Na₂ZnFe-PBAs shorter than expected. In addition, the copper hexacyanoferrate $Cu_3[Fe(CN)_6]_2$ material synthesized by replacing Fe with Cu element has a specific capacity of 44 mA h g$^{-1}$ at a current density of 20 mA g$^{-1}$ [112].

Figure 3.43 shows the charge/discharge curves of three typical Na₂MFe-PBAs materials (M = Ni, Cu and Zn). These PBAs can store, at the most, one Na ion per molecular unit, so they can only provide a specific capacity of 40–70 mA h g$^{-1}$. However, the introduction of nonredox active elements into the PBAs can improve structural stability and significantly reduce the volume change and lattice distortion during intercalation/deintercalation, thereby increasing cycle stability. However, the introduction of these elements will also reduce the sodium storage capacity [113]. It is still unclear how to replace Fe atoms by adding electrochemically inert elements to design a "zero-strain" lattice. Therefore, further research is needed to understand the influence of element substitution on structural stability and explore highly stable PBAs framework materials on this basis.

Cui et al. first reported NiFe-PBAs synthesized by the self-precipitation method of aqueous solution and used them as a cathode material for sodium-ion batteries. The chemical composition of this NiFe-PBAs material is $K_{0.6}Ni_{1.2}Fe(CN)_6 \cdot 3.6H_2O$, the theoretical specific capacity is 85 mA h g$^{-1}$, and the specific capacity is 60 mA · h · g$^{-1}$ at 0.83 C. When using 1 M NaNO₃ as the electrolyte, this material has a medium average

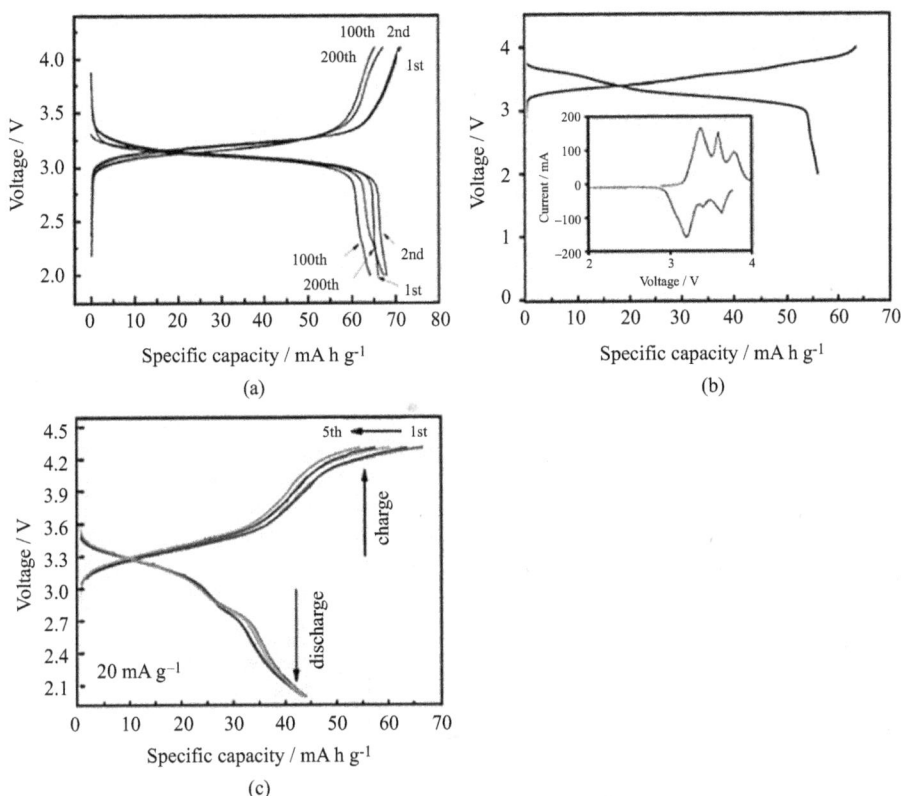

**Figure 3.43:** Three typical single Na$^+$ site in PBA: (a) Ni-Fe-PBA [110]; (b) Zn-Fe-PBA [111]; and (c) Cu-Fe-PBA [112].

working potential of 0.59 V. This material has the best rate performance. It can be charged and discharged at a density of 41.7 C and has a long cycle life of more than 5,000 cycles at a density of 8.3 C (Figure 3.44). [114]. Although NiFe-PBAs have an appropriate specific capacity and strong cycle stability as cathode of an aqueous sodium-ion batteries, the redox potential of this material is relatively low. In order to solve this problem, the researchers replaced part of the Fe element with Cu element, and synthesized a CuFe-PBAs framework with a chemical composition of $K_{0.71}Cu[Fe(CN)_6]_{0.72} \cdot 3.7H_2O$. The average working potential of this material is relatively high and can reach 1.0 V; it has a specific capacity of 60 mA h g$^{-1}$. The material also exhibits good rate performance. When cycled at 41.7 C, its specific capacity is about 34% of the specific capacity at 8.3 C after 500 cycles. As researchers expected, the redox potential of the material can be adjusted in the range of 0.6–1.0 V. The researchers placed $Cu_xNi_{1-x}$-PBAs electrode in 1 mol L$^{-1}$ NaNO$_3$ aqueous electrolyte. With the increase of Cu content, the redox potential increased from 0.6 to 1.0 V, while the reversible specific capacity remained stable, without significant loss. In current research,

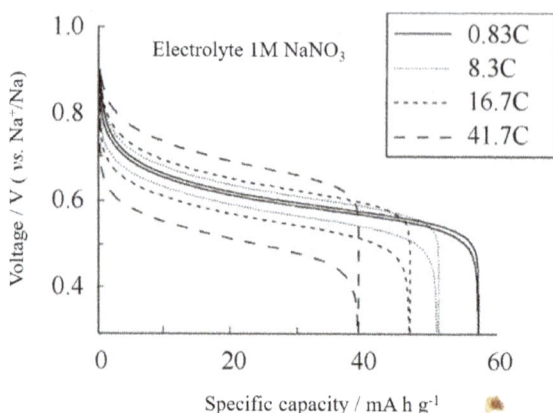

**Figure 3.44:** Rate performance of NiFE -PBA material [114].

the $Cu_{0.56}Ni_{0.44}$-PBAs framework has the best cycle performance and still has a 100% capacity retention after 200 cycles (Figure 3.45) [115].

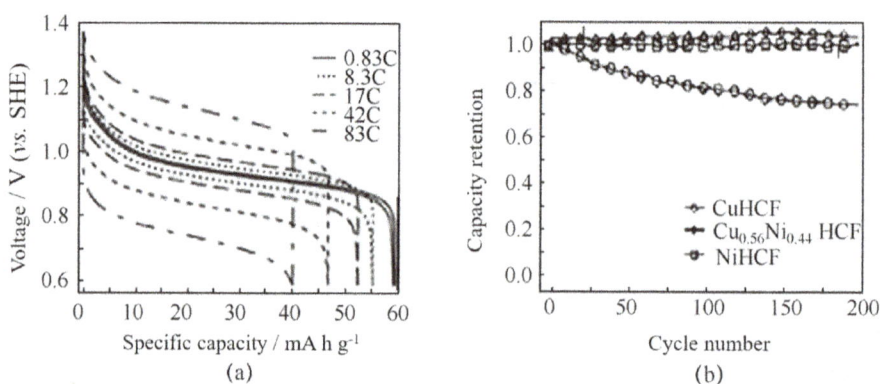

**Figure 3.45:** $Cu_{0.56}Ni_{0.44}$-PBA material: (a) rate performance and (b) cycle performance [115].

Although the abovementioned NiFe-PBAs and CuFe-PBAs materials show reliable high rate capability and cycle stability, they are all sodium-poor materials when they are not charged and cannot match the traditional sodium-free anode to construct practical sodium-ion batteries. To this end, the researchers prepared sodium-rich NiFe-PBAs and CuFe-PBAs frameworks, and their chemical compositions were $Na_2Ni$ $[Fe(CN)_6]$ and $Na_2Cu[Fe(CN)_6]$. Sodium storage performance is shown in Figure 3.46. The NiFe-PBAs material is immersed in 1 mol $L^{-1}$ $Na_2SO_4$ aqueous electrolyte, the reversible specific capacity at 1 C is 65 mA h $g^{-1}$, the capacity retention is 100% after 500 cycles, and the cycle stability is good [116]. Matching the sodium-rich NiFe-PBAs cathode ($Na_2NiFe(CN)_6$) with the sodium-poor $NaTi_2(PO_4)_3$ anode and using $Na_2SO_4$

aqueous electrolyte, the researchers assembled the first "rocking chair"-type aqueous sodium-ion batteries [103]. The average working voltage of this water-based full battery is 1.27 V, and the energy density is 42.5 W h kg$^{-1}$. After 250 cycles at 5 C, it can maintain 88% of the initial capacity (Figure 3.55(a)). Due to the use of water-based electrolytes, these water-based sodium-ion batteries show low cost, are environmentally friendly, show high safety, and are suitable for future large-scale energy storage applications, and have attracted widespread attention.

Similarly, CuFe-PBAs also exhibit good rate performance. They still have a reversible specific capacity of 60 mA h g$^{-1}$ after 500 cycles at 5 C, close to 70% capacity retention, and excellent cycle performance (Figure 3.46) [117].

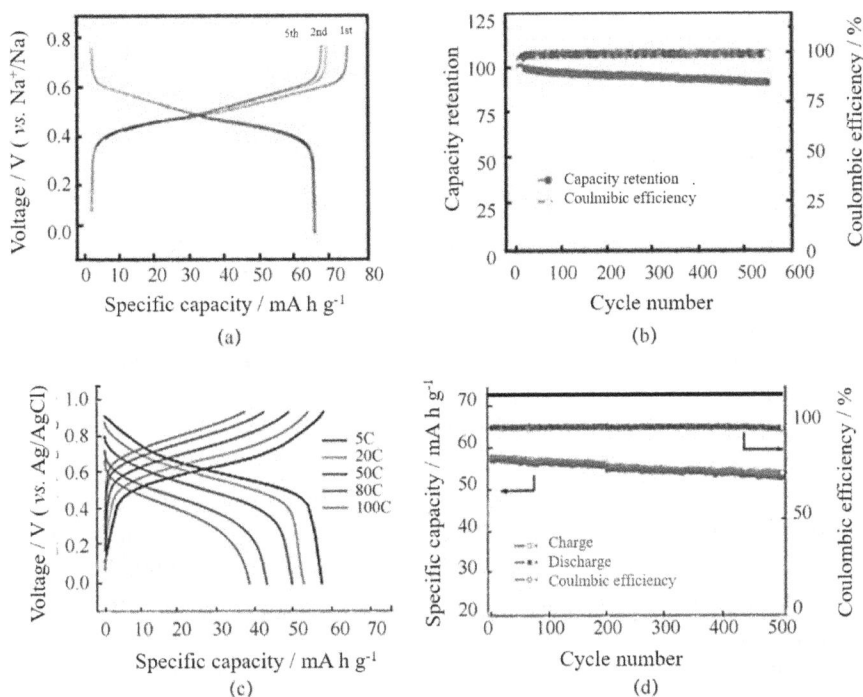

**Figure 3.46:** NiFe -PBA materials: (a) charge and discharge curves; and (b) cycle performance [116].

## 3.3.2 Prussian blue based on double-points reaction

Prussian blue compound (Na$_x$Fe[Fe(CN)$_6$], FeFe-PB) is one of the first coordination materials to be studied. It has strong electrochemical activity and has been widely used in electroanalysis and electrocatalysis applications. Since two different coordinated Fe atoms can be accompanied by the intercalation/deintercalation of alkali metal cations in the lattice for electron transfer reactions, the PB compound can theoretically be used as cathode material for the double Na-ion reaction, making full use of its two-electron oxidation. The theoretically high specific capacity brought by the reduction reaction is shown in Figure 3.47 [118]. However, the FeFe-PB material synthesized in the early research has relatively low reversible capacity and poor cycle stability. According to reports by Yu et al., $NaFeFe(CN)_6$ can provide a specific capacity of 118 mA h $g^{-1}$ at 5 mA $g^{-1}$, but after 20 cycles, the capacity retention is only 85% of the second cycle [119]. The contradiction between theoretical expectations and experimental results has always been a puzzling problem that has hindered the development of PBAs cathodes for many years.

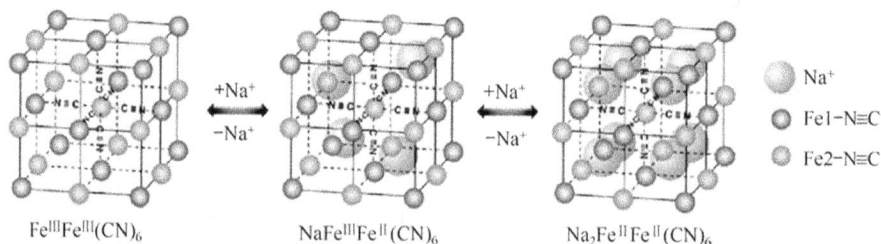

**Figure 3.47:** Schematic diagram of the crystal structure of the Prussian blue compound in the double Na$^+$ reaction [118].

In order to understand the mechanism of poor electrochemical performance of PB compounds, researchers synthesized single crystal and FeFe-PB without vacancies and used them as model compounds to study the relationship between PBAs crystal structure and electrochemical properties. This highly crystalline FeFe-PB material has a specific sodium storage capacity of 120 mA h $g^{-1}$, a high coulombic efficiency of 100%, significant rate performance, and excellent cycle stability at 20 C, and after 500 cycles have a capacity retention of 87% (Figure 3.48) [118]. The study found that lattice vacancies and the resulting coordination water are the main reasons for the electrochemical deactivation of FeFe-PB. This is because water molecules will occupy and block the redox-active sites and ion channels of Na ion, affecting the intercalation and transportation of Na ions. Therefore, the FeFe-PB material with high crystallinity, low defects, and no crystal water can improve the specific capacity and cycle life.

Based on the above judgments, several research groups used different synthesis strategies to minimize the lattice vacancy and water content, thereby controlling the crystallinity of the FeFe-PB material. Guo et al. used $Na_4Fe(CN)_6$ as a single iron source precursor to prepare $Na_{0.61}$FeFe-PB with high quality lattice [120]. This kind of FeFe-PB can theoretically undergo two Na-ion redox reactions and obtain a high specific capacity of 170 mA h $g^{-1}$ in the experiment, excellent rate performance, and 100% high coulombic efficiency. The stability in the cycle is good, and there will be no obvious capacity attenuation (Figure 3.49). The electrochemical performance of FeFe-PB cathode is also attributed to its high structural regularity. This material has

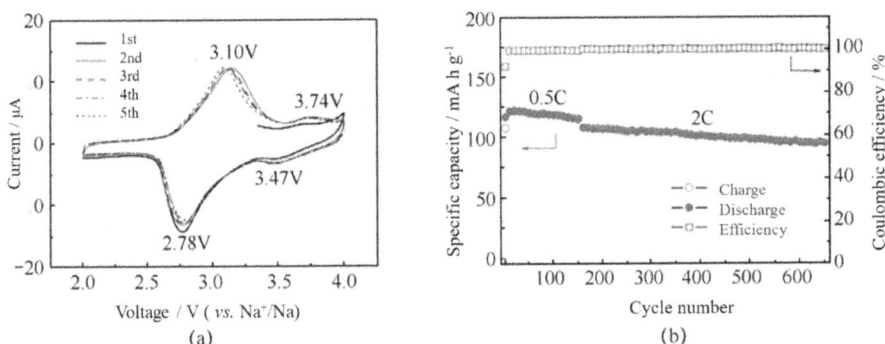

**Figure 3.48:** Highly crystalline FeFe-PB material: (a) CV curves; and (b) cycle and rate performance [118].

a very low vacancy content (6%) and water content (15.7%), making Na ion closer to the active site, which is more conducive to the electrochemical activation of iron ions with redox activity in the lattice of PB. However, the $Na_{0.61}$FeFe-PB material exists in a sodium-poor state, with only 0.6 Na atoms per unit molecule, which is not conducive to the practical application.

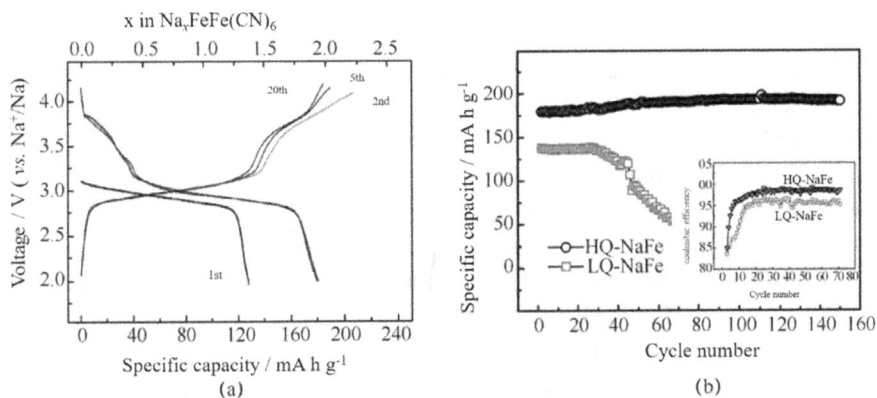

**Figure 3.49:** High-quality FeFe-PB: (a) charge and discharge curves; and (b) cycle performance [120].

Chou et al. used $Na_4$Fe(CN)$_6$ as the precursor to synthesize the sodium-rich $Na_{1.56}$FeFe (CN)$_6$ material in a saturated NaCl solution. The high sodium content in the PB lattice can not only reduce the amount of vacancies and coordination water, but also improve the stability of the material structure (Figure 3.50) [121]. Guo et al. further studied and found that the sodium-rich $Na_{1.63}Fe_{1.89}$(CN)$_6$ skeleton prepared under the conditions of ascorbic acid as a reducing agent and $N_2$ protection can further increase the Na content to 1.63 Na per molecule [122]. This sodium-rich PB sample can provide a reversible

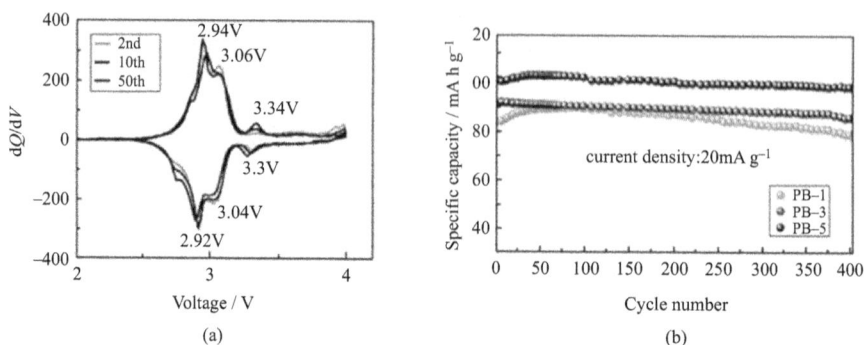

**Figure 3.50:** $Na_{1.56}FeFe(CN)_6$ material: (a) CV curves; and (b) cycle performance [121].

specific capacity of 150 mA h g$^{-1}$ in the first cycle and maintain 90% of the initial capacity after 200 cycles. In addition, Huang et al. synthesized similar sodium-rich, low-defect materials, such as $Na_{1.70}FeFe(CN)_6$ and $Na_{1.73}Fe[Fe(CN)_6]_{0.98}$. Compared to the same type of sodium-poor materials, these two materials have significantly improved electrochemical performance. In fact, it can be seen from the charge/discharge curve that these low-defect PB frameworks still contain a certain amount of crystal water. Although they are smaller than the theoretical specific capacity of the perfect defect-free PB lattice, the capacity decays very slowly during long-term cycles [123].

In order to further eliminate lattice defects and interstitial water, Goodenough et al. prepared $Na_{1.92}FeFe-PB$ ($Na_{1.92}FeFe(CN)_6$) rhombohedral lattice material, which almost completely removes the interstitial water in the lattice, and there is only 0.08 $H_2O$, so its water content is negligible [124]. As expected, the prepared $Na_{1.92}FeFe-PB$ showed excellent electrochemical performance as cathodes, with a high specific capacity of 160 mA h g$^{-1}$ and excellent cycle stability. After 800 cycles, its capacity retention is 80%, and shows a good rate performance of 100 mA h g$^{-1}$ at 15 C (Figure 3.51). It can also be seen from the X-ray spectrum results that during the charge and discharge process of the $Na_{1.92}FeFe-PB$ material, the two slightly split voltage plateaus come from the redox reactions of different Fe-3d states, at the C and N sites, respectively. This method of synthesizing dehydrated FeFe-PB lattice is simple and easy to be applied to the synthesis of other PB compounds.

### 3.3.3 Prussian Blue analogues based on double-point reaction

Many PBAs such as $Na_xMFe(CN)_6$ (M = Co, Mn, V, Ti) have a similar lattice structure, and have two redox sites (M, Fe) similar to FeFe-PB. The existence of this kind of crystal lattice makes it feasible for these materials to become the cathode of two Na-ion-reaction sodium storage. From an electrochemical point of view, these PBAs can be regarded as the formation of $Fe^{2+/3+}$, which replaces half of the redox-active $M^{2+/3+}$.

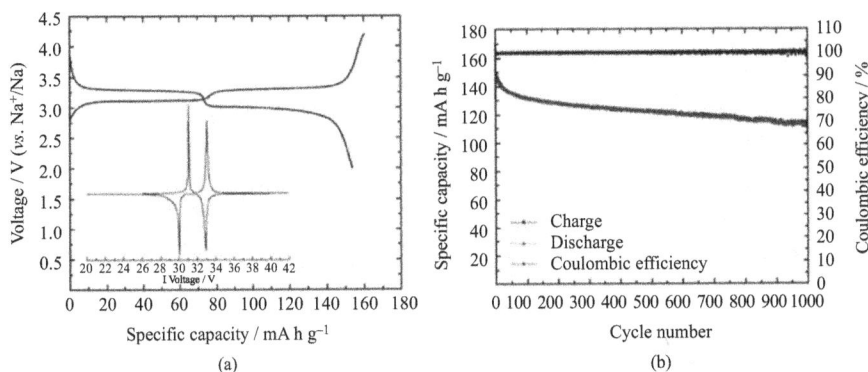

**Figure 3.51:** Dehydrated $Na_{1.92}$FeFe-PB: (a) charge and discharge curves; and (b) cycle performance [121].

This substitution may lead to different electrochemical reactions, thereby providing a wider selection of PBAs for better design of high-performance sodium storage.

Among the various redox pairs of transition metal ions, $Mn^{2+/3+}$ is a good choice for replacing $Fe^{2+/3+}$ in the PBAs lattice due to its low cost and reversible redox properties [125]. Moritomo et al. synthesized $Na_{1.32}Mn[Fe(CN)_6]_{0.83} \cdot 3.5H_2O$ thin-film electrodes by electrodeposition method [126]. There are two voltage platforms of 3.2 and 3.6 V during the charge and discharge process, which correspond to the oxidation-reduction processes of $Fe^{2+/3+}$ and $Mn^{2+/3+}$, respectively. The thin-film electrode has a discharge-specific capacity of 109 mA h $g^{-1}$ at 0.5 C and a specific capacity of 80 mA h $g^{-1}$ at 20 C. The higher specific capacity is because the film has a shorter ion diffusion path of 1 μm, which provides faster reaction kinetics. Goodenough et al. reported a sodium-rich rhombohedral lattice structure MnFe-PBAs ($Na_{1.72}Mn[Fe(CN)_6]_{0.99} \cdot 2.0H_2O$) material. This material shows a high specific capacity of 130 mA h $g^{-1}$ and a good rate performance at a high potential of 3.5 V (Figure 3.52(b)) [127]. They prepared high-purity MnFe-PBAs material after removing the water in the crystal lattice by vacuum drying. TGA and FT-IR confirmed that after sufficient drying, no water in the crystal lattice is decomposed again, which means that the material is anhydrous. The obtained anhydrous MnFe-PBAs material exhibits a single voltage plateau and extremely smooth charge/discharge curve, which may be due to the reversible phase transition of the tetragonal lattice and the rhombohedral lattice during the intercalation/deintercalation process. Anhydrous MnFe-PBAs material can maintain 75% of the initial discharge capacity after 500 cycles at 0.7 C (Figure 3.52) [128]. It can be seen that for MnFe-Mn materials, the reduction of the water content of crystallization significantly improves the cycle performance of the materials. At the same time, their research shows that the presence of crystal water will change the crystal structure of the material, thereby affecting the electrochemical performance. In the comparison of Figure 3.52(a) and (c), it can be clearly seen that the difference in

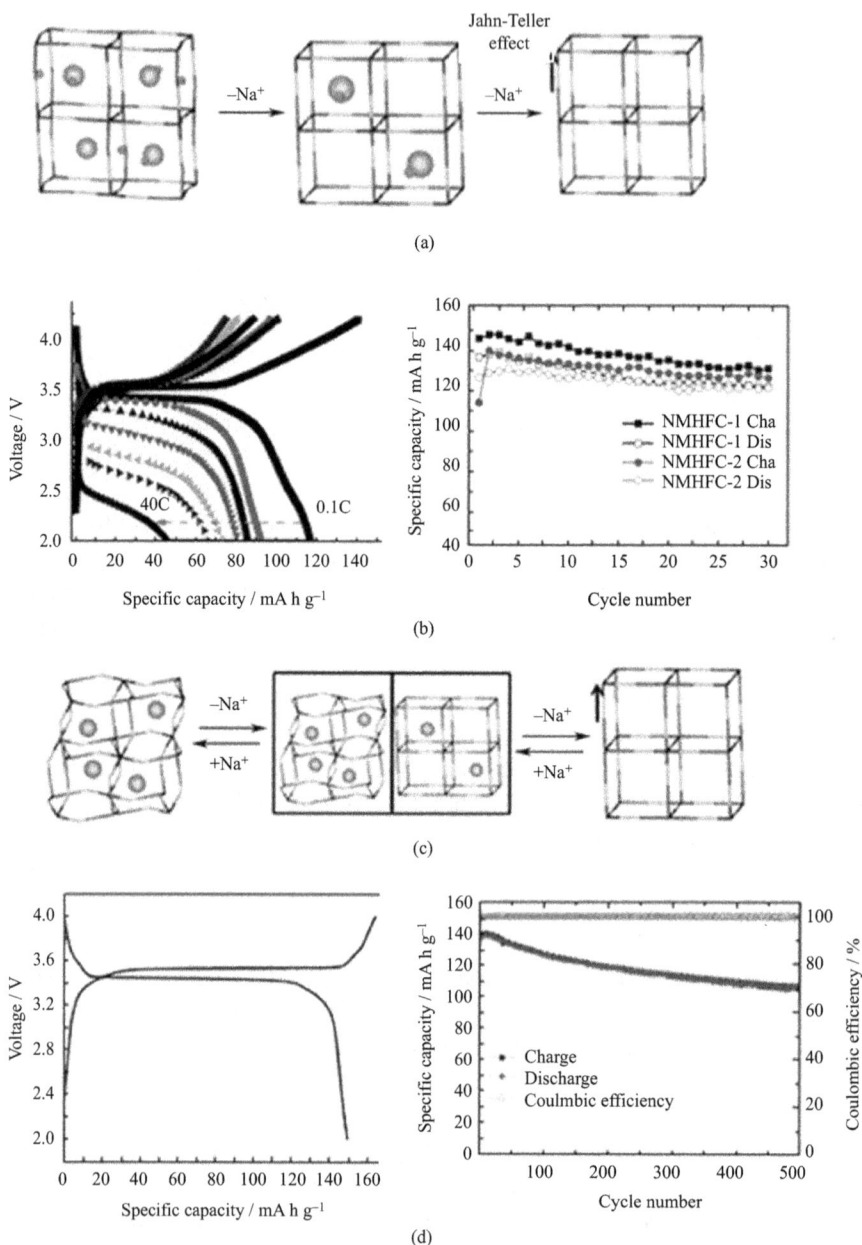

(a)

(b)

(c)

(d)

**Figure 3.52:** $Na_{1.72}Mn[Fe(CN)_6]_{0.99} \cdot 2H_2O$ material: (a) schematic diagram of crystal structure [127]; (b) cycle performance [128]; anhydrous MnFe-PBA material: (c) schematic diagram of crystal structure; and (d) cycle performance [129].

crystal water content will significantly affect the changes in the crystal structure of the material during the intercalation/deintercalation process. When the crystal water content is high, the crystal structure will undergo monoclinic, cubic, and tetragonal changes. When there is no crystal water, the crystal structure of the sodium storage state is a rhombus structure [129].

In addition, the $Na_2MnMnFe(CN)_6$ material with an open-frame structure exhibits an ultra-high-specific sodium storage capacity of 209 mA h g$^{-1}$ at 40 mA g$^{-1}$, and can still maintain 75% of the capacity, after 100 cycles (Figure 3.53) [130].

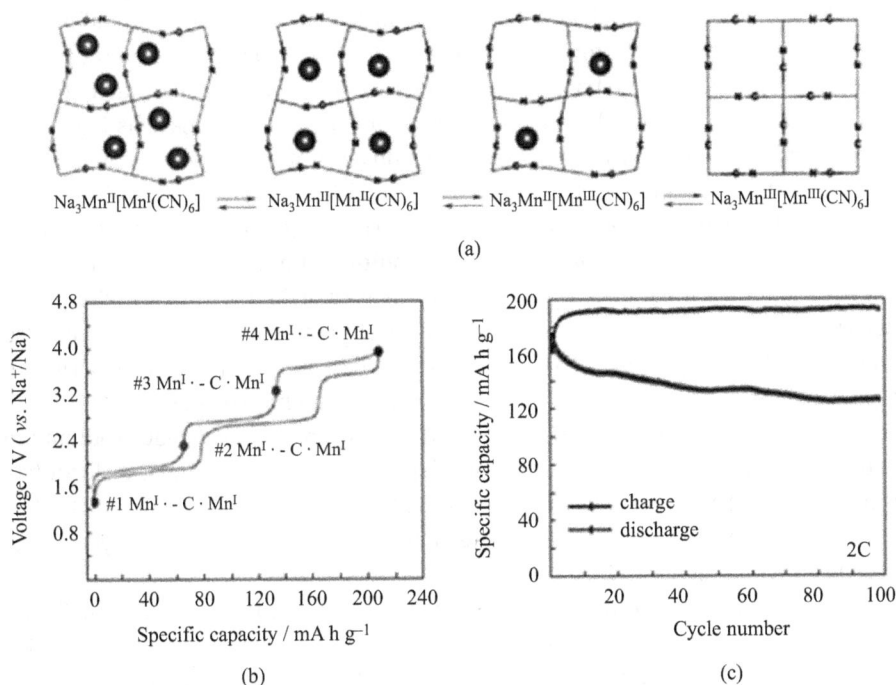

$Na_3Mn^{II}[Mn^I(CN)_6]$ ⇌ $Na_3Mn^{II}[Mn^{II}(CN)_6]$ ⇌ $Na_3Mn^{II}[Mn^{III}(CN)_6]$ ⇌ $Na_3Mn^{III}[Mn^{III}(CN)_6]$

(a)

(b)

(c)

**Figure 3.53:** $Na_2MnMnFe(CN)_6$ material: (a) schematic diagram of crystal structure; (b) charge and discharge curves; and (c) cycle performance [130].

Although researchers have made great progress in improving the sodium storage performance of the Mn-based PBAs framework, there are still some problems to be solved. For example, the Jahn-Teller effect of $Mn^{3+}$ often causes lattice deformation and dissolves Mn into the electrolyte, which affects the cycle stability of such materials (Figure 3.52(a)). After disassembling the cycled $Na/Na_xMnFeCN_6$, Chou et al. found that the entire separator turned brown, indicating that Mn was dissolved in the electrolyte [127]. Since increasing the electrolyte concentration can increase the electrochemical window, Okada and colleagues constructed a MnFe-PBAs/$NaTi_2(PO_4)_3$ full battery in a 17 mol L$^{-1}$ $NaClO_4$ aqueous electrolyte. The battery shows three stepped discharge platforms at 1.3 V, 5 V, and 1.8 V. Although the crystallinity of the MnFe-PBAs cathode is not complete, it still has good rate performance and cycle stability (Figure 3.55(d)) [131]. Since this full battery inhibits the precipitation reaction of oxygen and hydrogen in the high-concentration electrolyte, the use of a higher-concentration electrolyte is more conducive to the electrochemical stability of the aqueous sodium-ion batteries.

In further research, Co element is often used as a cathode structure stabilizer and conductivity enhancer and can replace part of Fe to improve the structural stability of PBA materials. For example, the $Na_xCoFe$-PBAs ($Na_{1.60}Co[Fe(CN)_6]_{0.90} \cdot 2.9H_2O$) thin-film electrode has a high specific capacity of 139 mA h g$^{-1}$ and two platforms of 3.4 V and 3.8 V, but the material will slowly decay after 100 cycles [132]. Yang et al. used the citric acid-assisted coprecipitation method to prepare the low-water content and low-defect $Na_2CoFe(CN)_6$ ($Na_2CoFe$-PBAs) material. This material has high crystallinity and low defects with two Na-ion reactions. During the process of extraction, the crystal structure is relatively regular, with a high specific capacity of 150 mA h g$^{-1}$ and capacity retention of 90% after 200 cycles (Figure 3.54) [133]. The researchers matched the CoFe-PBAs cathode with the modified graphene anode to assemble aqueous sodium-ion batteries with an energy density of 34.4 W h kg$^{-1}$ and a power density of up to 25 kW kg$^{-1}$ (Figure 3.55 (b)). Almost in the same period, Cui et al. also reported a PBAs full battery with CuFe-PBAs cathode and MnFe-PBAs anode. The energy density at 1 C is 27 W h kg$^{-1}$ and the average working voltage is 0.95 V [134]. Soon after, Yang and his colleagues prepared rocking chair-type aqueous sodium-ion batteries. The CoFe-PBAs cathode matches the $NaTi_2(PO_4)_3$ anode, and the electrolyte is a 1 mol L$^{-1}$ $Na_2SO_4$ aqueous solution. This type of aqueous sodium-ion battery has an average voltage of 1.45 V, an energy density of 67 W h kg$^{-1}$, and a long cycle life of 200 cycles (Figure 3.55(c)) [135].

Inspired by Ti-based phosphate framework materials, researchers used $Ti^{3+/4+}$ pairs with redox activity to replace $Fe^{2+/3+}$ for preparing Ti-based PBAs frameworks. The $Na_{0.7}TiFe$-PBAs material ($Na_{0.7}Ti$ [Fe(CN)$_6$]$_{0.9}$) showed a moderate specific capacity of 90 mA h g$^{-1}$ and two pairs of well-separated charge and discharge platforms, 3.0/2.6 V and 3.4/3.2 V [136].

The study found that when all Fe atoms are completely replaced by Mn atoms in the PBAs lattice, the resulting $Na_2Mn[Mn(CN)_6]$ is still a perfect PB lattice with a specific capacity of 200 mA h g$^{-1}$ and excellent cycle capacity. This high capacity is due to the three-electron redox reaction of Mn$^{ii}$-CN-Mn$^{ii/i}$, Mn$^{ii}$-CN-Mn$^{iii/ii}$, and Mn$^{iii/ii}$-CN-Mn$^{iii}$. This indicates that carbon-coordinated Mn atoms are electrochemically active between 1, 2, and 3 valence states, while N-coordinated Mn atoms can only switch between 2 and 3 valence states. However, this high capacity is gradually released within a wide potential range of 4.0–1.0 V, which is not conducive to applications. However, this new discovery reveals a new redox mechanism that can be used to explore new PBAs framework materials with higher sodium storage capacity.

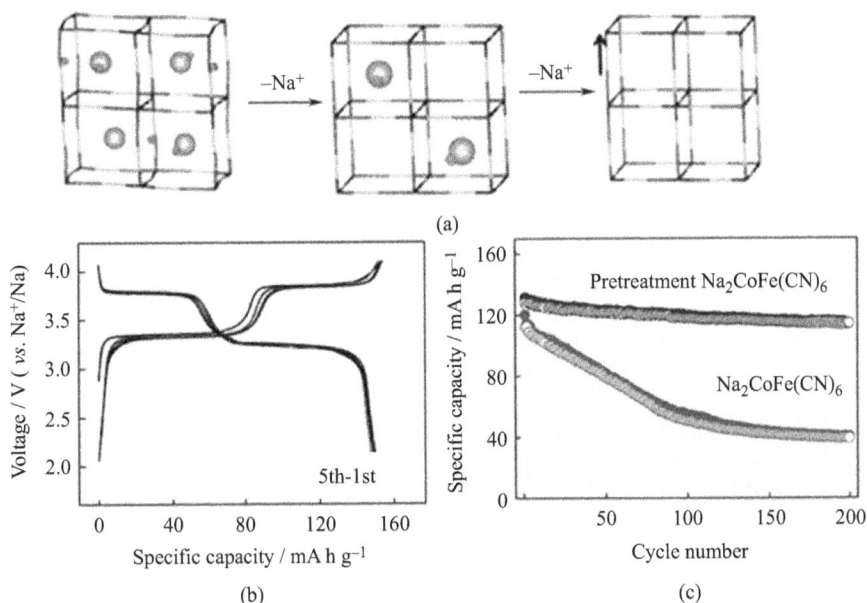

Figure 3.54: Na$_2$CoFe-PBA material: (a) schematic diagram of crystal structure; (b) charge and discharge curves; and (c) cycle performance [133].

Theoretically, a PBAs framework with two redox active centers, such as Na$_2$MFe (CN)$_6$ (M = Fe, Co, Mn, V), should all have two-electron redox capabilities. In the early studies, NaFeFe(CN)$_6$ and Co$_3$[Fe(CN)$_6$]$_2$ were placed in 1 mol L$^{-1}$ Na$_2$SO$_4$ aqueous electrolyte, and both had a low specific capacity of 65–70 mA h g$^{-1}$. It is much lower than the theoretical value of the two-electron redox reaction [137, 138]. There are three main reasons for this phenomenon when PBAs material is used as the cathode of sodium-ion batteries: first, only Fe coordinated with C is electrochemically active, which means that only four of the eight available Na-ion intercalation sites can improve the total specific capacity; second, the PBAs lattice synthesized by the conventional coprecipitation method contains about 30% lattice vacancies, which will also significantly reduce the specific capacity of the cathode material; finally, the PB or PBAs structure, in which the large amount of coordinated water in the slab blocks the intercalation reaction of Na active sites and sodium ions.

In order to overcome these problems, Yang et al. synthesized low-defect FeFe-PB nanocrystals and studied their sodium storage performance in aqueous electrolytes [139]. The low-defect FeFe-PB material prepared by the researchers has a higher specific capacity under water system conditions, with a specific capacity of 125 mA h g$^{-1}$. This highly crystalline FeFe-PB material has a specific capacity of 102 mA h g$^{-1}$ at 20 C. In addition to high rate performance, it also has long cycle capability. After 500 cycles, the capacity retention exceeds 83%. As shown in Figure 3.56 (a), the test

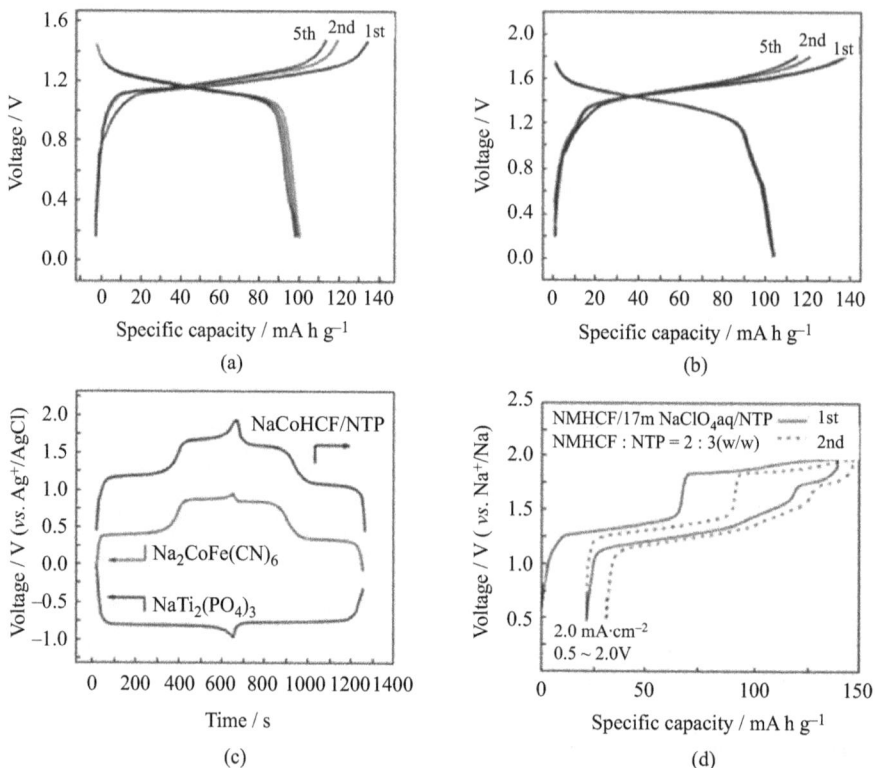

**Figure 3.55:** Electrochemical performance of the full cells: (a) charge and discharge curves of NiFe-PBA/NaTi$_2$(PO4)$_3$ [103]; (b) charge and discharge curves of CoFe-PBA/GO; (c) charge and discharge curves of CoFe-PBA/NaTi$_2$(PO$_4$)$_3$ [135]; and (d) charge and discharge curves of MnFe-PBA/NaTi$_2$(PO$_4$)$_3$ [134].

results show that this material has the ability to serve as a cathode for aqueous sodium-ion batteries.

Wu et al. also obtained similar results. They synthesized a high-quality FeFe-PB with a vacancy content as low as 9% (Na$_{1.29}$Fe[Fe(CN)$_6$]$_{0.91}$□$_{0.09}$). The FeFe-PB material has a specific capacity as high as 107 mA h g$^{-1}$ at 0.5 A g$^{-1}$. There is no obvious capacity loss after 1,100 cycles in 0.5 mol L$^{-1}$ Na$_2$SO$_4$ aqueous solution [140]. This research also provides new ideas for the design of sodium storage matrix, and all the experience and conclusions can be used to develop cathodes for high-capacity aqueous sodium-ion batteries.

Subsequently, Yang et al. further synthesized a vacancy-free CoFe-PBAs crystal with a chemical composition of Na$_{1.85}$Co[Fe(CN)$_6$]$_{0.99}$ · 2.5H$_2$O through a controlled crystallization reaction [135] and tested this material in a water system. As this material is a very low-defect lattice with two redox centers, it has an ultra-high specific capacity of 130 mA h g$^{-1}$, which is equivalent to 85% of the theoretical capacity of the two-site reaction of PBAs materials. As shown in Figure 3.56(b), there are two flat charge-discharge platforms of 0.92 V and 0.4 V in the charge/discharge curve. This phenomenon clearly proves that this is the intercalation process of two Na ions, corresponding to $Fe^{2+}/Fe^{3+}$ and $Co^{2+}/Co^{3+}$ redox reactions. The reversible specific capacity and working voltage of this material are the highest among the currently reported cathode materials for aqueous sodium-ion batteries. In addition, CoFe-PBA nanocrystals also show excellent high rate performance at 20 °C and high temperature performance. CoFe-PBA nanocrystals can maintain 90% capacity after 800 cycles, which is a high specific capacity and long cycle life for research and development in aqueous sodium-ion batteries.

MnFe-PBAs ($Na_2MnFe(CN)_6$) have two redox pairs: $Fe^{2+}/Fe^{3+}$ and $Mn^{2+}/Mn^{3+}$. Okada et al. used a simple coprecipitation method to prepare MnFe-PBAs powder with a chemical composition of $Na_{1.24}Mn[Fe(CN)_6]_{0.81} \cdot 1.28H_2O$, and studied the influence of electrolyte concentration on electrochemical performance, as well as application in sodium-ion batteries [131]. The researchers found that the electrochemical performance of the MnFe-PBAs cathode changes significantly with the change of the salt concentration in the aqueous electrolyte. In the concentration unit $NaClO_4$ aqueous electrolyte, the electrochemical window is narrowed due to the low electrolyte concentration, which is only 1.9 V, resulting in poor electrochemical performance such as reversible capacity and cycle stability. When the electrolyte concentration increases to 17 M, the electrochemical window of the electrolyte widens to 2.8 V. In this case, the MnFe-PBAs cathode works well, and the initial charge/discharge specific capacity is 124 /116 mA h g$^{-1}$. The charge/discharge process of this material is based on a two-site reaction, so the MnFe-PBAs cathode shows two charge-discharge platforms at 0.64/0.51 V and 1.2/1.03 V, respectively. The high concentration of electrolyte enables the MnFe-PBAs cathode to carry out more than 100 cycles at 5.0 mA cm$^{-2}$, with capacity retention of 92% (Figure 3.56(c)). This means that researchers can stabilize the electrochemical performance of PBAs materials by increasing the electrolyte concentration in the aqueous solution to provide a direction for future research.

Cui et al. proposed a new method to increase the specific capacity of PBAs, while maintaining its cycle stability. They separately studied the correlation between the electrochemical behavior and structural characteristics of MnFe-PBAs, CoFe-PBAs, and MnCoFe-PBAs materials in aqueous electrolytes. It is found that Fe-coordinated materials can maintain the crystal structure and good cycle life during cycling. In addition, since the N-coordinated transition metal will cause structural deformation of the overall material, the N-coordinated Co and Mn ions show a slower reaction kinetic state, but can still significantly increase the total capacity of the material. These results help people understand how to meet the needs of large-scale grid applications with extremely high cycle life and fast reaction kinetics, and promote the future development of PBAs-type cathode materials for sodium-ion batteries.

Vanadium (V) ions, with a variety of oxidation states from +2 to +5, show stable electrochemical properties within the electrochemical window of the aqueous

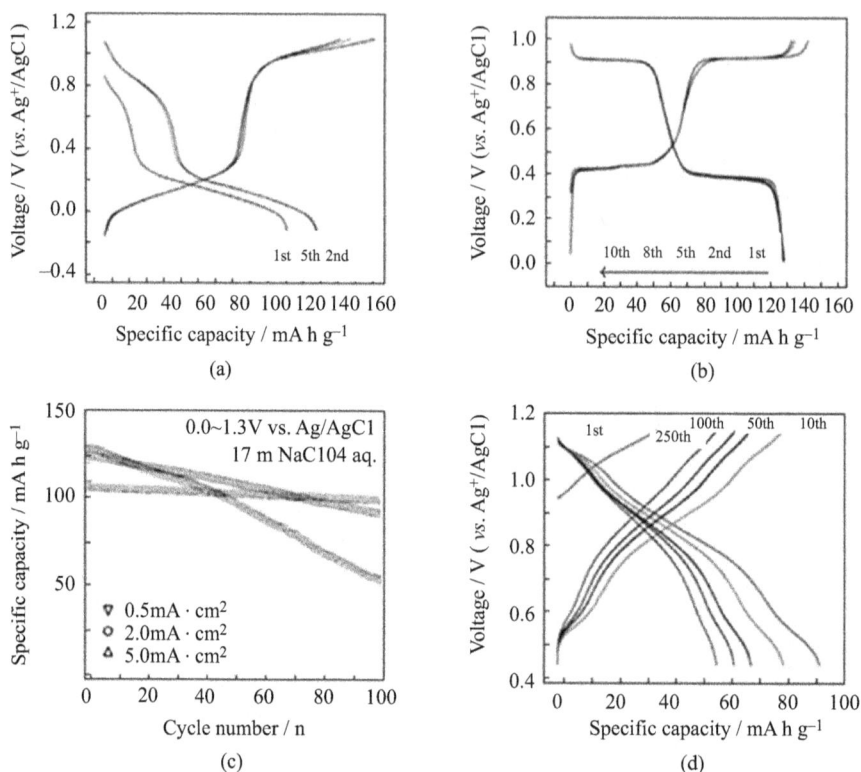

**Figure 3.56:** Electrochemical performance of PBA at multiple sites: (a) charge and discharge curves of FeFe-PB materials with low defects [139]; (b) charge and discharge curves of no-vacancy CoFe-PBA materials [135]; (c) cycle performance of MnFe-PBA materials [136]; and (d) charge and discharge curves of VFe-PBA materials [141].

electrolyte. In the VFe-PBAs framework, both V and Fe ions have redox activity, and both can provide redox properties for the material. Chung and colleagues first proposed VFe-PBAs ($V_3[Fe(CN)_6]_2$) as the cathode of water-based sodium-ion batteries. VFe-PBAs has a specific capacity of 91 mA h $g^{-1}$ at 110 mA $g^{-1}$. Under the extremely high 3,520 mA $g^{-1}$, the VFe-PBAs material still has a sufficiently high specific discharge capacity of 54 mA h $g^{-1}$(Figure 3.56(d)), because it is open-frame structure and 3D hydrogen bond network [141]. In contrast, Bandarenka and colleagues synthesized a new sodium-rich $VO_xFe$-PBAs framework ($Na_2VO_x[Fe(CN)_6]$) with an average working potential of 0.91 V. That shows a three-stage redox mechanism in the process of inserting sodium ions into the lattice. This material displays a specific capacity that can exceed 80 mA h $g^{-1}$ at 30 C when it is placed in a mixed solution of 3.6 mol $L^{-1}$ $H_2SO_4$ and 3 mol $L^{-1}$ $NaNO_3$ [142].

In the current research, many PBAs materials have a sufficiently high reversible specific capacity of 120 mA h $g^{-1}$ and cycle stability within 500 cycles, which meets the requirements of actual battery applications. However, further research is still needed to verify the factors that affect the structural stability of the PBAs framework under the high positive potential in the aqueous electrolyte, and use this material as a low-cost and large-scale energy storage application to prepare chemical composition and electrochemically stable PBAs cathodes material.

Since Fe element can be replaced by other transition metals in large amounts, without destroying the PB lattice, it is easy to design a lattice with better performance through transition metal doping. Ma et al. reported an Ni-doped MnFe-PBAs $(CN)_6$, which can alleviate the redox lattice distortion of $Mn^{2+/3+}$ pairs caused by nonredox active Ni doping [143]. The Ni-doped MnFe-PBAs$(CN)_6$ framework can achieve higher capacity and better cyclability. It showed a discharge-specific capacity of 118.2 mA h g$^{-1}$ at 10 mA g$^{-1}$ and a capacity retention rate of 83.8% after 800 cycles. Subsequently, Wu and his colleagues synthesized a series of Ni-doped CoFe-PBAs with different ratios of Ni-Co. They found that the best doping ratio in the PBAs lattice is 0.4 Ni, and its initial discharge specific capacity is 92 mA h g$^{-1}$, and the specific capacity exceeds 80 mA · h · g$^{-1}$ after 100 cycles [144]. Jiang and his colleagues studied the beneficial effects of Ni doping on the electrochemical performance of FeFe-PBAs. They found that Ni-induced low-spin $Fe^{2+}/Fe^{3+}$ coupling resulted in a higher discharge-specific capacity. Therefore, Ni-doped FeFe-PBAs can achieve a specific discharge capacity of 106 mA h g$^{-1}$, with a capacity retention rate of 96% after 100 cycles [145]. These studies show that the real reason for the limited electrochemical performance of PB-based materials is its lattice defects, and further research work should focus on solving this problem.

### 3.3.4 Prussian Blue analogues compound with optimized structure

With the in-depth study of the new PBA lattice structure, researchers have also actively explored various structure and morphological optimization methods in recent years, such as surface modification, nanostructure design, element doping, and morphology control, to further improve PBAs. The sodium storage performance of the material is shown in Figure 3.57.

In order to shorten the diffusion path of Na ions in the crystal lattice, the researchers synthesized several kinds of nano-PBA particles to enhance the high rate performance. For example, the double-textured KFeFe-PB nanocubes are prepared by a simple acid etching process, which is a combination of porous Prussian blue compound and nonporous Prussian blue compound (Figure 3.57(a)). The materials show reversible specific capacities of 85, 65, and 52 mA h g$^{-1}$, at 10, 240, and 480 mA g$^{-1}$, respectively [146]. Da et al. studied mesoporous/macroporous NiFe-PBAs in 2014 and found that larger pores on the surface can promote more Na-ion transfer to the inside of the PBAs framework. Hu et al. used polymer surfactants to prepare FeFe-PB crystals formed by the aggregation of nanocubes. Compared with PB single crystal, this

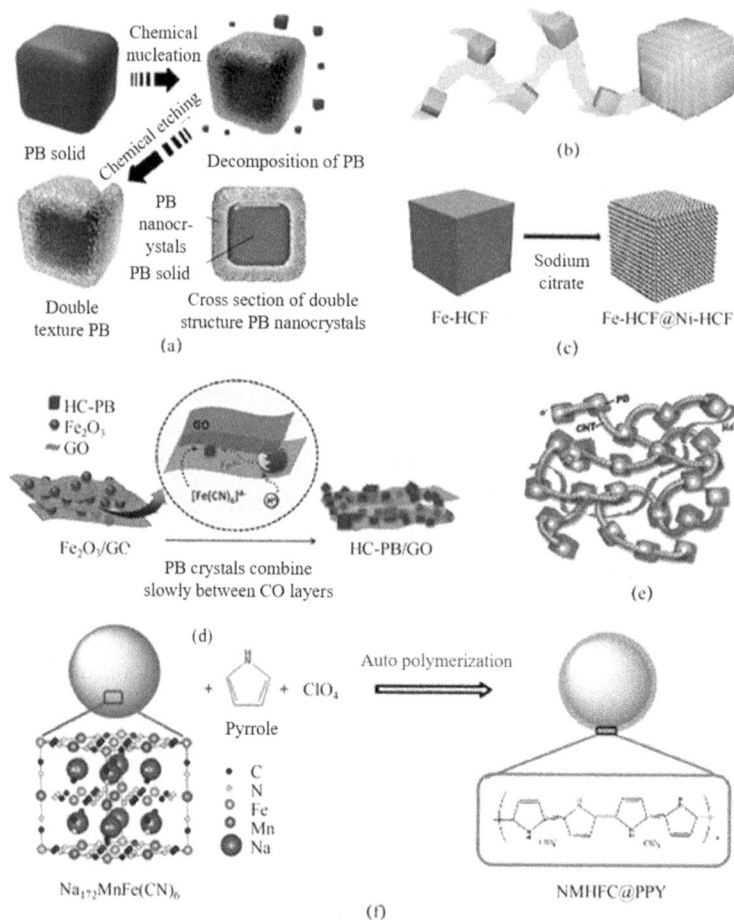

**Figure 3.57:** Structurally optimized Prussian blue material: (a) structurally optimized KFeFe-PB prepared by etching [146]; (b) Prussian blue with rich boundaries [147]; (c) core–shell structure FeFe-PBA [148]; (d) PB@GO nanocomposites [149]; (e) FeFe-PBPB/CNT composites [150]; and (f) NMHFC@PPY composites.

structure improves the reversible specific capacity and phase change behavior of PB. Recently, researchers have developed a PB material with many crystal boundaries (BR-FeHCF) (Figure 3.57(b)) by a simple and environmentally friendly coprecipitation method [147]. The $Na_4Fe(CN)_6 \cdot 10H_2O$ solution and $FeSO_4 \cdot 7H_2O$, sodium citrate and polyvinylpyrrolidone (PVP, K30, MW ≈ 40,000) solution are stirred and mixed vigorously, and reacted at 0 °C for 6 h, where the low temperature can effectively limit water molecules in the frameworks. This material has a high initial specific capacity of 120 mA h $g^{-1}$, and it still has a specific capacity of 95 mA h $g^{-1}$ after 280 cycles. As the abundant interface area provides good wettability and fast ion transport channel, the material also shows good rate performance of 60 mA h $g^{-1}$ at 1,600 mA $g^{-1}$ (Figure 3.58).

However, from the perspective of applications, most of the PBAs materials with commercial prospects have now been prepared in sub-micron to a few microns in size, which can achieve the perfect combination of volume energy density and high-rate performance of the material. The boundary-rich PBA cathode material has a larger specific surface area, can provide more contact interfaces with the electrolyte, and help increase the reactive sites of Na ions on the electrode. At the same time, this kind of material belongs to the sodium-rich cathode material, which has a rhombohedral crystal structure. As opposed to the traditional cubic phase structure, the rhombohedral phase structure has higher electronic conductivity, a more stable sodium storage structure, and a lower Na-ion transmission energy barrier. The rhombohedral phase will be transformed into a cubic phase after sodium removal, and this process is highly reversible. In addition, the rough interface also improves the contact area with the conductive agent and bonding agent in the electrode, so that the material has better conductivity and adhesion. However, due to the increase of the interface area, the boundary-rich PBAs cathode material has significantly increased interface side reactions. It is necessary to select appropriate additives to effectively improve the stability of the interface passivation layer, prevent the cracking and dissolution of the SEI film, and realize long cycle stability.

Surface modification is a widely used effective method, which can construct a favorable surface layer and protect the bulk to maintain the structural stability. The surface modification methods used in the structural optimization of PBAs materials can basically be divided into two categories: core–shell structure and surface coating. Talham et al. first reported a CuFe-PBAs@NiFe-PBAs material with a core–shell structure. It uses high-capacity CuFe-PBAs as the core and low-capacity, but highly stable, NiFe-PBAs as the shell. This material shows high capacity and long cycle stability because the shell protects the reaction interface of the CuFe-PBAs core. Huang et al. developed a core–shell structure FeFe-PBAs@NiFe-PBAs (Figure 3.58(c)), which can provide a fairly high capacity of 102 mA h g$^{-1}$ at 200 mA g$^{-1}$. After more than 800 cycles, it still retains a specific capacity of 79.7 mA h g$^{-1}$ [148]. What is impressive is that even at a current density as high as 2,000 mA g$^{-1}$, this core–shell structured PBAs material can still provide a specific capacity of about 60 mA h g$^{-1}$. Another type of core–shell structure PBAs material is the dual-texture PBAs nanocube synthesized by Kim et al. Through acid etching of FeFe-PBAs, the mixed microstructure of FeFe-PB nanocubes is composed of nonporous core and porous shell, which can significantly increase the reversible specific capacity and have stable cycle performance and rate performance [146]. The researchers used a simple and effective chemical co-precipitation method to prepare a new heterostructure with manganese-based Prussian blue material (PBM) as the core and nickel-based Prussian blue material (PBN) as the shell (Figure 3.59) [152].

The material effectively reduces the defect content and coordination water content of the PBA material by slowing down the rate of the nucleation reaction, and improves the structural stability of the material. Using PBM material with higher

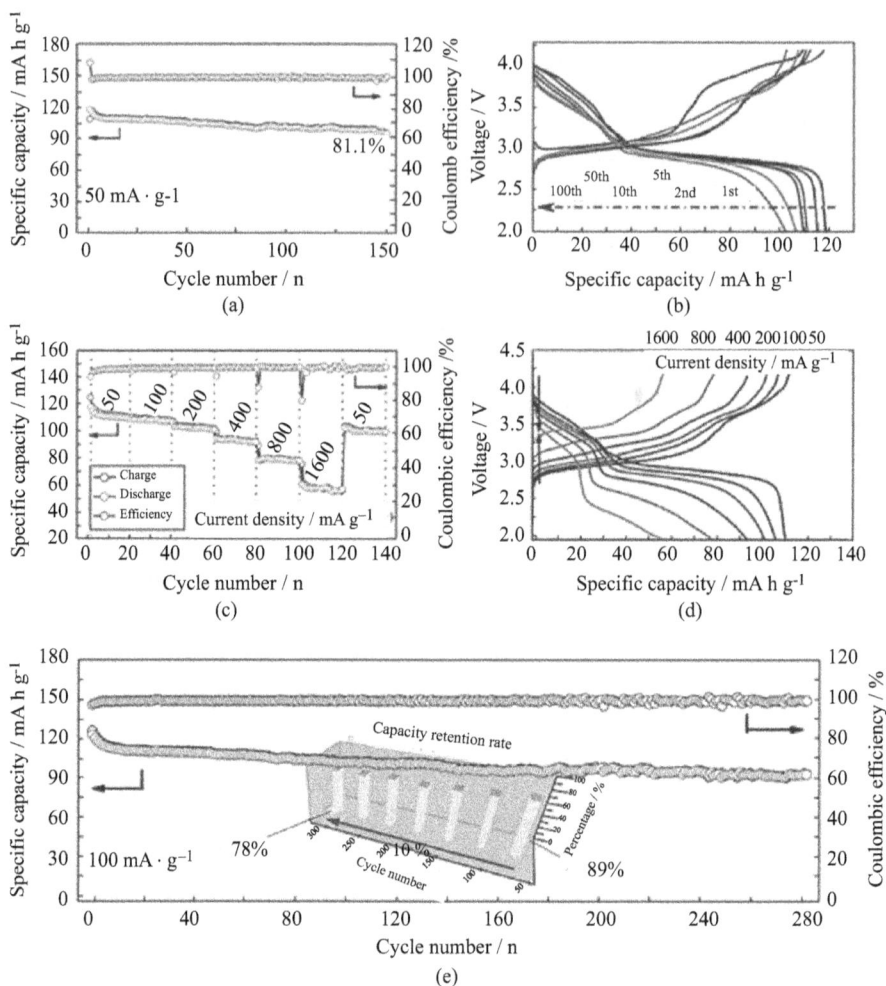

**Figure 3.58:** Electrochemical performances of boundary-rich Prussian blue electrode: (a) specific capacity and coulombic efficiency at 50 mA g$^{-1}$; (b) corresponding charge/discharge curves at different cycle numbers; (c) specific capacity and Coulombic efficiency at different rates; (d) corresponding charge/discharge curves at different rates; and (e) long cycle at 100 mA g$^{-1}$ [147].

specific capacity as the core and PBN material with higher structural stability as the shell, a core–shell structure FeFe-PBAs@NiFe-PBAs electrode was prepared. This not only improves the cycle performance of the electrode, but also improves the interface compatibility of the electrode. As shown in Figure 3.60, the core–shell structure of the electrode exhibits an initial specific capacity of 120 mA h g$^{-1}$, and still maintains 82.3% of the initial capacity after 600 cycles. Even at a current density of 3,200 mA g$^{-1}$, it still has a specific capacity of 52 mA h g$^{-1}$. When assembled with

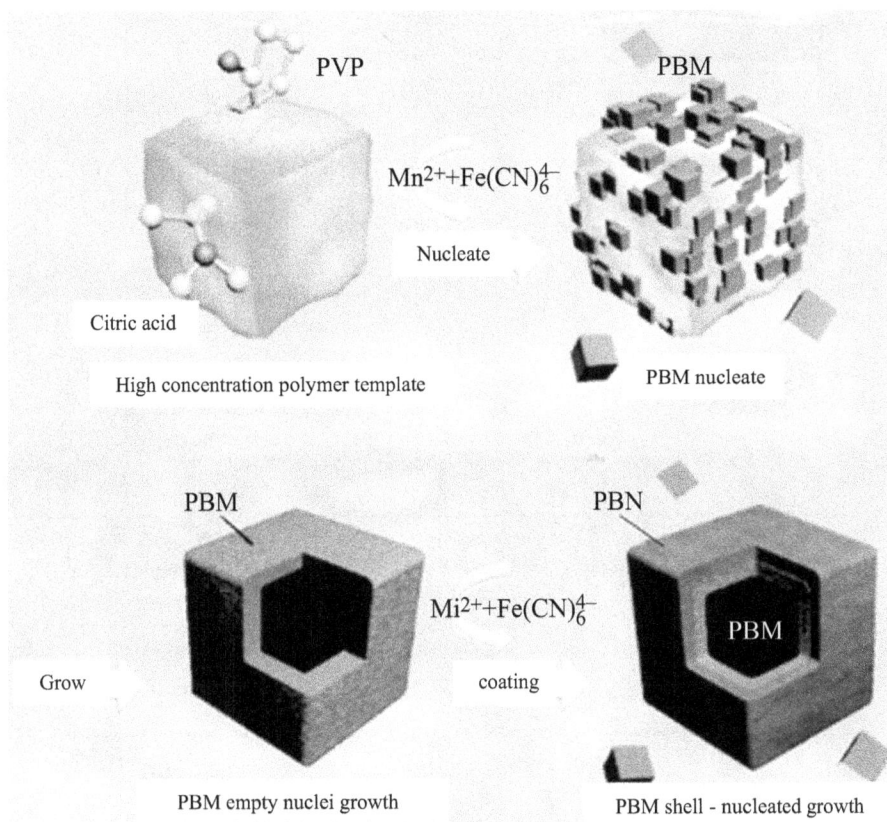

**Figure 3.59:** Schematic diagram of core–shell structure FeFe-PBA@NiFe-PBA [152].

the hard carbon anode and the NTP anode to form a full battery, both exhibit a reversible specific capacity of 120 mA h g$^{-1}$.

PBAs materials are usually combined with carbon coatings or highly conductive carbons, such as carbon nanotubes (CNTs), graphene oxide (GO), and conductive polymers (CPs), to enhance electronic conductivity. Pyo et al. prepared FeFe-PB/GO composites by embedding highly crystalline FeFe-PB particles in GO nanosheets and tested their sodium storage capacity (Figure 3.57(d)). Compared to the original PB particles without GO, the FeFe-PB/GO composite shows a significantly higher reversible specific capacity of 113 mA h g$^{-1}$ and a higher energy density of 338 W h kg$^{-1}$. In addition, the material has a significant improvement in its high-rate performance, showing a high energy density of 280 W h kg$^{-1}$ at 2,000 mA g$^{-1}$ [149].

Dou et al. prepared FeFe-PB@C composites by evenly embedding PB cubes in a conductive carbon matrix (Ketjen Black) to ensure the close electronic contact between the PB crystals and carbon. Compared to the bare FeFe-PB sample (with a specific capacity of 90 mA h g$^{-1}$), the FeFe-PB@C composite material has a higher

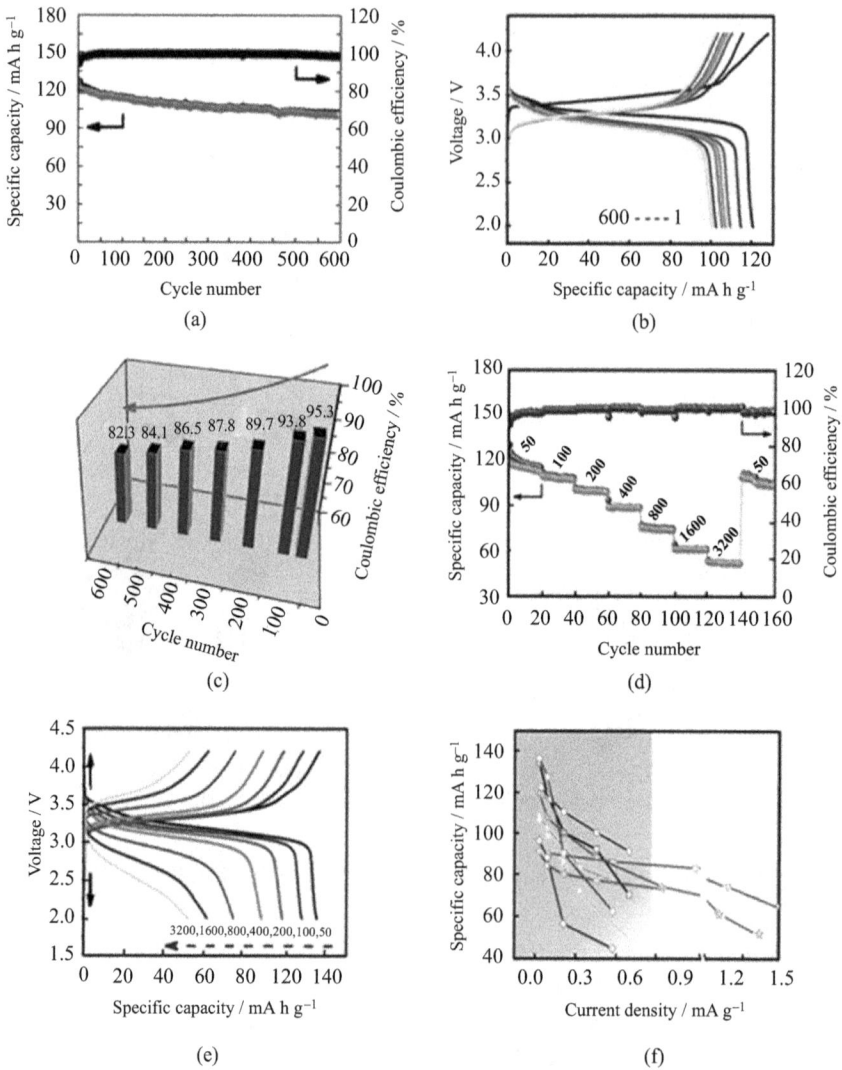

**Figure 3.60:** Core–shell structure FeFe-PBA@NiFe-PBA at 50 mA g$^{-1}$: (a) Long cycle performance; (b) charge/discharge curves; (c) capacity retention rates at different cycles; (d) specific capacity and Coulombic efficiency at different rates; (e) corresponding charge and discharge curves; and (f) comparison of rate properties with previously reported Prussian blue cathode materials [152].

reversible specific capacity of 120 mA h g$^{-1}$. The material achieves an initial specific capacity is 90 mA h g$^{-1}$, and it still has a specific capacity of 77.5 mA h g$^{-1}$ after 2,000 cycles at 90 C, with capacity retention rate of 90% at 20 °C. Guo et al. prepared FeFe-PB nanocubes by directly nucleating and growing on the CNT network, and synthesized a strong and flexible FeFe-PB/CNT composite material (Figure 3.57(e)). The

FeFe-PB/CNT composite exhibits a high specific discharge capacity of 142 mA h g$^{-1}$, an energy density of 408 W h kg$^{-1}$, a coulomb efficiency exceeding 99.4%, and a capacity retention rate of 86% after 1,000 cycles [150].

In contrast to the carbon coating, coating the conductive polymer on the PBAs particles can not only improve the surface conductivity, but also provide a flexible surface layer to adapt to the lattice strain generated during the intercalation/deintercalation of Na ions. Dou et al. developed a method to synthesize MnFe-PBAs@PPy composite material by coating a ClO$_4$ doped polypyrrole layer on MnFe-PBAs (Figure 3.57(f)) [151]. Through this work, the researchers found that the P-doped PPy layer can not only serve as a conductive network, but also prevent the Mn element from dissolving into the electrolyte. In addition, the P-doped PPy polymer can act as an active material to improve the redox ability of the composite material. The results show that the MnFe-PB@PPy composite maintains 67% of the initial capacity after 200 cycles and can provide a discharge-specific capacity of 55.6 mA h g$^{-1}$ at 40 C. Similarly, the FeFe-PB prepared by Huang et al. was compounded with conductive PPy to obtain FeFe-PB@PPy composite material. When used as a cathode material for sodium-ion batteries, this material has a reversible specific capacity of 113 mA h g$^{-1}$ at a current density of 25 mA g$^{-1}$ and can maintain 79% of the reversible capacity after 500 cycles [153].

Since 2013, researchers have developed a large number of PBAs framework materials, some of which have proven to have competitive advantages over other Na-ion transition metal oxides and phosphates. From the perspective of battery applications, Mn and Fe-based PBAs materials have great potential for commercial development of sodium-ion batteries due to their low cost and excellent electrochemical performance. Considering all the factors that affect the sodium storage capacity, the perfection of the crystal lattice is the decisive factor that will realize the high capacity and long cycle life of the PBAs compounds. Therefore, the development of a new synthesis method of low-cost and defect-free PBAs lattices that can be prepared on a large scale is very important for the development of practical sodium-ion batteries.

In short, we tried to make a comprehensive summary of the development of the PBAs framework as a cathode material for sodium-ion batteries, and mainly introduce the groundbreaking research of using PBAs framework for other monovalent and multivalent ion storage cathodes. In terms of material cost, environment friendliness and electrochemical performance, PBAs frameworks based on Mn and Fe (MnMn-PBAs, FeFe-PBAs, and MnFe-PBAs) are the most promising candidates for practical sodium-ion batteries. PBAs materials based on Ni, Co, and V lack competitive advantages due to their high cost, potential toxicity, and limited applications. In addition, due to its inherent low mass density, all PBA frameworks are not suitable as cathode materials to construct high specific energy batteries for portable and mobile applications. Therefore, using the PBAs framework to build low-cost sodium-ion batteries seems to be the best choice for large-scale stationary energy storage applications.

Compared to organic electrolytes, water-based sodium-ion batteries can well meet the demanding requirements of high rate, safety, and long cycle performance, and meet the requirements of various large-scale energy storage applications. However, the development of PBAs cathode materials is still at an early stage, so a lot of research is needed to design and manufacture new PBAs frameworks with specific structures and shapes, improve the electrochemical performance of sodium storage, and realize the manufacture of commercial sodium-ion batteries. From the performance characteristics of PBAs-based sodium-ion batteries reported in recent years, we can clearly see that some water-based sodium-ion batteries exhibit an energy density of 45 W h kg$^{-1}$ and a cycle life of more than 1,000 cycles. These performance characteristics of ion batteries make it more competitive than traditional Pb/PbO$_2$, Ni/Cd, and Ni/MH batteries, and are more suitable for large-scale applications.

In recent years, the demonstration of PBAs framework materials for reversible multivalent ion storage has opened up a new research field for the development of new materials and batteries, which will undoubtedly contribute to the promotion of large-scale energy storage applications in the future. At present, this research field is still in the initial stage, and a lot of research and supporting evidence are still needed in the future development of electrochemical energy storage technology.

## 3.4 Other cathode materials for sodium-ion batteries

### 3.4.1 Transition metal fluorides

Transition metal fluorides are different from general transition metal oxides and sulfide anode materials. Metal fluorides have unique coordination effects between metal ions and fluoride ions, and can be used as a cathode for sodium-ion batteries. The biggest problem with metal fluoride as an electrode material is that the stable chemical bond composed of metal and F makes the material have high resistivity. It will have a great impact on the rate performance of the material, and it may also make the capacity of the material differ greatly from the theoretical capacity [154]. At the same time, Okada et al. through the study of NaMF$_3$ (M = Fe, V, Ti, V, Mn) showed that these materials all realize the insertion/extraction of Na ions through the redox of the valence metal ion. However, except for NaFeF$_3$, which has a larger specific capacity (128 mA h g$^{-1}$), other materials exhibit poor electrochemical performance [155]. Although NaFeF$_3$ can only conduct a single-electron reaction through Fe$^{2+/3+}$, there are sufficient Na ions migration channels in the crystal structure. The theoretical specific capacity can reach 197 mA h g$^{-1}$, and the discharge platform is close to 3 V (Figure 3.61(a)). The key to improve the electrochemical performance of NaFeF$_3$ is mainly to improve the electronic conductivity of the material. It can directly increase the conductivity or reduce the transmission path of Na ions by carbon composite or nanomaterials.

Although graphene, carbon nanotubes, and other materials can improve the stability and rate performance of $FeF_3$ materials [154, 156], the use of high-cost carbon materials does not meet the original intention of choosing low-cost halogen element F as the electrode material for sodium-ion batteries. Some studies have found that $FeF_3 \cdot xH_2O$ has crystal water, which broadens the Na-ion transmission channel [157]. One way to further stabilize the crystal structure and cycle stability is the doping of $K^+$ [158] and $Cr^{2+}$ [159]. However, these ions themselves cannot contribute to the capacity and also have the disadvantage of reducing the discharge platform of the material. Among iron-based fluoride materials, $Fe_2F_5 \cdot H_2O$ is currently the only material with a similar structure to anionic groups. It is very rare for $Fe^{2+}$ and $Fe^{3+}$ to appear in a compound of such a small molecular weight at the same time. Wang et al. made a series of modifications to $Fe_2F_5 \cdot H_2O$, and its initial discharge-specific capacity as high as 251 mA h g$^{-1}$ (Figure 3.61(b)) will also be the same for iron-based fluoride. Materials have entailed a lot of research and development work and brought high expectations [160, 161].

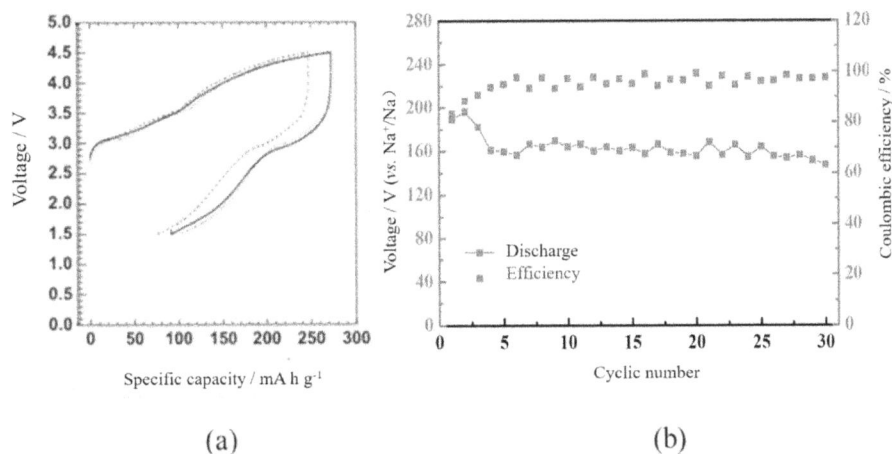

(a)  (b)

Figure 3.61: Electrochemical performance of Fe-based fluoride: (a) charge and discharge curves of $NaFeF_3$ at 0.01 C [155]; and (b) cycling performance of $Fe_2F_5 \cdot H_2O$ at 20 mA g$^{-1}$ in 1.0–4.0 V [160].

## 3.4.2 Sodium-containing organic compound cathode materials

Organic electrodes have the advantages of easy synthesis, rich variety, and safety and are a class of electrode materials that may be used in large-scale production in the future. At the same time, the raw materials required for the synthesis of organic electrodes are various biomass materials, and biomass is basically produced by using solar energy and consuming $CO_2$ through photosynthesis of plants. Therefore, organic electrodes are highly renewable and are environment-friendly electrode materials. The most notable feature of organic electrode materials is their rich variety, with the presence of various single and double bonds, functional groups, heteroatoms, etc. At the same time, polymerization can connect small molecules to synthesize polymer materials. The rich composition and structure of organic materials have created various types of sodium storage cathode materials. Their main active groups include carbonyl groups, disulfide bonds, and imine groups. The main reaction mechanism is shown in Figure 3.62 [162].

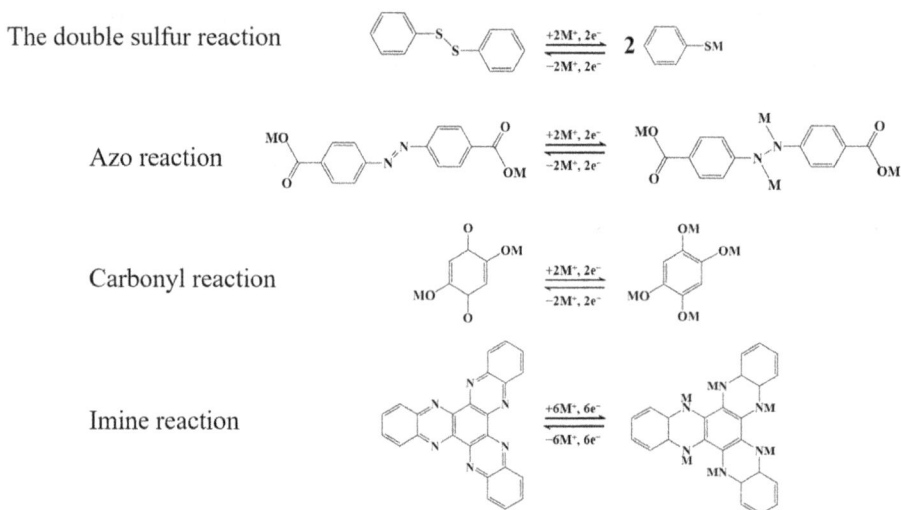

**Figure 3.62:** Several reaction mechanisms of organic electrode materials [162].

Organic electrode voltage is generally not high. Based on the above reaction mechanism, there are many organic materials that can be used in alkali metal-ion batteries. In the field of cathodes for sodium-ion batteries, high energy density sodium rose palmitate ($Na_2C_6O_6$) has received widespread attention. Theoretically, $Na_2C_6O_6$ (Figure 3.63(a)) can react reversibly with up to four Na ions, and the theoretical specific capacity is as high as 501 mA h $g^{-1}$. However, the electrochemical test result of Okada et al. on $Na_2C_6O_6$ is that it has a reversible specific capacity of about 180 mA h $g^{-1}$ in the voltage range of 1.5–2.8 V, and only about two Na ions reversibly participate in the reaction. And, the materials have problems of capacity loss at the first cycle and poor cycle stability [163]. Bao et al. analyzed the reason for the large difference between the actual capacity and the theoretical capacity. As shown in Figure 3.63(b) and (c), the general $Na_2C_6O_6$ will change from the α phase to the γ phase during the reaction, which inhibits the Na-ion reaction, but if the electrolyte type and the size of the active material particles are limited, it is possible to maintain the stability of the material structure and realize the four Na-ion reactions. The researchers synthesized a nano-sized electrolyte solution that matches diethylene

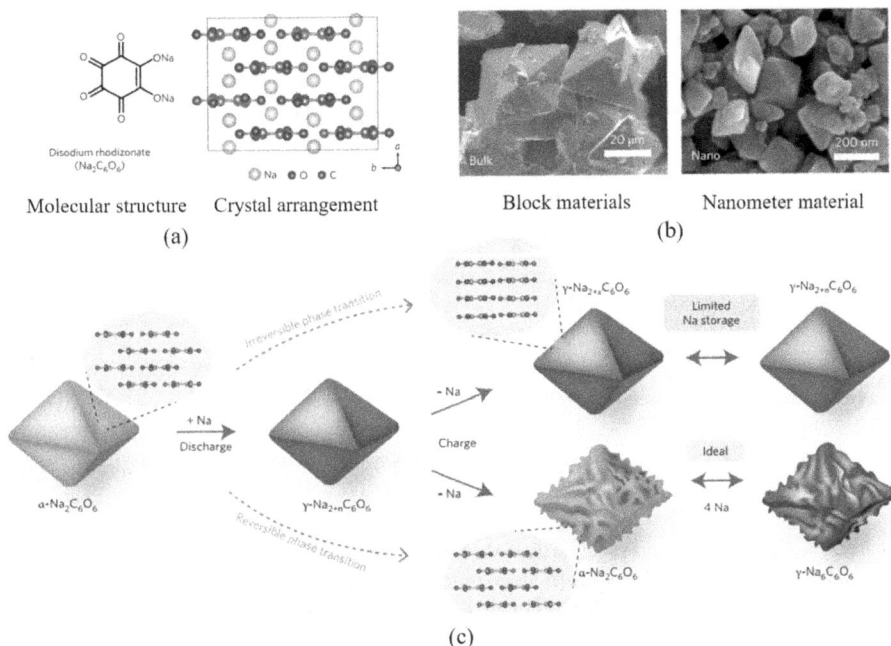

**Figure 3.63:** Structural characteristics of $Na_2C_6O_6$: (a) molecular structure and crystal structure of $Na_2C_6O_6$; (b) comparison of particle size of $Na_2C_6O_6$; (c) whether $Na_2C_6O_6$ can undergo reversible reactions with four $Na^+$ in different phases [164].

glycol dimethyl ether. The electrode material has a reversible specific capacity of 484 mA h g$^{-1}$ in the voltage range of 0.5–3.3 V, and maintains a capacity of 90.6% after 50 cycles [164].

In order to increase the initial sodium content, Chen et al. synthesized 2,5-dihydroxyterephthalic acid ($Na_4C_8H_2O_6$) electrode material with two groups of active functional groups. The reaction mechanism of this material is shown in Figure 3.64(a), and it can be found that this material has two active regions at 1.6–2.8 V and 0.1–1.8 V, respectively. The materials based on the two Na-ion reactions have a reversible specific capacity of 180 mA h g$^{-1}$. The electrochemical performance of the assembled organic symmetrical full battery is shown in Figure 3.64(b–d). The working voltage of the full battery is 1.8 V, the energy density reaches 65 W h kg$^{-1}$, and it can be cycled stably for 100 cycles [165].

Organic cathode materials have two significant shortcomings: ① poor conductivity, resulting in low Na ions or electron migration rate, slow electrochemical reactions, and poor rate performance; ② easy to dissolve in organic electrolyte, resulting in rapid capacity attenuation. The main modification processes of organic electrodes include: ① carbon coating the material to improve conductivity, while inhibiting the dissolution of electrode materials; ② synthesizing polymer materials through

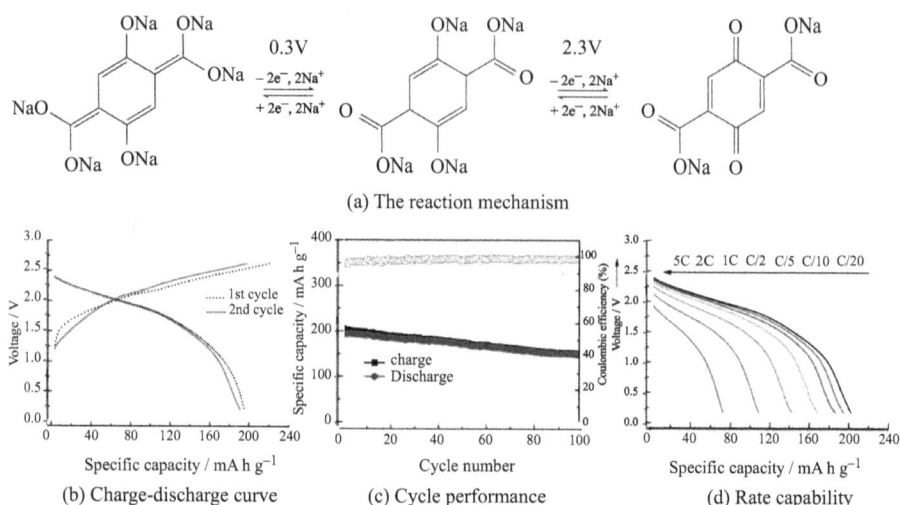

(a) The reaction mechanism

(b) Charge-discharge curve

(c) Cycle performance

(d) Rate capability

**Figure 3.64:** Reaction mechanism of $Na_4C_8H_2O_6$ electrode and electrochemical performance of assembled symmetrical full batteries [165]: (a) reaction mechanism; (b) charge and discharge curves of the full batteries; (c) cycle performance; and (d) rate performance.

polymerization to inhibit the dissolution of electrode materials. For example, the Calix[4] quinone synthesized by Huang et al. (C4Q for short, chemical formula is $C_{28}H_{16}O_8$) can be coated with 50% mesoporous carbon and carbon nanotubes and has a stable reversible specific capacity of 290 mA h g$^{-1}$, after 100 cycles at 0.1 C [166]. Liu et al. used RGO and carbon nanotubes to clad perylene-3,4,9,10-tetraformic dianhydride (PTCDA/RGO/CNT). According to the step reaction of one Na ion on two anhydrides in the voltage range of 1.5–3.5 V, the initial specific capacity of the electrode could reach 126 mA h g$^{-1}$ at the current density of 10 mA g$^{-1}$. Even at a current density of 200 mA g$^{-1}$, after 500 cycles, the capacity retention rate of the electrode material is still as high as 99%. For polymer materials, Shaijumon et al. used perylene-3,4,9,10-tetracarboxylic dianhydride as the raw material and reacted with hydrazine and imidazole to generate N, N'-diamino-3,4,9,10-perylenetetramine methyl polyimide (PI) material. PI generally has high electronic conductivity and high thermal stability. When PI is used as cathode material for sodium-ion batteries, it can contribute 126 mA h g$^{-1}$ specific capacity and high rate performance in the voltage range of 1.5–3.5 V [167].

The research on organic electrode materials is currently in a stage of rapid development. Since the alkali Li$^+$ and K$^+$ have electrochemical properties similar to Na ions, the development of many materials is not limited to one battery system. In addition to organic electrode materials with carbonyl active sites, research on other reactive types of organic electrode materials has just started and is mainly used in lithium-ion batteries. The problems of low sodium, low conductivity, and easy solubility in the electrolyte of organic electrodes can be solved by discovering new materials, or can be overcome by choosing aqueous electrolytes, ionic liquid electrolytes, or solid electrolytes according to the use environment. Although organic electrodes still have many shortcomings, their environmental friendliness, low cost, and ease of manufacturing make them worthy of further research.

# References

[1]    Delmas C, Braconnier J J, Fouassier C. et al., Electrochemical Intercalation of Sodium in $Na_xCoO_2$ Bronzes[J]. Solid State Ionics, 1981, 3–4: 165–169.
[2]    彭正军, 李法强, 诸葛芹等. 共沉淀法制备正极材料$LiCoO_2$及其电化学性能研究[J]. 盐湖研究, 2008, 03: 25–29.
[3]    Delmas C, Fouassier C, Hagenmuller P. Structural Classification and Properties of the Layered Oxides[J]. Physica B+C, 1980, 99(1): 81–85.
[4]    Yabuuchi N, Komaba S. Recent Research Progress on Iron- and Manganese-Based Positive Electrode Materials for Rechargeable Sodium Batteries[J]. Science and Technology of Advanced Materials, 2014, 15(4): 43501.
[5]    Yabuuchi N, Yoshida H, Komaba S. Crystal Structures and Electrode Performance of Alpha-$NaFeO_2$ for Rechargeable Sodium Batteries[J]. Electrochemistry, 2012, 80(10): 716–719.
[6]    Takeda Y, Nakahara K, Nishijima M. et al., Sodium Deintercalation from Sodium Iron Oxide[J]. Materials Research Bulletin, 1994, 29(6): 659–666.
[7]    Zhao J, Zhao L, Dimov N. et al., Electrochemical and Thermal Properties of -$nafeo_2$ Cathode for Na-Ion Batteries[J]. Journal of the Electrochemical Society, 2013, 160(5): A3077–A3081.
[8]    Yoshida H, Yabuuchi N, Komaba S. $NaFe_{0.5}Co_{0.5}O_2$ as High Energy and Power Positive Electrode for Na-Ion Batteries[J]. Electrochemistry Communications, 2013, 34, 60–63.
[9]    Ma X, Chen H, Ceder G. Electrochemical Properties of Monoclinic $NaMnO_2$[J]. Journal of the Electrochemical Society, 2011, 158(12): A1307–A1312.
[10]   Mendiboure A, Delmas C, Hagenmuller P. Electrochemical Intercalation and Deintercalation of $Na_xMnO_2$ Bronzes[J]. Journal of Solid State Chemistry, 1985, 57, 323–331.
[11]   Nur K S, Norashikin K, Norlida K. Studies of $NaCoO_2$: Synthesis and Characterization[J]. Aiaa Journal, 2009, 36(9): 1763–1765.
[12]   Shacklette L W. Rechargeable Electrodes from Sodium Cobalt Bronzes[J]. Journal of the Electrochemical Society, 1988, 135(11): 2669–2674.
[13]   Molenda J, Delmas C, Dordor P. et al., Transport Properties of $Na_xCoO_{2-y}$[J]. Solid State Ionics, 1984, 12: 473–477.
[14]   Braconnier J J, Delmas C, Hagenmuller P. Etude Par Desintercalation Electrochimiq Ue Des Systemes $Na_xCrO_2$ Et $Na_xNiO_2$[J]. Materials Research Bulletin, 1982, 17, 993–1000.
[15]   Li G, Yue X, Luo G. et al., Electrode Potential and Activation Energy of Sodium Transition-Metal Oxides as Cathode Materials for Sodium Batteries: A First-Principles Investigation[J]. Computational Materials Science, 2015, 106: 15–22.
[16]   Komaba S, Takei C, Nakayama T. et al., Electrochemical Intercalation Activity of Layered $NaCrO_2$ Vs. $LiCrO_2$[J]. Electrochemistry Communications, 2010, 12(3): 355–358.
[17]   Xia X, Dahn J R. $NaCrO_2$ Is a Fundamentally Safe Positive Electrode Material for Sodium-Ion Batteries with Liquid Electrolytes[J]. Electrochemical and Solid-State Letters, 2012, 15(1): A1.
[18]   Chen C, Matsumoto K, Nohira T. et al., Electrochemical and Structural Investigation of $NaCrO_2$ as a Positive Electrode for Sodium Secondary Battery Using Inorganic Ionic Liquid NaFSA–KFSA[J]. Journal of Power Sources, 2013, 237: 52–57.

[19] Fukunaga A, Nohira T, Kozawa Y. et al., Intermediate-Temperature Ionic Liquid NaFSA-KFSA and Its Application to Sodium Secondary Batteries[J]. Journal of Power Sources, 2012, 209: 52–56.

[20] Vassilaras P, Ma X, Li X. et al., Electrochemical Properties of Monoclinic $NaNiO_2$[J]. Journal of the Electrochemical Society, 2012, 160(2): A207–A211.

[21] Hamani D, Ati M, Tarascon J M. et al., $Na_xVO_2$ as Possible Electrode for Na-ion Batteries[J]. Electrochemistry Communications, 2011, 13(9): 938–941.

[22] Maazaz A, Delmas C, Hagenmuller P. A Study of the $Na_xTiO_2$ System by Electrochemical Deintercalation[J]. Journal of Inclusion Phenomena, 1983, 1, 45–51.

[23] Yu H, Guo S, Zhu Y. et al., Novel Titanium-Based O3-Type $NaTi_{0.5}Ni_{0.5}O_2$ as a Cathode Material for Sodium Ion Batteries[J]. Chemistry Communication, 2014, 50(4): 457–459.

[24] Gao Y, Wang Z, Lu G. Atomistic Understanding of Structural Evolution, Ion Transport and Oxygen Stability in Layered $NaFeO_2$[J]. Journal of Materials Chemistry A, 2019, 7(6): 2619–2625.

[25] Mortemard D B B, Carlier D, Guignard M. et al, Structural and Electrochemical Characterizations of P2 and New O3-$Na_xMn_{1-y}Fe_yO_2$ Phases Prepared by Auto-Combustion Synthesis for Na-Ion Batteries[J]. Journal of the Electrochemical Society, 2013, 160(4): A569–A574.

[26] Li C, Reid A, Saunders S. Nonstoichiometric Alkali Ferrites and Aluminates in the Systems $NaFeO_2$-$TiO_2$, $KFeO_2$-$TiO_2$ $KAlO_2$-$TiO_2$ and $KAlO_2$-$SiO_2$[J]. Journal of Solid State Chemistry, 1971, 3, 614–620.

[27] Wang X, Liu G, Iwao T. et al., Role of Ligand-to-Metal Charge Transfer in O3-Type $NaFeO_2$–$NaNiO_2$ Solid Solution for Enhanced Electrochemical Properties[J]. The Journal of Physical Chemistry C, 2014, 118(6): 2970–2976.

[28] Komaba S, Yabuuchi N, Nakayama T. et al., Study on the Reversible Electrode Reaction of $Na_{1-x}Ni_{0.5}Mn_{0.5}O_2$ for a Rechargeable Sodium-Ion Battery[J]. Inorganic Chemistry, 2012, 51(11): 6211–6220.

[29] Sathiya M, Hemalatha K, Ramesha K. et al., Synthesis, Structure, and Electrochemical Properties of the Layered Sodium Insertion Cathode Material: $NaNi_{1/3}Mn_{1/3}Co_{1/3}O_2$[J]. Chemistry of Materials, 2012, 24(10): 1846–1853.

[30] Mu L, Xu S, Li Y. et al., Prototype Sodium-Ion Batteries Using an Air-Stable and Co/Ni-Free O3-Layered Metal Oxide Cathode[J]. Advanced Materials, 2015, 27(43): 6928–6933.

[31] Tamaru M, Wang X, Okubo M. et al., Layered $Na_2RuO_3$ as a Cathode Material for Na-Ion Batteries[J]. Electrochemistry Communications, 2013, 33: 23–26.

[32] Chen M, Liu Q, Wang S. et al., High-Abundance and Low-Cost Metal-Based Cathode Materials for Sodium-Ion Batteries: Problems, Progress, and Key Technologies[J]. Advanced Energy Materials, 2019, 9: 1803609.

[33] Delmas C, Maazaz A, Fouassier C. et al., Influence de L'environnement de L'ion Alcalin Sur Sa Mobilite Dans Les Structures a Feuillets $A_x(L_xM_{1-x})O_2$[J]. Material Research Bulletin, 1979, 14: 329–335.

[34] Lee D H, Xu J, Meng Y S. An Advanced Cathode for Na-Ion Batteries with High Rate and Excellent Structural Stability[J]. Physical Chemistry Chemical Physics, 2013, 15(9): 3304–3312.

[35] Chou F C, Abel E T, Cho J H. et al., Electrochemical De-Intercalation, Oxygen Non-Stoichiometry, and Crystal Growth of $Na_xCoO_2-\delta$[J]. Journal of Physics and Chemistry of Solids, 2005, 66(1): 155–160.

[36] Zhu Y E, Qi X, Chen X. et al., P2-$Na_{0.67}Co_{0.5}Mn_{0.5}O_2$ Cathode Materials with Excellent Rate Capability and Cycling Stability for Sodium Ion Batteries[J]. Journal of Materials Chemistry A, 2016, 4: 11103–11109.

[37] Berthelot R, Carlier D, Delmas C. Electrochemical Investigation of the P2–NaxCoO$_2$ Phase Diagram[J]. Nature Materials, 2011, 10(1): 74–80.

[38] Lotfabad E M, Ding J, Cui K. et al, High-Density Sodium and Lithium Ion Battery Anodes from Banana Peels[J]. ACS Nano, 2014, 8(7): 7115–7129.

[39] Ma Y, Doeff M M, Visco S J. et al., Rechargeable Na/Na$_x$CoO$_2$ and Na$_{15}$Pb$_4$/Na$_x$CoO$_2$ Polymer Electrolyte Cells[J]. cheminform, 1993, 140(10): 2726–2731.

[40] Parant J P, Olazcuaga R, Devalette M. et al., Sur Quelques Nouvelles Phases de Formule NaxMnO$_2$ (X≤1)[j]. Journal of Solid State Chemistry, 1971, 3(1): 1–11.

[41] Caballero A, Hernan L, Morales J. et al., Synthesis and Characterization of High-Temperature Hexagonal P2-Na$_{0.6}$MnO$_2$ and Its Electrochemical Behaviour as Cathode in Sodium Cells[J]. Journal of Materials Chemistry, 2002, 12(4): 1142–1147.

[42] Yabuuchi N, Kubota K, Dahbi M. et al., Research Development on Sodium-Ion Batteries[J]. Chemical Reviews, 2014, 114(23): 11636–11682.

[43] Yabuuchi N, Kajiyama M, Iwatate J. et al., P2-Type Na$_x$[Fe$_{1/2}$Mn$_{1/2}$]O$_2$ Made from Earth-Abundant Elements for Rechargeable Na Batteries[J]. Nature Materials, 2012, 11(6): 512–517.

[44] Singh G, Amo J, Galceran M. et al., Structural Evolution during Sodium Deintercalation/Intercalation in Na$_{2/3}$[Fe$_{1/2}$Mn$_{1/2}$]O$_2$[J]. Journal of Materials Chemistry A, 2015, 3: 6954–6961.

[45] Carlier D, Cheng J H, Berthelot R. et al., The P2-Na$_{2/3}$Co$_{2/3}$Mn$_{1/3}$O$_2$ Phase: Structure, Physical Properties and Electrochemical Behavior as Positive Electrode in Sodium Battery[J]. Dalton Transactions, 2011, 40(36): 9306.

[46] Yang P, Zhang C, Li M. P2-NaCo$_{0.5}$Mn$_{0.5}$O$_2$ as a Positive Electrode Material for Sodium-Ion Batteries[J]. Chemphyschem, 2015. 16(16): 3408–3412.

[47] Chen X, Zhou X, Hu M. et al., Stable Layered P3/P2 Na Co Mn O Cathode Materials for Sodium-Ion Batteries[J]. Journal of Materials Chemistry A, 2015, 3(41): 20708–20714.

[48] Lu Z, Dahn J. In Situ X-ray Diffraction Study of P2-Na$_{2/3}$[Ni$_{1/3}$Mn$_{2/3}$]O$_2$[J]. Journal of the Electrochemical Society, 2001, 148(11): A1225.

[49] Zhou Q, Cai W, Zhang Y. et al., Electricity Generation from Corn Cob Char though a Direct Carbon Solid Oxide Fuel Cell[J]. Biomass & Bioenergy, 2016, 91: 250–258.

[50] Singh G, Tapia N, Lopez D A. et al., High Voltage Mg-Doped Na$_{0.67}$Ni$_{0.3-x}$Mg$_x$Mn$_{0.7}$O$_2$ (X=0.05, 0.1) Na-Ion Cathodes with Enhanced Stability and Rate Capability[J]. Chemistry of Materials, 2016, 28(14): 5087–5094.

[51] Clément R J, Xu J, Middlemiss D S. et al., Direct Evidence for High Na$^+$ Mobility and High Voltage Structural Processes in P2-Na$_x$[Li$_y$Ni$_z$Mn$_{1-y-z}$]O$_2$ (X, Y, Z ≤ 1) Cathodes from Solid-State NMR and DFT Calculations[J]. Journal of Materials Chemistry A, 2017, 5: 4129–4143.

[52] Singh G, Acebedo B, Cabanas M C. et al., An Approach to Overcome First Cycle Irreversible Capacity in P2-Na$_{2/3}$[Fe$_{1/2}$Mn$_{1/2}$]O$_2$[J]. Electrochemistry Communications, 2013, 37: 61–63.

[53] Nazar L F, Talaie E, Duffort V. et al., Structure of the High Voltage Phase of Layered P2-Na$_{2/3-x}$[Mn$_{1/2}$Fe$_{1/2}$]O$_2$ and the Positive Effect of Ni Substitution on Its Stability[J]. Energy & Environmental Science, 2015, 8(8): 2512–2523.

[54] Yabuuchi N, Hara R, Kubota K. et al., New Electrode Material for Rechargeable Sodium Batteries: P2-type Na$_{2/3}$[Mg$_{0.28}$Mn$_{0.72}$]O$_2$ with Anomalously High Reversible Capacity[J]. Journal of Materials Chemistry A, 2014, 2(40): 16851–16855.

[55] Yabuuchi N, Hara R, Kajiyama M. et al., New O2/P2-Type Li-Excess Layered Manganese Oxides as Promising Multi-Functional Electrode Materials for Rechargeable Li/Na Batteries[J]. Advanced Energy Materials, 2014, 4(13): 13072–13072.

[56] Guo S, Liu P, Yu H. et al., A Layered P2- and O3-Type Composite as A High-Energy Cathode for Rechargeable Sodium-Ion Batteries[J]. Angewandte Chemie International Edition, 2015, 54(20): 5894–5899.

[57]  Komaba S, Mikumo T, Yabuuchi N. et al., Electrochemical Insertion of Li and Na Ions into Nanocrystalline $Fe_3O_4$ and $\alpha$-$Fe_2O_3$ for Rechargeable Batteries[J]. Journal of the Electrochemical Society, 2010, 1(157): A60–A65.

[58]  Komaba S, Mikumo T, Ogata A. Electrochemical Activity of Nanocrystalline $Fe_3O_4$ in Aprotic Li and Na Salt Electrolytes[J]. Electrochemistry Communications, 2008, 10(9): 1276–1279.

[59]  Wei W, Cui X, Chen W. et al., Manganese Oxide-Based Materials as Electrochemical Supercapacitor Electrodes[J]. Chemical Society Reviews, 2011, 40(3): 1697–1721.

[60]  Brousse T, Toupin M, Dugas R. et al., Crystalline $MnO2$ as Possible Alternatives to Amorphous Compounds in Electrochemical Supercapacitors[J]. Journal of the Electrochemical Society, 2006, 153(12): A2171–A2179.

[61]  Sauvage F, Laffont L, Tarascon J M. et al., Study of the Insertion/Deinsertion Mechanism of Sodium into $Na_{0.44}MnO_2$[J]. Cheminform, 2007, 38(28): 3289–3294.

[62]  Kim H, Kim D, Hwa D. Ab Initio Study of the Sodium Intercalation and Intermediate Phases in $Na_{0.44}MnO_2$ for Sodium-Ion Battery[J]. Chemistry of Materials, 2012, 24(6): 1205–1211.

[63]  Guo S, Yi J, Sun Y. et al., Recent Advances in Titanium-Based Electrode Materials for Stationary Sodium-Ion Batteries[J]. Energy & Environmental Science, 2016, 9(10): 2978–3006.

[64]  Whitacre J F, Tevar A, Sharma S. $Na_4Mn_9O_{18}$ as a Positive Electrode Material for an Aqueous Electrolyte Sodium-Ion Energy Storage Device[J]. Electrochemistry Communications, 2010, 12(3): 463–466.

[65]  Kim D J, Ponraj R, Kannan A G. et al., Diffusion Behavior of Sodium Ions in $Na_{0.44}MnO_2$ in Aqueous and Non-Aqueous Electrolytes[J]. Journal of Power Sources, 2013, 244: 758–763.

[66]  Spahr M E, Novák P, Haas O. et al., Electrochemical Insertion of Lithium, Sodium, and Magnesium in Molybdenum(VI) Oxide[J]. Journal of Power Sources, 1995, 54(2): 346–351.

[67]  Epavcevic S, Xiong H, Stamenkovic V R. et al., Nanostructured Bilayered Vanadium Oxide Electrodes for Rechargeable Sodium-Ion Batteries[J]. ACS Nano, 2011, 6(1): 530–538.

[68]  Ni Q, Bai Y, Wu F. et al., Polyanion-Type Electrode Materials for Sodium-Ion Batteries[J]. Advanced Science, 2017, 4(3): 1600275.

[69]  Li C, Miao X, Chu W. et al., Retracted Article: Hollow Amorphous $NaFePO_4$ Nanospheres as a High-Capacity and High-Rate Cathode for Sodium-Ion Batteries[J]. Journal of Materials Chemistry A, 2015, 3(16): 8265–8271.

[70]  Zaghib K, Trottier J, Hovington P. et al., Characterization of Na-Based Phosphate as Electrode Materials for Electrochemical Cells[J]. Journal of Power Sources, 2011, 196(22): 9612–9617.

[71]  Moreau P, Guyomard D, Gaubicher J. et al., Structure and Stability of Sodium Intercalated Phases in Olivine $FePO_4$[J]. Chemistry of Materials, 2010, 22(14): 4126–4128.

[72]  Fernandez A J, Saurel D, Acebedo B. et al., Electrochemical Characterization of $NaFePO_4$ as Positive Electrode in Aqueous Sodium-Ion Batteries[J]. Journal of Power Sources, 2015, 291: 40–45.

[73]  Lee K T, Ramesh T N, Nan F. et al., Topochemical Synthesis of Sodium Metal Phosphate Olivines for Sodium-Ion Batteries[J]. Chemistry of Materials, 2011, 23(16): 3593–3600.

[74]  Jian Z, Zhao L, Pan H. et al., Carbon Coated $Na_3V_2(PO_4)_3$ as Novel Electrode Material for Sodium Ion Batteries[J]. Electrochemistry Communications, 2012, 14(1): 86–89.

[75]  Fang Y, Xiao L, Ai X. et al., Hierarchical Carbon Framework Wrapped $Na_3V_2(PO_4)_3$ as a Superior High-Rate and Extended Lifespan Cathode for Sodium-Ion Batteries[J]. Advanced Materials, 2015, 27(39): 5895.

[76]  Aragón M J, Lavela P, Alcántara R. et al., Effect of Aluminum Doping on Carbon Loaded $Na_3V_2(PO_4)_3$ as Cathode Material for Sodium-Ion Batteries[J]. Electrochimica Acta, 2015, 180: 824–830.

[77] Barpanda P, Nishimura S I, Yamada A. High-Voltage Pyrophosphate Cathodes[J]. Advanced Energy Materials, 2012, 2(7): 1–19.

[78] Barpanda P, Ye T, Nishimura S. et al., Sodium Iron Pyrophosphate: A Novel 3.0V Iron-Based Cathode for Sodium-Ion Batteries[J]. Electrochemistry Communications, 2012, 24: 116–119.

[79] Barpanda P, Oyama G, Nishimura S. et al., A 3.8V Earth-Abundant Sodium Battery Electrode[J]. Nature Communications, 2014, 5: 1.

[80] Barpanda P, Chotard J, Recham N. et al., Structural, Transport, and Electrochemical Investigation of Novel $AMSO_4F$ (A = Na, Li; M = Fe, Co, Ni, Mn) Metal Fluorosulphates Prepared Using Low Temperature Synthesis Routes[J]. Inorganic Chemistry, 2010, 49(16): 7401–7413.

[81] Wang H, Hu P, Yang J. et al., Renewable-Juglone-Based High-Performance Sodium-Ion Batteries[J]. Advanced Materials, 2015, 27(14): 2348–2354.

[82] Wood S M, Eames C, Kendrick E. et al., Sodium Ion Diffusion and Voltage Trends in Phosphates $Na_4M_3(PO_4)_2P_2O_7$ (M = Fe, Mn, Co, Ni) for Possible High-Rate Cathodes[J]. The Journal of Physical Chemistry C, 2015, 119(28): 15935–15941.

[83] Kim H, Park I, Lee S. et al., Understanding the Electrochemical Mechanism of the New Iron-Based Mixed-Phosphate $Na_4Fe_3(PO_4)_2(P_2O_7)$ in a Na Rechargeable Battery[J]. Chemistry of Materials, 2013, 25(18): 3614–3622.

[84] Nose M, Nakayama H, Nobuhara K. et al., $Na_4Co_3(PO_4)_2P_2O_7$: A Novel Storage Material for Sodium-Ion Batteries[J]. Journal of Power Sources, 2013, 234: 175–179.

[85] Nose M, Shiotani S, Nakahara K. et al., $Na_4Co_{2.4}Mn_{0.3}Ni_{0.3}(PO_4)_2P_2O_7$: High Potential and High Capacity Electrode Material for Sodium-Ion Batteries[J]. Electrochemistry Communications, 2013, 34: 266–269.

[86] Sauvage F, Quarez E, Tarascon J M. et al., Crystal Structure and Electrochemical Properties Vs. $Na^+$ of the Sodium Fluorophosphate $Na_{1.5}VOPO_4F_{0.5}$[J]. Solid State Sciences, 2006, 8(10): 1215–1221.

[87] Chihara K, Kitajou A, Gocheva I D. et al., Cathode Properties of $Na_3M_2(PO_4)_2F_3$ [M=ti, Fe, V] for Sodium-Ion Batteries[J]. Journal of Power Sources, 2013, 227: 80–85.

[88] Yakubovich O, Karimova O, Mel'nikov O. The Mixed Anionic Framework in the Structure of $Na_2\{MnF[PO_4]\}$[J]. Inorganic Compounds, 1997(C53): 395–397.

[89] Chen H, Hao Q, Zivkovic O. et al., Sidorenkite ($Na_3mnpo_4co_3$): A New Intercalation Cathode Material for Na-Ion Batteries[J]. Chemistry of Materials, 2013, 25(14): 2777–2786.

[90] Sun Q, Ren Q Q, Fu Z W. NASICON-type $Fe_2(MoO_4)_3$ Thin Film as Cathode for Rechargeable Sodium Ion Battery[J]. Electrochemistry Communications, 2012, 2, 145–148.

[91] Uebou Y, Okada S, Yamaki J I. Electrochemical Insertion of Lithium and Sodium into $(Moo_2)_2p_2o_7$[j]. Journal of Power Sources, 2003, 115(1): 119–124.

[92] Itaya K, Uchida I, Neff V D. Electrochemistry of Polynuclear Transition Metal Cyanides: Prussian Blue and Its Analogues[J]. Accounts of Chemical Research, 1986, 19(6): 162–168.

[93] Lee H, Wang R Y, Pasta M. et al., Manganese Hexacyano manganate Open Framework as a High-capacity Positive Electrode Material for Sodium-ion Batteries[J]. Nature Communications, 2014, 5(1): 5280.

[94] Lu Y, Wang L, Cheng J. et al., Prussian Blue: A New Framework of Electrode Materials for Sodium Batteries[J]. Chemical Communications, 2012, 48(52): 6544.

[95] Qian J, Zhou M, Cao Y. et al., Nanosized $Na_4Fe(CN)_6$/C Composite as a Low-cost and High-rate Cathode Material for Sodium-ion Batteries[J]. Advanced Energy Materials, 2012, 2(4): 410–414.

[96] Buser H, Schwarzenbach D, Petter W. et al., The Crystal Structure of Prussian Blue: $Fe_4[Fe(CN)_6]_3*xH_2O$[J]. Inorganic Chemistry, 1977, 16(11): 2705.

[97] Shibata T, Moritomo Y. Ultrafast Cation Intercalation in Nanoporous Nickel Hexacyanoferrate[J]. Chemical Communications (Cambridge, England), 2014, 50(85): 12941–12943.

[98] Takachi M, Fukuzumi Y, Moritomo Y. $Na^+$ Diffusion Kinetics in Nanoporous Metal-Hexacyanoferrates[J]. Dalton Transactions, 2016, 45(2): 458–461.

[99] Wu X, Wu C, Wei C. et al., Highly Crystallized $Na_2CoFe(CN)_6$ with Suppressed Lattice Defects as Superior Cathode Material For sodium-ion Batteries[J]. ACS Applied Materials & Interfaces, 2016, 8(8): 5393–5399.

[100] Zhang H, Yang J, Gao J. Effects of Transverse Profile of Pump Field on Second Harmonic Generation in Periodic Nonlinear Materials[J]. Chinese Physics, 2003, 12(5): 518–523.

[101] Asakura D, Okubo M, Mizuno Y. et al., Fabrication of a Cyanide-bridged Coordination Polymer Electrode for Enhanced Electrochemical Ion Storage Ability[J]. The Journal of Physical Chemistry C, 2012, 116(15): 8364–8369.

[102] You Y, Wu X, Yin Y. et al., A Zero-strain Insertion Cathode Material of Nickel Ferricyanide for Sodium-ion Batteries[J]. Journal of Materials Chemistry A, 2013, 1: 14061.

[103] Wu X, Cao Y, Ai X. et al., A Low-cost and Environmentally Benign Aqueous Rechargeable Sodium-ion Battery Based on $NaTi_2(PO4)_3$–$Na_2NiFe(CN)_6$ Intercalation Chemistry[J]. Electrochemistry Communications, 2013, 31: 145–148.

[104] Wessells C D, Huggins R A, Cui Y. Copper Hexacyanoferrate Battery Electrodes with Long Cycle Life and High Power[J]. Nature Communications, 2011, 2(1): 1–5.

[105] Yu S, Shokouhimehr M, Hyeon T. et al., Iron Hexacyanoferrate Nanoparticles as Cathode Materials for Lithium and Sodium Rechargeable Batteries[J]. ECS Electrochemistry Letters, 2013, 2(4): A39–A41.

[106] Imanishi N, Morikawa T, Kondo J. et al., Lithium Intercalation Behavior into Iron Cyanide Complex as Positive Electrode of Lithium Secondary Battery[J]. Journal of Power Sources, 1999, 79: 215–219.

[107] Neff V D. Electrochemical Oxidation and Reduction of Thin Films of Prussian Blue[J]. Journal of the Electrochemical Society, 1978, 125(6): 886.

[108] Ware M. Prussian Blue: Artists' pigment and Chemists' sponge[J]. Journal of Chemical Education, 2008, 85(5): 612.

[109] Pramudita J, Schmid S, Godfrey T. et al., Sodium Uptake in Cell Construction and Subsequent in Operando Electrode Behaviour of Prussian Blue Analogues, $Fe[Fe(CN)_6]_{1-x}·yH_2O$ and $FeCo(CN)_6$[J]. Physical Chemistry Chemical Physics, 2014, 16(44): 24111–24636.

[110] You Y, Wu X, Yin Y. et al., A Zero-strain Insertion Cathode Material of Nickel Ferricyanide for Sodium-ion Batteries[J]. Journal Materials Chemistry A, 2013, 1(45): 14033–14410.

[111] Lee H, Kim Y, Park J. et al., Sodium Zinc Hexacyanoferrate with a Well-defined Open Framework as a Positive Electrode for Sodium Ion Batteries[J]. Chemical Communications, 2012, 48(67): 8416.

[112] Tuo J, Xie H, Cai Z. et al., The Electrochemical Performance of $Cu_3[Fe(CN)_6]_2$ as a Cathode Material for Sodium-ion Batteries[J]. Materials Research Buttetin, 2016, 86: 194–200.

[113] Wu X, Deng W, Qian J. et al., Single-crystal $FeFe(CN)_6$ Nanoparticles: A High Capacity and High Rate Cathode for Na-ion Batteries[J]. Journal of Materials Chemistry A, 2013, 1(35): 10130.

[114] Wessells C D, Peddad S V, Huggins R A. et al., Nickel Hexacyanoferrate Nanoparticle Electrodes for Aqueous Sodium and Potassium Ion Batteries[J]. Nano Letters, 2011, 11(12): 5421–5425.

[115] Wessells C D, Mcdowell M T, Peddada S V. et al., Tunable Reaction Potentials in Open Framework Nanoparticle Battery Electrodes for Grid-scale Energy Storage[J]. ACS Nano, 2012, 6(2): 1688–1694.

[116] Doucette R T, Mcculloch M D. A Comparison of High-speed Flywheels, Batteries, and Ultracapacitors on the Bases of Cost and Fuel Economy as the Energy Storage System in A Fuel Cell Based Hybrid Electric Vehicle[J]. Journal of Power Sources, 2011, 196(3): 1163–1170.

[117] Wu X, Sun M, Shen Y. et al., Energetic Aqueous Rechargeable Sodium-ion Battery Based on $Na_2CuFe(CN)_6$-$NaTi_2(PO4)_3$ Intercalation Chemistry[J]. ChemSusChem, 2014, 7(2): 407–411.

[118] Yu S H, Shokouhimehr M, Hyeon T. et al., Iron Hexacyanoferrate Nanoparticles as Cathode Materials for Lithium and Sodium Rechargeable Batteries[J]. ECS Electrochemistry Letters, 2013, 2(4): A39–A41.

[119] Slater M D, Kim D, Lee E. et al., Sodium-ion Batteries[J]. Advanced Functional Materials, 2013, 23(8): 947–958.

[120] Fidalgo L, Maerkl J. High-quality Prussian Blue Crystals as Superior Cathode Materials for Room-temperature Sodium-ion Batteries[J]. Energy & Environmental Science, 2010, 1, 1–6.

[121] Li W, Chou S, Wang J. et al., Facile Method to Synthesize Na-enriched $Na_{1+x}FeFe(CN)_6$ Frameworks as Cathode with Superior Electrochemical Performance for Sodium-ion Batteries[J]. Chemistry of Materials, 2015, 27(6): 1997–2003.

[122] You Y, Yu X, Yin Y. et al., Sodium Iron Hexacyanoferrate with High Na Content as a Na-rich Cathode Material for Na-ion Batteries[J]. Nano Research, 2015, 8(1): 117–128.

[123] Liu Y, Qiao Y, Zhang W. et al., Sodium Storage in Na-rich $Na_xFeFe(CN)_6$ Nanocubes[J]. Nano Energy, 2015, 12: 386–393.

[124] Wang L, Song J, Qiao R. et al., Rhombohedral Prussian White as Cathode for Rechargeable Sodium-ion Batteries[J]. Journal of the American Chemical Society, 2015, 137(7): 2548–2554.

[125] Sottmann J, Bernal F L M, Yusenko K V. et al., In Operando Synchrotron XRD/XAS Investigation of Sodium Insertion into the Prussian Blue Analogue Cathode Material $Na_{1.32}Mn[Fe(CN)_6]_{0.83} \cdot 2H_2O$[J]. Electrochimica Acta, 2016, 200: 305–313.

[126] Matsuda T, Takachi M, Moritomo Y. ChemInform Abstract: A Sodium Manganese Ferrocyanide Thin Film for Na-ion Batteries.[J]. ChemInform, 2013, 44, 23.

[127] Wang L, Lu Y, Liu J. et al., A Superior Low-cost Cathode for A Na-Ion Battery[J]. Angewandte Chemie International Edition, 2013, 52(7): 1964–1967.

[128] Song J, Wang L, Lu Y. et al., Removal of Interstitial $H_2O$ in Hexacyanometallates for a Superior Cathode of a Sodium-ion Battery[J]. Journal of the American Chemical Society, 2015, 137(7): 2658–2664.

[129] Pasta M, Wessells C D, Liu N. et al., Full Open-framework Batteries for Stationary Energy Storage[J]. Nature Communications, 2014, 5(1): 3007.

[130] Li W, Chou S, Wang J. et al., Multifunctional Conducing Polymer Coated $Na_{1+x}MnFe(CN)_6$ Cathode for Sodium-ion Batteries with Superior Performance via a Facile and One-step Chemistry Approach[J]. Nano Energy, 2015, 13: 200–207.

[131] Nakamoto K, Sakamoto R, Ito M. et al., Effect of Concentrated Electrolyte on Aqueous Sodium-ion Battery with Sodium Manganese Hexacyanoferrate Cathode[J]. Electrochemistry, 2017, 85(4): 179–185.

[132] Takachi M, Matsuda T, Moritomo Y. Cobalt Hexacyanoferrate as Cathode Material for $Na^+$ Secondary battery[J]. Applied Physics Express, 2013, 6, 25802.

[133] Song J, Wang L, Lu Y. et al., Removal of Interstitial $H_2O$ in Hexacyanometallates for a Superior Cathode of a Sodium-ion Battery[J]. Journal of the American Chemical Society, 2015, 137(7): 2658–2664.

[134] Yang Y, Brownell C, Sadrieh N. et al., Quantitative Measurement of Cyanide Released from Prussian Blue[J]. Clinical Toxicology, 2010, 45(7): 776–781.

[135] Wu X, Sun M, Guo S. et al., Vacancy-free Prussian Blue Nanocrystals with High Capacity and Superior Cyclability for Aqueous Sodium Ion Batteries[J]. Chemistry Nano Materials, 2015, 1: 188–193.

[136] Senguttuvan P, Rousse G, Arroyo Y. et al., Low-Potential Sodium Insertion in a NASICON-Type Structure through the Ti(III)/Ti(II) Redox Couple[J]. Journal of the American Chemical Society, 2013, 135(10): 3897–3903.

[137] Fernández A J, Piernas M J, Castillo E. et al., Electrochemical Characterization of $NaFe_2(CN)_6$ Prussian Blue as Positive Electrode for Aqueous Sodium-ion Batteries[J]. Electrochimica Acta, 2016, 210: 352–357.

[138] Zhao F, Wang Y, Xu L. et al., Cobalt Hexacyanoferrate Nanoparticles as a High-rate and Ultra-stable Supercapacitor Electrode Material[J]. ACS Applied Materials & Interfaces, 2014, 6(14): 11007–11012.

[139] Wu X, Luo Y, Sun M. et al., Low-defect Prussian Blue Nanocubes as High Capacity and Long Life Cathodes for Aqueous Na-ion Batteries[J]. Nano Energy, 2015, 13: 117–123.

[140] Zhou L, Yang Z, Li C. et al., Prussian Blue as Positive Electrode Material for Aqueous Sodium-ion Capacitor with Excellent Performance[J]. RSC Advances, 2016, 6(111): 109340–109345.

[141] Lee J, Ali G, Kim D H. et al., Metal-organic Framework Cathodes Based on a Vanadium Hexacyanoferrate Prussian Blue Analogue for High-performance Aqueous Rechargeable Batteries[J]. Advanced Energy Materials, 2017, 7(2): 1601491.

[142] Paulitsch B, Yun J, Bandarenka A S. Electrodeposited $Na_2VO_x$ [Fe(cn)$_6$] Films as a Cathode Material for Aqueous Na-ion Batteries[J]. ACS Applied Materials & Interfaces, 2017, 9(9): 8107–8112.

[143] Yang D, Xu J, Liao X. et al., Structure Optimization of Prussian Blue Analogue Cathode Materials for Advanced Sodium Ion Batteries[J]. Chemistry Communication, 2014, 50(87): 13377–13380.

[144] Xie M, Xu M, Huang Y. et al., $Na_2Ni_xCo_{1-x}Fe(CN)_6$: A Class of Prussian Blue Analogs with Transition Metal Elements as Cathode Materials for Sodium Ion Batteries[J]. Electrochemistry Communications, 2015, 59: 91–94.

[145] Yu S, Li Y, Lu Y. et al., A Promising Cathode Material of Sodium Iron-nickel Hexacyanoferrate for Sodium Ion Batteries[J]. Journal of Power Sources, 2015, 275(1): 45–49.

[146] Kim D, Zakaria M, Park M. et al., Dual-textured Prussian Blue Nanocubes as Sodium Ion Storage Materials[J]. Electrochimica Acta, 2017, 54: 4.

[147] Huang Y, Xie M, Wu F. et al., A Novel Border-rich Prussian Blue Synthetized by Inhibitor Control as Cathode for Sodium Ion Batteries[J]. Nano Energy, 2017, 39: 273–283.

[148] Wan M, Tang Y, Wang L. et al., Core-shell Hexacyanoferrate for Superior Na-ion Batteries[J]. Journal of Power Sources, 2016, 329: 290–296.

[149] Prabakar S J R, Jeong J, Pyo M. Highly Crystalline Prussian Blue/Graphene Composites for High-rate Performance Cathodes in Na-ion Batteries[J]. RSC Advances, 2015, 5(47): 37545–37552.

[150] You Y, Yao H, Xin S. et al., Subzero-Temperature Cathode for a Sodium-ion battery[J]. Advanced Materials, 2016, 28(33): 7243–7248.

[151] Chou S, Wang J, Liu H. et al., Rapid Synthesis of $Li_4Ti_5O_{12}$ Microspheres as Anode Materials and Its Binder Effect for Lithium-ion Battery[J]. The Journal of Physical Chemistry C, 2011, 115 (32): 16220–16227.

[152] Huang Y, Xie M, Wang Z. et al., A Chemical Precipitation Method Preparing Hollow-core-shell Heterostructures Based on the Prussian Blue Analogs as Cathode for Sodium-ion Batteries[J]. Small, 2018, 14(28): 1801246.

[153] Tang Y, Zhang W, Xue L. et al., Polypyrrole-promoted Superior Cyclability and Rate Capability in $Na_xFe[Fe(CN)_6]$ Cathode for Sodium-ion Batteries[J]. Journal of Materials Chemistry A, 2016.

[154] Li L, Zhu J, Xu M. et al., In-Situ Engineering toward Core Regions: A Smart Way to Make Applicable $FeF_3$@Carbon Nanoreactor Cathodes for Li-Ion Batteries[J]. ACS Applied Materials & Interfaces, 2017, 9(21): 17992–18000.

[155] Nishijima M, Gocheva I D, Okada S. et al., Cathode Properties of Metal Trifluorides in Li and Na Secondary Batteries[J]. Journal of Power Sources, 2009, 190(2): 558–562.

[156] Wei S, Wang X, Liu M. et al., Spherical $FeF_3 \cdot 0.33H_2O$/MWCNTs Nanocomposite with Mesoporous Structure as Cathode Material of Sodium Ion Battery[J]. Journal of Energy Chemistry, 2017, 27(2): 578–581.

[157] Zhang R, Wang X, Wang X. et al., Iron Fluoride Packaged into 3D Order Mesoporous Carbons as High-Performance Sodium-Ion Battery Cathode Material[J]. Journal of the Electrochemical Society, 2018, 165(2): A89–A96.

[158] Cao D, Yin C, Shi D. et al., Cubic Perovskite Fluoride as Open Framework Cathode for Na-Ion Batteries[J]. Advanced Functional Materials, 2017, 27: 1701130.

[159] Liu M, Wang X, Wei S. et al., Cr-Doped $Fe_2F_5 \cdot H_2O$ with Open Framework Structure as a High Performance Cathode Material of Sodium-Ion Batteries[J]. Electrochimica Acta, 2018, 269: 479–489.

[160] Jiang M, Wang X, Shen Y. et al., New Iron-Based Fluoride Cathode Material Synthesized by Non-Aqueous Ionic Liquid for Rechargeable Sodium Ion Batteries[J]. Electrochimica Acta, 2015, 186: 7–15.

[161] Jiang M, Wang X, Hu H. et al., In Situ Growth and Performance of Spherical $Fe_2F_5$ Center Dot $H_2O$ Nanoparticles in Multi-Walled Carbon Nanotube Network Matrix as Cathode Material for Sodium Ion Batteries[J]. Journal of Power Sources, 2016, 316: 170–175.

[162] Shea J J, Luo C. Organic Electrode Materials for Metal Ion Batteries[J]. ACS Applied Materials & Interfaces, 2020, 12(5): 5361–5380.

[163] Chihara K, Chujo N, Kitajou A. et al., Cathode Properties of $Na_2C_6O_6$ for Sodium-Ion Batteries[J]. Electrochimica Acta, 2013, 110: 240–246.

[164] Lee M, Hong J, Lopez J. et al., High-Performance Sodium–Organic Battery by Realizing Four-Sodium Storage in Disodium Rhodizonate[J]. Nature Energy, 2017.

[165] Wang S, Wang L, Zhu Z. et al., All Organic Sodium-Ion Batteries with $Na_4C_8H_2O_6$[J]. Angewandte Chemie International Edition, 2014, 53(23): 5892–5896.

[166] Zhou G, Miao Y E, Wei Z. et al., Bioinspired Micro/Nanofluidic Ion Transport Channels for Organic Cathodes in High-Rate and Ultrastable Lithium/Sodium-Ion Batteries[J]. Advanced Functional Materials, 2018, 28(52): 1804629.

[167] Harish B, Dijo D, Kalaivanan N. et al., A Polyimide Based All-Organic Sodium Ion Battery[J]. Journal of Materials Chemistry A, 2015, 3(19): 10453–10458.

# Chapter 4
# Anode material for sodium-ion battery

Electrode materials directly determine the electrochemical performance of Na-ion batteries, and the development of electrode materials with excellent performance and low price is the key to promote the early commercialization of sodium-ion batteries. At present, the research on sodium storage cathode materials has made some progress in the direction of layered oxides, polyanionic compounds, Prussian blue, and analogues [1–5]. In the case of sodium storage anode materials, the criteria for the selection of anode materials are also more stringent due to the lower voltage requirements (for $Na^+/Na$) of being not too close to the voltage of $Na^+/Na$, to avoid dendrite generation. Most of the current literature on the electrochemical performance of the negative electrode focuses more on the discharge process than on the charging process. However, considering the possibility of practical applications, the charging voltage of the half-cell, composed of sodium storage cathode material as the positive electrode and sodium metal, cannot be too high; too high a charging voltage will result in a very low voltage of the full cell and thus is not appropriate for practical applications [6]. The sodium storage anode materials studied so far are mainly classified into carbon-based anode materials, titanium-based anode materials, conversion-responsive anode materials, and intermetallic compound anode materials, as shown in Figure 4.1. The average voltage and specific capacity relationships of each type of materials are shown in Figure 4.2 [7]. Among them, the carbon-based anode materials and titanium-based anode materials store sodium through the occurrence of an embedding reaction, and their specific capacities are low, but the structural stability and cycling stability of the materials are excellent during the cycling process. Conversion-responsive anode materials first undergo conversion reactions and continue to undergo alloying reactions if the central transition metal atom is active, while intermetallic compound anode materials undergo alloying reactions [8]. Currently, hard carbon anode materials have relatively high compatibility with both ester and ether electrolytes and have been commercially applied, but their specific capacity needs to be further improved. Therefore, researchers have shifted their attention to conversion-responsive anode materials and intermetallic compound anode materials with high specific capacities. Electrode materials, based on conversion reaction and alloying reaction, have high specific capacity because they can make full use of the valence state of metal elements or can form intermetallic compounds, but they are accompanied by huge volume expansion during cycling, which causes crushing and collapse of the electrode materials [9]. The reaction kinetics of these two electrode materials are also relatively sluggish; so further improvement of sodium storage kinetic processes is needed. Besides, some low-cost and easy-to-synthesize organic electrode materials have been developed, and such organic materials are made into flexible electrodes, which can be used in wearable devices and other fields in the future. In this chapter, the research

https://doi.org/10.1515/9783110749069-004

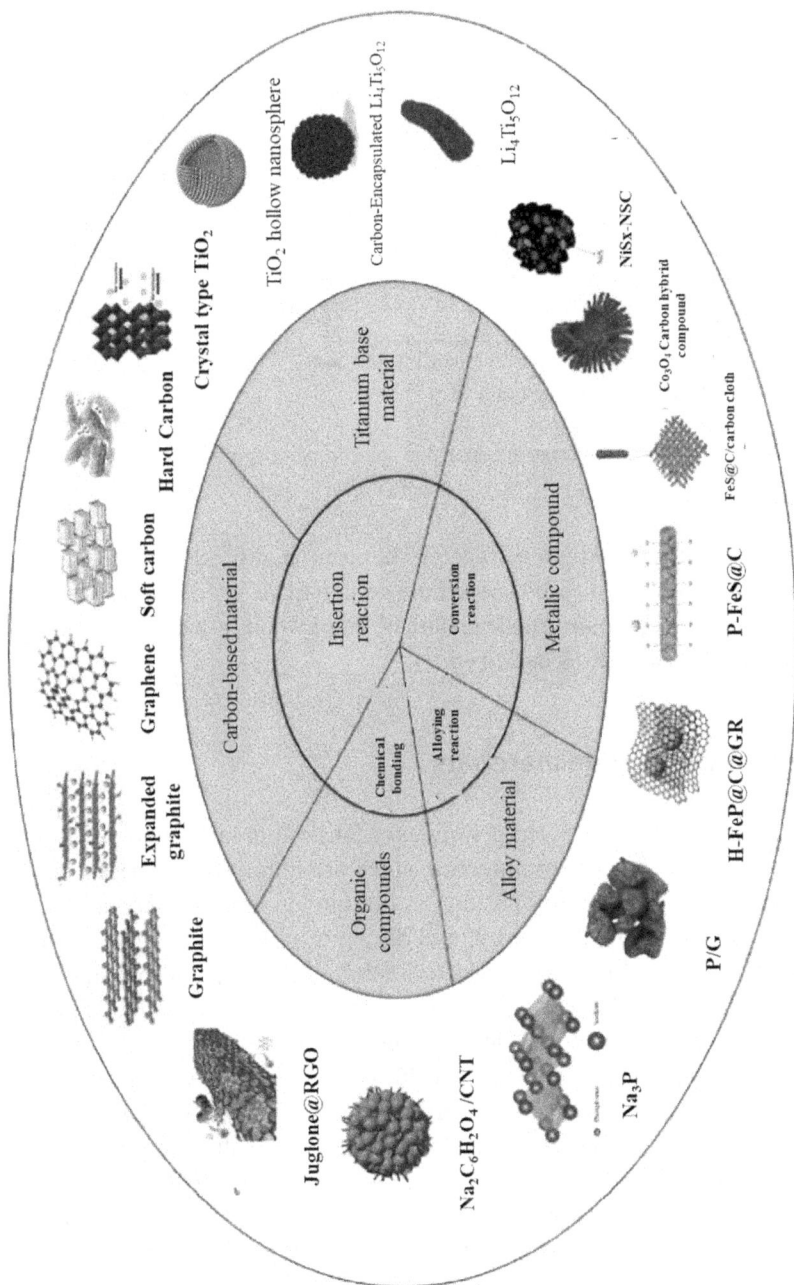

**Figure 4.1:** Classification of anode materials for sodium-ion batteries.

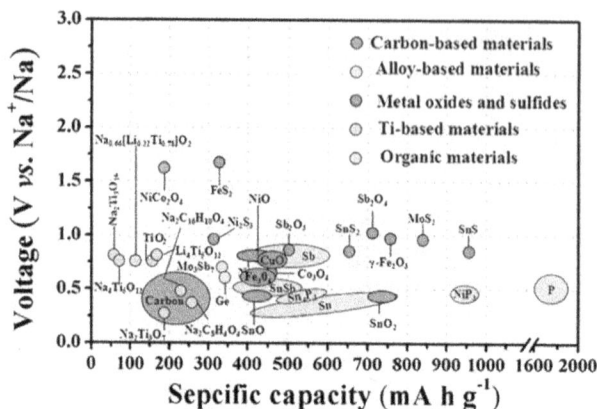

**Figure 4.2:** Relationship between average potential and specific capacity of anode materials in sodium-ion batteries [7].

progress of the structure, sodium storage mechanism, modification methods, and electrochemical properties of carbon-based anode materials, titanium-based anode materials, conversion-responsive anode materials, intermetallic anode materials, and other anode materials are mainly described.

## 4.1 Carbon-based anode materials

As an anode material with excellent performance for lithium-ion batteries, carbon-based anode materials have been attracting much attention. Carbon-based anode materials occupy a very important position in lithium-ion battery anode materials due to their wide availability, low price, and easy preparation, and hence become the first choice of anode materials for sodium-ion batteries [10]. According to the different microstructures of materials, carbon-based anode materials are mainly divided into graphitized carbon and non-graphitized carbon. Graphitized carbon, such as graphite, has a high volumetric specific capacity and good cycling performance and is therefore used as anode material for commercial lithium-ion batteries. However, the large ionic radius of sodium ions makes it difficult to be embedded in graphite; so graphite is still difficult to be used as an anode material in sodium-ion batteries [11]. Non-graphitized carbon, such as graphene, can provide abundant sodium storage sites for sodium ions due to its large specific surface area and more surface defects; however, the use of graphene in sodium-ion battery anode materials is limited by its expensive price, difficulty in preparation, and higher reaction potential. Currently, the ideal carbon-based anode material is hard carbon material. Hard carbon material belongs to non-graphitized carbon, which has higher capacity and a lower working potential, but it still needs to be studied thoroughly due to its debatable sodium storage mechanism.

### 4.1.1 Graphite

Graphite is an isomer of carbon, gray-black, opaque solid, with a density of 2.25 g · cm$^{-3}$, a melting point of 3,652 °C and a boiling point of 4,827 °C. It is chemically stable, corrosion-resistant, does not react easily with acids and alkalis, etc. It burns in oxygen at 687 °C to produce carbon dioxide. It can be oxidized by strong oxidizing agents, such as concentrated nitric acid, potassium permanganate, etc. It can be used as an anti-wear agent, lubricant, etc. Graphite with high purity can be used as neutron reducer in an atomic reactor, and also used in manufacturing crucible, electrode, electric brush, dry cell, graphite fiber, heat exchanger, cooler, electric arc furnace, arc lamp, pencil core of pencil, etc. Graphite structure is the crystal structure of mineral graphite in the form of hexagonal layers. In graphite crystals, carbon atoms in the same layer form covalent bonds by sp$^2$ hybridization, and each carbon atom is connected to three other atoms by three covalent bonds, as shown in Figure 4.3, and the typical layer spacing of graphite is 3.4 Å.

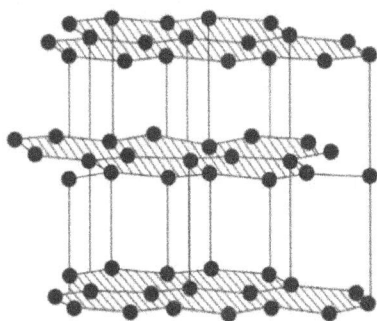

**Figure 4.3:** Structure of graphite.

Graphite is widely used in anode materials for lithium-ion battery, and its theoretical specific capacity is 372 mA h g$^{-1}$ [12]. During the electrochemical reduction process, Li$^+$ are embedded between graphite layers to form the intergraphite compound, LiC$_x$, where $x \leq 6$. Although the physicochemical properties of sodium and lithium are similar, the capacity of graphite-embedded sodium is actually very low and cannot be used directly as anode materials in sodium-ion batteries because sodium cannot form a stable binary compound, Na-C, with graphite. The already reported graphite sodium storage product, NaC$_{70}$, corresponds to a theoretical specific capacity of 31 mA h g$^{-1}$ [13]. First-principles calculations of the formation energy of sodium-graphite interlayer compounds (Na-GIC) show that Na is difficult to embed in graphite due to Na-GIC energy instability, and Na-GIC thermodynamic instability does not favor the formation of

NaC$_6$ and NaC$_8$; thus limiting the capacity of graphite-embedded sodium [14–16]. The large radius of sodium ions and the small graphite layer spacing are not favorable for the reversible deintercalation of sodium ions under the corresponding electrochemical window. Due to this characteristic, many researchers have mainly promoted sodium-ion embedding in graphite by solventizing sodium ions and increasing the graphite layer spacing to improve the sodium storage performance of graphite. Most of the methods related to solventized enlarged layer spacing mainly use ether-based electrolytes because ether-based electrolytes can inhibit the decomposition of electrolytes and form a thin SEI film on the graphite surface, which facilitates the migration of solventized sodium ions to the graphite lattice. The diethylene glycol dimethyl ether electrolyte can significantly increase the layer spacing of graphite and ensure the reversible embedding/detachment of sodium ions between the graphite layers. The graphite layer spacing can also be increased by redox, where graphite is first oxidized to obtain graphene oxide, followed by partial reduction to increase the graphite layer spacing to 4.3 Å [17]. In situ TEM studies revealed that the intercalation/deintercalation of Na$^+$ during the electrochemical reaction was reversible, exhibiting a specific capacity of about 284 mA h g$^{-1}$ at a current density of 20 mA g$^{-1}$, and the capacity remained well-maintained after 2,000 weeks of cycling. Although both solventization and redox methods can improve the sodium storage performance of graphite to some extent, the sodium storage performance of graphite is still far from its lithium storage performance. This potential can be extended to 0.38 V by adjusting the relative stability of the ternary graphite embedding compound and the solvent activity in the electrolyte, thus reducing the solvent dependence of the embedding voltage, as shown in Figure 4.4 [18]. The feasibility of graphite negative electrode in sodium-ion batteries was demonstrated by optimizing the electrolyte and matching it with the Na$_{1.5}$VPO$_{4.8}$F$_{0.7}$ cathode. By increasing the operating voltage of this sodium-ion full

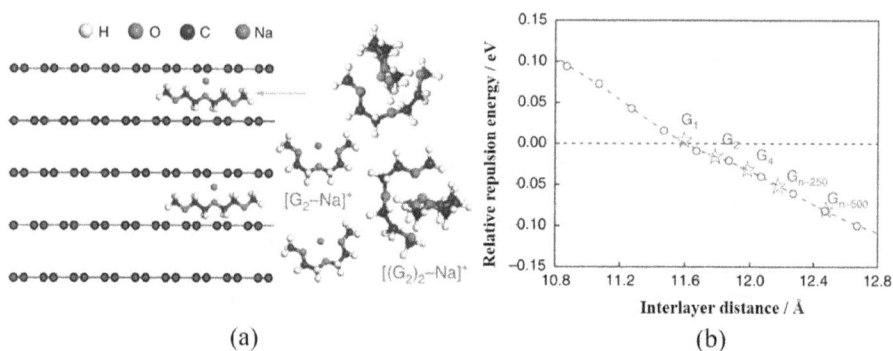

**Figure 4.4:** Solvent dependence of co-embedded voltage: (a) schematic diagram of solventized sodium-ion co-embedded graphite layer; (b) normalized repulsion energy as a function of the interlayer distance to clarify the correlation between the Na storage potential in graphite and ether in electrolyte [18].

cell to 3.1 V, the power density calculated based on the active material of the anode and cathode can be as high as 3,863 W kg$^{-1}$, with an average weekly capacity decay rate of only 0.007% during a long-term cycle of up to 1,000 weeks and a good capacity retention over a wide temperature range (0–60 °C).

### 4.1.2 Graphene

Graphene is a two-dimensional dotted structure with a hexagonal honeycomb lattice formed by carbon atoms, hybridized with sp$^2$ electron orbitals, and is the basic building block of carbon isomers, such as graphite, carbon nanotubes, and fullerenes (as shown in Figure 4.5). The unique atomic crystal structure and electron arrangement characteristics of graphene determine its excellent mechanical, electrical, thermal, optical, and chemical properties, making it suitable for wide range of applications in scientific research and industrial technology. The large theoretical specific surface area of graphene (about 2,600 m$^2 \cdot$g$^{-1}$) endows it with great adsorption properties and surface activity, which can be used as an excellent material for adsorption and energy storage devices, and allows it to be used in a wide range of promising applications in chemisorption, environmental remediation, capacitors, batteries, etc. [19].

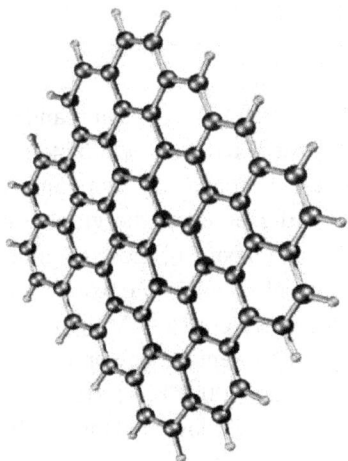

**Figure 4.5:** Graphene structure diagram.

Graphene has a very large specific surface area and a high number of defects, while its high electronic conductivity and excellent chemical stability can provide abundant storage sites for sodium ions. For lithium-ion and sodium-ion batteries, graphene or reduced graphene oxide (RGO) can help improve the storage capacity of lithium and sodium ions in their composite anode materials. Graphene materials are prepared by mechanical exfoliation, epitaxial growth, chemical vapor deposition (CVD), microwave plasma chemical vapor deposition (MPCVD), chemical intercalation exfoliation, and graphite oxide reduction [20]. The basic principle of the Hummer's method of chemical oxidation is to oxidize, intercalate, and swell the scaled graphite powder by strong oxidizing agents (concentrated sulfuric acid, concentrated nitric acid, potassium permanganate, etc.), and subsequently exfoliate it by a strong ultrasonic action to form a water-soluble and stably dispersed graphene oxide (GO) solution [21]. The GO prepared by this method is grafted with a large number of hydrophilic oxygen-containing groups on the surface, which makes the GO have good water solubility. It can be compounded with other materials in an aqueous solution to construct large-scale derivative products and multifunctional macro-assembly materials to form graphene. These oxygen-containing groups also provide reaction sites for further chemical modification and functionalization of graphene, making it easier for graphene to be uniformly dispersed in matrix polymers during the preparation of the functional composites. Although the chemical oxidation process will inevitably introduce defects and grafted oxygen-containing functional groups to the graphene flakes, resulting in a partial or complete loss of many properties of graphene, such as electrical and thermal conductivity, the subsequent heat treatment or chemical repair can fully restore the excellent thermal and electrical conductivity and other properties of graphene. Sodium ions have reversible storage properties in RGO [22].RGO has a higher electronic conductivity and more active sites as well as a larger layer spacing and disordered structure, and thus can store more $Na^+$, as shown in Figure 4.6(a).

The XRD results show that RGO consists of disordered graphite nanocrystals and stacked graphene sheets. As shown in Figure 4.6(b), the Raman spectra of RGO show two peaks at about 1334.8 $cm^{-1}$ and 1603.4 $cm^{-1}$, corresponding to the typical D and G bands of carbon materials. The intensity ratio (ID/IG) of the D band to the G band reflects the degree of disorder in RGO. A large ID/IG value (1.22) and a very pronounced D-band and G-band indicate the loss of long-range order between the graphene sheets, which is consistent with the XRD results. XRD and Raman spectroscopy indicate that RGO was successfully synthesized after treatment by graphene oxide. The RGO negative electrode achieved a specific capacity of 141 mA·h·g$^{-1}$. A highly ordered structure with a layer spacing of 0.388 nm and good Na + embedding properties can be obtained even at higher carbonization temperatures (1,100 °C) [23]. However, these graphene defects also pose some serious problems, such as low Coulombic efficiency. Therefore, graphene is still difficult to be considered as an ideal choice for the commercialization of anode materials in sodium-ion batteries due to its high price and complicated preparation. It also has the disadvantages of low first-period efficiency and high reaction potential.

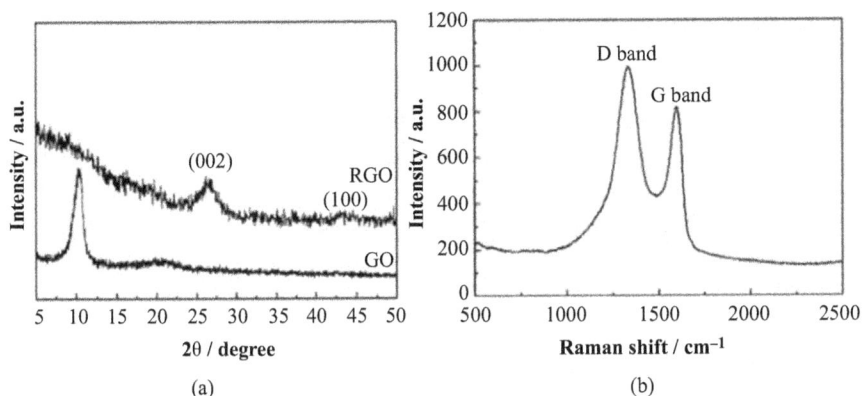

Figure 4.6: Physical characterization of RGO: (a) XRD patterns of GO and RGO; and (b) Raman atlas of RGO [22].

### 4.1.3 Soft carbon

Non-graphitized carbon, such as soft and hard carbon, has been widely investigated for use as anode materials in sodium-ion batteries.

The difference between the two mainly lies in the ease of graphitization and the arrangement of graphite microcrystals. Soft carbon can be transformed into graphitized carbon above 2,000 °C, while hard carbon keeps its amorphous structure. As shown in Figure 4.7, the graphite microcrystals inside the soft carbon are relatively ordered, and the thickness and width of the microchip layer are larger, while the graphite microcrystals inside the hard carbon are disordered and heterogeneously arranged.

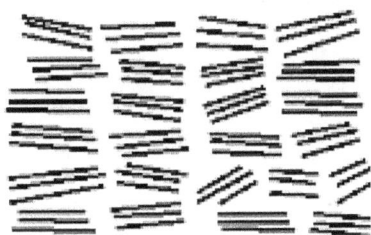

Figure 4.7: Schematic diagram of soft carbon structure.

Common soft carbons are petroleum coke, needle coke, carbon fibers, and carbon microspheres. In 1993, Doeff et al. first reported the sodium storage properties of disordered soft carbons prepared from petroleum coke pyrolysis, discussing the extent of $Na^+$ embedding/detachment reactions in soft carbons and their potential application in SIBs [13]. In 2017, the storage mechanism of soft carbons was revealed experimentally and computationally for the first time. Sodium ions were found inserted between graphite layers, forming an irreversible quasi-platform at about 0.5 V [24].

The irreversible expansion can be observed by in situ transmission electron microscopy (In situ TEM) and X-ray diffraction (XRD). This high potential plateau is related to the high binding energy of the internal local defects of the soft carbon, associated with Na$^+$. In addition, during the first sodization, the soft carbon exhibits a long ramp region, above and below the quasi-plateau, and the ramp region shows a highly reversible behavior. Total neutron scattering and correlation pair distribution function studies reveal that this is caused by the high number of defects contained in soft carbon. Although soft carbon has some sodium storage properties, the soft carbon material has a low sodium storage specific capacity of about 200 mA $\cdot$ h $\cdot$ g$^{-1}$, while the high charging potential of soft carbon – higher than 0.5 V – limits its application as a carbon-based anode material in -ion batteries.

Microporous soft carbon porous nanosheet electrode materials (SC-NS) were prepared from conventional soft carbon compounds, obtained by the pyrolysis of 3,4,9,10-perylenetetracarboxylic acid dianhydride by a simple microwave-assisted exfoliation process, and it was found that the surface area increased significantly after exfoliation – the microporous volume increased more than 100-fold, and the defects at the microporous edges synergistically enhanced the kinetic properties and provided additional sodium storage sites, enhancing the sodium storage capacity [25]. In addition, a soft carbon anode material (PA) with excellent sodium storage performance was obtained by simple crushing and one-step carbonization, using low-cost anthracite as a precursor [26]. As shown in Figure 4.8(a), the presence of carbon is well demonstrated by the broad diffraction peaks of carbon that can be seen in the XRD pattern, and the obvious D and G peaks are also detected in the Raman pattern (Figure 4.8(b)). The specific sodium storage capacity of the material reaches 220 mA $\cdot$ h $\cdot$ g$^{-1}$, which indicates the excellent cyclic stability of the electrochemical performance of the anode material. The energy density of the soft pack battery, made by using this soft carbon material as the negative electrode and

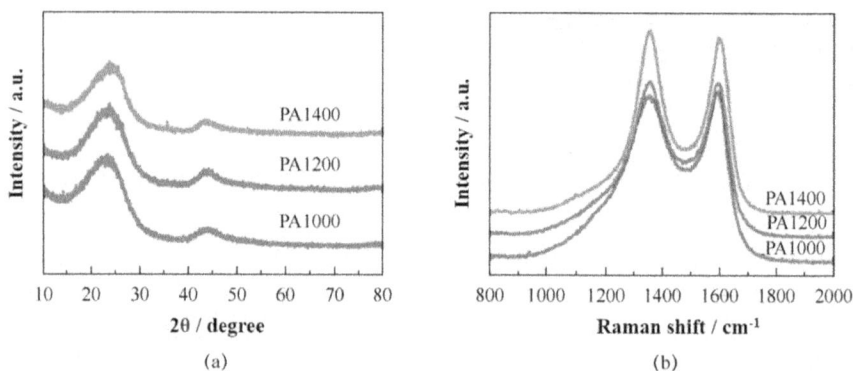

Figure 4.8: Physical characterization of PA at different carbonization temperatures: (a) XRD pattern; and (b) Raman spectroscopy.

Cu base layer oxide as the positive electrode, reached $100 \ W \cdot h \cdot kg^{-1}$, and the capacity retention rate was 80% at 1 C charge/discharge multiplier, and it passed a series of safety experiments, relevant to sodium-ion batteries, offering good application prospects. The research of anthracite soft carbon cathode material and its application in low-speed electric vehicles will greatly promote the development of low-cost sodium-ion batteries.

### 4.1.4 Hard carbon

Hard carbon, also known as hard graphitized carbon material, is difficult to graphitize even in a high temperature environment of above 2,500 °C. It is generally prepared by calcination at high temperature (1,000 °C, inert gas environment) of carbon-containing precursors. As shown in Figure 4.9, hard carbon has a larger layer spacing and disordered microporosity, possessing more sodium storage sites. Mature hard carbon materials are widely used in various tests of sodium-ion batteries and are called the "first generation sodium-ion batteries carbon-based materials", which can significantly improve the electrochemical performance of sodium-ion batteries. Common hard carbons include resin carbons (such as phenolic resins, epoxy resins, and polyfurfuryl alcohols), organic polymer pyrolysis carbons (such as tetrafluoroethylene perfluoroalkyl vinyl ether, polyvinyl chloride PVC, polyvinylidene fluoride PVDF, and polyacrylonitrile PAN), carbon black (acetylene black) and biomass carbon. As early as 2000, hard carbon anodes were synthesized using glucose pyrolysis, demonstrating a reversible specific capacity of $300 \ mA \cdot h \cdot g^{-1}$, far exceeding that of graphite and soft carbon materials [27]. The microstructure of carbon materials has a very important influence on their sodium storage performance.

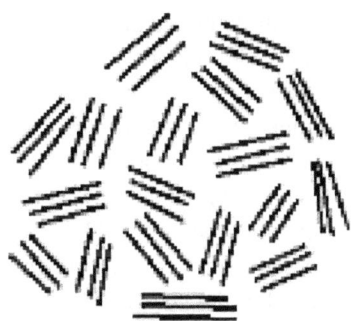

**Figure 4.9:** Schematic diagram of hard carbon structure.

Researchers continue to have different views on the mechanism of sodium storage in hard carbon materials. There are two main views on the mechanism of sodium storage in hard carbon materials: "intercalation-adsorption" mechanism and "adsorption-intercalation" mechanism. Two regions of the discharge curve for sodium storage at hard carbon: the high-potential ramp region (2–0.1 V) and the low-potential plateau region (0.1–0 V). As shown in Figure 4.10, Jahn et al. first proposed that the capacity of the slope region is mainly derived from the deintercalation of $Na^+$ between graphite-like layers, while the capacity of the plateau region is derived from the filling or deposition of $Na^+$ in the micropores (referred to as the "embedding-adsorption" mechanism here) [28], while Cao et al. first proposed that the slope-zone capacity is mainly derived from the adsorption of $Na^+$ on the carbon surface and edge defects, while the platform-zone capacity is mainly reflected in the embedding and de-embedding of $Na^+$ between the graphite-like layers, similar to the embedding and de-embedding behavior of $Li^+$ in graphite (referred to as the "adsorption-embedding" mechanism here) [29].

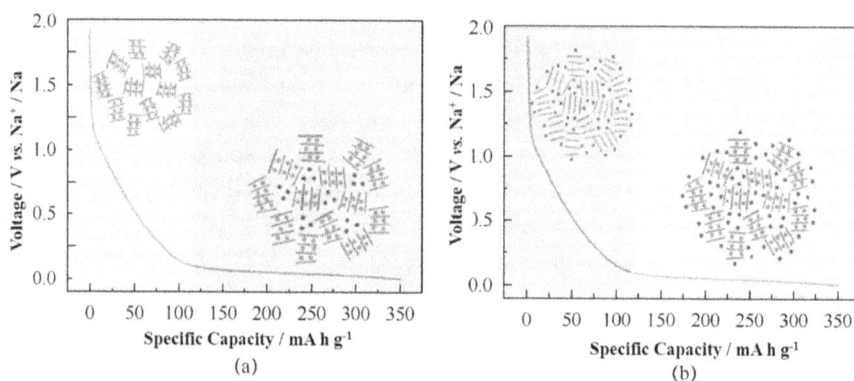

**Figure 4.10:** Hard carbon sodium storage mechanism: (a) "Imbibed–adsorbed" mechanism; and (b) "adsorption-embed" mechanism [30].

The electrochemical properties of hollow carbon nanowires (HCNWs) as sodium and lithium storage anode materials are shown in Figure 4.11. The sodium storage behavior of HCNWs in the low potential region is very similar to the lithium storage behavior of graphite, reflecting the embedding of alkali metal ions between the graphite layers and the formation of carbon-embedded compounds (MCx) [31]. The sodium storage behavior in the high-voltage region, on the other hand, corresponds to the charge transfer on the surface of the graphite microcrystals. To further confirm the mechanism, researchers performed theoretical simulations of the $Na^+$ embedding/detachment mechanism, based on the van der Waals forces between carbon layers and the repulsive forces between $Na^+$ and carbon, and found that the spacing between the carbon layers, suitable for sodium embedding, is about 0.37 nm. In addition, it was revealed that hard carbon nanoparticles undergo phase transition behavior from graphitic carbon to the intercalation compound, $NaC_x$.

Heteroatoms doped into the carbon structure, which readily form defective sites for $Na^+$ absorption, can significantly improve the electrical conductivity of carbon-based electrodes and improve the electrode-electrolyte interactions by functionalizing the carbon surface. The doping of carbon materials with heterogeneous atoms (e.g., N, B, S, and P) can improve the electrochemical properties of the carbon anode materials [33–37]. In general, N and B doping into the carbon structure significantly improves the ion transport and charge transfer processes, while S and P are electrochemically active elements that can react reversibly with Na, so the introduction of S into the carbon structure can provide additional reaction sites for $Na^+$ and thus increase the specific capacity [38, 39].

The three-dimensional interconnected structure of freestanding flexible films, composed of N-doped porous nanofibers and N-doped porous carbon materials, can significantly improve the sodium storage properties of carbon materials [34, 39]. N doping increases the electrical conductivity while generating notches on the exterior of the material, increasing the specific surface area of the material and thus the specific capacity. In addition, the interaction of electrons and N vacancies on the surface of the material also provides a certain capacity. In a recent study, some researchers have proposed the use of S-doped disordered carbon as anode material for sodium-ion batteries. S doping into the carbon structure increases the interlayer distance and provides additional reaction sites to facilitate the diffusion of $Na^+$ ions. Li et al. synthesized an S-doped disordered carbon with a high S doping rate (about 26.9%) and a unique coral-like three-dimensional structure. The prepared sulfur-doped carbon has a high reversible capacity of 516 mA h $g^{-1}$ with excellent multiplicative properties and good reversibility after 1,000 cycles [36].

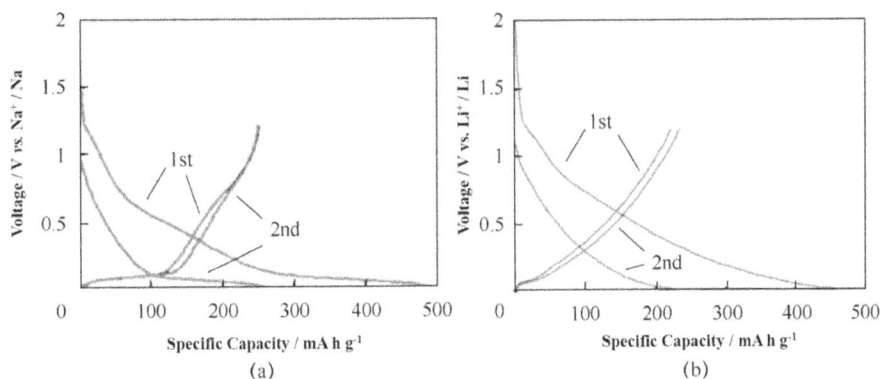

**Figure 4.11:** Comparison of sodium and lithium storage performance of HCNWs: (a) charge and discharge curves of HCNWs as the negative electrode of sodium-ion battery in the first two weeks; and (b) charge and discharge curves of HCNWs as anode of lithium-ion battery in the first 2 weeks [32].

The mechanism of sodium embedding in hard carbon is still debatable, and further theoretical and experimental studies are needed to elucidate the relevant reaction mechanism in order to promote the further development of hard carbon anode materials, for a better understanding of the sodium storage mechanism. At the same time, hard carbon anode faces many challenges in practical applications. The selection of suitable carbon sources and fabrication processes can significantly increase the sodium storage capacity of hard carbon materials and improve their cycling and multiplicity performances. Based on previous studies, the effects of factors, such as particle size, additives, vacancy defects and porosity measurements on the reversible capacity and multiplicative performance of hard carbon materials should be considered. In addition, the relationship between the electrochemical properties of hard carbon and solid electrolyte membrane (SEI) should be investigated, and theoretical calculations should be applied to predict and verify the experiments.

### 4.1.5 Other carbon materials

Graphyne, a porous two-dimensional all-carbon isomer, was synthesized artificially and controllably for the first time in 2010. As a new material, graphite alkynes have received a lot of attention worldwide. There has been research in the fields of theory and synthesis methods, and applications have been rapidly developed. A graphite alkyne is defined as an all-carbon network structure, with alkyne bonds and double bonds periodically connected in the two-dimensional plane, as shown in Figure 4.12. The periodic structural unit of graphite alkyne can be precisely controlled by the design of precursors, and the all-carbon skeleton construction not only inherits the excellent physical and chemical stability of carbon materials perfectly, but also the alkyne-rich structure gives graphite alkyne good electron and hole conductivity. As a potential lithium/sodium storage anode material, its theoretical predictions indicate that graphyne has a high theoretical storage density (-ion storage specific capacity up to 744 mA h g$^{-1}$) and three-dimensional ion transport pore channels [40]. The two-dimensional planar spacing of 0.365 nm in graphene is larger than the graphene layer spacing (0.34 nm), which facilitates the fast transport of lithium ions within the graphene interlayer. Also, graphene shows excellent storage performance in the storage of sodium ions.

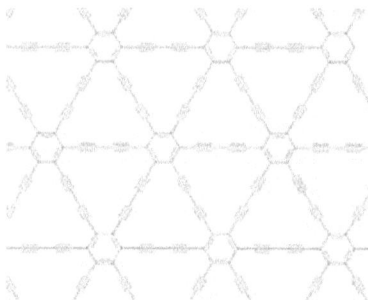

**Figure 4.12:** Schematic diagram of the structure for graphite acetylene.

It is theoretically and experimentally demonstrated that graphidiyne (GDY)-based materials can be used as the negative electrode of sodium-ion batteries. The binding of Na on monolayer GDY and monolithic GDY is investigated by DFT calculations, and it is pointed out that the insertion of Na into GDY results in a stable configuration of $NaC_3$, which allows for the most efficient Na storage. Also, the lower potential barrier of 0.4 eV facilitates the rapid migration of $Na^+$ [41]. In addition, the maximum specific capacity of a single GDY layer of sodium atoms ($NaC_{2.57}$, equivalent to 497 mA h $g^{-1}$) and a bulk GDY layer ($NaC_{5.14}$, equivalent to 316 mA h $g^{-1}$) could be determined using a bottom-up synthesis approach [40]. In the search for better performance, heteroatom doping was explored for the preparation of new GDY-based sodium-ion batteries. GDY, synthesized by boron doping (BGDY), with high conductivity and relatively low band gap, exhibited excellent electrochemical performance in -ion batteries [42].

Hollow carbon nanotubes, as one-dimensional nanomaterials with light mass and perfectly connected hexagonal structure, have many excellent mechanical, electrical, and chemical properties. In recent years, as the research of carbon nanotubes and nanomaterials has been intensified, their broad application prospects have been continuously shown. Hollow carbon nanospheres have similar great application prospects as carbon nanotubes in the fields of hydrogen storage and anode materials for secondary batteries. Hollow carbon nanotubes and hollow carbon nanospheres, as new carbon-based anode materials for sodium-ion batteries, can diffuse and migrate sodium ions more effectively; so this material can perform the sodium-ion embedding/exfoliation reaction more efficiently [43]. The hollow carbon nanosphere material has a suitable specific surface area and porosity due to the characteristic of disordered carbonaceous structure. When a PC solution of 1 M $NaClO_4$ was used as the electrolyte, the reversible specific capacity of hollow carbon nanotubes reached 200 mA h $g^{-1}$ after 100 weeks of charge/discharge cycles at a current density of 50 mA $g^{-1}$ in the voltage range of 0.001–3.0 V. Hollow carbon nanospheres were charged and discharged at the same current density. The hollow carbon nanotubes were cycled for 400 weeks at the same current density in the voltage range of 0.01–1.2 V. The reversible charge/discharge capacity of 250 mA h $g^{-1}$ was still achieved after cycling, and the capacity retention rate was 82%.

Biomass materials have attracted much attention due to their low production cost and low energy consumption in the synthesis process. Biomass materials can be directly prepared by the pyrolysis of amorphous carbon materials, which are mainly composed of C, H, and O elements, and have low polluting properties, wide distribution, and are abundantly available. The biomass materials that can be used as hard carbon precursors are currently reported as mullein [44], oak leaves [45], corn cobs [46], and peanut shells [47]. A carbon material prepared by the pyrolysis of banana peels has a low surface area and high vibrational density, with a specific capacity of 336 mA h $g^{-1}$ at a current density of 0.1 A $g^{-1}$ and a capacity retention of 89% after 300 weeks of cycling [48]. A recent study showed that the use of biomass by-product soybean residue as a biochar anode material can be sustainable [49]. Soybean residue can increase the carbon layer spacing to obtain a high specific surface area during the process of being carbonized and stripped, which facilitates easier embedding and detachment of sodium ions in the carbon structure during the cycling process. As shown in Figure 4.13, researchers investigated the overall structure, size, and morphology of the nitrogen-doped carbon flakes using transmission electron microscopy (TEM), and the high-resolution TEM (HRTEM) showed that the products consisted of graphite microcrystals with a layer spacing of 0.39 nm observed at the edges. It is shown that graphite microcrystalline NDCS can provide enough electrons for the redox reaction, thus improving the multiplicative performance of sodium-ion batteries. At the same time, this unique structure and large specific surface area give it a high specific capacity of about 292.2 mA h $g^{-1}$ and a good cycling stability for 2,000 weeks. From the perspective of green and recycling, the development of electrode materials using biomass is of great importance for future sustainable development.

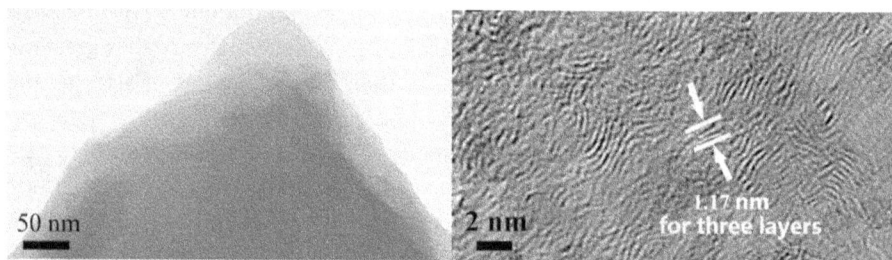

**Figure 4.13:** TEM images of NDCS at different resolutions: (a) TEM images with low resolution; and (b) high-resolution TEM images [49].

Initial progress has been made in the study of other carbon materials. These carbon materials have different structures, resulting in different physical and chemical properties, different synthesis methods, and different electrochemical properties being exhibited in their application when used in sodium-ion battery anode materials. In practical battery applications, there is still a need to solve the problems of initial irreversible capacity, low capacity, low multiplicative performance, and low mass loading of the battery.

## 4.2 Titanium-based negative electrode materials

Titanium-based materials have been widely investigated in the field of rechargeable lithium-ion batteries due to their structural stability, excellent cycling performance, abundant crystal resources, low cost, and non-toxicity [50, 51]. Due to the similar working mechanism of sodium-ion batteries and lithium-ion batteries, titanium-based materials have also been extensively used in the research of electrode materials in sodium-ion batteries. In addition, titanium-based materials have a low charge/discharge voltage plateau in sodium-ion batteries, which has led to a great deal of attention in the research of their application in the field of negative electrodes. At present, the research hotspots in titanium-based materials are mainly focused on titanium dioxide [52–54], lithium titanate [55–58], sodium titanate [59–62], and sodium titanium phosphate. In recent years, the work of researchers has focused on exploring the oxidation/reduction mechanism and optimization of electrochemical properties of titanium-based anode materials. In this section, the basic properties and crystal structures of several titanium-based anode materials are presented, and the modification studies, electrochemical properties, and sodium storage mechanisms of the materials are described.

### 4.2.1 TiO$_2$ electrode material

Titanium dioxide has received widespread attention as a typical embedded anode material in lithium-ion batteries due to its structural stability, non-toxicity, low price, and abundant availability [63–65]. TiO$_2$ consists of TiO$_6$ octahedra and Ti$^{4+}$ connections and the polycrystalline type of TiO$_2$ is related to the connection of TiO$_6$ octahedra. The current research types mainly include anatase [53, 54, 66–69], rutile [70–74], bronze [75], and slate, and the crystal structures are shown in Figure 4.14. Rutile TiO$_2$ is a tetragonal crystal system with the Ti atoms located at the center of the lattice and six oxygen atoms located above the prongs of the octahedral; each octahedron is connected to the surrounding ten octahedra, eight of which share corners and two share edges, and every two TiO$_2$ molecules form a crystal cell. The structure of anatase TiO$_2$ is also tetragonal, in which each octahedron and the surrounding eight octahedra are connected to each other, of which four co-angles and four TiO$_2$ form one crystal cell. Platytitanite TiO$_2$ belongs to the rhombohedral crystal system, where six TiO$_2$ form one cell, and the structure is unstable; so platytitanite-type TiO$_2$ exists relatively rarely in nature. Among these crystalline phases, rutile phase is the most stable, and anatase and slate TiO$_2$ can be transformed into the rutile phase by an irreversible exothermic reaction after a high-temperature heating. Although the radius of Na$^+$ is larger than that of Li$^+$, most research results are centered on anatase TiO$_2$ because the activation potential of Na embedded in the anatase lattice is similar to that of Li. Moreover, among the various crystalline forms of TiO$_2$, natural anatase TiO$_2$ is one of the earliest Li-embedded host structures studied [51].

Although TiO$_2$ possesses various advantages, such as low price and stable structure, there are still many shortcomings in its use as the negative electrode in sodium-ion batteries. For example, in TiO$_2$, all TiO$_2$ crystals are electron insulators, because Ti$^{4+}$ has no d electrons, which largely limits the electrochemical performance of TiO$_2$

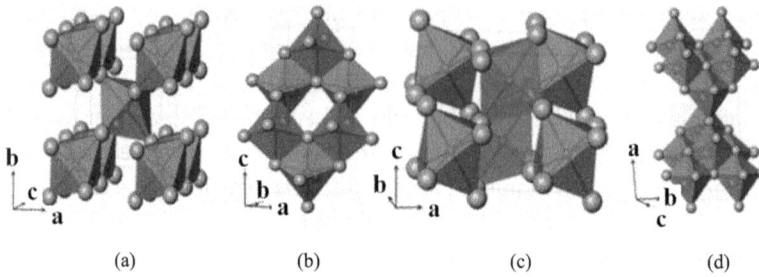

Figure 4.14: TiO₂ crystal structure: (a) rutile; (b) anatase; (c) bronze; and (d) platelet [50, 51].

in -ion batteries. Also, due to some defects of $TiO_2$ itself, coupled with the large radius of $Na^+$ and slow transport rate, it leads to a very unsatisfactory sodium storage performance. Therefore, many studies on $TiO_2$ are currently focused on improving the enhancement of electronic conductivity and ionic conductivity of $TiO_2$ by material nanosizing, interfacial modification, ion doping, using different types of C composite, and morphological structure optimization.

$TiO_2$ mainly stores sodium in the form of pseudo-capacitance, and there is no obvious charging and discharging plateau in the constant-current charging and discharging curves (Figure 4.15). The sodium storage form of pseudo-capacitance ensures that $TiO_2$ has good electrochemical kinetics and also ensures good multiplicative performance of $TiO_2$. Because of the diverse crystal structures of $TiO_2$, the electrochemical properties exhibited under its different crystal structures are also very different.

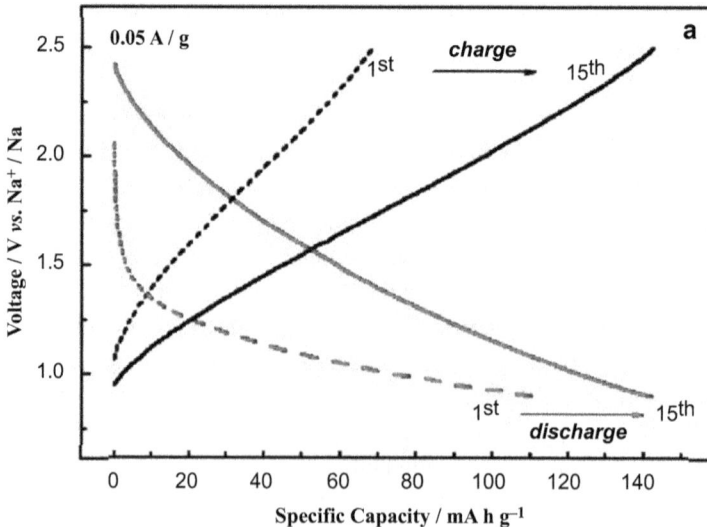

Figure 4.15: Constant-current charge–discharge curve of TiO₂ [76].

Moreover, in addition to the above-mentioned crystal structure mechanisms, the amorphous structure of $TiO_2$ has also been studied. Xiong et al. first reported the use of amorphous $TiO_2$ as anode material in sodium-ion batteries by growing amorphous $TiO_2$ nanotubes directly on Ti foil and using the obtained material as the anode of sodium-ion batteries, with a reversible specific capacity up to 150 mA h $g^{-1}$ and excellent cycling performance of the battery [54]. The studies on the different crystalline types of $TiO_2$ mainly focused on the rutile and anatase types. Nanocluster-type rutile $TiO_2$ microspheres [73] showed capacity retention of 83.1% after 200 cycles at a current density of 0.1 C. The sodium storage performance of $TiO_2$ is affected by the slow sodium kinetics due to the large size of $Na^+$. To solve this problem, researchers proposed a strategy to modify the nanostructures with metallic Ti and highly conductive carbon additives. Compared to rutile $TiO_2$, $Na^+$ has a lower potential barrier to de-embedding in anatase crystalline form and a faster ion transport rate; so anatase $TiO_2$ type is the most widely studied among the many $TiO_2$ electrode materials.

For example, by synthesizing anatase $TiO_2$ nanorods by the hydrothermal method and coating a thin layer of carbon on its surface, this material not only has high specific surface area and good electron conductivity, but the nanorod structure also provides good electron/ion transport channels. The capacity and multiplicity performances of the material are also significantly improved. The reversible specific capacity is 190 mA h $g^{-1}$ in the voltage interval of 0–3 V and still a specific capacity of 82 mA h $g^{-1}$ at 30 C [69]. Moreover, in addition to the optimization of the material structure, some studies have been carried out with regard to the optimized structure of the cladding, using different carbon sources. Anatase $TiO_2$ nanorods, coated with bituminous carbon, have a high specific capacity of 193 mA h $g^{-1}$ and excellent multiplicative properties [69]. The $TiO_2$/C hierarchical structure, synthesized by the hydrothermal method, has a discharge specific capacity of 277.5 mA h $g^{-1}$ at 50 mA $g^{-1}$ and 153.9 mA h $g^{-1}$ at 5,000 mA $g^{-1}$ [77]. The changes in the crystal structure, before charging and discharging for two weeks, were analyzed by in situ XRD characterization, as shown in Figure 4.16. The results indicate that the $TiO_2$/C structure is stable and the electrochemical cycle is reversible during the electrochemical reaction, which further proves that the $TiO_2$ sodium storage mechanism is an embedding/detachment reaction mechanism.

In addition to compounding with different carbon sources to improve the electrical conductivity of the material, some researchers have proposed to improve the reversible sodium storage performance of titanium dioxide by doping with different elements to reduce the average oxidation state of titanium, thereby creating more defects and vacancies, and increasing the layer spacing of the material. Wang et al. prepared B-doped $TiO_2$ by a facile hydrothermal method with a reversible specific capacity of 150 mA h $g^{-1}$ at 2C current density and stable cycling for 400 weeks [78]. In addition, the highly conductive self-supported S-doped $TiO_2$ nanotubes showed a high specific capacity of 320 mA h $g^{-1}$ at a current density of 33.5 mA $g^{-1}$ and the stable capacity retention after 4,400 weeks of cycling at a high current density of 3.35 A g$^{-1}$ was 91% [79].

Of course, researchers do not limit themselves to one approach to modify materials; elemental doping and structural optimization can improve the electrochemical properties of materials. Xu et al. designed a novel layer-less graphene-coated cobalt and nitrogen double-doped $TiO_2$/C multilevel pore skeleton structure (CN-TC@FG) using cobalt-doped, aminated titanium-based metal organic framework (CNH-T) as a multifunctional precursor [80]. On the macroscopic structural scale, CN-TC@FG has a continuous macroporous network structure, which facilitates the electron transfer and electrolyte penetration. On the microscopic structural scale, the ultrafine $TiO_2$ nanoparticles are uniformly embedded into the mesoporous carbon skeleton, forming a good three-dimensional conductive network framework. This not only promotes the diffusive transport of $Na^+$, but also enhances the electron conduction rate of $TiO_2$ particles. At the atomic scale, the double doping of cobalt and nitrogen greatly enhances the inherent electrical conductivity of $TiO_2$. When the material is used in the anode of sodium-ion batteries, the CN-TC@FG electrode exhibits high specific capacity, good cycle life, and fast pseudocapacitive sodium-ion storage capacity, with a reversible specific capacity of 174 mA h g$^{-1}$ after 5,000 weeks of cycling at 6 C and 121 mA h g$^{-1}$ current density after 10,000 weeks of cycling at 15 C.

Therefore, for materials like $TiO_2$, which exhibit pseudocapacitive behavior in sodium-ion batteries, the pseudocapacitive behavior of $TiO_2$ materials should be given full play in the synthesis method and structure design, and thus the sodium-ion storage performance of the materials should be improved. These methods can further promote the application of $TiO_2$ in the anode materials of sodium-ion batteries by improving the synthesis method, selecting suitable elements to dope them, or by designing a reasonable structure, combining the advantages of different structural scale designs to achieve multi-scale design.

### 4.2.2 Li$_x$TiO$_y$-type negative electrode materials

The spinel structure of $Li_4Ti_5O_{12}$ has a flat high potential of about 1.5 V during charge and discharge, with low volume change and a good cycle life, making it a typical "zero stress" anode material in Li-ion batteries [81]. One quarter of the lithium ions and all of the titanium ions are located in octahedral sites in this spinel-type skeleton structure. The octahedral sites share edges and form three-dimensional tunnels in the structure. The remaining lithium ions are located in the tetrahedral positions of the spinel skeleton, as shown in Figure 4.17. When $Li^+$ is embedded in $Li[Li_{1/3}Ti_{5/3}]O_4$, the lithium ions in the tetrahedral sites migrate to the adjacent octahedral sites, and $Ti^{4+}$ is reduced to $Ti^{3+}$, eventually forming $Li_2[Li_{1/3}Ti_{5/3}]O_4$.

Also, spinel-type $Li_4Ti_5O_{12}$ is an anode material that allows the reversible de-embedding of sodium ions. The reversible specific capacity of spinel $Li_4Ti_5O_{12}$,

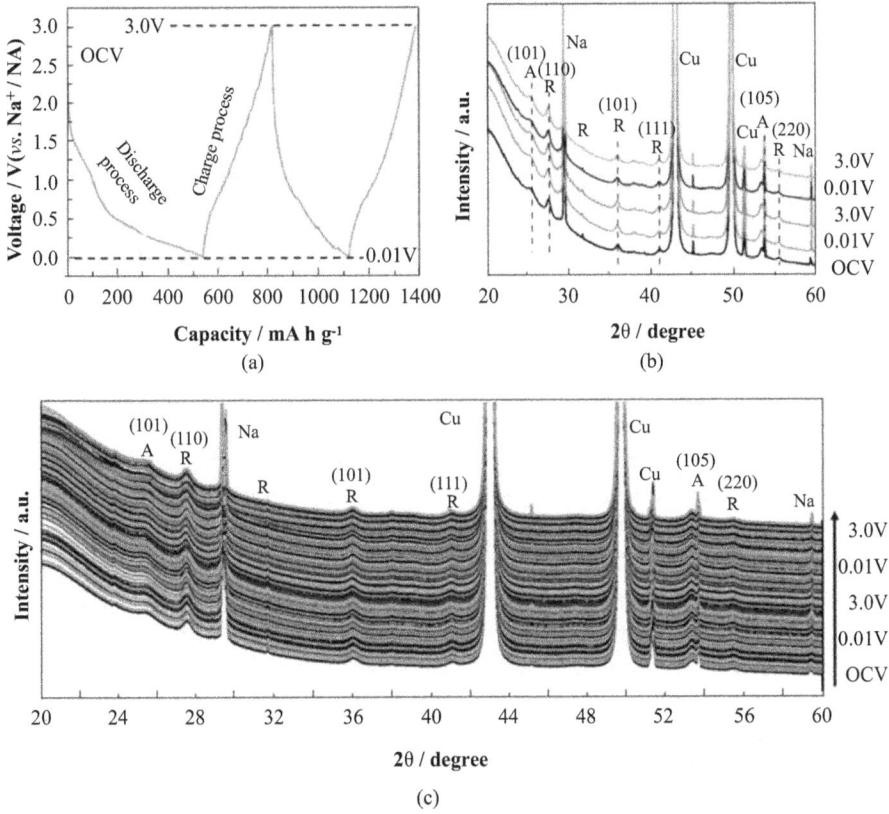

**Figure 4.16:** Physical phase changes during the first 2 weeks of TiO$_2$/C charging and discharging: (a) charging and discharging curves of TiO$_2$/C for the first 2 weeks; (b) high-energy XRD patterns of TiO$_2$/C at the charging and discharging cutoff voltages for the first two weeks; (c) in situ XRD patterns during the first 2 weeks of charging and discharging (test conditions: voltage range 0.01–3.0 V, current density 50 mA g$^{-1}$) [77].

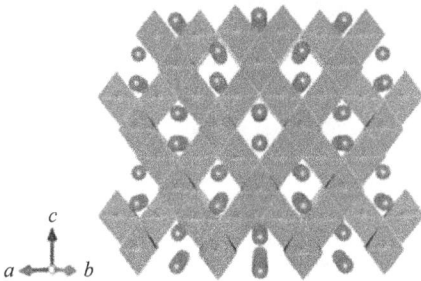

**Figure 4.17:** Spinel-type Li$_4$Ti$_5$O$_{12}$.

used as a substrate material for Na$^+$ embedding in sodium-ion batteries, was first reported by Zhao et al. to be about 145 mA h g$^{-1}$ with relatively low embedding/de-embedding voltages and good recyclability [57]. The reduction products of $Li_4Ti_5O_{12}$ in sodium-ion batteries were considered by researchers as a mixture of $Li_7Ti_5O_{12}$ and $Na_6LiTi_5O_{12}$ [51], with the following chemical reaction equation:

$$2Li_4Ti_5O_{12} + 6Na^+ + 6e^- \quad Li_7Ti_5O_{12} + Na_6LiTi_5O_{12} \tag{4.1}$$

The sodification process is shown in Figure 4.18. In addition, the electrochemical properties of $Li_4Ti_5O_{12}$ can be improved by changing the electrolyte and binder. Compared with $Li^+$, $Na^+$ is larger in size and requires a larger potential barrier for diffusion and phase transition in the spinel skeleton structure, which hinders its high multiplicity performance. To solve this problem, combining conductive additives with $Li_4Ti_5O_{12}$ is an effective approach. Graphene is an efficient conductive additive, and porous $Li_4Ti_5O_{12}$ nanofibers encapsulated in a graphene framework were used as anode materials in sodium-ion batteries. It was found by in situ XRD (Figure 4.19) that the new phase of $Na_6[LiTi_5]O_{12}$ appeared on the discharge and disappeared on charge, corresponding to the $Na^+$ embedding/detachment of $Li_4Ti_5O_{12}$ ($Li_4Ti_5O_{12}$ $Na_6$ $[LiTi_5]O_{12}$). As seen from the XPS pattern, the sodiuming/desalting process is accompanied by a redox reaction between $Ti^{4+}/Ti^{3+}$. The highly conductive 3D graphene wrapping provides a highly conductive network for electron transport, providing a shorter channel for $Na^+$ diffusion, with a high reversible specific capacity of 195 mA h g$^{-1}$ at 0.2 C [82]. In addition, compared to the expensive graphene, pitch is an inexpensive conductive agent, and coating $Li_4Ti_5O_{12}$ nanowires with pitch carbon significantly improved the electronic conductivity, with an electronic conductivity of about $3 \times 10^{-1}$ S cm$^{-1}$ and a 168 mA hg$^{-1}$ at 0.2 C, with high reversible specific capacity and significantly better electrochemical properties than blank $Li_4Ti_5O_{12}$ [83].

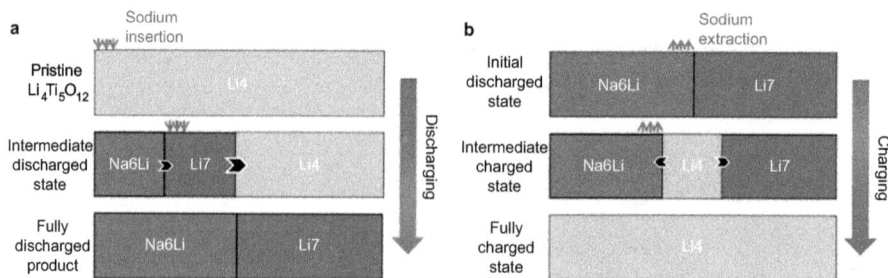

**Figure 4.18:** Natrifying process of $Li_4Ti_5O_{12}$.

Because $Li_xTiO_y$-like anode materials initially exhibit excellent electrochemical performance in Li-ion batteries, the mechanism study related to their use in Li-ion batteries can be transferred to sodium-ion batteries. Since $Li_xTiO_y$-like materials contain $Li^+$, the sodium-ion embedding process will have corresponding lithium-containing sodium

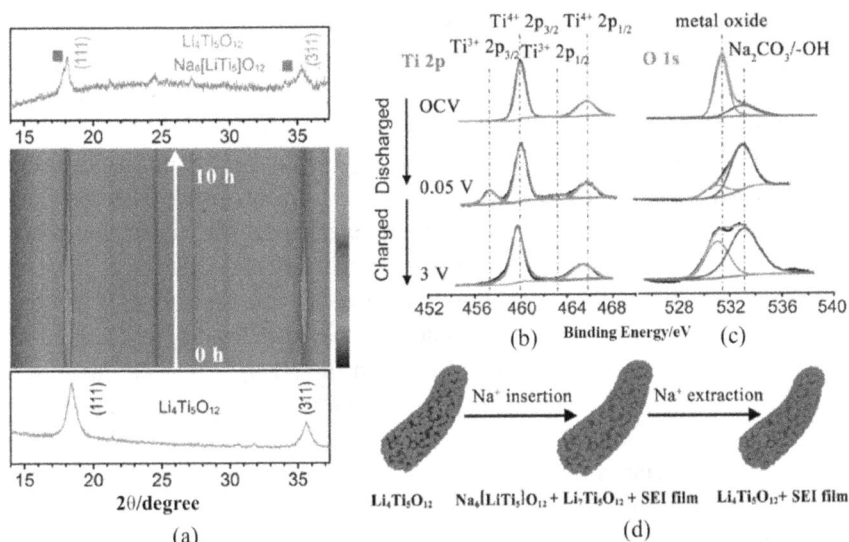

**Figure 4.19:** Physical characterization of G-PLTO electrode reaction process: (a) in situ XRD results; (b) and (c) in situ high-resolution XPS spectra of G-PLTO electrodes during discharge and subsequent charging; and (d) diagram of Na$^+$ embedding/stripping process in G-PLTO electrode [82].

products, leading to the emergence of a new sodium-ion storage mechanism, which can be studied in depth with the technical means of high-end characterization.

### 4.2.3 Na$_x$TiO$_y$-type negative electrode materials

The compound formed by Na$_2$O, an oxide of sodium and TiO$_2$, can be expressed as $x$Na$_2$O–$n$TiO$_2$, which is considered a promising electrode material due to the abundance of elements on the earth's crust, its non-toxicity, and structural stability. Among them, Na$_2$Ti$_3$O$_7$ is considered as one of the most promising anode materials for SIBs. Na$_2$Ti$_3$O$_7$ has a low working potential, high specific capacity (177 mA h g$^{-1}$), stable crystal structure, and high energy density, among all titanium-based anode materials. It is an anode material in sodium-ion batteries, with excellent electrochemical performance at room temperature. It is anode material for sodium-ion batteries, with excellent electrochemical performance at room temperature.

Na$_2$Ti$_3$O$_7$ has been used in other applications, including sensors, catalysts, and clean-up agents for toxic pollutants [55]. Since the operating voltage of Na$_2$Ti$_3$O$_7$ is about 0.3 V, which is much lower than that of the other current titanium-based anode materials, research work on the electrochemical properties of Na$_2$Ti$_3$O$_7$ has also been initiated. Na$_2$Ti$_3$O$_7$ has a unique Z-shaped layered crystal structure, with cells consisting of two-dimensional (Ti$_3$O$_7$)$^{2-}$ sheet layers, and Na$^+$ is distributed between the $(Ti_3O_7)^{2-}$ sheet layers. As shown in Figure 4.20, researchers found that the $Na_2Ti_3O_7$ material can perform two $Na^+$ embedding/de-embedding reactions during the charge/discharge cycles, with a plateau voltage of about 0.3 V (for $Na/Na^+$), and exhibits a reversible specific capacity of about 200 mA h $g^{-1}$ [59]. In-depth analysis by in situ XRD revealed that the sodium in $Na_2Ti_3O_7$ occupied two different sites in the $(Ti_3O_7)^{2-}$ sheet layer, specifically as Na(1) and Na(2). The reversible de-embedding of the two sodium ions by $Na_2Ti_3O_7$ allows the theoretical specific capacity to reach 177 mA h $g^{-1}$, but in the actual charging and discharging process, the actual capacity of the material was much lower than the theoretical specific capacity [84]. The essence of this phenomenon is due to the smaller lamellar layer spacing and larger $Na^+$ radius, which makes one of the Na(1) and Na(2) in the $(Ti_3O_7)^{2-}$ lamellar layer, the active sodium site Na(1), while the other one is the inert sodium site Na(2). During the discharge process of the cell, it is always difficult for the Na(2) to be reversibly embedded/deposited. This is the reason why it is difficult to increase the actual capacity of the material. Therefore, researchers used a few small molecules to replace the sodium out of the $(Ti_3O_7)^{2-}$ sheet layer and subsequently peeled off the sheet by centrifugation to expand the layer spacing. This not only increases the migration rate of $Na^+$ but also effectively reduces the inert sodium sites in the $(Ti_3O_7)^{2-}$ sheet layer, increasing the actual capacity [59].

The charging and discharging mechanism, based on $Na_2Ti_3O_7$, has been studied accordingly. During the charging process of $Na_2Ti_3O_7$, two additional $Na^+$ are embedded in one $(Ti_3O_7)^{2-}$ sheet layer; it is difficult to embed four $Na^+$ on one $(Ti_3O_7)^{2-}$ sheet layer at the same time, due to the limitation of the $(Ti_3O_7)^{2-}$ sheet ground layer spacing. Therefore, in order that more $Na^+$ can be stably accommodated, the $Na_4Ti_3O_7$ material structure undergoes rearrangement of the crystal structure to expand its layer spacing to ensure that more $Na^+$ can be stably accommodated. During the discharge phase, the nano-sized $Na_2Ti_3O_7$ does not produce an intermediate phase during the $Na^+$ de-uteration process, and $Na_4Ti_3O_7$ is directly converted to

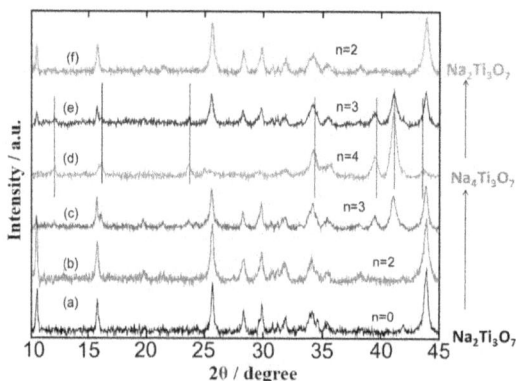

Figure 4.20: In situ XRD of $Na_2Ti_3O_7$/Na cells cycling in the range of 2.5–0.01 V at C/50 [59].

$Na_2Ti_3O_7$ [85]. It is also because of this that during the long cycle, the crystal structure changes continuously by the continuous de-embedding of $Na^+$ inside the material, leading to a final irreversible distortion of the structure and a rapid capacity decline, as a result.

The main factors that currently limit the performance of $Na_2Ti_3O_7$ materials include the low electronic conductivity of the material, the slow $Na^+$ transport rate, and the unstable SEI film formed. Therefore, compounding $Na_2Ti_3O_7$ with carbon materials can effectively suppress the surface side reactions and significantly improve the electrical conductivity of the materials, thus improving the electrochemical performance of $Na_2Ti_3O_7$ electrode materials. A novel $Na_2Ti_3O_7$@C composite was prepared by the rheological phase method, and it was found that the carbon coating could significantly improve the electrochemical performance of $Na_2Ti_3O_7$, with a specific capacity of 111.8 mA h $g^{-1}$ even after 100 weeks of cycling at 1 C, while the uncoated carbon material only had a specific capacity of 48.6 mA h $g^{-1}$ after 100 weeks of cycling. The composites also showed relatively stable sodium storage performance during long-term cycling at a larger current multiplication rate (5 C) [86]. This is because the uniformly distributed carbon forms a good conductive network between $Na_2Ti_3O_7$ particles, and the $Na_2Ti_3O_7$ active material can obtain electrons from all directions, which significantly improves the specific capacity, multiplicative performance, and cycling performance of the material. Balaya et al. prepared $Na_2Ti_3O_7$ by the solid-phase method and ball-milled it with carbon black, and it was still maintained after 90 weeks of cycling at 1 C with a specific capacity of more than 100 mA h $g^{-1}$, while the long-term cycling performance was relatively stable at different current multiplication rates [61]. Although $Na_2Ti_3O_7$ is considered to be the most promising anode material for sodium-ion batteries, it still suffers from poor capacity retention performance. In order to improve the electrochemical performance of $Na_2Ti_3O_7$ electrodes, further studies, such as the use of active materials as protective coatings, are needed.

More studies on $Na_xTi_yO_z$-type materials have also been carried out as potential anode materials for sodium-ion batteries [87–90]. Among them, $Na_2Ti_6O_{13}$ and $Na_4Ti_5O_{12}$ materials have received the most attention for lithium-ion battery applications, and many scientists have speculated that this type of material can be applied to sodium-ion battery anodes. Woo et al. studied the performance of $Na_4Ti_5O_{12}$ material on sodium-ion batteries under low temperature conditions and found that the material can efficiently perform sodium-ion embedding/detachment reactions [91]. However, the reversible charge/discharge specific capacity was only 50 mA h $g^{-1}$. This may be due to the poor kinetic performance during $Na^+$ migration, resulting in a potential resistance effect. The structures of $Na_2Ti_3O_7$ and $Na_2Ti_6O_{13}$ have similarities [92], and both are composed of basic $(Ti_3O_7)^{2-}$ units. $Na_2Ti_6O_{13}$ can accommodate 0.85 mol of Na and has a discharge performance of more than 65 mA h $g^{-1}$ at a voltage plateau of 0.8 V. The proposed mechanism of $Na^+$-embedding of $Na_2Ti_6O_{13}$ by non-in situ XRD tests is [61]

$$Na_2Ti_6O_{13} + xNa^+ + xe^- \rightarrow Na_{2+x}Ti_6O_{13} (x = 0.85) \tag{4.2}$$

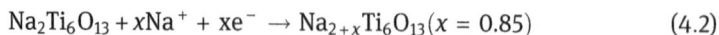

The $Na_2Ti_6O_{13}$ material was investigated by a combination of theoretical calculations and experiments, and it was found that the discharge cutoff potential has an important effect on the specific capacity of the material [93]. By reducing the cutoff voltage from 0.3 to 0 V, the specific capacity of the $Na_2Ti_6O_{13}$ anode material could be increased from 49.5 mA h $g^{-1}$ ($Na_{2+1}Ti_6O_{13}$) to 196 mA h $g^{-1}$ ($Na_{2+4}Ti_6O_{13}$).

In recent years, the Li/Na-mixed titanate, $Na_{0.66}[Li_{0.22}Ti_{0.78}]O_2$, has also been used as an embedded host for $Na^+$. Hu et al. designed a zero-strain anode material, P2-$Na_{0.66}[Li_{0.22}Ti_{0.78}]O_2$, with a layered structure, based on the spinel structure $Li_4Ti_5O_{12}$ [89]. Experimental results (Figure 4.21(a)) revealed that the crystal structure of $Na0.66 [Li_{0.22}Ti_{0.78}]O_2$ is a P2-type layered structure, with a space group of P63/mmc, with Li\Ti occupying the transition metal sites and sodium occupying two positions between the alkali metal layers (2b, 2d), forming a trigonal structure with the upper and lower oxygen; the particle size of the material is 10–15 μm. The P2-$Na_{0.66}[Li_{0.22}Ti_{0.78}]O_2$ has a reversible specific capacity of 116 mA h $g^{-1}$ at an average storage voltage of 0.75 V. In addition, the strain of the P2-$Na_{0.66}[Li_{0.22}Ti_{0.78}]O_2$ material is zero during the sodium embedding/de-embedding process, which ensures a lifetime issue of the material at long cycles, with a volume change of about 0.77% and an actual usable specific capacity of 100 mA h $g^{-1}$. By the in situ XRD patterns during the charge/discharge process (Figure 4.21(b)), it can be found that the P2-layered material, with lithium-doped transition metal layer, exhibits a nearly single-phase behavior during the embedding/desalination process, unlike the reaction mechanism of the conventional P2-layered material, in which multiple phase transitions occur during the embedding/desalination reaction. The zero-stress characteristics of P2-$Na0.66 [Li_{0.22}Ti_{0.78}]O_2$ materials are fully exploited, which can further promote the research of anode materials for rechargeable sodium-ion batteries.

### 4.2.4 NaTi₂(PO₄)₃-type negative electrode materials

$NaTi_2(PO_4)_3$ with the NASICON-type three-dimensional mesh structure is a typical representative of polyanionic titanium-based materials, which can be used as electrode materials in sodium-ion batteries. The three-dimensional channels contained in the crystal structure of $NaTi_2(PO_4)_3$ facilitate the rapid migration of $Na^+$. The theoretical specific capacity of $NaTi_2(PO_4)_3$ material is 133 mA h g $^{-1}$ and its specific capacity is low. The specific capacity of $NaTi_2(PO_4)_3$ material is 133 mA h $g^{-1}$, which is low and has not been extensively studied in organic sodium-ion batteries systems before, but it has received extensive attention again due to the ability of $NaTi_2(PO_4)_3$ material to be used in aqueous sodium-ion batteries with a stable structure. 2011, Park et al. investigated $NaTi_2(PO_4)_3$ in organic and aqueous systems – the electrochemical behavior of sodium at 2.1 V (for Na/$Na^+$), embedded in $NaTi_2(PO_4)_3$

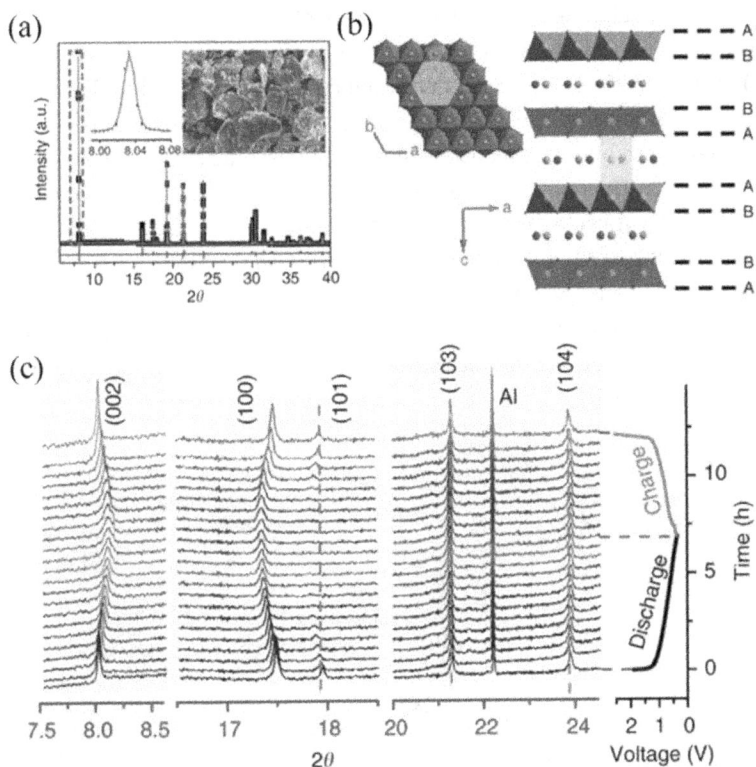

**Figure 4.21:** Characterization of $P_2$-$Na_{0.66}$ [$Li_{0.22}Ti_{0.78}$]$O_2$: (a) XRD pattern and morphology of $P_2$-$Na_{0.66}$ [$Li_{0.22}Ti_{0.78}$]$O_2$; (b) structure diagram of $P_2$-$Na_{0.66}$ [$Li_{0.22}Ti_{0.78}$]$O_2$ material samples; (c) electrochemical embedding /structural changes of $P_2$-$Na_{0.66}$ [$Li_{0.22}Ti_{0.78}$]$O_2$ during sodium removal: in situ XRD patterns by synchrotron radiation.

through a two-phase reaction mechanism [94]. The reversible cyclic specific capacity of $NaTi_2(PO_4)_3$ material is 120 mA h $g^{-1}$ in the organic system and 123 mA h $g^{-1}$ in the aqueous solution system, both exceeding 90% of its theoretical capacity.

$NaTi_2(PO_4)_3$ is considered an ideal anode material for aqueous sodium-ion batteries because the sodium-ion embedding/detachment voltage is close to the minimum voltage that the aqueous system can withstand. As shown in Figure 4.22(a), the CV curves of $NaTi_2(PO_4)_3$ exhibit a pair of very symmetric redox peaks at –0.96 V and –0.68 V, corresponding to the reversible $Na^+$ embedding/detachment reactions in the $NaTi_2(PO_4)_3$ lattice:

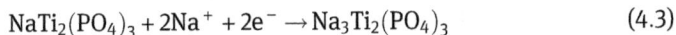

$$NaTi_2(PO_4)_3 + 2Na^+ + 2e^- \rightarrow Na_3Ti_2(PO_4)_3 \qquad (4.3)$$

Figure 4.22(b) shows the charge and discharge curves of $NaTi_2(PO_4)_3$ negative electrode at a high current multiplication rate of 5 C in an aqueous electrolyte, indicating that $NaTi_2(PO_4)_3$ has good electrochemical performance. The Prussian blue-like cathode material in sodium-ion batteries ($Na_2NiFe(CN)_6$) was used as the cathode and $NaTi_2(PO_4)_3$ material as the anode, which was matched to become a full battery in an aqueous solution system. The battery had a discharge voltage of 1.27 V and an energy density of 42.5 W h $kg^{-1}$. It cycled for 250 cycles at a multiplication rate of 5 C. The capacity retention was 88%. When $Na_2CuFe(CN)_6$ was used as the anode, the average voltage reached 1.4 V, the energy density increased to 48 W h $kg^{-1}$, and the capacity retention rate reached 90%, at a high multiplier of 1,000 cycles. These studies provide effective technical support for the industrial application of sodium-ion batteries in an aqueous solution system.

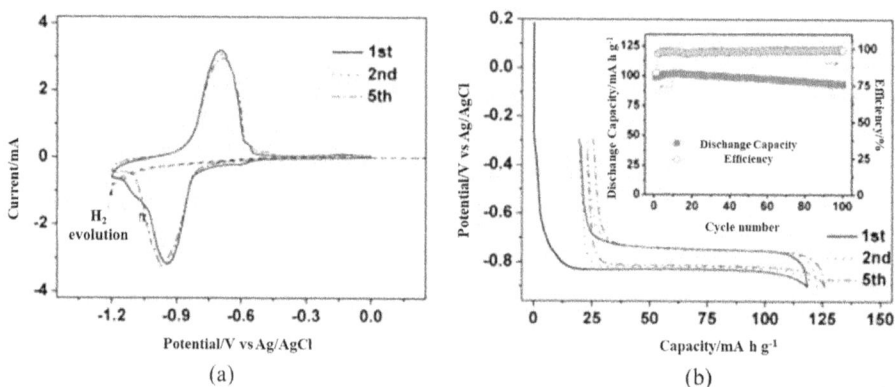

**Figure 4.22:** Electrochemical REDOX behavior of $NaTi_2(PO_4)_3$ electrode: (a) CV curve at scanning rate of 5 mV · s $^{-1}$, dotted line represents electrochemical window of electrolyte; (b) charging and discharging at high current double speed of 5 C electric curve, illustration shows reversible specific capacity and Coulomb efficiency at high rate (5 C) cycles [94].

The inset gives the reversible specific capacity and Coulombic efficiency at high multiplicity (5 C) cycles [94] Currently, the modification of $NaTi_2(PO_4)_3$ has been widely studied, and the compounding of $NaTi_2(PO_4)_3$ with carbon materials is the most effective way to modify $NaTi_2(PO_4)_3$, because the electronic and ionic conductivity of $NaTi_2(PO_4)_3$ can be significantly improved by compounding with carbon materials. The "double carbon coating" $NaTi_2(PO_4)_3$ electrode material, prepared by soft chemical method, is a double carbon coated NTP@C@PC material by embedding NTP@C particles into a porous carbon (PC) matrix, which still has a reversible specific capacity of 103 mA h $g^{-1}$ after 5,000 weeks of cycling at 5 C, and a reversible specific capacity with 88 mA h $g^{-1}$ of after 5,000 cycles of cycling at 5 C, and 76 mA h $g^{-1}$ after 6,000 weeks of cycling at 10 C [95]. The high specific surface area, porous structure characteristics, and high electronic and ionic conductivity of this material are the keys to its excellent multiplicative and cycling properties. In addition, compounding with other materials that have good electrical conductivity can significantly improve the electrical conductivity of $NaTi_2(PO_4)_3$ material, and further enhance the electrochemical properties of the material.

## 4.3 Transformation-responsive anode materials

Metal compounds have received a lot of attention as they follow the conversion mechanism when used as anode materials in -ion batteries with high theoretical specific capacity. Unlike the embedding and alloying reaction mechanisms, the conversion reaction mechanism involves the breaking of chemical bonds and the generation of new bonds to form compounds with sodium. In essence, conversion reactions can be described as replacement reactions, but the structure of the metal compound may change.

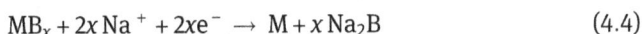

$$MB_x + 2x\,Na^+ + 2xe^- \rightarrow M + x\,Na_2B \tag{4.4}$$

where B stands for non-metallic elements, such as O, S, Se, and P, and M stands for metallic elements, such as Mn, Fe, Co, Ni, and Cu. Conversion reactions generally occur in the potential interval below 1 V. Materials generally do not undergo conversion reactions in the high potential interval. Although this kind of metal compound, based on multi-electron conversion mechanism, is used as anode material in sodium-ion batteries with high theoretical specific capacity, there are also many problems. First, the poor conductivity of the electrode material itself leads to poor overall electrochemical performance, and the generation of SEI film during cycling causes the reduction of the first-period coulombic efficiency and irreversible capacity loss, resulting in poor multiplier performance and cycle stability of the battery. Meanwhile, the destruction of the material structure during charging and discharging can cause serious volume expansion problems, resulting in poor contact between the active substances and deteriorating the electrochemical performance of the battery. In order to solve the above problems, researchers have increased the sodium storage capacity of metal compounds and improved their cycling performance by constructing composite electrodes, with conductive carbon, and designing electrodes with nanostructures. In this section, the research progress of several conversion-responsive anode materials, such as metal oxides [96–98], metal sulfides [99–101], and metal phosphides [102–104], is summarized.

### 4.3.1 Metal oxide cathode materials

In 2002, Alcantara et al. first utilized $NiCo_2O_4$ spinel oxide as an anode material in sodium-ion batteries and introduced the concept of conversion mechanism [105], initiating the use of metal oxide materials as anode materials in sodium-ion batteries, and the researchers concluded the conversion mechanism of $NiCo_2O_4$ as:

$$NiCo_2O_4 + 8Na \rightarrow Ni + 2Co + 4Na_2O \tag{4.5}$$

Thereafter, researchers concluded that the transformation mechanism of metal oxides is:

$$MO_x + 2xNa^+ + 2xe^- \rightarrow M + xNa_2O \tag{4.6}$$

where M stands for Mn, Fe, Co, N, and Cu. The metal oxides studied so far mainly include iron oxide, tin oxide, copper oxide, and cobalt oxide.

Iron-based oxides are a typical anode material in sodium-ion batteries because iron oxide has excellent chemical stability, high capacity, costs less, demonstrates easy preparation, is non-toxic, and has a high theoretical specific capacity. The currently reported iron-based oxides include $Fe_2O_3$ and $Fe_3O_4$.

$Fe_2O_3$ has a high theoretical specific capacity (1,007 mA h g$^{-1}$); however, the low electronic conductivity ($\sim 7 \times 10^{-3}$ S cm$^{-1}$) causes a large volume expansion during sodium-ion embedding/detachment, leading to agglomeration and polarization of the active material, which eventually leads to a rapid capacity loss, finally leading to a rapid capacity decline. In order to solve these problems, methods such as synthesis of nanostructures with different morphologies and compounding with carbon can be used. Among them, nanostructures can provide abundant active sites, large specific surface area, and short electron/ion diffusion distance. Compounding with carbon materials can significantly improve the electrical conductivity of metal oxides. The sodium storage process of $Fe_2O_3$ materials is [106]:

$$Fe_2O_3 + 6Na^+ + 6e \rightarrow 2Fe + 3Na_2O \tag{4.7}$$

The sodium-ion embedding/detachment behavior of $Fe_2O_3$@GNS can be analyzed by cyclic voltammetry (CV). Figure 4.23 shows the CV curves of the $Fe_2O_3$@GNS electrode for the first 5 weeks. The currents in the first cathodic scan, starting at 0.6 V, are higher than those in the subsequent scans, due to the irreversible formation of SEI. The curves almost overlap, except for the first week of cycling, indicating good stability of $Fe_2O_3$@GNS cycling. Beyond the first cycle, two weak redox peaks, located at 0.6 V (cathode) and 0.7 V (anode), can be observed, which may correspond to the embedded detachment from the $Fe_2O_3$ lattice.

A reasonable structural design can significantly improve the electrochemical properties of the material. For example, the yolk-shell-like $Fe_2O_3$@C composite is attached to the surface of multi-walled carbon nanotubes (MWNTs) [107]. On the one hand, MWNTs and carbon layers enhance the electrical conductivity of the material, and on the other hand, the egg yolk shell-like structure can buffer the volume expansion of $Fe_2O_3$ during the battery cycle, and this electrode has good sodium storage performance. The $Fe_2O_3$ nanocrystals were homogeneously immobilized on graphene nanoflakes ($Fe_2O_3$@ GNS) by the nanocasting technique, and the composite exhibited

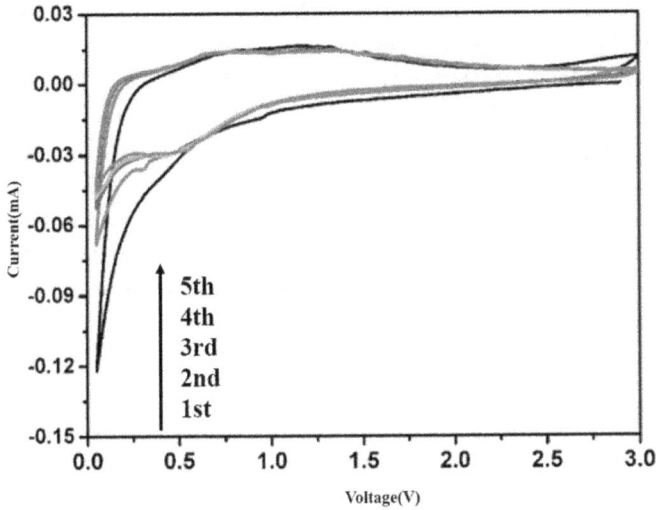

**Figure 4.23:** CV curve of $Fe_2O_3@GNS$ at a scanning rate of $0.05 \, mV \cdot s^{-1}$ in the range of 0.05–3 V compared with $Na^+/Na$.

excellent cycling and multiplicative performance when used as the anode of sodium-ion batteries [106].

As early as 2009, Omaba et al. reported that the iron-based oxide $Fe_3O_4$ can be used as a sodium electronegative electrode to achieve an embedded reaction sodium storage process in the voltage interval of 1.2 to 4.0 V [108].The $Fe_3O_4$ samples crystallized into spinel-type structures, with space group symmetry. For the industrial $Fe_3O_4$, ball-milled industrial $Fe_3O_4$ and precipitated samples, the calculated lattice parameters were $a = 8.394$, 8.37, and 8.35 Å, respectively. The methanol wet ball milling resulted in a significant broadening of the Bragg peak profile, indicating a significant reduction in the grain size of the industrial $Fe_3O_4$ powder. Half-cells composed as anode materials in sodium-ion batteries undergo conversion reactions at discharge voltages of 0.04–3 V [109]:

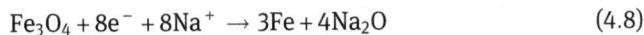

$$Fe_3O_4 + 8e^- + 8Na^+ \rightarrow 3Fe + 4Na_2O \tag{4.8}$$

The experimentally measured discharge specific capacity at the first cycle was 643 mA h $g^{-1}$, which is lower than the theoretical specific capacity of $Fe_3O_4$ of 926 mA h $g^{-1}$. To improve the electrical conductivity of metal oxides, γ-$Fe_2O_3@C$ nanocomposites were prepared for use as anode materials in sodium-ion batteries [110]. Three-dimensional porous nanocomposites were prepared using spray pyrolysis, and consisted of interconnected nanochannels and γ-$Fe_2O_3$ nanoparticles uniformly embedded in a porous carbon matrix inside. γ-$Fe_2O_3@C$ nanocomposites exhibited excellent electrochemical performance when used as anode in sodium-ion batteries, cycling at a current density of 2,000 mA g$^{-1}$. A specific capacity of 358 mA h g$^{-1}$ was still available after 1,400 weeks.

SnO$_2$ is abundantly available, has low operating voltage, and has high theoretical specific capacity. The reaction with sodium ions occurs in two steps: conversion reaction and alloying reaction. Using in situ transmission electron microscopy and DFT calculations to explore the structure and chemical change process of the negative electrode of SnO$_2$ nanowires in sodium-ion batteries, and by quantitatively comparing the results with those of lithium-ion batteries, it was found that the replacement reaction occurs when Na is embedded in SnO$_2$, forming amorphous Na$_x$Sn nanoparticles dispersed in the Na$_2$O matrix [111]. With further embedding of Na, Na$_x$Sn crystallizes into Na$_{15}$Sn$_4$ ($x = 3.75$). After Na is taken off, Na$_x$Sn transforms into Sn nanoparticles:

$$(4 + x)Na + SnO_2 \rightarrow Na_xSn + 2Na_2O \tag{4.9}$$

$$(15 - 4x)Na + 4 Na_xSn \rightarrow Na_{15}Sn_4 \tag{4.10}$$

As the de-alloying proceeds, some pores are formed and the Sn particle structure is confined to the Na$_2$O hollow matrix. These pores greatly increase the electrical impedance and lead to poor recyclability of SnO$_2$. The average migration velocity of Na$^+$ in SnO$_2$ is about $(0.63 \pm 0.2)$ nm · s$^{-1}$, which is about 1/20 of that of Li$^+$. How to improve the kinetics of SnO$_2$ materials in sodium-ion batteries is the key to drive the development of SnO$_2$ materials in sodium ion. The main modification methods used by researchers include the introduction of porous carbon, MWCNTs, and graphene to prepare composite anode. Nitrogen-doped graphene nanohybridized SnO$_2$ (SnO$_2$/NG) was prepared by the in situ hydrothermal method, and the grain size of SnO$_2$ in SnO$_2$/NG (1.6 nm) was similar to that of SnO$_2$ in undoped nitrogen graphene nanohybridized SnO$_2$ (SnO$_2$/G) (1.4 nm) [112]. The microstructure of nitrogen-doped graphene (NG) was investigated by Raman spectroscopy, and two characteristic peaks of 1,323 cm$^{-1}$ and 1,589 cm$^{-1}$ could be observed in the range of 800–2,000 cm$^{-1}$, corresponding to the D and G bands of NG, respectively, which are increased compared to SnO$_2$/G. This is probably due to the distortion of the graphene structure caused by the different C-C and C-N bond distances. In addition, the $I_D/I_G$ (intensity ratio of D-band to G-band) value of SnO$_2$/NG (1.33) is higher than that of the SnO$_2$/G composite (1.18), indicating that the structure of NG is more disordered than that of graphene due to the introduction of N atoms in the graphene network. The initial reversible specific capacity of this composite anode was 339 mA h g$^{-1}$ at a current density of 20 mA g$^{-1}$, and a specific capacity of 640 mA h g$^{-1}$ was also maintained at high current densities.

CuO materials have also been investigated due to their wide availability, stability, chemical properties, and high theoretical capacity. The specific transformation mechanism of CuO nanowires during sodization is as follows [97]:

$$2CuO + 2Na^+ + 2e^- \rightarrow Cu_2O + Na_2O \tag{4.11}$$

$$Cu_2O + Na_2O \rightarrow 2NaCuO \tag{4.12}$$

$$7NaCuO + Na^+ + e^- \rightarrow Na_6Cu_2O_6 + Na_2O + 5Cu \tag{4.13}$$

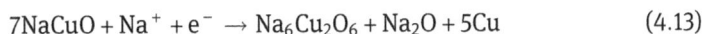

Similar to other metal oxides, copper oxides also suffer from poor electrical conductivity and large volume changes during cycling. In order to improve the electronic conductivity and regulate the volume change during cycling, researchers have used compounding methods with carbon and optimized the structure. For example, micro- and nanostructured CuO/C was prepared by spray pyrolysis, which has a specific capacity of 402 mA h $g^{-1}$ after 600 cycles at a current density of 200 mA $g^{-1}$ [113]. CuO/rGO composites were synthesized in one step using copper acetate and GO as synthetic raw materials and isopropyl alcohol as solvent and a reducing agent for the reaction in a water bath at 83 °C. The final specific capacity after 100 cycles at a current density of 50 mA $g^{-1}$ was 237.6 mA h $g^{-1}$ [114]. Copper oxide nanoarrays were prepared by the micro emulsion method, and the copper oxide array structure was grown directly on the conductive copper sheet, which facilitates electron transport and accelerates the electrochemical reaction rate. The presence of open spaces in the copper oxide array structure provides a buffering effect for the volume expansion caused by $Na^+$ insertion/extrusion, which can significantly improve the cycling and multiplicity performance of the battery at high multiplicity when applied to the negative electrode of sodium-ion batteries [115].

Spinel $Co_3O_4$, with a theoretical specific capacity of up to 890 mA h $g^{-1}$, is a promising anode material for sodium-ion batteries. However, it has some problems at present. The poor conductivity of $Co_3O_4$, as well as the large volume change during charging and discharging, lead to the actual specific capacity of $Co_3O_4$ anode being much lower than the theoretical specific capacity, and the cycling performance is not satisfactory. Researchers used the method of combining with highly conductive carbon materials or designing reasonable $Co_3O_4$ nanostructures to solve the problem of poor electrochemical performance of $Co_3O_4$ materials. Cu-doped $Co_3O_4$ anode materials, with internal voids and nitrogen-doped carbon layer cladding, were prepared using a stepwise cladding-etching method, which alleviated the volume expansion problem of $Co_3O_4$ due to the hollow structure, internal voids of the material, and external nitrogen-doped carbon layer [116]. Meanwhile, it is observed that the copper doping of the internal $Co_3O_4$ improves the electronic conductivity of the material and enhances the multiplicative properties of the material. This is further confirmed by the theoretical calculations as well. In addition, the results of theoretical calculations show that the band gap of pure $Co_3O_4$ is 1.56 eV, and after replacing $Co^{2+}$ with $Cu^{2+}$, the band gap of Cu-doped $Co_3O_4$ reduced to 1.33 eV, the Fermi energy level guide band was shifted, and an impurity energy level appeared at the top of the valence band, indicating that the electrical conductivity of $Co_3O_4$ is enhanced after Cu-ion doping. The material still has a specific capacity of 312 mA h $g^{-1}$ after 300 cycles of cycling at 2 A $g^{-1}$ when used as an anode material for sodium-ion batteries. The rhodopsin-like carbon-matrix-limited $Co_3O_4$ nanoparticle hollow sphere complex (R-$Co_3O_4$/C) was prepared by a simple one-step hydrothermal and one-step annealing process, and the material exhibited excellent cycling and multiplicative performance when used as an anode material for sodium-ion batteries. At a current density of 0.5 A $g^{-1}$, the R-$Co_3O_4$/C hybrid hollow sphere composite electrode can still maintain 74.5% of its initial capacity after 500 cycles of cycling, and even at a high current density of 5 A $g^{-1}$, the R-$Co_3O_4$/C hybrid hollow sphere electrode maintains a reversible specific capacity of 223 mA h $g^{-1}$ [117]. The ultrafine $Co_3O_4$ nanoparticles are uniformly embedded in the amorphous carbon matrix, which can effectively improve the reactivity, shorten the ion diffusion distance, and enhance the electrical conductivity and structural stability. Meanwhile, the carbon material also acts as a buffer matrix to mitigate the volume change and electrode chalking during charging and discharging.

### 4.3.2 Metal sulfide anode materials

Metal sulfide materials have attracted much attention from researchers because they undergo transformation reactions, similar to those of metal oxides, during electrochemical cycling, while having a high theoretical capacity. The M-S bond in metal sulfides is weaker than the corresponding M-O bond in metal oxides, which kinetically facilitates the conversion reaction with $Na^+$. However, similar to metal oxide materials, metal sulfides also produce large volume changes while undergoing conversion reactions, which affect the cyclic stability of the material. Depending on the metal element, the sodium storage mechanism of metal sulfides can be divided into conversion reactions, combined with embedding reactions or alloying reactions. Researchers have investigated the electrochemical properties of various metal sulfides, including tin sulfide (SnS, $SnS_2$), molybdenum sulfide ($Mo_2S$, $MoS_2$), iron sulfide (FeS, $FeS_2$), cobalt sulfide (CoS, $CoS_2$), and nickel sulfide (NiS).

Tin-based sulfides (SnS, $SnS_2$) are of interest because of their high theoretical capacities, with SnS having a theoretical specific capacity of 1,136 mA h $g^{-1}$ and $SnS_2$ having a theoretical specific capacity of 1,022 mA h $g^{-1}$. The reaction mechanism for sodium storage in tin-based sulfides is:

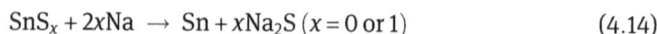

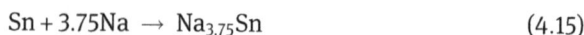

$$SnS_x + 2xNa \rightarrow Sn + xNa_2S \ (x = 0 \text{ or } 1) \tag{4.14}$$

$$Sn + 3.75Na \rightarrow Na_{3.75}Sn \tag{4.15}$$

SnS has a distorted NaCl-type structure and the reaction mechanism for the exfoliation/embedding of sodium in nano-carbon composites (SnS-C), composed of well-crystallized SnS nanoparticles, is [118]

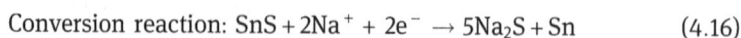

$$\text{Conversion reaction: } SnS + 2Na^+ + 2e^- \rightarrow 5Na_2S + Sn \tag{4.16}$$

$$\text{Alloying reaction: } Sn + 3.75Na^+ + 3.75e^- \rightarrow Na_{3.75}Sn \qquad (4.17)$$

The prepared SnS-C composite has a high sodium storage performance, with a specific capacity of 568 mA h g$^{-1}$ at a current density of 20 mA g$^{-1}$, and at the same time, it has a good multiplicative performance.

SnS$_2$ is a typical CdI-2-type structure with a layer spacing of 0.59 nm. In order to improve the electrochemical performance, researchers usually introduce carbon additives, such as graphene and carbon nanotubes, into the active material, thus effectively alleviating the volume expansion/shrinkage problem while improving the electronic conductivity and Na$^+$ transport rate. For example, a thin-layer SnS$_2$/graphene material was prepared using a solvothermal method, which has high capacity, ultra-long cycle life, and excellent multiplicative performance when used as the anode of sodium-ion batteries, with a high specific capacity of 649 mA h g$^{-1}$ at a current density of 100 mA g$^{-1}$ and 12.8 A g$^{-1}$ current density, and still has a reversible specific capacity of 330 mA h g$^{-1}$ [119]. The sodium storage of "multi-space" MoS$_2$ and graphene-composite microspheres synthesized by the one-step spray drying method can effectively alleviate the volume expansion during sodium storage and maintain a specific capacity of 322 mA h g$^{-1}$ at a high current density of 1.5 A g$^{-1}$ [120]. In addition, Sb$_2$S$_3$ particles loaded on S-doped graphene, which give full play to the strong adsorption capacity of polysulfides, can effectively improve the cycle life and maintain more than 83% of the initial capacity after 900 cycles of cycling.

Layered MoS$_2$ has also been reported for sodium-ion battery anode materials. The unique sandwich-type structure (S-Mo-S) is connected by van der Waals forces, with a typical layer spacing of 0.62 nm. Depending on the order of layer stacking, MoS$_2$ can be classified into two-layer stacked hexagonal polycrystalline 2 H-MoS$_2$, single-layer stacked triangular 1T-MoS$_2$, and three-layer stacked oblique hexahedral 3 R-MoS$_2$, as shown in Figure 4.24. In each layer, Mo and S exhibit trigonal coordination (2 H-MoS$_2$ and 3 R-MoS$_2$) or octahedral coordination (1T-MoS$_2$). Among them, 2 H-MoS$_2$ is a product that can be obtained by some common methods (e.g., solvothermal reaction, microwave-assisted solvothermal reaction, spray pyrolysis, CVD, and electrostatic spinning method) that can exist stably at room temperature.

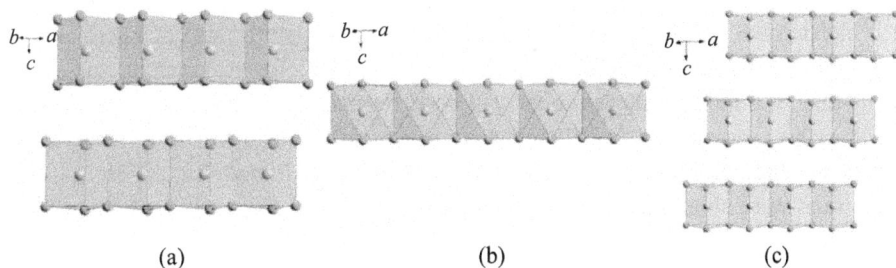

**Figure 4.24:** Crystal structures of MoS$_2$: (a) 2 H-MoS$_2$; (b) 1T-MoS$_2$; and (c) 3 R-MoS$_2$ [121].

$MoS_2$ was found to be an excellent anode material for sodium-ion batteries by theoretical calculations using first principles approach [122]. Sodium ions in $MoS_2$ materials have low embedding reaction energy barriers, fast migration rates, and their insertion/extrusion reactions have little effect on the crystal structure of the material. Moderate strength forces are formed between Na and $MoS_2$. In addition, the process of Na embedding into $MoS_2$ is a phase transition from hexagonal to tetragonal crystalline phase, which gives the material a theoretical specific capacity of 146 mA h $g^{-1}$ and a low voltage plateau of 0.75 to 1.25 V. $MoS_2$ can store $Na^+$ by embedding and transformation reactions within the working voltage window [123]. The electrochemical reaction of $MoS_2$ is divided into two steps

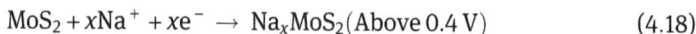

$$MoS_2 + xNa^+ + xe^- \rightarrow Na_xMoS_2 (\text{Above } 0.4 \text{ V}) \tag{4.18}$$

$$Na_xMoS_2 + (4-x)Na^+ + (4-x)e^- \rightarrow 2Na_2S + Mo (\text{Below } 0.4 \text{ V}) \tag{4.19}$$

Due to the irreversible reaction and shuttle effect of the generated polysulfides, $MoS_2$ electrodes suffer from serious cycling problems and large voltage polarization. The effective solutions considered so far include: 1. improving the electrical conductivity of the material by carbon modification; 2. improving the structural stability of the material during cycling by structural design. The monolayer of ultra-small $MoS_2$ dispersed in carbon fibers by electrostatic spinning, has a reversible specific capacity of up to 854 mA h $g^{-1}$ at a current density of 0.1 A $g^{-1}$, and after 100 cycles at a current density of 10 A $g^{-1}$, it still has a specific capacity of 253 mA h $g^{-1}$. $MoS_2$ nanoflowers, with a high discharge specific capacity of 350 mA h $g^{-1}$ at a current density of 50 mA $g^{-1}$, were prepared by increasing the planar layer spacing and could be stably cycled for more than 1,500 cycles [124]. The heterogeneous structure formed between $MoS_2$ and C in the $MoS_2$/C micron tube, with a layered structure prepared by a simple template method, can prevent the accumulation and agglomeration of $MoS_2$, in addition to providing more sodium-embedded active sites, while facilitating the improvement of the electrical conductivity and $Na^+$ diffusion ability of $MoS_2$ and inhibiting the volume expansion of $MoS_2$ during electrochemical cycling. The prepared $MoS_2$/C micron tubes have excellent electrochemical performance as the anode of sodium-ion batteries, with a reversible specific capacity of up to 563.5 mA h $g^{-1}$ after 100 cycles at a current density of 200 mA $g^{-1}$. The specific capacity remained at 484.9 mA h $g^{-1}$ after 1,500 cycles at a high current density [125]. The in situ TEM confirmed that the $MoS_2$/C micron tubes were able to maintain structural stability during the electrochemical cycling, and at the same time, the $MoS_2$/C micron tubes had good reversibility during the charging and discharging process. The specific capacity remained at 484.9 mA h $g^{-1}$ after 1,500 cycles at a high current density [125]. The in situ TEM confirmed that the $MoS_2$/C micron tubes were able to maintain structural stability and had good reversibility during the charge/discharge process.

Pyrite $FeS_2$ is a natural mineral and has the advantages of abundant resources, low cost, and a theoretical specific capacity up to 894 mA h $g^{-1}$; so it can be used as an anode material for sodium-ion batteries. $FeS_2$ is a cubic structure belonging to the $Pa\bar{3}$ space group, in which Fe occupies the $(FeS_6)$ octahedral site and sulfur atoms are present as $S^{2-}$. Ahn et al. first reported a $Na/FeS_2$ cell, and the XRD patterns and SEM images of the $FeS_2$ electrode are shown in Figure 4.25(a) and (b) [126]. The battery had a first discharge specific capacity of 447 mA h $g^{-1}$ at room temperature and maintained a specific capacity of 70 mA h $g^{-1}$ after 50 cycles. The nanoscale $FeS_2$ material [127] was applied to the negative electrode of the sodium-ion battery and still had a high capacity of 500 mA h $g^{-1}$ after 400 cycles at a current density of 1 A $g^{-1}$. The conversion mechanism of $FeS_2$ is

$$FeS_2 + 2Na^+ + 2e^- \rightarrow Na_2FeS_2 \tag{4.20}$$

$$Na_2FeS_2 + 2Na^+ + 2e^- \rightarrow 2Na_2S + Fe \tag{4.21}$$

Ferrous sulfide (FeS) has a lower theoretical specific capacity of 609 mA h $g^{-1}$, compared to $FeS_2$. FeS belongs to the hexagonal structure of the P63/mmc space group. The sodium storage mechanism is

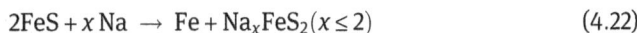

$$2FeS + x\,Na \rightarrow Fe + Na_xFeS_2 (x \leq 2) \tag{4.22}$$

$$Na_xFeS_2 + (4-x)\,Na \rightarrow 2Na_2S + Fe \tag{4.23}$$

$FeS_2$ materials suffer from poor cycling stability due to the large volume expansion rate, and researchers have used methods such as tuning the microscopic size and morphology of the materials to improve the electrochemical performance. Flexible self-supporting carbon cloth thin film carbon-coated FeS materials were prepared (Figure 4.26(c), (d)), which have high reversible capacity and excellent multiplicity performance when used as anode materials for sodium-ion batteries [128]. FeS/C nanocomposites are dominated by pseudocapacitive behavior with fast electrochemical reaction kinetics and excellent electrochemical performance, especially sodium storage performance at high multiplicity. The nanocomposites have reversible specific capacities of 547.1 mA h $g^{-1}$ and 228 mA h $g^{-1}$ at current densities of 0.05 A $g^{-1}$ and 80 A $g^{-1}$.

In addition, cobalt-based sulfides ($CoS$, $CoS_2$, $Co_3S_4$, and $Co_9S_8$) and nickel-based sulfides ($NiS_2$ and $Ni_3S_2$) have been investigated as anode materials for sodium-ion batteries because of their high theoretical specific capacities. The theoretical specific capacity of CoS is 589 mA h g-1, of $CoS_2$ is 872 mA h g-1, of $Co_3S_4$ is 702 mA h g-1, and of $Co_9S_8$ is 544 mA h g-1. Among them, $CoS_2$ is a cubic phase Pa space group and the reaction mechanism in the sodium-ion battery is

$$CoS_2 + xNa \rightarrow Na_xCoS_2 \text{ (above 1.0 V)} \tag{4.24}$$

$$Na_xCoS_2 + (4-x)Na \rightarrow Co + 2Na_2S \text{ (below 1.0V)} \tag{4.25}$$

The nickel-based sulfides include $NiS_2$ in the Pa space group cubic phase and $Ni_3S_2$ in the rhombohedral structure, with theoretical specific capacities of 873 mA h $g^{-1}$

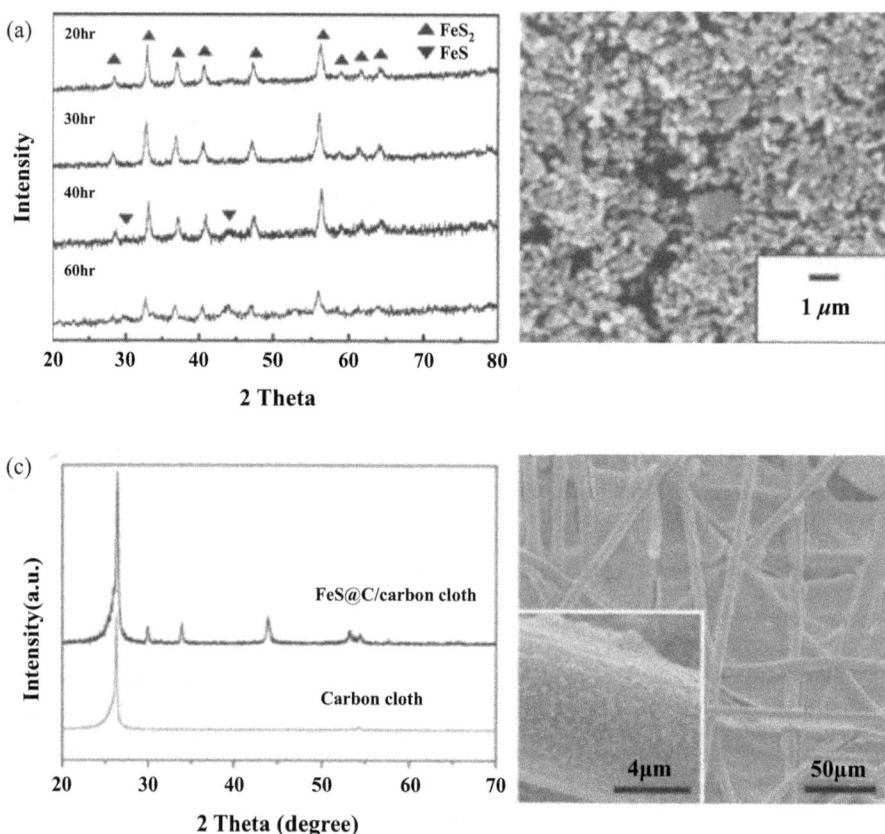

**Figure 4.25:** Physical characterization of FeS$_2$ and FeS@C/carbon cloth: (a) XRD patterns of FeS$_2$ powders at different mechanical alloying times; (b) SEM image of FeS$_2$ electrode [126]; (c) FeS@C/ XRD patterns of carbon cloth and carbon cloth; and (d) SEM images and enlarged SEM images of FeS@C/carbon cloth [128].

and 642 mA h g$^{-1}$, respectively. Although the theoretical specific capacities of cobalt-based sulfides and nickel-based sulfides are high, the electrical conductivity is low. At the same time, a large volume expansion is generated during the cycling process, leading to agglomeration of the active material and a very obvious capacity decline. Similar to other metal sulfide modification methods, researchers mostly design a new nanostructure and prepare composite nanomaterials using various carbon additives to improve their sodium storage performance. Hybrid nanocomposites were obtained by immobilizing cobalt sulfide (CoS) nanoplates on reduced graphene oxide (RGO), which has a high specific capacity of 540 mA h g$^{-1}$ at a current density of 1 A g$^{-1}$ and a capacity retention of 88% after 1,000 cycles [129]. The preparation of carbon-coated NiS$_x$ complexes by N/S co-doping as anode materials for sodium-ion batteries can effectively improve the electronic and ionic conductivity of the materials, with a reversible specific capacity of up to 338.4 mA h g$^{-1}$ at a current density of 2 A g$^{-1}$ [130]. Nickel sulfide@nitrogen-sulfur co-doped carbon nanotube materials (Ni$_3$S$_2$@NS-CNTs) were fabricated at a heat treatment temperature of 1,000 °C using glucose as the carbon source to achieve in situ filling of metal sulfide nanoparticles during the growth of carbon nanotubes, while uniformly doping N, S, and O into the carbon nanotube walls [131]. As shown in Figure 4.26, Ni$_3$S$_2$@NS-CNTs were able to form an ultrathin stable SEI film on the surface in ether electrolyte, and the dense and stable inorganic composition and partially highly reversible chemical composition of this SEI film ensured the electrochemical performance of this electrode material. The cyclic voltage curves show that after the first cycle, the second and third CV curves have obvious overlapping redox peaks, indicating that the reaction is reversible. Ni$_3$S$_2$@NS-CNTs were applied to sodium-ion batteries with a high initial coulombic efficiency of 93% in ether electrolyte and a specific capacity of 463 mA h g$^{-1}$ at a current density of 0.2 A g$^{-1}$. The above results indicate that the conductivity of the material can be improved when compounded with carbon materials, while the unique structure can alleviate the volume expansion during the cycling process, so that the anode material of sodium sulfide ion battery with good cycling performance can be obtained.

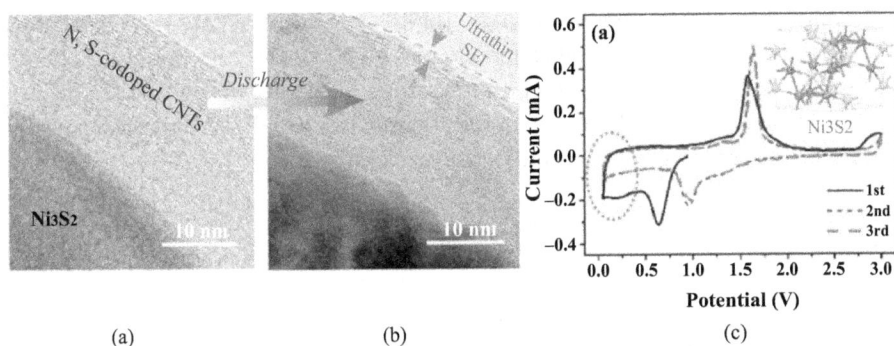

(a)  (b)  (c)

**Figure 4.26:** HRTEM and CV curves of Ni$_3$S$_2$@NS-CNTs in different states: (a) original Ni$_3$S$_2$@NS-CNTs; (b) HRTEM image of edge after discharge; and (c) CV curve of Ni$_3$S$_2$@NS-CNTs electrode scanning rate of 0.1 mV·s$^{-1}$ in the first three cycles [131].

### 4.3.3 Metal selenide anode materials

Metal selenides are one of the current research hotspots for anode materials due to their easily controlled morphology and high theoretical specific capacity, as well as their higher initial coulombic efficiency and more stable cycling performance than oxides and sulfides. Compared to MoS$_2$, MoSe$_2$ has a larger layer spacing and smaller band gap, with higher coulombic efficiency and electronic conductivity. However, MoSe$_2$ has lower ionic conductivity, larger volume expansion, and higher mechanical stress during Na$^+$ insertion/extrusion, which limits its electrochemical performance, and has poor cycling stability and multiplicative performance. As shown in Figure 4.27, it was found that the two reduction peaks found near 0.6 V and 0.3 V during the initial discharge of $MoS_2$ correspond to the sodium embedding into $MoSe_2$ crystals and the conversion of $MoSe_2$ to Mo and NaSe, respectively [132–136]. However, there are higher reduction peaks above 1.3 V and they originate from the sodium embedding process [137–140]. The reduction peak near 1.8 V is common during charging, but its corresponding reaction mechanism still has no unified answer. One theory suggests that this reduction peak corresponds to the formation of selenium, while the other suggests that this peak reflects the formation of $MoSe_2$.

In order to improve the electrochemical properties of $MoSe_2$, researchers have modified $MoSe_2$. The modification methods carried out mainly include carbon cladding, porous structure design, and crystallization. The $MoSe_2$/reduced graphene oxide (rGO) composites, synthesized by the hydrothermal method, showed better cycling performance, with the material having a specific capacity of 462 mA h $g^{-1}$ and 456 mA h $g^{-1}$ at the 10th and 100th cycle at a current density of 200 mA $g^{-1}$ [134]. $MoSe_2$/C composites synthesized the amorphous carbon on the surface of $MoSe_2$ using the oleic acid-assisted reflow method, resulting in a more stable structure, and the material exhibited a reversible specific capacity of 445 mA h $g^{-1}$ after 100 cycles at a current density of 1A $g^{-1}$ [136]. $MoSe_2$, with a core-shell structure, was synthesized using the selenization method for $MoO_3$ [137]. Its charge/discharge curves showed that there was a phase transition occurring in the first two cycles and the multiplicative performance of the material improved, discharging a specific capacity of 345 mA h $g^{-1}$ at a current density of 1.5 A $g^{-1}$.

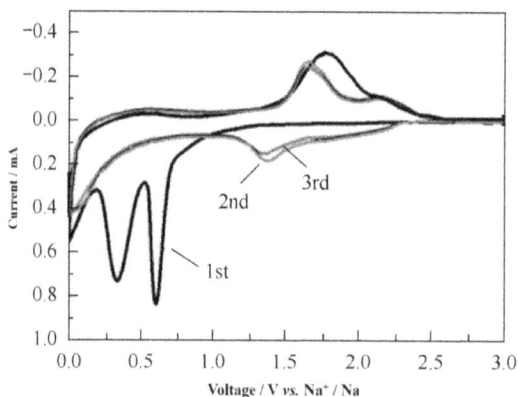

Figure 4.27: CV curve of $MoSe_2$ in the process of sodium-ion implantation/exhumation [133].

$FeSe_2$ is an orthorhombic crystal system, with a space group of Pnnm and a theoretical specific capacity of 500 mA h $g^{-1}$. The synthesis method includes solvothermal reaction or spray pyrolysis, combined with the sulfide method. During the discharge process, FeSe$_2$ is first reduced to NaxFeSe$_2$ and then FeSe is generated. Finally, iron atoms and Na$_2$Se are detected in the fully discharged state. Completely different from iron sulfide, the reaction is reversible during the charging process and the final charging product is still FeSe$_2$ [141]. The specific reaction equation for the charging process is

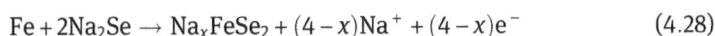

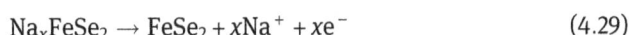

$$FeSe_2 + xNa^+ + xe^- \rightarrow Na_xFeSe_2 \qquad (4.26)$$

$$Na_xFeSe_2 + (2-x)Na^+ + (2-x)e^- \rightarrow FeSe + Na_2Se \qquad (4.27)$$

The specific reaction equation for the charging process is

$$Fe + 2Na_2Se \rightarrow Na_xFeSe_2 + (4-x)Na^+ + (4-x)e^- \qquad (4.28)$$

$$Na_xFeSe_2 \rightarrow FeSe_2 + xNa^+ + xe^- \qquad (4.29)$$

Zhang et al. synthesized FeSe$_2$ for the first time by a solvothermal reaction using ammonium iron sulfate and selenium [141]. As shown in Figure 4.28, the FeSe$_2$ microspheres are 2–5 μm in size, and each microsphere is assembled from several nano-octahedra. FeSe$_2$, using an ether-based electrolyte, can put out a stable specific capacity of 403 mA h g$^{-1}$ at 1 A g$^{-1}$ current density, and has a capacity retention of 89.0% after 2,000 cycles. The material multiplicative performance is better due to the contribution originating from the pseudocapacitance, with a discharge specific capacity of 226 mA h g$^{-1}$ at 25 A g$^{-1}$. It has been shown that carbon modification can also improve the cycling performance of such materials. The FeSe$_2$/amorphous carbon composite, synthesized by spray drying and selenization method, discharged a specific capacity of 383 mA h g$^{-1}$ at a current density of 500 mA g$^{-1}$, but the capacity loss was as high as 69% after 150 cycles [142]. The FeSe$_2$/rGO composite obtained using the GO-FeSe$_2$ composite method discharged a specific capacity of 408 mA h g$^{-1}$ after 100 cycles, with a capacity retention of 90% (compared to the capacity of the second cycle) [143].

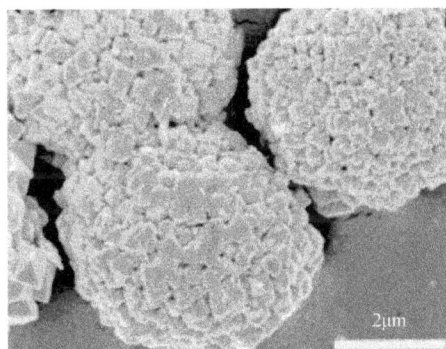

**Figure 4.28:** SEM images of FeSe$_2$ microspheres [141].

CoSe$_2$ is an orthorhombic albite phase, with a theoretical specific capacity of 494 mA h g$^{-1}$ and a reaction mechanism similar to that of FeSe$_2$ [144]. CoSe$_2$ exhibits good cycling and multiplicative properties when used as anode material for lithium-ion batteries, but less research has been carried out with respect to sodium-ion batteries. At the early stage of discharge, Na$_x$CoSe$_2$ and CoSe are produced. In the fully discharged state, the products, Co and Na$_2$Se, are obtained. During charging, all reactions are reversible. CoSe$_2$ has received much attention from researchers because of its high sodium storage capacity and suitable reaction potential. However, the poor conductivity of metallic selenides and the severe volume expansion during sodium storage lead to their poor multiplicity and cycling stability performance. It is particularly important to know how to enhance the sodium storage performance of metalloselenides through the modulation of the structure and components. Sea urchin-like CoSe$_2$ is synthesized by the solvothermal reaction, as shown in Figure 4.29. The material is assembled from nanorods, with lengths ranging from 20 to 100 nm and a size of about 1 μm [144]. If 1 mol NaCF$_3$SO$_3$/diethylene glycol dimethyl ether is used as the electrolyte, the material is able to discharge specific capacity of 434 mA h g$^{-1}$ and 354 mA h g$^{-1}$ at current densities of 0.1 A g$^{-1}$ and 10 A g$^{-1}$, respectively. Even at a current density of 1 A g$^{-1}$, the material was able to discharge a specific capacity of about 410 mA h g$^{-1}$ after 30 cycles and maintained it until the 1,800th cycle. In conclusion, the use of redox graphene or porous carbon, as a support for active materials, can improve the electrochemical properties of the materials by using more gentle and facile synthesis methods, such as hydrothermal or solvent heat [145, 146].

**Figure 4.29:** TEM images of sea urchin-like CoSe$_2$ [144].

### 4.3.4 Metal phosphide anode materials

Metal phosphide is a promising anode material for sodium-ion batteries. Similar to metal oxides and metal sulfides, metal phosphides undergo conversion reactions with metal Na to produce metal monomers and $Na_3P$. The uniform distribution of metal monomers in $Na_3P$ speeds up the kinetic process of the $Na_3P$ oxidation reaction, giving metal phosphides good electrochemical performance when used as anode materials for sodium-ion batteries. However, the defects of metal phosphides also limit their own development. On the one hand, the low electrical conductivity of metal phosphide makes its active material utilization in sodium-ion batteries low, showing a capacity much lower than its theoretical capacity, and on the other hand, during the charging and discharging alloying process, the large volume change destroys the original microscopic morphology of the electrode material, resulting in poor cycling stability. So far, the improvement scheme mainly focuses on the compounding of phosphorus-based materials with carbon materials and the construction of nanostructures. The carbon material, as a material with good electrical conductivity, can improve the overall stability of the electrode material, while the nanoscale particles can reduce the ion/electron transport path, and the reasonable nanostructure construction can protect the microscopic morphology of the metal phosphide from being destroyed during the charging and discharging process.

These metal phosphides are broadly divided into two types based on whether the metals in the metal phosphides are electrochemically active: inactive metal phosphides and active metal phosphides. Inactive metal phosphides include Co-P (CoP), Cu-P ($CuP_2$, $Cu_3P$), Fe-P (FeP, $FeP_4$), and Ni-P ($NiP_3$). The use of electrochemically inert metals, such as Cu, Fe or Ni, does not contribute to the capacity of these metal phosphides, and also reduces the theoretical capacity of the metal phosphides. However, metals have high electrical conductivity, which helps to fully utilize the P capacity. At the same time, these metal inert phosphides have a relatively small volume expansion. Their charging and discharging mechanism is

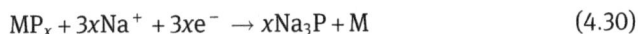

$$MP_x + 3xNa^+ + 3xe^- \rightarrow xNa_3P + M \qquad (4.30)$$

$NiP_3$/carbon nanotubes (CNTs) composites were synthesized by a simple ball milling method, which consists of $NiP_3$ particles chemically bonded to functionalized CNTs [147]. The addition of conductive carbon nanotubes increased the $Na^+$ diffusion coefficient by two orders of magnitude, from $9.15 \times 10^{-17}$ $cm^2 \cdot S^{-1}$ to $2.91 \times 10^{-15}$ $cm^2 \cdot S^{-1}$, and the charge transfer resistance was reduced to 1/6th of the original one. The $NiP_3$/CNTs composite negative electrode has a high initial reversible specific capacity of 853 mA h $g^{-1}$ and maintains more than 80% capacity after 120 cycles at a current density of 200 mA $g^{-1}$, and the high multiplicative capacity of 363.8 mA h $g^{-1}$ remains after 200 cycles at a current density of 1,600 mA $g^{-1}$. Density functional theory (DFT) calculations, combined with molecular dynamics (AIMD), and simulations reveal strong chemical interactions between the red P in $NiP_3$ and the functional groups on the carbon nanotubes through ball milling, to form P-C and P-O-C bonds. Figure 4.30 shows the X-ray diffraction (XRD) patterns of $NiP_3$, CNTs and $NiP_3$/CNT composites. All sharp peaks match well with the crystalline $NiP_3$ (JCPDS: 98–152-6485, cubic), and the broad peak at ~ 25.9° corresponds to CNTs in the composites, demonstrating the presence of CNTs. The absence of the Ni peak confirms the complete formation of the phosphorus-rich $NiP_3$ phase. FeP has the advantage of abundant resources and low price, while its theoretical specific capacity can reach 926 mA h $g^{-1}$. Wang et al. interconnected hollow FeP (H-FeP) nanospheres with GR to form a three-dimensional laminar structure, which improved the electronic conductivity of the material and accelerated the electrode reaction kinetic process [148]. The material still has a specific capacity of 400 mA h $g^{-1}$ after 250 cycles at a current density of 0.1 A $g^{-1}$, when used as anode material for sodium-ion batteries, with excellent electrochemical performance.

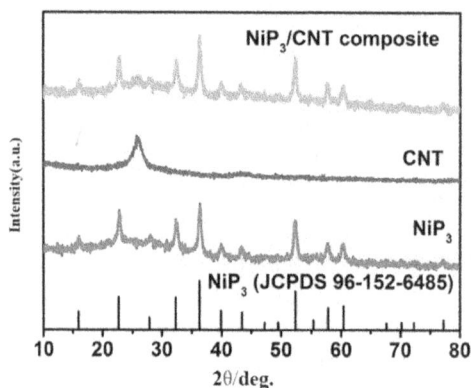

**Figure 4.30:** XRD patterns of $NiP_3$, CNTs, and $NiP_3$/CNT composites [147].

The metals in the metal-active phosphides are electrochemically active (Se-P, Sn-P, and Ge-P). In addition to conversion reactions, these reactive metal phosphides undergo alloying reactions to form $Na_yM$:

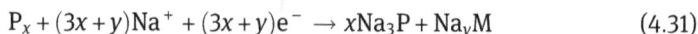

$$P_x + (3x+y)Na^+ + (3x+y)e^- \rightarrow xNa_3P + Na_yM \tag{4.31}$$

Since both metals and P react with $Na^+$, active metal phosphides usually have a higher theoretical capacity than inactive metal phosphides, and their theoretical capacity depends on the type and content of the metal in the metal phosphide. Among them, tin-based phosphides ($Sn_4P_3$, $SnP_3$) are widely reported as anode materials for sodium-ion batteries. The theoretical capacity of $Sn_4P_3$ is 1,133 mA h $g^{-1}$ and the theoretical specific capacity of $SnP_3$ is 1,616 mA h $g^{-1}$. High-energy mechanical ball milling is considered a relatively simple method for the synthesis of phosphorus-based materials, and the $Sn_4P_3$ intermetallic compound is prepared by high-energy mechanical ball milling, which has a reversible specific capacity of 718 mA h $g^{-1}$ with stable cycling performance and almost zero capacity decay at 100 cycles [149]. Current approaches to improve the electrochemistry of tin-based phosphides include conductive carbon cladding and nanostructure design. Homogeneous yolk-shell $Sn_4P_3@C$ nanosphere structures can alleviate volume expansion, and when used as anode materials for sodium-ion batteries exhibit a specific capacity of 360 mA h $g^{-1}$ at 1.5C for 400 cycles [103]. In addition, other metal phosphides have been extensively studied as anode materials for sodium-ion batteries. Similar to tin phosphide, selenophosphide was also proposed as a novel metal-active phosphide by Chen et al. in 2017 [150]. The sodization process of $Se_4P_4$ consists of three steps: first, $Se_4P_4$ is converted to amorphous $Na_xSe_4P_4$; then $Na_xSe_4P_4$ generates $Na_2Se$ and elemental P; finally, monomeric P further reacts with $Na^+$ to form $Na_3P$. The reaction equation is:

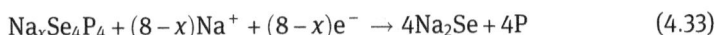

$$Se_4P_4 + xNa^+ + xe^- \rightarrow Na_xSe_4P_4 \tag{4.32}$$

$$Na_xSe_4P_4 + (8-x)Na^+ + (8-x)e^- \rightarrow 4Na_2Se + 4P \tag{4.33}$$

$$4P + 12Na^+ + 12e^- \rightarrow 4Na_3P \tag{4.34}$$

The XRD and Raman spectra of $Se_4P_4$, P, and Se are shown in Figure 4.31. During the sodization process, 20 $Na^+$ were involved in the reaction, with a theoretical capacity of 1217 mA h $g^{-1}$. The reversible specific capacity of $Se_4P_4$ synthesized by this ball mill was 1,048 mA h $g^{-1}$ (about 86% of the theoretical specific capacity), and 804 mA h $g^{-1}$ was still retained after 60 cycles. Its excellent electrochemical properties are attributed to the fact that element P prevents the aggregation of $Na_2Se$ during the sodization process, while the semiconductor, $Na_2Se$, provides the conductive pathway and facilitates the reaction. So far, some progress has been made in metal phosphides such as $Sn_4P_3$ [103], FeP [148], $Cu_3P$ [104], and $NiP_3$ [151], but for practical application in sodium-ion batteries, researchers still need to make improvements to material design and electrode matching.

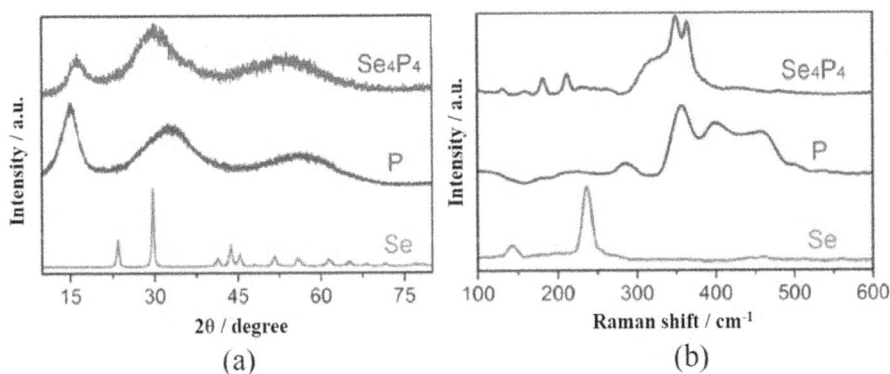

**Figure 4.31:** Physical characterization of $Se_4P_4$, P, and Se:(a) XRD pattern; and (b) Raman atlas [150].

### 4.3.5 Metal fluoride anode materials

Other transforming anode materials, such as transition metal fluorides, have also been used by researchers as anode materials for sodium-ion batteries. Transition metal fluorides have received increasing attention as promising anode and cathode materials for lithium-ion and sodium-ion batteries due to their abundant resources, low cost, and high specific capacity. The currently studied transition metal fluorides include $NaMF_3$, among which, the low electrical conductivity due to the large energy band gap of M = Fe, Mn, Ni and V, as well as the large volume expansion during charging and discharging, reduce the cycling stability and multiplicative performance of the transition metal fluorides and limit their practical applications. In addition, the transition metal fluorides exhibit relatively poor sodium storage performance due to the large $Na^+$ radius. In order to improve the electrical conductivity of the transition metal fluorides, researchers have prepared carbon-based transition metal fluorides by compounding different carbon materials (carbon black, graphene, carbon nanotubes) with them. Most synthesis methods of metal fluorides involve ball milling and fluorination. The materials obtained by ball milling generally have micron-level structures, and the splitting during the cycling process reduces the cycle life of the materials; while the fluorination method adds a large amount of fluorine source, which results in the waste of fluorine source, on the one hand, and generates toxic gas during the process and pollutes the environment. Conventional synthesis of carbon-based transition metal compounds generally requires two steps: first, synthesizing the transition metal compound, and then, compounding it with carbon materials; the compounding method is mostly a ball milling method. Therefore, it is extremely critical to develop an environmentally friendly method to prepare carbon-based transition metal fluorides with nanostructures.

Using $SnF_2$ and acetylene black as raw materials, a composite material of $SnF_2$ and C was prepared as a high-performance anode material for sodium-ion batteries by a simple ball milling method, and its electrochemical properties and related energy storage mechanism were studied as

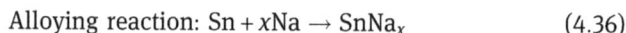

$$\text{Transformation reaction: } SnF_2 + 2Na \rightarrow Sn + 2NaF \tag{4.35}$$

$$\text{Alloying reaction: } Sn + xNa \rightarrow SnNa_x \tag{4.36}$$

The nanocomposite electrode has a high reversible specific capacity of 563 mA h $g^{-1}$, which can still reach 191 mA h $g^{-1}$ at a high current density of 1 C, with a high multiplicative performance [152]. $NH_4FeF_3$/carbon nanosheet ($NH_4FeF_3$/CNS) composites can be prepared by the pyrolysis reaction, using iron acetylacetonate and $NH_4F$ as raw materials. $NH_4FeF_3$ has an open skeleton structure, with a chalcogenide topology, in which $FeF_6$ octahedral monomers are interconnected by F-anion to form cavities, while cations are located in the structure. $NH_4FeF_3$/CNS has a specific capacity of 504 mA h g$^{-1}$ when used as the anode of sodium-ion batteries, with good multiplicative performance and cycling stability, as shown in Figure 4.32 [153].

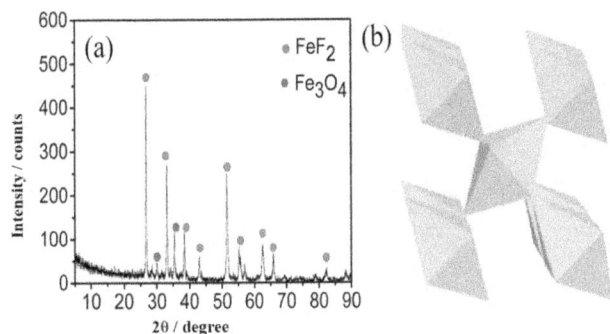

**Figure 4.32:** XRD patterns of FeF$_2$/CNS and crystal structure of FeF$_2$: (a) XRD patterns of FeF$_2$/CNS obtained by annealing NH$_4$FeF$_3$/CNS composite at 500 °C for 3 h; and (b) crystal structure.

## 4.4 Intermetallic compound anode materials

Intermetallic compound materials have received a lot of attention from researchers because of their high specific capacity and relatively low working potential (below 1.0 V), as well as the advantages of simple processing and non-pollution. The reactions of Na + with intermetallic compounds in alloying/de-alloying reactions can be summarized as

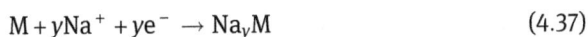

$$M + yNa^+ + ye^- \rightarrow Na_yM \tag{4.37}$$

Metals (Sn, Bi), quasimetals (Si, Ge, As, Sb), and polyatomic nonmetallic compounds (P) are all promising anode materials for sodium-ion batteries, as shown in Figure 4.33 [154]. In addition, intermetallic compound materials have relatively high electron conductivity, which facilitates electron transport in the electrode material. However, the alloying reaction of Na$^+$ with larger ionic radius causes large volume expansion, which is restricted to produce huge stress when assembled into a battery, leading to pulverization and flaking of the electrode material. Alloyed electrodes are subjected to huge mechanical stresses during charge/discharge reactions and inevitably become pulverized; thus, gradually losing electrochemical activity [154]. In order to alleviate the volume expansion of alloy materials during cycling, researchers usually use modification methods, such as improving binders, using electrolyte additives, preparing composites by compounding with conductive carbon, and preparing nanostructured alloy materials, to improve the specific capacity, cycling stability, and multiplicative performance of alloy materials. This section introduces the research progress and current status of various alloy materials.

(a)

(b)

(c)

(d)

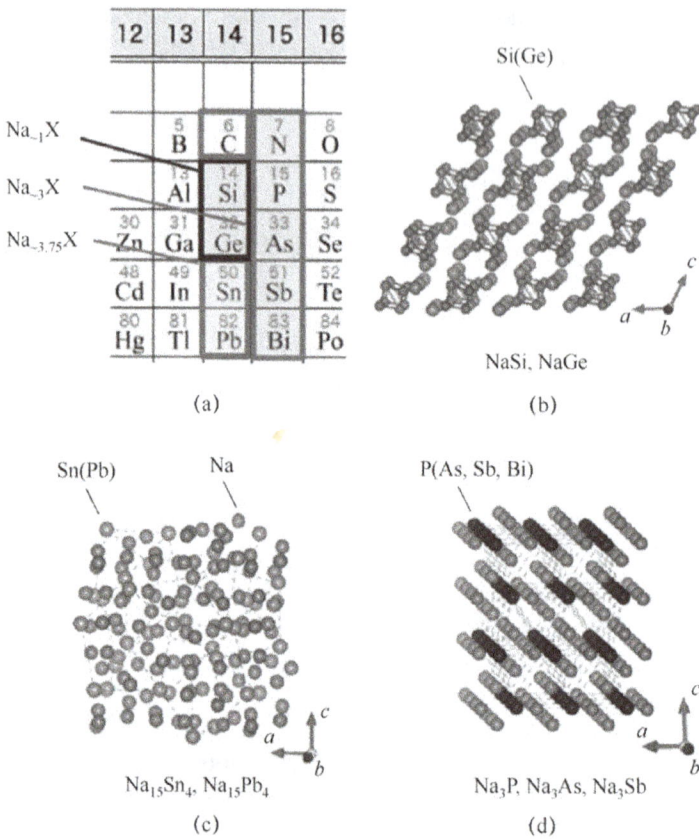

**Figure 4.33:** Intermetallic compounds used as anode materials for sodium-ion batteries: (a) group I4, I5 in the periodic table of elements; (b) Si, Ge; (c) Sn, Pb; and (d) schematic diagram of the structure of P, As, Sb embedded sodium [154].

## 4.4.1 Metal anode materials (Sn, Pb, Bi)

Tin-based materials are a very promising material for sodium storage anode, and the theoretical specific capacity of Sn, calculated according to the formation of $Na_{15}Sn_4$ alloy between Sn and Na, is 847 mA h $g^{-1}$. The discharge curves and volume change process curves for the elements of group I4 and I5 of the periodic table were calculated and obtained by means of DFT, as shown in Figure 4.34 [155].

The course of Na-Sn alloying reaction needs to go through four phases of phase change, $NaSn_5$, $NaSn$, $Na_9Sn_4$, and $Na_{15}Sn_4$, and the electrochemical reaction steps of Na and Sn are shown in Figure 4.35:

$$\text{Platform 1: } Na + 3Sn \rightarrow NaSn_3^* \tag{4.38}$$

$$\text{Platform 2: } 2Na + NaSn_3^* \rightarrow 3(a - NaSn) \tag{4.39}$$

$$\text{Platform 3: } 5Na + 4(a - NaSn) \rightarrow Na_9Sn_4 \tag{4.40}$$

$$\text{Platform 4: } 6Na + Na_9Sn_4^* \rightarrow Na_{15}Sn_4 \tag{4.41}$$

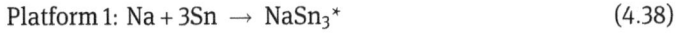

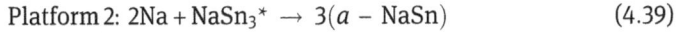

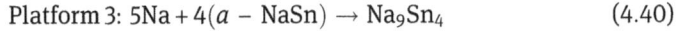

where $a$ is the amorphous phase; * is the new crystalline phase.

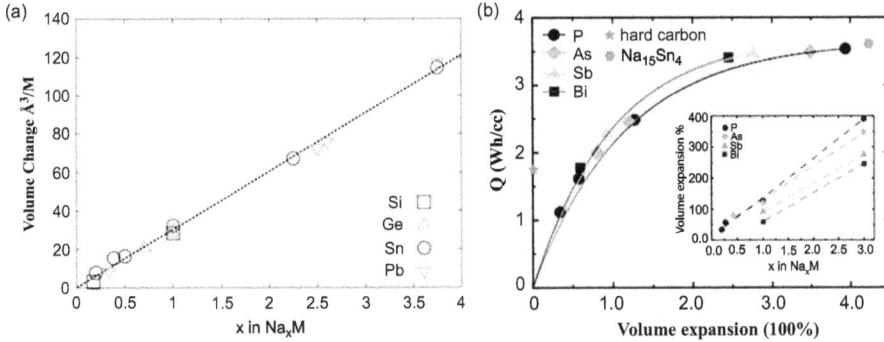

**Figure 4.34:** Discharge curves and volume changes of main group elements of $I_4$ and $I_5$: (a) main group elements of $I_4$; and (b) $I_5$ main group elements [155].

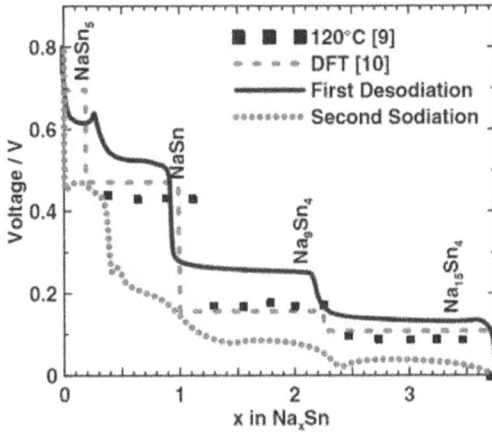

**Figure 4.35:** Reaction mechanism of Sn as anode material of sodium-ion battery.

Tin undergoes a two-step sodium embedding process in the first step to form amorphous $NaSn_2$ (56% expansion), and in the second step to form amorphous $Na_9Sn_4$, $Na_3Sn$ (336% expansion), and crystalline $Na_{15}Sn_4$ (420% expansion), in that order [154, 156, 157]. The large volume changes that exist in the tin-based anode during Na atom insertion and extraction can lead to chalking of the electrode and accelerated capacity decay during cycling. Thus, most of the work of researchers has focused on dealing with the volume change during the alloying/de-alloying reaction, and composite formation with conductive carbon is a very effective way to buffer the volume change. Liu et al. introduced Sn@C composites with ultrafine Sn nanoparticles [158, 159]. Firstly, ultra-small Sn nanoparticles ($\approx$8 nm) were uniformly embedded in spherical carbon using an aerosol spray pyrolysis method for use as anode materials for sodium-ion batteries. The nanocomposite has an initial reversible specific capacity of 493.6 mA h g$^{-1}$ at a current density of 200 mA g$^{-1}$, and still provides high multiplicative specific capacity of 349 mA h g$^{-1}$ at a current density of 4,000 mA g$^{-1}$. The remarkable electrochemical performance of the material is attributed to the synergistic effect between the well-dispersed ultrasmall Sn nanoparticles and the conductive carbon network. The unique structure of the ultrasmall Sn nanoparticles, embedded in the porous carbon network, can effectively suppress the volume fluctuation and particle aggregation during cycling, effectively solving the major problems of electrode crushing and poor electrical contact of the Sn electrodes. Later, they also uniformly encapsulate ultrasmall Sn nanodots (1–2 nm) in porous N-doped carbon nanofibers by a simple electrostatic spinning method, to form flexible self-supporting films, which can be directly used as anode materials for sodium-ion batteries with high reversible capacity, excellent multiplicative performance, and ultra-long cycle life.

In addition to improving the electrochemical performance of Sn by constructing composite electrodes, the reversibility of the alloying/de-alloying process of Sn electrodes can be effectively improved by using polyacrylate (PAA) binders [160]. The use of PAA binder, instead of PVdF, can also effectively improve the specific capacity and coulombic efficiency, but the poor capacity retention of the Sn-PAA electrode can be attributed to the volume change of the electrode material and the electrolyte decomposition on the surface of the Na–Sn alloy. The performance of the Sn-PAA electrode was further improved, when a small amount of FEC was added to the electrolyte, and the Sn electrode still had a reversible specific capacity of 700 mA h g$^{-1}$ for more than 20 cycles.

In addition, bismuth (Bi), due to its layered crystal structure with large layer spacing, reacts with Na to form Na$_3$Bi, with a theoretical specific capacity of 385 mA h g$^{-1}$. NaBi is more readily converted to c-Na$_3$Bi (c: cubic) crystal structure, as opposed to h-Na$_3$Bi (h: hexagonal) crystal structure [161]. The phase fractions of Na-Bi in the 100th cycle in the charged (2 V) and discharged (0 V) states indicate that c-Na$_3$Bi is formed on the crystalline surface. And, it was calculated by DFT simulations that Bi can provide sites for Na$^+$ diffusion, according to the intercalation mechanism rather than the alloying mechanism [162]. Bi has a layered crystal structure with a large layer spacing along the c-axis (d(003) = 3.95 Å) that can accommodate Na$^+$ and, therefore, can exhibit good multiplicative properties.

## 4.4.2 Metal-like anode materials (Si, Ge, As, Sb)

The quasi-metallic elements contained in IVA and VA groups are Si, Ge, As, and Sb. Silicon, as the material with the highest theoretical lithium storage capacity, has been successfully used in commercial lithium-ion batteries and is one of the current research hotspots as the anode material for sodium-ion batteries; however, silicon has been considered to have no sodium storage activity. Relevant theoretical calculations show that sodium storage in amorphous silicon is promising due to the fact that sodium ions need to consume a lot of energy to embed in crystalline silicon materials, while the energy consumed in amorphous silicon is much less than that of crystalline silicon. It has been calculated that a single amorphous Si can store 0.76 Na, corresponding to a theoretical specific capacity of 725 mA h g$^{-1}$. In recent years, there have been experimental studies on sodium storage in silicon materials, in which silicon as the active material is basically partially or fully amorphous, confirming the sodium storage activity of amorphous Si. The Na-Si binary phase diagram was first presented from the results of differential thermal analysis and X-ray diffraction by Moritoet, as shown in Figure 4.36, which confirmed the fully embedded sodium form of Na-Si as monoclinic structure of NaSi [163]. The diffusion kinetics of Na in Si are slow, and theoretical calculations indicate that the structurally modified Si is more suitable as an ideal electrode for sodium-ion batteries, such as amorphous Si, that can bind better to Na [164]. The reversible electrochemical properties of Na$^+$ in Si were first demonstrated experimentally by Xu et al. The prepared nanoparticles of amorphous Si and crystalline Si exhibited 10 mA g$^{-1}$ current density with a reversible specific capacity of 279 mA h g$^{-1}$ and a stable specific capacity of 248 mA h g$^{-1}$ after 100 cycles at a current density of 20 mA g$^{-1}$ [165]. Also, through various analytical techniques, they proposed a possible mechanism for Na storage. The reaction that occur during the sodification process is

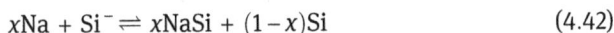

$$x\text{Na} + \text{Si}^- \rightleftharpoons x\text{NaSi} + (1-x)\text{Si} \qquad (4.42)$$

The reaction that occurs during the desodification process is

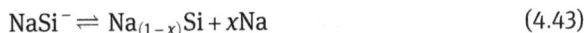

$$\text{NaSi}^- \rightleftharpoons \text{Na}_{(1-x)}\text{Si} + x\text{Na} \qquad (4.43)$$

Germanium is chemically similar to silicon, and is bonded to at most one sodium atom [166, 167]. Sodium can be alloyed with germanium to form sodium-germanium compounds, with a theoretical specific capacity of 369 mA h g$^{-1}$ [168]. However, during the charge and discharge of the battery, the huge volume change effect of germanium-based materials can lead to electrode pulverization and flaking, which can seriously affect the cycling performance of Li-ion batteries. Also, similar to silicon, higher activation energy is required for sodium to jump between the germanium lattices. Novel electrode designs, such as amorphous germanium films and amorphous germanium nanowires, can address the slow kinetics of Na$^+$ in germanium. Mullins et al. prepared germanium nanofilms [168] using evaporative deposition and investigated them as anode materials for sodium-ion batteries. The reversible specific capacity of the germanium nanofilms was 430 mA h g$^{-1}$, which was higher than the theoretical specific capacity of 369 mA h g$^{-1}$, and after 100 cycles at C/5, the germanium nanofilms retained 88% of the initial specific capacity. In addition, the germanium thin film electrode, prepared by Baggetto et al,. had a high discharge specific capacity of 350 mA h g$^{-1}$, which is close to the theoretical value, and it was found that the use of fluoroethylene carbonate (FEC) electrolyte additive could promotes the formation of a thinner SEI, which could effectively improve the capacity retention [169].

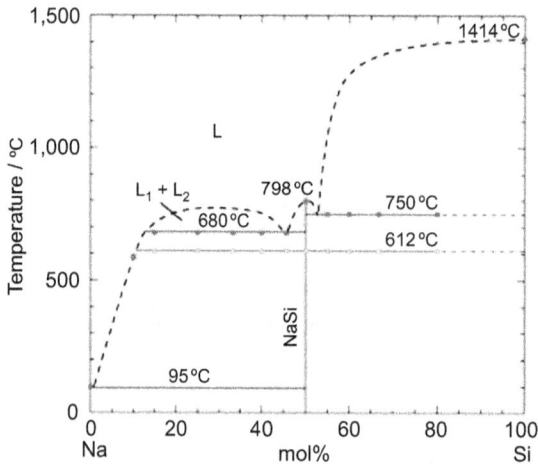

**Figure 4.36:** Na–Si binary phase diagram [163].

Antimony-based anode materials have become a focus of attention for researchers because of their inherent advantages of high theoretical specific capacity (660 mA h g$^{-1}$), low operating voltage (1 V), high electronic conductivity, non-toxic and low cost, and highly reversible alloying/de-alloying. However, it is necessary to construct composites with a structural diversity of antimony-based materials, based on the problems of volume expansion, low cycling coulombic efficiency and poor cycling performance accompanying the charging and discharging process of batteries. Qian et al. prepared Sb/C composite electrodes by the mechanical ball milling of Sb powder and conductive carbon (Figure 4.37(a)), and using Scherrer's formula, the average grain size of Sb particles in the composite was calculated to have an average grain size of about 10 nm [170]. The SEM images are shown in Figure 4.37(b), and the morphological characteristics of the Sb/C composite exhibit irregular aggregates of hundreds of nanometers in size. The Sb/C electrode has a reversible specific capacity of 610 mA h g$^{-1}$, and the electrode has a good multiplicative performance, with 50% capacity retention at a high current density of 2,000 mA g$^{-1}$. The Sb/C electrode has a reversible specific capacity of 610 mA h g$^{-1}$ and good multiplicity performance, with 50% capacity retention at a high current density of 2,000 mA g$^{-1}$ and good long-term cycling stability performance. Although the Sb/C electrode could only cycle stably for the first 50 cycles in the FEC-free electrolyte, the cycling performance was greatly improved by adding 5% FEC to the electrolyte, which was similar to that of the Sn electrode. In the optimized electrolyte, the Sb/C electrode has long-term cycling stability, with 94% capacity retention over 100 cycles [170]. Wu et al. prepared Sb/C composites that exhibited excellent cycling performance by maintaining a reversible specific capacity of 630 mA h g$^{-1}$ even after 100 cycles. The material has a unique fibrous structure and the nano-Sb particles are uniformly distributed in the carbon fibers, which play a key role in maintaining the reversibility of the material during cycling [171]. In addition, the rod-in-tube Sb@N-C composites were produced by in situ high-temperature carbon reduction using an in situ polymerization method to coat polypyrrole (PPy) on Sb$_2$S$_3$ [172]. The cavities formed inside the carbon tubes during calcination can buffer the volume expansion of Sb during electrochemical cycling, while PPy is converted into amorphous nitrogen-doped carbon hollow tubes during high-temperature carbonization, and the abundant nitrogen doping significantly enhances the electronic conductivity of the material while adding more intrinsic defects and active sites for charge transport. The Sb@N-C composite has good cycling performance and multiplicative performance when used in the anode of sodium-ion batteries, maintaining a reversible specific capacity of 345.6 mA h g$^{-1}$ after 3,000 cycles at a current density of 2.0 A g$^{-1}$.

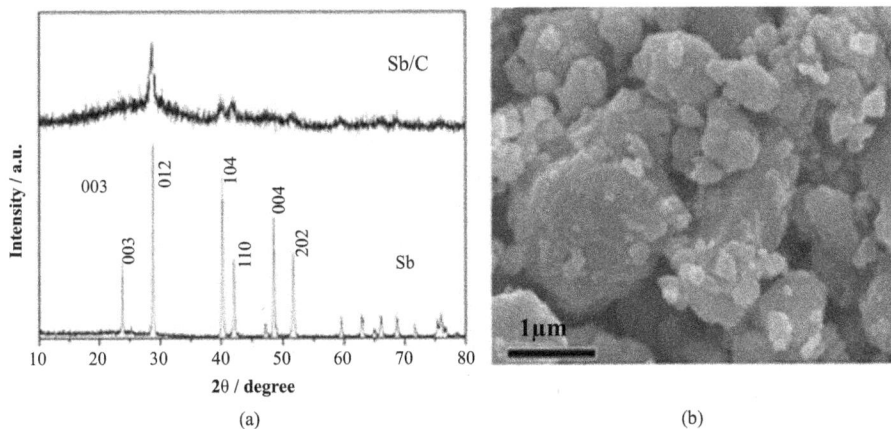

**Figure 4.37:** Physical characterization of Sb/C composites: (a) XRD patterns of Sb/C nanocomposites and metal Sb; and (b) SEM images of Sb/C anode composites [170].

### 4.4.3 Non-metallic (P)

Phosphorus is a nonmetallic element in group VA, and it has three main isomers: white phosphorus, red phosphorus, and black phosphorus, as shown in Figure 4.38.

White phosphorus is volatile and unstable, and spontaneous combustion occurs when exposed to the natural environment. Red phosphorus is usually amorphous and is widely used commercially. Black phosphorus is a crystalline phase that converts to red phosphorus at higher temperatures [173]. Amorphous red P and rhombohedral black P have been extensively studied as anode materials for SIBs. However, the electrochemical properties of red P and black P are reduced due to the effect of large volume changes (490%) during electrochemical sodalization/desodalization [174]. However, alloy-based anodes with suitable sodium-embedded potential can effectively improve the electrochemical performance of alloy anodes if certain modifications are used, such as carbon-based material cladding, to effectively improve the electrical conductivity, synthesis of three-dimensional frameworks to limit the volume expansion, and the use of material nanosizing to shorten the ion transport distance. These methods can provide improvements in the performance of P-based electrode materials. By preparing an amorphous red phosphorus/carbon composite anode, the material was found to have a redox potential of 0.4 V and a reversible specific capacity of 1,890 mA h $g^{-1}$, with a specific capacity of 1,540 mA h $g^{-1}$ at a high current density of 2.86 A $g^{-1}$ and a multiplicative good performance [175]. The P@RGO composite can be obtained by the homogeneous growth of red phosphorus on reduced graphene oxide using physical vapor deposition (PVD), which effectively alleviates the problems of poor electrical conductivity, poor sodium-ion transport kinetics, and large volume changes during cycling [176]. The composite has high specific capacity, excellent multiplicative performance, and cycling stability when used as the anode of sodium-ion batteries.

White phosphorus          Red phosphorus          Black phosphorus

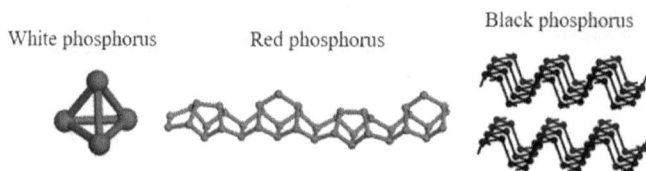

**Figure 4.38:** Schematic diagram of white phosphorus, red phosphorus, and black phosphorus [177].

Black phosphorus has many other special advantages over red phosphorus. Black phosphorus has a graphite-like layered structure with larger layer spacing, and its interlayer channel size can reach 3.08 Å, implying that $Na^+$ (1.04 Å) can be stored between the phosphorus monolayers [173]. Meanwhile, the layered crystal structure of orthorhombic black phosphorus is the most thermodynamically stable and has a higher bulk conductivity, compared to red phosphorus, which provides a greater electrochemical advantage in sodium-ion batteries [178, 179]. An in-depth study of the structure and surface evolution of black phosphorus during charging and discharging revealed that black phosphorus belongs to the sub-stable polycrystalline type. When fully discharged, Na$_3$P is produced, and then reverse charging obtains an amorphous P [180]. It was also demonstrated that VC additives can effectively improve the interfacial properties of black phosphorus and prolong the cycle life of electrodes. Therefore, the development of black phosphorus will be one of the next technological routes for high specific energy sodium-ion batteries. In the nanophosphorus-graphene hybrid material, with a small amount of phosphorus lamellae sandwiched between graphene lamellae, a two-step sodium embedding and alloying mechanism was discovered using in situ transmission electron microscopy and X-ray diffraction techniques [173]. The graphene lamellae in this nanostructure provide an elastic buffer layer to accommodate the anisotropic volume expansion during sodium embedding. Meanwhile, the phosphorus lamellae with increased interlayer distance shorten the Na$^+$ diffusion length. The electrode has a high specific capacity of 2,440 mA h g$^{-1}$ at a current density of 0.05 A g$^{-1}$ and maintains 83% capacity after 100 cycles, in the voltage range of 0 to 1.5 V. Qian et al. reported for the first time an amorphous P/C composite with 140 cycles of stable cycling, maintaining a reversible specific capacity of 1,750 mA h g$^{-1}$ [170]. Subsequently, they prepared Sn$_4$P$_3$/C nanocomposites by mechanical ball milling, with a specific capacity of 850 mA h g$^{-1}$ in the first cycle and a capacity retention of 86% after 150 cycles [181]. In this composite, a synergistic effect was produced between Sn and P, in which P served to disperse the Sn particles, while the high electrical conductivity of Sn improved the electrical insulation of P, thus serving to increase the electronic conductivity and inhibit the volume expansion during the intercalation/deintercalation process.

## 4.5 Other anode materials

### 4.5.1 MXene

MXene materials are made by etching MAX materials with different concentrations of HF, NH$_4$HF$_2$, and fluorinated salts (LiF, NaF, KF, NH$_4$F) with a mixture of concentrated HCl, etc. The schematic diagram of MXene preparation from MAX is shown in Figure 4.39 [182]. MXene is a new two-dimensional material that has been hotly researched in recent years and is used in electrochemical energy storage because of its excellent electronic conductivity, low ion diffusion energy barrier, and unique structure. They are used in electrochemical energy storage. However, the electrochemical performance of MXene is limited by the unavoidable interlayer stacking phenomenon and the low theoretical specific capacity. In order to utilize MXene effectively, researchers have used structural design and compounding with other materials to improve the electrochemical properties of MXene. The template method was used to process 2D MXene sheets into hollow spheres and 3D structures, and due to the good contact between the spheres and the metallic conductivity of MXene, the 3D macroporous MXene has high electrical conductivity and greatly improved capacity, multiplicity, and cycling stability when used as the anode of sodium-ion batteries [183].When MXene is compounded with other materials, the laminar structure of MXene can provide more channels for ion movement, and other materials are uniformly attached or grown between MXene lamellae to extend the lamellar space of MXene. The MXene@SnS composites are synthesized by a simple hydrothermal method. On the one hand, the SnS nanoparticles expand the layer spacing of MXene and provide high reversible capacity, and on the other hand, MXene improves the electrical conductivity of the SnS materials and the composites, showing better multiplicative performances at different current densities [184]. In addition, MXene materials not only have metallic properties but also have hydrophilic properties, and these two advantages can be fully exploited in the material design. The three-dimensional network, composed of conductive hard carbon and MXene 2D nanosheets, can effectively stabilize the electrode structure, alleviate the volume expansion of hard carbon during charging and discharging, and improve the electrode capacity and cycling performance of sodium-ion batteries [185].

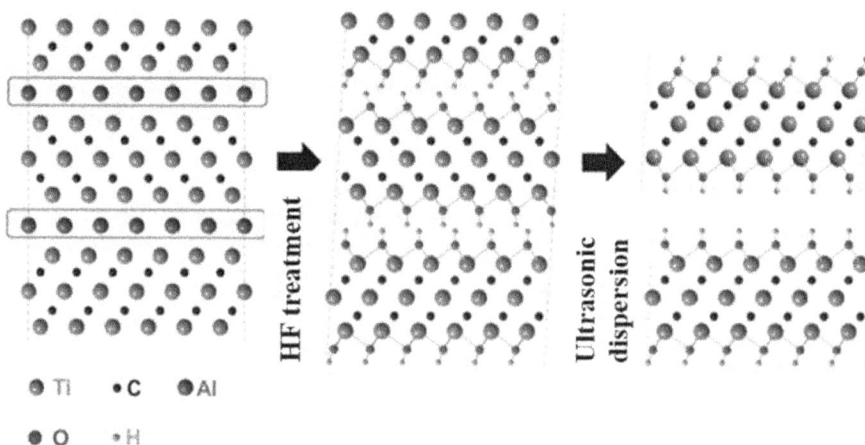

**Figure 4.39:** Ti$_3$AlC$_2$ stripping process: (a) Ti$_3$AlC$_2$ structure; (b) after HF treatment, Al atom is replaced by OH; and (c) stripping of MXene nanosheets [182].

### 4.5.2 Organic compound anode materials

Compared with inorganic materials, organic materials used as anode materials for sodium-ion batteries have received less attention. With the new requirements of portable devices for secondary battery flexibility, low cost, and environmental protection, organic compounds are gradually gaining attention as electrode materials. Organic compounds are renewable energy sources with many advantages such as light weight, flexibility, and cost-effectiveness, and have a wide range of application prospects in batteries [186, 187]. However, at the same time, organic compounds have some problems: firstly, the extremely low electronic conductivity of organic compounds can lead to slow kinetics; secondly, the large volume changes in the sodium-ion intercalation/deintercalation process can cause particle crushing. In addition, organic solvents are unstable during cycling [188]. These problems can lead to severe capacity degradation and loss of active material during the cell cycling process. Carbonyl compounds mainly refer to a class of organic materials that contain carbonyl groups (C = O) in their molecular structure. Among them, the carbonyl group is electrochemically active and its redox mechanism is the enolization reaction of C = O, i.e., the carbonyl group can undergo a reduction reaction to gain an electron to generate free radical negative ions. In the redox reaction of the carbonyl group, the cation only plays a role in compensating the charge; so it is not sensitive to the cation radius. In addition, the reaction kinetics are very fast; so carbonyl compounds usually have excellent charge and discharge reversibility. The more studied organic carbonyl compounds include conjugated carboxyl, imine and quinone groups, and Schiff bases. This section mainly summarizes the sodium storage properties of various organic compounds and the current status of the related research.

Zhao et al. first applied $Na_2C_8H_4O_4$/KB (Cochin black) to the anode of sodium-ion battery, exhibiting a specific capacity of 250 mA h $g^{-1}$, corresponding to a two-electron transfer reaction. In addition, $Al_2O_3$ was encapsulated on the electrode surface, improving the dissolution problem and increasing the cycle life of the battery [186].Wang et al. prepared the organic tetrasodium salt 2,5-dihydroxyterephthalate $Na_4C_8H_2O_6$ by a green one-pot method and found that the material could be cycled at two potential windows, from 1.6 to 2.8 V and from 0.1 to 1.8 V, under reversible cycling, providing an equivalent stable specific capacity of >180 mA h $g^{-1}$ and a coulombic efficiency close to 100%. Electrochemical tests, XPS, and computational studies showed that the reversible embedding and deintercalation of both $Na^+$ corresponded to reversible redox reactions between phenoxy and quinone carbonyls, and carboxycarbonyl, and enol groups [189]. The average operating voltage of the cells assembled using $Na_4C_8H_2O_6$, as both positive and negative initial active materials, was 1.8 V, with an actual energy density of about 65 W h $kg^{-1}$.

Due to the low electronic conductivity of organic compound materials, the volume change of $Na^+$ during insertion/extraction causes particle fragmentation, and in order to avoid massive dissolution of the active material during sodium-ion battery cycling, researchers have proposed many modification methods, such as the preparation of composite electrodes, by combining organic matter with conductive additives. Biomolecule-based electrodes, prepared by a simple self-assembly process, can be grown on reduced graphene oxide (rGO) nanoflakes due to the strong π-π interactions between the aromatic structure and the carbon scaffold [190]. XPS results showed that the redox molecules were immobilized by π-π interactions on the rGO carbon scaffold, which inhibited the dissolution of organic matter and improved the electrode conductivity, and facilitated $Na^+$ migration. It shows good electrochemical performance in reversible Na$^+$ transport. Electrochemical tests showed that the electrode had a high specific capacity of 305 mA h g$^{-1}$ after 100 cycles, with good cycling stability. Then, the disodium 2,5-dihydroxy-1,4-benzoquinone (Na$_2$C$_6$H$_2$O$_4$)/CNT nanocomposite, prepared by the spray drying method, was found to have an average sodium storage voltage of 1.4 V, which can effectively prevent the formation of solid electrolyte intermediate phase layer, thus ensuring high safety, high first-cycle coulombic efficiency, and excellent multiplicative performance [191]. The reversible specific capacity of this composite when used as the negative electrode of sodium-ion batteries reaches 259 mA h g$^{-1}$, with a first-cycle coulombic efficiency of 88% and a sodium storage specific capacity of 142 mA h g$^{-1}$ at 7 C. In addition, the spray drying method can be further extended to water- or oil-soluble materials to improve the electrode's electron conductivity, which can be further extended for studies of other materials. However, organic compounds with electrochemical activity are very soluble in organic electrolytes, resulting in poor cycling performance. At present, the research and development of organic compounds as electrode materials for sodium-ion batteries is still in its initial stage, and more organic compounds are pending further research and development by researchers.

### 4.5.3 Sodium metal anode materials

In recent years, the sodium metal anode has received much attention, again for its high specific capacity (1,166 mAh g$^{-1}$) and low overpotential (−2.714 V, compared with the standard hydrogen electrode), and meanwhile, the study of sodium metal anode is important for the research of high specific energy Na-S and Na-O$_2$ battery systems. The study of sodium metal anode is important for the research of high energy Na-S and Na-O$_2$ battery systems. However, due to the uneven ion deposition during the cycling process, a large number of sodium dendrites are generated in the sodium cathode, and with the increase of cycling times, the dendrites gradually grow and produce a large amount of sodium that loses its electrochemical activity ("dead sodium"), which can cause short-circuiting and battery accidents, in serious cases. In addition, due to the characteristics of sodium metal anode without a skeleton structure during the cycling process, the electrode undergoes unlimited volume expansion, which makes the SEI film rupture, and the exposed new sodium reacts with the electrolyte, reducing the cycling efficiency, and the nucleation potential at the rupture site is lower, further increasing the non-uniformity of nucleation. In recent years, many research results on sodium-metal anode have been reported, and the research hotspots are mainly focused on the formation of SEI film, the construction of three-dimensional conducting skeleton, and the application of solid electrolyte selection for sodium-metal anode, because solving the problems of dendrite growth, volume expansion, and suppression of cell short circuit are the keys to realize the practical application of sodium-metal anode.

During the charging and discharging reaction of the battery, a SEI film is formed on the surface of the sodium metal, and the main components of this film are $Na_2CO_3$, $Na_2O$, and $NaOH$. This electron-insulating and ion-conducting SEI film is mainly produced at the early stage of the electrochemical reaction and can effectively avoid the constant side reactions between the sodium metal and the electrolyte. So far, the composition of the SEI film has been impossible to be precisely determined. In sodium metal anode, as the dendrite growth during the cycling process, it will puncture the SEI film and form a short circuit, which in turn will cause capacity decay and safety accidents. In order to effectively suppress the dendrite growth and volume expansion, and reduce the occurrence of short circuit, the formed SEI film must have excellent chemical stability and mechanical strength to ensure uniform line nucleation of sodium ions during the deposition and detachment process. However, the SEI films, formed by conventional methods, cannot meet the requirements of long cycling of metal anodes, in terms of mechanical properties and compositional homogeneity; so, finding simple and feasible methods to improve the stability of SEI films or constructing artificial SEI films by modulating the electrolyte additives is one of the important directions to improve the electrochemical performance of sodium metal anodes.

Due to the low melting point of sodium metal (98 °C), there are many difficulties in constructing artificial SEI films using conventional methods. As shown in Figure 4.40, Hu et al. constructed nanoscale $Al_2O_3$ films on the surface of sodium metal using low-temperature atomic layer deposition at around 70 °C. The $Al_2O_3$ films have high elastic modulus and stability, which can physically inhibit the growth of sodium dendrites and block the side reactions between sodium metal and electrolyte [192]. Not only that, this artificial interfacial film can promote the transport rate of sodium ions between the sodium metal and the electrolyte and make the sodium ions uniformly distributed, inhibiting the growth of sodium dendrites. After assembling the sodium counter electrode, there was still no significant overpotential change after cycling for 450 h at higher current densities.

The use of three-dimensional conductive frameworks is one of the most effective ways to mitigate the volume expansion during the cycling of sodium metal anode secondary cells. Volume expansion during plating/stripping is a persistent problem in the application of sodium metal anodes, and many new and feasible solutions have been proposed in recent years to address this problem. Among them, the introduction of a three-dimensional porous conducting skeleton and the infiltration of molten sodium metal into the nano-pores of the main skeleton structure to form a three-dimensional composite sodium-metal conducting anode is an effective method. The main reason for dendrites in sodium metal cycling is the uneven ion deposition; so, improving the uniform ion distribution in the cycling process is a very important modification tool, of which the most direct method is to directly increase the actual contact area and thus reduce the actual current density. The three-dimensional porous skeleton has abundant specific surface area, which can effectively reduce the actual current density, alleviate the dendrite growth, and improve the multiplicative performance of the cell. At the same time, the 3D conductive skeleton also provides a large number of line nucleation sites, which effectively regulates the deposition and exfoliation of sodium metal, and the abundant pores suppress the volume expansion problem generated during the cycling process.

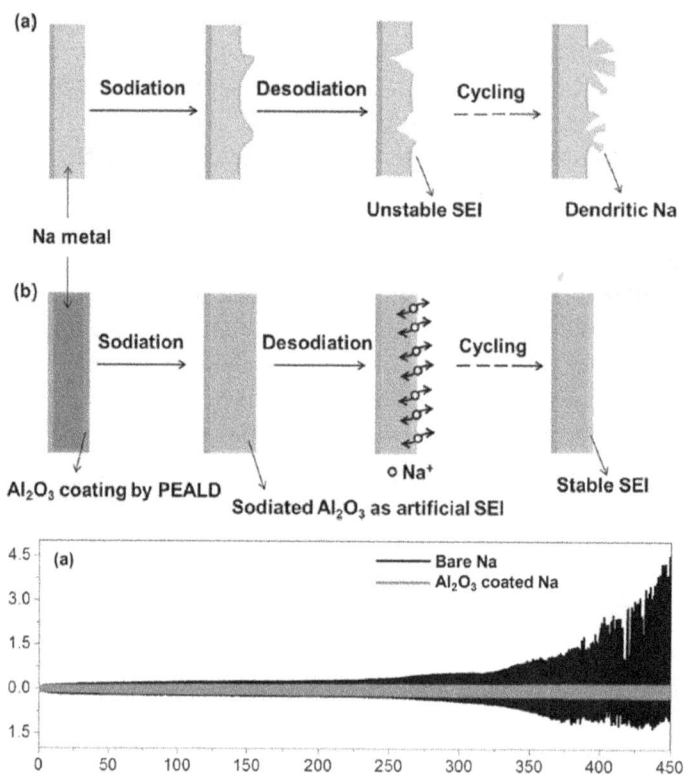

**Figure 4.40:** Deposition mechanism diagram and electrochemical performance of metal sodium anode modified by $Al_2O_3$.

It is important to select a 3D conductive material that is wettable, and has stable chemical and mechanical properties. Feng Wu's team used a hydrothermal self-assembly method to obtain a graphene aerogel conductive framework. As shown in Figure 4.41, the skeleton material has good flexibility and mechanical strength. It can effectively store sodium metal and mitigate volume expansion, and is compounded with molten sodium metal at 300 °C to form a graphene aerogel-sodium metal composite electrode, which effectively suppresses volume expansion and dendrite growth generated during sodium-metal cycling [193]. By modulating the wettable functional groups on the surface, such as -COOH, -OH, and C = O, the composite electrode that can directly react with alkali metals was constructed, and the molten sodium metal was allowed to climb onto the skeleton material by capillary action. This method solves the essential problem of sodium metal without supporting skeleton, and the introduction of skeleton material further increases the active site for sodium-ion deposition and reduces the formation of sodium dendrites, while having better adaptability at high current density.

**Figure 4.41:** Graphene aerogel skeleton showing good flexibility and mechanical strength.

The development of advanced solid-state electrolytes is very effective in suppressing short circuits and side reactions in batteries. Among them, the use of solid polymer electrolytes is still one of the hot directions to solve the alkali metal problem. In terms of adhesion to electrodes, solid polymer electrolytes are better than liquid and ceramic electrolytes, and most solid polymer electrolytes have good flexibility and processability, which provides strong support for the development of flexible wearable devices in the future.

# References

[1]    Qian J F, Zhou M, Cao Y L. et al., Nanosized $Na_4Fe(CN)_6$/C Composite as a Low-Cost and High-Rate Cathode Material for Sodium-Ion Batteries[J]. Advanced Energy Materials, 2012, 2(4): 410–414.

[2]    Yuan D D, Liang X M, Wu L. et al., A Honeycomb-Layered $Na_3Ni_2SbO_6$: A High-Rate and Cycle-Stable Cathode for Sodium-Ion Batteries[J]. Advanced Materials, 2014, 26(36): 6301–6306.

[3]    Wu X Y, Luo Y, Sun M. et al., Low-defect Prussian Blue Nanocubes as High Capacity and Long Life Cathodes for Aqueous Na-ion Batteries[J]. Nano Energy, 2015, 13: 117–123.

[4]    Fang Y J, Xiao L F, Qian J F. et al., Mesoporous Amorphous $FePO_4$ Nanospheres as High-Performance Cathode Material for Sodium-Ion Batteries[J]. Nano Letters, 2014, 14(6): 3539–3543.

[5]    Fang Y J, Xiao L F, Ai X P. et al., Hierarchical Carbon Framework Wrapped $Na_3V_2(PO_4)_3$ as a Superior High-Rate and Extended Lifespan Cathode for Sodium-Ion Batteries[J]. Advanced Materials, 2015, 27(39): 5895–5900.

[6]    Delmas C. Sodium and Sodium-Ion Batteries: 50 Years of Research[J]. Advanced Energy Materials, 2018, 8(17): 1703137.

[7]    Yabuuchi N, Kubota K, Dahbi M. et al., Research Development on Sodium-Ion Batteries[J]. Chemical Reviews, 2014, 114(23): 11636–11682.

[8]    Klein F, Jache B, Bhide A. et al., Conversion Reactions for Sodium-Ion Batteries[J]. Physical Chemistry Chemical Physics, 2013, 15(38): 15876–15887.

[9]     Kim Y J, Ha K H, Oh S M. et al., High-Capacity Anode Materials for Sodium-Ion Batteries[J]. Chemistry-A European Journal, 2014, 20(38): 11980–11992.

[10]    Hou H S, Qiu X Q, Wei W F. et al., Carbon Anode Materials for Advanced Sodium-Ion Batteries [J]. Advanced Energy Materials, 2017, 7(24): 1602898.

[11]    Kim H, Kim H, Ding Z. et al, Recent Progress in Electrode Materials for Sodium-Ion Batteries [J]. Advanced Energy Materials, 2016, 6(19): 1600943.

[12]    Hwang J Y, Myung S T, Sun Y K. Sodium-Ion Batteries: Present and Future[J]. Chemical Society Reviews, 2017, 46(12): 3529–3614.

[13]    Thomas P, Billaud D. Electrochemical Insertion of Sodium into Hard Carbons[J]. Electrochimica Acta, 2002, 47(20): 3303–3307.

[14]    Wang Z H, Selbach S M, Grande T. Van der Waals Density Functional Study of the Energetics of Alkali Metal Intercalation in Graphite[J]. RSC Advances, 2014, 4(8): 4069–4079.

[15]    Nobuhara K, Nakayama H, Nose M. et al., First-Principles Study of Alkali Metal-Graphite Intercalation Compounds[J]. Journal of Power Sources, 2013, 243: 585–587.

[16]    Jache B, Adelhelm P. Use of Graphite as a Highly Reversible Electrode with Superior Cycle Life for Sodium-Ion Batteries by Making Use of Co-Intercalation Phenomena[J]. Angewandte Chemie International Edition, 2014, 53(38): 10169–10173.

[17]    Wen Y, He K, Zhu Y J. et al., Expanded Graphite as Superior Anode for Sodium-Ion Batteries[J]. Nature Communications, 2014, 5: 4033.

[18]    Xu Z L, Yoon G, Park K Y. et al., Tailoring Sodium Intercalation in Graphite for High Energy and Power Sodium Ion Batteries[J]. Nature Communications, 2019, 10(1): 1–10.

[19]    Berger C, Song Z M, Li X B. et al., Electronic Confinement and Coherence in Patterned Epitaxial Graphene[J]. Science, 2006, 312(5777): 1191–1196.

[20]    Evans K E. Auxetic Polymers: A New Range of Materials[J]. Endeavour, 1991, 15(4): 170–174.

[21]    Zhang Q Q, Xu X, Li H. et al., Mechanically Robust Honeycomb Graphene Aerogel Multifunctional Polymer Composites[J]. Carbon, 2015, 93: 659–670.

[22]    Wang Y X, Chou S L, Liu H K. et al., Reduced Graphene Oxide with Superior Cycling Stability and Rate Capability for Sodium Storage[J]. Carbon, 2013, 57: 202–208.

[23]    Ding J, Wang H L, Li Z. et al, Carbon Nanosheet Frameworks Derived from Peat Moss as High Performance Sodium Ion Battery Anodes[J]. ACS Nano, 2013, 7(12): 11004–11015.

[24]    Jian Z L, Bommier C, Luo L L. et al., Insights on the Mechanism of Na-Ion Storage in Soft Carbon Anode[J]. Chemistry of Materials, 2017, 29(5): 2314–2320.

[25]    Yao X H, Ke Y J, Ren W H. et al., Defect-Rich Soft Carbon Porous Nanosheets for Fast and High-Capacity Sodium-Ion Storage[J]. Advanced Energy Materials, 2019, 9(6): 1803260.

[26]    Hu Y S, Li H M, Qi X G. et al., Advanced Sodium-Ion Batteries Using Superior Low Cost Pyrolyzed Anthracite Anode[J]. Energy Storage Materials, 2016, 5: 191–197.

[27]    Stevens D A, Dahn J R. High Capacity Anode Materials for Rechargeable Sodium-Ion Batteries [J]. Journal of the Electrochemical Society, 2000, 147(4): 1271–1273.

[28]    Stevens D A, Dahn J R. High Capacity Anode Materials for Rechargeable Sodium-Ion Batteries [J]. Journal of the Electrochemical Society, 2000, 147(4): 1271.

[29]    Cao Y L, Xiao L F, Sushko M L. et al., Sodium Ion Insertion in Hollow Carbon Nanowires for Battery Applications[J]. Nano Letters, 2012, 12(7): 3783–3787.

[30]    Qiu S, Xiao L F, Sushko M L. et al., Manipulating Adsorption-Insertion Mechanisms in Nanostructured Carbon Materials for High-Efficiency Sodium Ion Storage[J]. Advanced Energy Materials, 2017, 7(17): 1700403.

[31]    Xiao L F, Cao Y L, Henderson W A. et al., Hard Carbon Nanoparticles as High-Capacity, High-Stability Anodic Materials for Na-Ion Batteries[J]. Nano Energy, 2016, 19: 279–288.

[32]    Cao Y L, Xiao L F, Sushko M L. et al., Sodium Ion Insertion in Hollow Carbon Nanowires for Battery Applications[J]. Nano Letters, 2012, 12(7): 3783–3787.

[33] Xu J T, Wang M, Wickramaratne N P. et al., High-Performance Sodium Ion Batteries Based on a 3D Anode from Nitrogen-Doped Graphene Foams[J]. Advanced Materials, 2015, 27(12): 2042–2048.

[34] Wang S Q, Xia L, Yu L. et al., Free-Standing Nitrogen-Doped Carbon Nanofiber Films: Integrated Electrodes for Sodium-Ion Batteries with Ultralong Cycle Life and Superior Rate Capability[J]. Advanced Energy Materials, 2016, 6(7): 1502217.

[35] Yang F H, Zhang Z A, Du K. et al., Dopamine Derived Nitrogen-Doped Carbon Sheets as Anode Materials for High-Performance Sodium Ion Batteries[J]. Carbon, 2015, 91: 88–95.

[36] Li W, Zhou M, Li H M. et al., A High Performance Sulfur-Doped Disordered Carbon Anode for Sodium Ion Batteries[J]. Energy & Environmental Science, 2015, 8(10): 2916–2921.

[37] Li Y M, Wang Z G, Li L L. et al., Preparation of Nitrogen and Phosphorous Co-Doped Carbon Microspheres and Their Superior Performance as Anode in Sodium-Ion Batteries[J]. Carbon, 2016, 99: 556–563.

[38] Wang Z H, Qie L, Yuan L X. et al., Functionalized N-doped Interconnected Carbon Nanofibers as an Anode Material for Sodium-Ion Storage with Excellent Performance[J]. Carbon, 2013, 55: 328–334.

[39] Shi X D, Zhang Z A, Fu Y. et al., Self-Template Synthesis of Nitrogen-Doped Porous Carbon Derived from Zeolitic Imidazolate Framework-8 as an Anode for Sodium Ion Batteries[J]. Materials Letters, 2015, 161: 332–335.

[40] Farokh Niaei A H, Hussain T, Hankel M. et al., Sodium-Intercalated Bulk Graphdiyne as an Anode Material for Rechargeable Batteries[J]. Journal of Power Sources, 2017, 343: 354–363.

[41] Xu Z M, Lv X J, Li J. et al., A Promising Anode Material for Sodium-Ion Battery with High Capacity and High Diffusion Ability: Graphyne and Graphdiyne[J]. RSC Advances, 2016, 6: 25594–25600.

[42] Wang N, Li X D, Tu Z Y. et al., Synthesis and Electronic Structure of Boron-Graphdiyne with an sp-Hybridized Carbon Skeleton and Its Application in Sodium Storage[J]. Angewandte Chemie International Edition, 2018, 57(15): 3968–3973.

[43] Tang K, Fu L J, White R J. et al., Hollow Carbon Nanospheres with Superior Rate Capability for Sodium-Based Batteries[J]. Advanced Energy Materials, 2012, 2(7): 873–877.

[44] Long C L, Chen X, Jiang L L. et al., Porous Layer-Stacking Carbon Derived from In-Built Template in Biomass for High Volumetric Performance Supercapacitors[J]. Nano Energy, 2015, 12: 141–151.

[45] Li Y M, Mu L Q, Hu Y S. et al., Pitch-Derived Amorphous Carbon as High Performance Anode for Sodium-Ion Batteries[J]. Energy Storage Materials, 2016, 2: 139–145.

[46] Zhou Q, Cai W Z, Zhang Y P. et al., Electricity Generation from Corn Cob Char though a Direct Carbon Solid Oxide Fuel Cell[J]. Biomass & Bioenergy, 2016, 91: 250–258.

[47] Ding J, Wang H L, Li Z. et al, Peanut Shell Hybrid Sodium Ion Capacitor with Extreme Energy–Power Rivals Lithium Ion Capacitors[J]. Energy & Environmental Science, 2015, 8(3): 941–955.

[48] Lotfabad E M, Ding J, Cui K. et al, High-Density Sodium and Lithium Ion Battery Anodes from Banana Peels[J]. ACS Nano, 2014, 8(7): 7115–7129.

[49] Yang T Z, Qian T, Wang M F. et al., A Sustainable Route from Biomass Byproduct Okara to High Content Nitrogen-Doped Carbon Sheets for Efficient Sodium Ion Batteries[J]. Advanced Materials, 2016, 28(3): 539–545.

[50] Aravindan V, Lee Y S, Yazami R. et al., $TiO_2$ Polymorphs in 'Rocking-chair' Li-Ion Batteries[J]. Materials Today, 2015, 18(6): 345–351.

[51] Guo S H, Yi J, Sun Y. et al., Recent Advances in Titanium-Based Electrode Materials for Stationary Sodium-Ion Batteries[J]. Energy & Environmental Science, 2016, 9(10): 2978–3006.

[52] Su D W, Dou S X, Wang G X. Anatase $TiO_2$: Better Anode Material than Amorphous and Rutile Phases of $TiO_2$ for Na-Ion Batteries[J]. Chemistry of Materials, 2015, 27(17): 6022–6029.

[53] Lunell S, Stashans A, Ojamae L. et al., Li and Na Diffusion in $TiO_2$ from Quantum Chemical Theory versus Electrochemical Experiment[J]. Journal of the American Chemical Society, 1997, 119(31): 7374–7380.

[54] Xiong H, Slater M D, Balasubramanian M. et al., Amorphous $TiO_2$ Nanotube Anode for Rechargeable Sodium Ion Batteries[J]. The Journal of Physical Chemistry Letters, 2011, 2(20): 2560–2565.

[55] Umebayashi T, Yamaki T, Itoh H. et al., Band Gap Narrowing of Titanium Dioxide by Sulfur Doping[J]. Applied Physics Letters, 2002, 81(3): 454–456.

[56] Jung H G, Myung S T, Yoon C S. et al., Microscale Spherical Carbon-Coated $Li_4Ti_5O_{12}$ as Ultra High Power Anode Material for Lithium Batteries[J]. Energy & Environmental Science, 2011, 4 (4): 1345–1351.

[57] Zhao L, Pan H L, Hu Y S. et al., Spinel Lithium Titanate ($Li_4ti_5o_{12}$) as Novel Anode Material for Room-Temperature Sodium-Ion Battery[J]. Chinese Physics B, 2012, 21(2): 0799017.

[58] Sun Y, Zhao L, Pan H L. et al., Direct Atomic-Scale Confirmation of Three-Phase Storage Mechanism in $Li_4Ti_5O_{12}$ Anodes for Room-Temperature Sodium-Ion Batteries[J]. Nature Communications, 2013, 4: 1870.

[59] Senguttuvan P, Rousse G, Seznec V. et al., $Na_2Ti_3O_7$: Lowest Voltage Ever Reported Oxide Insertion Electrode for Sodium Ion Batteries[J]. Chemistry of Materials, 2011, 23(18): 4109–4111.

[60] Xu J, Ma C, Balasubramanian M. et al., Understanding $Na_2Ti_3O_7$ as an Ultra-Low Voltage Anode Material for a Na-Ion Battery[J]. Chemical Communications, 2014, 50(83): 12564–12567.

[61] Rudola A, Saravanan K, Mason C W. et al., $Na_2Ti_3O_7$: An Intercalation Based Anode for Sodium-ion Battery Applications[J]. Journal of Materials Chemistry A, 2013, 1(7): 2653.

[62] Rudola A, Sharma N, Balaya P. Introducing a 0.2 V Sodium-Ion Battery Anode: The $Na_2Ti_3O_7$ to $Na_{3-x}Ti_3O_7$ Pathway[J]. Electrochemistry Communications, 2015, 61: 10–13.

[63] Reddy M, Subba Rao G, Chowdari B V. Metal Oxides and Oxysalts as Anode Materials for Li Ion Batteries[J]. Chemical Reviews, 2013, 113(7): 5364–5457.

[64] Senguttuvan P, Rousse G, Vezin H. et al., Titanium(III) Sulfate as New Negative Electrode for Sodium-Ion Batteries[J]. Chemistry of Materials, 2013, 25(12): 2391–2393.

[65] Chen J S, Luan D, Li C M. et al., $TiO_2$ and $SnO_2@TiO_2$ Hollow Spheres Assembled from Anatase $TiO_2$ Nanosheets with Enhanced Lithium Storage Properties[J]. Chemical Communications, 2009, 46(43): 8252.

[66] Xu Y, Lotfabad E M, Wang H. et al., Nanocrystalline Anatase $TiO_2$: A New Anode Material for Rechargeable Sodium Ion Batteries[J]. Chemical Communications, 2013, 49(79): 8973.

[67] Wu L M, Buchholz D, Bresser D. et al., Anatase $TiO_2$ Nanoparticles for High Power Sodium-Ion Anodes[J]. Journal of Power Sources, 2014, 251: 379–385.

[68] Gonzalez J R, Alcantara R, Nacimiento F. et al., Microstructure of the Epitaxial Film of Anatase Nanotubes Obtained at High Voltage and the Mechanism of Its Electrochemical Reaction with Sodium[J]. Crystengcomm, 2014, 16(21): 4602–4609.

[69] Kim K T, Ali G, Chung K Y. et al., Anatase Titania Nanorods as an Intercalation Anode Material for Rechargeable Sodium Batteries[J]. Nano Letters, 2014, 14(2): 416–422.

[70] Hong Z S, Hong J X, Xie C B. et al., Hierarchical Rutile $TiO_2$ with Mesocrystalline Structure for Li-Ion and Na-Ion Storage[J]. Electrochimica Acta, 2016, 202: 203–208.

[71] Hong Z S, Zhou K Q, Zhang J W. et al., Facile Synthesis of Rutile $TiO_2$ Mesocrystals with Enhanced Sodium Storage Properties[J]. Journal of Materials Chemistry, 2015, 3(33): 17412–17416.

[72] Usui H, Yoshioka S, Wasada K. et al., Nb-Doped Rutile TiO$_2$: A Potential Anode Material for Na-Ion Battery[J]. ACS Applied Materials & Interfaces, 2015, 7(12): 6567–6573.

[73] Zhang Y, Pu X L, Yang Y C. et al., An Electrochemical Investigation of Rutile TiO$_2$ Microspheres Anchored by Nanoneedle Clusters for Sodium Storage[J]. Physical Chemistry Chemical Physics, 2015, 17(24): 15764–15770.

[74] Zhang Y, Foster C W, Banks C E. et al., Graphene-Rich Wrapped Petal-Like Rutile TiO$_2$ Tuned by Carbon Dots for High-Performance Sodium Storage[J]. Advanced Materials, 2016, 28(42): 9391–9399.

[75] Søndergaard M, Dalgaard K J, Bøjesen E D. et al., In-situ Monitoring of TiO$_2$(B)/Anatase Nanoparticles Formation and Application in Li-Ion and Na-Ion Batteries[J]. Materials Chemistry A, 2012, 3: 18667–18674.

[76] Xiong H, Slater M D, Balasubramanian M. et al., Amorphous TiO$_2$ Nanotube Anode for Rechargeable Sodium Ion Batteries[J]. The Journal of Physical Chemistry Letters, 2011, 2(20): 2560–2565.

[77] He H N, Gan Q M, Wang H Y. et al., Structure-Dependent Performance of TiO$_2$/C as Anode Material for Na-Ion Batteries[J]. Nano Energy, 2018, 44: 217–227.

[78] Wang B F, Zhao F, Du G D. et al., Boron-Doped Anatase TiO$_2$ as a High-Performance Anode Material for Sodium-Ion Batteries[J]. ACS Applied Materials & Interfaces, 2016, 8(25): 16009–16015.

[79] Ni J F, Fu S D, Wu C. et al., Self-Supported Nanotube Arrays of Sulfur-Doped TiO$_2$ Enabling Ultrastable and Robust Sodium Storage[J]. Advanced Materials, 2016, 28(11): 2259–2265.

[80] Xu H, Liu Y T, Qiang T T. et al., Boosting Sodium Storage Properties of Titanium Dioxide by a Multiscale Design Based on MOF-Derived Strategy[J]. Energy Storage Materials, 2019, 17: 126–135.

[81] Jung H G, Myung S T, Yoon C S. et al., Microscale Spherical Carbon-Coated Li$_4$Ti$_5$O$_{12}$ as Ultra High Power Anode Material for Lithium Batteries[J]. Energy & Environmental Science, 2011, 4 (4): 1345.

[82] Chen C J, Xu H H, Zhou T F. et al., Integrated Intercalation-Based and Interfacial Sodium Storage in Graphene-Wrapped Porous Li$_4$Ti$_5$O$_{12}$ Nanofibers Composite Aerogel[J]. Advanced Energy Materials, 2016, 6(13): 1600322.

[83] Kim K T, Yu C Y, Yoon C S. et al., Carbon-Coated Li$_4$Ti$_5$O$_{12}$ Nanowires Showing High Rate Capability as an Anode Material for Rechargeable Sodium Batteries[J]. Nano Energy, 2015, 12: 725–734.

[84] Tsiamtsouri M A, Allan P K, Pell A J. et al., Exfoliation of Layered Na-Ion Anode Material Na$_2$Ti$_3$O$_7$ for Enhanced Capacity and Cyclability[J]. Chemistry of Materials, 2018, 30(5): 1505–1516.

[85] Pan H L, Lu X, Yu X Q. et al., Sodium Storage and Transport Properties in Layered Na$_2$Ti$_3$O$_7$ for Room-Temperature Sodium-Ion Batteries[J]. Advanced Energy Materials, 2013, 3(9): 1186–1194.

[86] Yan Z C, Liu L, Shu H B. et al., A Tightly Integrated Sodium Titanate-Carbon Composite as an Anode Material for Rechargeable Sodium Ion Batteries[J]. Journal of Power Sources, 2015, 274: 8–14.

[87] Rudola A, Saravanan K, Devaraj S. et al., Na$_2$Ti$_6$O$_{13}$: A Potential Anode for Grid-Storage Sodium-Ion Batteries[J]. Chemical Communications, 2013, 49(67): 7451.

[88] Shirpour M, Cabana J, Doeff M. New Materials Based on a Layered Sodium Titanate for Dual Electrochemical Na and Li Intercalation Systems[J]. Energy & Environmental Science, 2013, 6(8): 2538.

[89] Wang Y S, Yu X Q, Xu S Y. et al., A Zero-Strain Layered Metal Oxide as the Negative Electrode for Long-Life Sodium-Ion Batteries[J]. Nature Communications, 2013, 4: 2365.

[90] Li H, Fei H L, Liu X. et al., In-situ Synthesis of $Na_2Ti_7O_{15}$ Nanotubes on a Ti Net Substrate as a High Performance Anode for Na-Ion Batteries[J]. Chemical Communications, 2015, 51(45): 9298–9300.

[91] Woo S H, Park Y, Choi W Y. et al., Trigonal $Na_4Ti_5O_{12}$ Phase as an Intercalation Host for Rechargeable Batteries[J]. Journal of the Electrochemical Society, 2012, 159(12): A2016–A2023.

[92] Andersson S, Wadsley A D. The Structures of $Na_2Ti_6O_{13}$ and $Rb_2Ti_6O_{13}$ and the Alkali Metal Titanates[J]. Topics in Catalysis, 1994, 1: 137–144.

[93] Shen K, Wagemaker M. $Na_{2+x}Ti_6O_{13}$ as Potential Negative Electrode Material for Na-Ion Batteries[J]. Inorganic Chemistry, 2014, 53(16): 8250–8256.

[94] Wu X Y, Cao Y L, Ai X P. et al., A Low-Cost and Environmentally Benign Aqueous Rechargeable Sodium-Ion Battery Based on $NaTi_2(PO_4)_3$-$Na_2NiFe(CN)_6$ Intercalation Chemistry[J]. Electrochemistry Communications, 2013, 31: 145–148.

[95] Jiang Y, Zeng L C, Wang J Q. et al., A Carbon Coated NASICON Structure Material Embedded in Porous Carbon Enabling Superior Sodium Storage Performance: $NaTi_2(PO_4)_3$ as an Example [J]. Nanoscale, 2015, 7(35): 14723–14729.

[96] Yuan S, Huang X L, Ma D L. et al., Engraving Copper Foil to Give Large-Scale Binder-Free Porous CuO Arrays for a High-Performance Sodium-Ion Battery Anode[J]. Advanced Materials, 2014, 26(14): 2273–2279.

[97] Liu H H, Cao F, Zheng H. et al., In-situ Observation of the Sodiation Process in CuO Nanowires [J]. Chemical Communications, 2015, 51(52): 10443–10446.

[98] Liu S H, Wang Y W, Dong Y F. et al., Ultrafine $Fe_3O_4$ Quantum Dots on Hybrid Carbon Nanosheets for Long-Life, High-Rate Alkali-Metal Storage[J]. ChemElectroChem, 2016, 3(1): 38–44.

[99] Hu Z, Zhu Z Q, Cheng F Y. et al., Pyrite $FeS_2$ for High-Rate and Long-Life Rechargeable Sodium Batteries[J]. Energy & Environmental Science, 2015, 8(4): 1309–1316.

[100] Liu J, Wu C, Xiao D D. et al., MOF-Derived Hollow $Co_9S_8$ Nanoparticles Embedded in Graphitic Carbon Nanocages with Superior Li-Ion Storage[J]. Small, 2016, 12(17): 2354–2364.

[101] Wang Y X, Yang J P, Chou S L. et al., Uniform Yolk-Shell Iron Sulfide-Carbon Nanospheres for Superior Sodium-Iron Sulfide Batteries[J]. Nature Communications, 2015, 6(1): 8689.

[102] Kim S O, Manthiram A. The Facile Synthesis and Enhanced Sodium-Storage Performance of a Chemically Bonded $CuP_2$/C Hybrid Anode[J]. Chemical Communications, 2016, 52(23): 4337–4340.

[103] Liu J, Kopold P, Wu C. et al., Uniform Yolk-Shell $Sn_4P_3$@C Nanospheres as High-Capacity and Cycle-Stable Anode Materials for Sodium-Ion Batteries[J]. Energy & Environmental Science, 2015, 8(12): 3531–3538.

[104] Fan M P, Chen Y, Xie Y H. et al., Half-Cell and Full-Cell Applications of Highly Stable and Binder-Free Sodium Ion Batteries Based on $Cu_3P$ Nanowire Anodes[J]. Advanced Functional Materials, 2016, 26(28): 5019–5027.

[105] Alcantara R, Jaraba M, Lavela P. et al., $NiCo_2O_4$ Spinel: First Report on a Transition Metal Oxide for the Negative Electrode of Sodium-Ion Batteries[J]. Chemistry of Materials, 2002, 14(7): 2847–2848.

[106] Jian Z L, Zhao B, Liu P. et al., $Fe_2O_3$ Nanocrystals Anchored onto Graphene Nanosheets as the Anode Material for Low-Cost Sodium-Ion Batteries[J]. Chemical Communications, 2014, 50(10): 1215–1217.

[107] Zhao Y, Feng Z X, Xu Z C. Yolk-Shell $Fe_2O_3$ Middle Dot in Circle C Composites Anchored on MWNTs with Enhanced Lithium and Sodium Storage[J]. Nanoscale, 2015, 7(21): 9520–9525.

[108] Komaba S, Mikumo T, Yabuuchi N. et al., Electrochemical Insertion of Li and Na Ions into Nanocrystalline Fe$_3$O$_4$ and alpha-Fe$_2$O$_3$ for Rechargeable Batteries[J]. Journal of the Electrochemical Society, 2010, 157(1): A60–A65.

[109] Hariharan S, Saravanan K, Ramar V. et al., A Rationally Designed Dual Role Anode Material for Lithium-Ion and Sodium-Ion Batteries: Case Study of Eco-Friendly Fe$_3$O$_4$[J]. Physical Chemistry Chemical Physics, 2013, 15(8): 2945–2953.

[110] Zhang N, Han X P, Liu Y C. et al., 3D Porous γ-Fe$_2$O$_3$@C Nanocomposite as High-Performance Anode Material of Na-Ion Batteries[J]. Advanced Energy Materials, 2015, 5(5): 1401123.

[111] Gu M, Kushima A, Shao Y Y. et al., Probing the Failure Mechanism of SnO$_2$ Nanowires for Sodium-Ion Batteries[J]. Nano Letters, 2013, 13(11): 5203–5211.

[112] Xie X Q, Su D W, Zhang J Q. et al., A Comparative Investigation on the Effects of Nitrogen-Doping into Graphene on Enhancing the Electrochemical Performance of SnO$_2$/Graphene for Sodium-Ion Batteries[J]. Nanoscale, 2015, 7(7): 3164–3172.

[113] Lu Y Y, Zhang N, Zhao Q. et al., Micro-Nanostructured CuO/C Spheres as High-Performance Anode Materials for Na-Ion Batteries[J]. Nanoscale, 2015, 7(6): 2770–2776.

[114] 韩季颖. 铜基氧化物/石墨烯复合材料的制备及其在钠离子电池中的应用[D]. 北京: 北京化工大学, 2017.

[115] 唐正, 何国强. 微乳液法制备氧化铜阵列应用于钠离子电池[J]. 电源技术, 2019, 43(5): 795–797.

[116] Kong H B, Wu Y S, Hong W Z. Structure-Designed Synthesis of Cu-Doped Co$_3$O$_4$@N-Doped Carbon with Interior Void Space for Optimizing Alkali-Ion Storage[J]. Energy Storage Materials, 2019, 24: 610–617.

[117] Xu M, Xia Q H, Yue J L. et al., Rambutan-Like Hybrid Hollow Spheres of Carbon Confined Co$_3$O$_4$ Nanoparticles as Advanced Anode Materials for Sodium-Ion Batteries[J]. Advanced Functional Materials, 2019, 29: 1807377.

[118] Wu L, Lu H Y, Xiao L F. et al., A Tin(ii) Sulfide-Carbon Anode Material Based on Combined Conversion and Alloying Reactions for Sodium-Ion Batteries[J]. Journal of Materials Chemistry A, 2014, 2(39): 16424–16428.

[119] Zhang Y D, Zhu P Y, Huang L L. et al., Few-Layered SnS$_2$ on Few-Layered Reduced Graphene Oxide as Na-Ion Battery Anode with Ultralong Cycle Life and Superior Rate Capability[J]. Advanced Functional Materials, 2015, 25(3): 481–489.

[120] Choi S H, Ko Y N, Lee J K. et al., 3D MoS$_2$ -graphene Microspheres Consisting of Multiple Nanospheres with Superior Sodium Ion Storage Properties[J]. Advanced Functional Materials, 2015, 25(12): 1780–1788.

[121] Hu Z, Liu Q N, Chou S L. et al., Advances and Challenges in Metal Sulfides/ Selenides for Next-Generation Rechargeable Sodium-Ion Batteries[J]. Advanced Materials, 2017, 29(48): 1700606.

[122] Mortazavi M, Wang C, Deng J. et al., Ab Initio Characterization of Layered MoS$_2$ as Anode for Sodium-Ion Batteries[J]. Journal of Power Sources, 2014, 268: 279–286.

[123] Hu Z, Wang L X, Zhang K. et al., MoS$_2$ Nanoflowers with Expanded Interlayers as High-Performance Anodes for Sodium-Ion Batteries[J]. Angewandte Chemie, 2014, 126(47): 13008–13012.

[124] Zhu C B, Mu X K, van Aken P A. et al., Single-Layered Ultrasmall Nanoplates of MoS$_2$ Embedded in Carbon Nanofibers with Excellent Electrochemical Performance for Lithium and Sodium Storage[J]. Angewandte Chemie-International Edition, 2014, 53(8): 2152–2156.

[125] Pan Q C, Zhang Q B, Zheng F H. et al., Construction of MoS$_2$/C Hierarchical Tubular Heterostructures for High-Performance Sodium Ion Batteries[J]. ACS Nano, 2018, 12(12): 12578–12586.

[126] Kim T B, Jung W H, Ryu H S. et al., Electrochemical Characteristics of Na/FeS$_2$ Battery by Mechanical Alloying[J]. Journal of Alloys and Compounds, 2008, 449(1–2): 304–307.

[127] Walter M, Zuend T, Kovalenko M V. Pyrite (Fes$_2$) Nanocrystals as Inexpensive High-Performance Lithium-Ion Cathode and Sodium-Ion Anode Materials[J]. Nanoscale, 2015, 7(20): 9158–9163.

[128] Wei X, Li W H, Shi J A. et al., FeS@C on Carbon Cloth as Flexible Electrode for Both Lithium and Sodium Storage[J]. ACS Applied Materials & Interfaces, 2015, 7(50): 27804–27809.

[129] Peng S J, Han X P, Li L L. et al., Unique Cobalt Sulfide/Reduced Graphene Oxide Composite as an Anode for Sodium-Ion Batteries with Superior Rate Capability and Long Cycling Stability [J]. Small, 2016, 12(10): 1359–1368.

[130] Tao H W, Zhou M, Wang K L. et al., N/S Co-Doped Carbon Coated Nickel Sulfide as a Cycle-Stable Anode for High Performance Sodium-Ion Batteries[J]. Journal of Alloys and Compounds, 2018, 754: 199–206.

[131] Chang X Q, Ma Y F, Yang M. et al., In-situ Solid-State Growth of N, S Codoped Carbon Nanotubes Encapsulating Metal Sulfides for High-Efficient-Stable Sodium Ion Storage[J]. Energy Storage Materials, 2019, 23: 359–366.

[132] Zhang Z A, Yang X, Fu Y. et al., Ultrathin Molybdenum Diselenide Nanosheets Anchored on Multi-Walled Carbon Nanotubes as Anode Composites for High Performance Sodium-Ion Batteries[J]. Journal of Power Sources, 2015, 296: 2–9.

[133] Wang H, Lan X Z, Jiang D L. et al., Sodium Storage and Transport Properties in Pyrolysis Synthesized MoSe$_2$ Nanoplates for High Performance Sodium-Ion Batteries[J]. Journal of Power Sources, 2015, 283: 187–194.

[134] Xie D, Tang W J, Wang Y D. et al., Facile Fabrication of Integrated Three-Dimensional C-MoSe$_2$/Reduced Graphene Oxide Composite with Enhanced Performance for Sodium Storage [J]. Nano Research, 2016, 9(6): 1618–1629.

[135] Yang X, Zhang Z A, Shi X D. Rational Design of Coaxial-Cable MoSe$_2$/C: Towards High Performance Electrode Materials for Lithium-Ion and Sodium-Ion Batteries[J]. Journal of Alloys and Compounds, 2016, 686: 413–420.

[136] Tang Y C, Zhao Z B, Wang Y W. et al., Carbon-Stabilized Interlayer-Expanded Few-Layer MoSe$_2$ Nanosheets for Sodium Ion Batteries with Enhanced Rate Capability and Cycling Performance[J]. ACS Applied Materials & Interfaces, 2016, 8(47): 32324–32332.

[137] Ko Y N, Choi S H, Park S B. et al., Hierarchical MoSe$_2$ Yolk-Shell Microspheres with Superior Na-Ion Storage Properties[J]. Nanoscale, 2014, 6(18): 10511.

[138] Choi S H, Kang Y C. Fullerene-Like MoSe$_2$ Nanoparticles-Embedded CNT Balls with Excellent Structural Stability for Highly Reversible Sodium-Ion Storage[J]. Nanoscale, 2016, 8(7): 4209–4216.

[139] Zhang Z A, Fu Y, Yang X. et al., Hierarchical MoSe$_2$ Nanosheets/Reduced Graphene Oxide Composites as Anodes for Lithium-Ion and Sodium-Ion Batteries with Enhanced Electrochemical Performance[J]. ChemNanoMat, 2015, 1(6): 409–414.

[140] Yang X, Zhang Z A, Fu Y. et al., Porous Hollow Carbon Spheres Decorated with Molybdenum Diselenide Nanosheets as Anodes for Highly Reversible Lithium and Sodium Storage[J]. Nanoscale, 2015, 7(22): 10198–10203.

[141] Zhang K, Hu Z, Liu X. et al., FeSe$_2$ Microspheres as a High-Performance Anode Material for Na-Ion Batteries[J]. Advanced Materials, 2015, 27(21): 3305–3309.

[142] Park G D, Kim J H, Kang Y C. Large-Scale Production of Spherical FeSe$_2$-Amorphous Carbon Composite Powders as Anode Materials for Sodium-Ion Batteries[J]. Materials Characterization, 2016, 120: 349–356.

[143] Zhang Z, Shi X, Yang X. et al., Nanooctahedra Particles Assembled FeSe$_2$ Microspheres Embedded into Sulfur-Doped Reduced Graphene Oxide Sheets as a Promising Anode for Sodium Ion Batteries[J]. ACS Applied Materials & Interfaces, 2016, 8(22): 13849–13856.

[144] Zhang K, Park M H, Zhou L M. et al., Urchin-Like CoSe$_2$ as a High-Performance Anode Material for Sodium-Ion Batteries[J]. Advanced Functional Materials, 2016, 26(37): 6728–6735.

[145] Park G D, Kang Y C. One-Pot Synthesis of CoSe$_x$-rGO Composite Powders by Spray Pyrolysis and Their Application as Anode Material for Sodium-Ion Batteries[J]. Chemistry-A European Journal, 2016, 22(12): 4140–4146.

[146] Cho J S, Won J M, Lee J. et al., Design and Synthesis of Multiroom-Structured Metal Compounds-Carbon Hybrid Microspheres as Anode Materials for Rechargeable Batteries[J]. Nano Energy, 2016, 26: 466–478.

[147] Ihsan-Ul-Haq M, Huang H, Cui J. et al., Chemical Interactions between Red P and Functional Groups in NiP$_3$/CNT Composite Anodes for Enhanced Sodium Storage[J]. Journal of Materials Chemistry A, 2018, 6(41): 20184–20194.

[148] Wang X J, Chen K, Wang G. et al., Rational Design of Three-Dimensional Graphene Encapsulated with Hollow FeP@Carbon Nanocomposite as Outstanding Anode Material for Lithium Ion and Sodium Ion Batteries[J]. ACS Nano, 2017, 11(11): 11602–11616.

[149] Kim Y, Kim Y, Choi A. et al., Tin Phosphide as a Promising Anode Material for Na-Ion Batteries [J]. Advanced Materials, 2014, 26(24): 4139–4144.

[150] Lu Y Y, Zhou P F, Lei K X. et al., Selenium Phosphide (Se$_4$P$_4$) as a New and Promising Anode Material for Sodium-Ion Batteries[J]. Advanced Energy Materials, 2017, 7(7): 1601973.

[151] Fullenwarth J, Darwiche A, Soares A. et al., NiP$_3$: A Promising Negative Electrode for Li- and Na-ion Batteries[J]. Journal of Materials Chemistry A, 2014, 2(7): 2050–2059.

[152] Ali G, Lee J, Oh S. et al., Elucidating the Reaction Mechanism of SnF$_2$@C Nanocomposite as a High-Capacity Anode Material for Na-Ion Batteries[J]. Nano Energy, 2017, 42: 106–114.

[153] Kong M H, Liu K H, Ning J Y. et al., Perovskite Framework NH$_4$FeF$_3$/Carbon Composite Nanosheets as a Potential Anode Material for Li and Na Ion Storage[J]. Journal of Materials Chemistry A, 2017, 5(36): 19280–19288.

[154] Yabuuchi N, Kubota K, Dahbi M. et al., Research Development on Sodium-Ion Batteries[J]. Chemical Reviews, 2014, 114(23): 11636–11682.

[155] Mortazavi M, Ye Q J, Birbilis N. et al., High Capacity Group-I5 Alloy Anodes for Na-Ion Batteries: Electrochemical and Mechanical Insights[J]. Journal of Power Sources, 2015, 285: 29–36.

[156] Ellis L D, Hatchard T D, Obrovac M N. Reversible Insertion of Sodium in Tin[J]. Journal of the Electrochemical Society, 2012, 159(11): A1801–A1805.

[157] Wang J W, Liu X H, Mao S X. et al., Microstructural Evolution of Tin Nanoparticles during In-situ Sodium Insertion and Extraction[J]. Nano Letters, 2012, 12(11): 5897–5902.

[158] Liu Y C, Zhang N, Jiao L F. et al., Tin Nanodots Encapsulated in Porous Nitrogen- Doped Carbon Nanofibers as a Free-Standing Anode for Advanced Sodium-Ion Batteries[J]. Advanced Materials, 2015, 27(42): 6702–6707.

[159] Liu Y C, Zhang N, Jiao L F. et al., Ultrasmall Sn Nanoparticles Embedded in Carbon as High-Performance Anode for Sodium-Ion Batteries[J]. Advanced Functional Materials, 2015, 25(2): 214–220.

[160] Komaba S, Matsuura Y, Ishikawa T. et al., Redox Reaction of Sn-Polyacrylate Electrodes in Aprotic Na Cell[J]. Electrochemistry Communications, 2012, 21: 65–68.

[161] Sottmann J, Herrmann M, Vajeeston P. et al., How Crystallite Size Controls Reaction Path in Non-Aqueous Metal Ion Batteries: The Example of Sodium Bismuth Alloying[J]. Chemistry of Materials, 2016, 28(8): 2750–2756.

[162] Su D, Dou S, Wang G X. Bismuth: A New Anode for the Na-Ion Battery[J]. Nano Energy, 2015, 12: 88–95.

[163] Morito H, Yamada T, Ikeda T. et al., Na-Si Binary Phase Diagram and Solution Growth of Silicon Crystals[J]. Journal of Alloys and Compounds, 2009, 480(2): 723–726.

[164] Zhang L, Hu X L, Chen C J. et al., In Operando Mechanism Analysis on Nanocrystalline Silicon Anode Material for Reversible and Ultrafast Sodium Storage[J]. Advanced Materials, 2017, 29: 16047085.

[165] Xu Y L, Swaans E, Basak S. et al., Reversible Na-Ion Uptake in Si Nanoparticles[J]. Advanced Energy Materials, 2016, 6(2): 1501436.

[166] Sangster J, Pelton A D. The Cs-Ge (Cesium-germanium) System[J]. Journal of Phase Equilibria, 1997, 18(3): 284–286.

[167] Yue C, Yu Y J, Sun S B. et al., High Performance 3D Si/Ge Nanorods Array Anode Buffered by TiN/Ti Interlayer for Sodium-Ion Batteries[J]. Advanced Functional Materials, 2015, 25(9): 1386–1392.

[168] Abel P R, Lin Y M, de Souza T. et al., Nanocolumnar Germanium Thin Films as a High-Rate Sodium-Ion Battery Anode Material[J]. The Journal of Physical Chemistry C, 2013, 117(37): 18885–18890.

[169] Baggetto L, Keum J K, Browning J F. et al, Germanium as Negative Electrode Material for Sodium-Ion Batteries[J]. Electrochemistry Communications, 2013, 34: 41–44.

[170] Qian J F, Chen Y, Wu L. et al., High Capacity Na-Storage and Superior Cyclability of Nanocomposite Sb/C Anode for Na-Ion Batteries[J]. Chemical Communications, 2012, 48(56): 7070.

[171] Liu Y H, Xu Y H, Zhu Y J. et al., Tin-Coated Viral Nanoforests as Sodium-Ion Battery Anodes[J]. ACS Nano, 2013, 7(4): 3627–3634.

[172] Luo W, Li F, Gaumet J J. et al., Bottom-Up Confined Synthesis of Nanorod-in- Nanotube Structured Sb@N-C for Durable Lithium and Sodium Storage[J]. Advanced Energy Materials, 2018, 8(19): 1703237.

[173] Sun J, Lee H W, Pasta M. et al., A Phosphorene-Graphene Hybrid Material as A High-Capacity Anode for Sodium-Ion Batteries[J]. Nature Nanotechnology, 2015, 10(11): 980–985.

[174] Song J X, Yu Z X, Gordin M L. et al., Chemically Bonded Phosphorus/Graphene Hybrid as a High Performance Anode for Sodium-Ion Batteries[J]. Nano Letters, 2014, 14(11): 6329–6335.

[175] Kim Y, Park Y, Choi A. et al., An Amorphous Red Phosphorus/Carbon Composite as a Promising Anode Material for Sodium Ion Batteries[J]. Advanced Materials, 2013, 25(22): 3045–3049.

[176] Liu Y H, Zhang A Y, Shen C F. et al., Red Phosphorus Nanodots on Reduced Graphene Oxide as a Flexible and Ultra-Fast Anode for Sodium-Ion Batteries[J]. ACS Nano, 2017, 11(6): 5530–5537.

[177] Sun J, Zheng G Y, Lee H W. et al., Formation of Stable Phosphorus-Carbon Bond for Enhanced Performance in Black Phosphorus Nanoparticle-Graphite Composite Battery Anodes[J]. Nano Letters, 2014, 14(8): 4573–4580.

[178] Qian J F, Wu X Y, Cao Y L. et al., High Capacity and Rate Capability of Amorphous Phosphorus for Sodium Ion Batteries[J]. Angewandte Chemie International Edition, 2013, 52(17): 4633–4636.

[179] Li W J, Chou S L, Wang J Z. et al., Simply Mixed Commercial Red Phosphorus and Carbon Nanotube Composite with Exceptionally Reversible Sodium-Ion Storage[J]. Nano Letters, 2013, 13(11): 5480–5484.

[180] Dahbi M, Yabuuchi N, Fukunishi M. et al., Black Phosphorus as a High-Capacity, High-Capability Negative Electrode for Sodium-Ion Batteries: Investigation of the Electrode/Electrolyte Interface[J]. Chemistry of Materials, 2016, 28(6): 1625–1635.

[181] Wu L, Hu X H, Qian J F. et al., Sb-C Nanofibers with Long Cycle Life as an Anode Material for High-Performance Sodium-Ion Batteries[J]. Energy Environment Science, 2014, 7(1): 323–328.

[182] Naguib M, Mochalin V N, Barsoum M W. et al., 25th Anniversary Article: MXenes: A New Family of Two-Dimensional Materials[J]. Advanced Materials, 2014, 26(7): 992–1005.

[183] Zhao M Q, Xie X Q, Ren C E. et al., Hollow MXene Spheres and 3D Macroporous MXene Frameworks for Na-Ion Storage[J]. Advanced Materials, 2017, 29(37): 1702410.

[184] Zhang Y Q, Guo B S, Hu L Y. et al., Synthesis of SnS Nanoparticle-Modified MXene ($Ti_3C_2t_x$) Composites for Enhanced Sodium Storage[J]. Journal of Alloys and Compounds, 2018, 732: 448–453.

[185] Sun N, Zhu Q Z, Anasori B. et al., MXene-Bonded Flexible Hard Carbon Film as Anode for Stable Na/K-Ion Storage[J]. Advanced Functional Materials, 2019, 29(51): 1906282.

[186] Zhao L, Zhao J M, Hu Y S. et al., Disodium Terephthalate ($Na_2C_8h_4O_4$) as High Performance Anode Material for Low-Cost Room-Temperature Sodium-Ion Battery[J]. Advanced Energy Materials, 2012, 2(8): 962–965.

[187] Park Y, Shin D S, Woo S H. et al., Sodium Terephthalate as an Organic Anode Material for Sodium Ion Batteries[J]. Advanced Materials, 2012, 24(26): 3562–3567.

[188] Haupler B, Wild A, Schubert U S. Carbonyls: Powerful Organic Materials for Secondary Batteries[J]. Advanced Energy Materials, 2015, 5(11): 1402034.

[189] Wang S W, Wang L J, Zhu Z Q. et al., All Organic Sodium-Ion Batteries with $Na_4C_8H_2O_6$[J]. Angewandte Chemie International Edition, 2014, 53(23): 5892–5896.

[190] Wang H, Hu P F, Yang J. et al., Renewable-Juglone-Based High-Performance Sodium-Ion Batteries[J]. Advanced Materials, 2015, 27(14): 2348–2354.

[191] Wu X Y, Ma J, Ma Q D. et al., A Spray Drying Approach for the Synthesis of A $Na_2C_6H_2O_4$/CNT Nanocomposite Anode for Sodium-Ion Batteries[J]. Journal of Materials Chemistry A, 2015, 3(25): 13193–13197.

[192] Luo W, Lin C F, Zhao O. et al., Ultrathin Surface Coating Enables the Stable Sodium Metal Anode[J]. Advanced Energy Materials, 2017, 7(2): 1601526.

[193] Wu F, Zhou J H, Luo R. et al., Reduced Graphene Oxide Aerogel as Stable Host for Dendrite-Free Sodium Metal Anode[J]. Energy Storage Materials, 2019, 22: 376–383.

# Chapter 5
# Sodium-ion battery electrolyte

One of the most vital components for batteries is the electrolyte. It impacts deep influence on the overall electrochemical performance of the battery for which it plays a role in connecting cathodes with anodes and constructing the path for ion transmission between them. Therefore, the performance of batteries, for example: cost, safety, and specific capacity, is affected by the selection of electrolytes.

For sodium-ion batteries, the feature of electrolytes is nonnegligible. According to their physical property, electrolytes are divided into liquid electrolyte and solid electrolyte, and it can also be classified into organic electrolyte and inorganic electrolyte as the main chemical composition of electrolytes. Due to the similarity between sodium-ion batteries and lithium-ion batteries, the performance of the former is usually consistent with the performance requirements of the latter, including ① chemical stability, no chemical reaction with cathodes and anodes; ② electrochemical stability, no decomposition occurs during charging and discharging; ③ thermal stability, the electrolyte state remains stable when the temperature increases; ④ ion conduction and electronic insulation, the electrolytes can transport ions, but cannot conduct electrons; and ⑤ low toxicity, low production cost [1]. In this chapter, according to the state and composition of electrolytes, the electrolytes of sodium-ion batteries are divided into five types: organic liquid electrolytes, ionic liquid electrolytes, polymer electrolytes, inorganic solid electrolytes, and aqueous electrolytes. Their physical and chemical properties, synthesis methods, and matching with electrode materials were discussed as follows.

## 5.1 Overview of electrolytes and their characteristics

The electrolyte plays an important part of batteries, which has unique physical and chemical properties. Not only a comprehensive study of the electrode materials should be made but also components such as electrolytes, current collectors, and separators, and the compatibility between them must be optimized in practical application. Although there are some similarities existing in characteristics between sodium-ion battery electrolytes and lithium-ion battery electrolytes, differences in transportation and storage of different ions, such as working potential range and ionic conductivity, have to be noticed. Therefore, this section will introduce the characteristics of electrolytes for sodium-ion batteries in chemical–electrochemical stability, thermal stability, ion transport, and other properties (such as toxicity and cost) in detail.

https://doi.org/10.1515/9783110749069-005

### 5.1.1 Chemical–electrochemical stability

The battery that involves a series of chemical and electrochemical reactions is a complex system. As for electrolytes, especially, the basic requirements for the electrolyte are negative to reacting with other components. Corrosion of collectors by electrolytes impacts great influence on cycling life and safety of the battery. Therefore, in order to obtain better performance, it is necessary to avoid the corrosion. Because the standard electrode potential of $Na^+/Na$ redox pair is higher than that of $Li^+/Li$ redox pair, the reaction of sodium is carried before decomposition of aluminum foil, so aluminum foil can be used as collectors of both the cathode and anode, which can not only reduce the cost but also increase the energy density of sodium-ion batteries. Furthermore, the ductility of aluminum foil can effectively adapt to the slight deformation of batteries. However, it is worth noting that the anionic groups of sodium salts and solvents may lead to pitting corrosion of aluminum foils under extreme conditions. In the same organic solvent (ethylene carbonate (EC):diethyl carbonate (DEC), v:v = 1:1), the increasing order of aluminum dissolution and anionic decomposition of different sodium salts is as follows: $NaPF_6 < NaClO_4 < NaTFSI < NaFTFSI < NaFSI$ [bis(trifluoromethylsulfonyl) imide (TFSI), bis(fluorosulfonyl) imide (FSI) and fluorosulfonyl-trifluoromethylsulfonyl) imide (FTFSI) anion]. The anions reacting with aluminum usually show a large irreversible oxidation current [2]. By adding a small amount of $NaPF_6$ (mass fraction 5%) to the electrolyte, an $AlF_3$ or $AlO_xF_y$ protective layer can be formed to improve the stability of aluminum foil in imide-based electrolytes.

As a result of solvent effect, the imide salt dissolved in the ionic liquids (ILs) to form an electrolyte can produce a stable passivation layer, whose active components include Al $(TFSI)_3$ and Al $(FTFSI)_3$, on the surface of current collectors to prevent it from further corrosion. Because the solubility of these aluminum perylene imides in ionic liquid-based electrolytes is lower than that in carbonate-based electrolytes, aluminum perylene imides are able to form a good protective layer in ionic liquid-based electrolytes [3]. Among these aluminum perylene imides, Al $(TFSI)_3$ is the most stable substance in ILs and carbonate-based electrolytes. Similar results have been confirmed by sodium semicells with $NaFe_{0.4}Ni_{0.3}Ti_{0.3}O_2$ layered oxides as positive electrodes. At 55 ℃, the oxidation decomposition voltage of aluminum foil in ionic liquid electrolyte based on NaFSI can reach 5 V. Under the same condition, however, the oxidation decomposition reaction of aluminum foil in carbonate electrolyte occurs at about 3.5 V. The ionic liquid electrolyte matched with $NaFe_{0.4}Ni_{0.3}Ti_{0.3}O_2$ electrode showed high capacity retention after 50 cycles. And no sign of aluminum dissolution was detected. A more effective way to restrain aluminum foil corrosion at high voltage is to use a high concentration of bis (fluorosulfonyl) imide (FSA) electrolyte [4]. As shown in Figure 5.1, although $AlF_3$ is hard to form on the electrode surface, no free solvent molecules can easily solvate $Al^{3+}$.

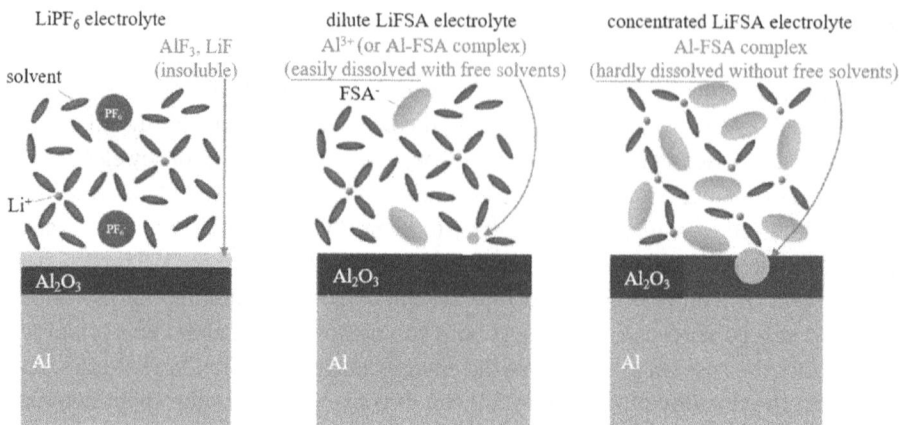

**Figure 5.1:** Schematic diagram of the behavior of diluted LiPF$_6$-based electrolytes in conventional Lithium bis(fluorosulfonyl)amide (LiFSA)-based electrolytes and Al electrodes in high-concentration LiFSA electrolytes [4].

For high-performance electrolytes, in addition to chemical stability, good electrochemical stability is also very necessary. The electrochemical stability of electrolytes usually refers to a wide electrochemical stability window (ESW) and a stable electrode/electrolyte interface. The electrochemical stability depends not only on the composition of the electrolyte, but also on the compatibility between electrodes and electrolytes. In general, the ESW of aqueous electrolyte is limited by hydrogen evolution/oxygen evolution reaction, while the ESW of inorganic solid electrolyte shows a wide range because of stable crystal structure. As for organic liquid electrolytes, the determinants of ESW are mainly related to composition and concentration. In the same organic solvent, all sodium salts except NaTFSI showed similar electrochemical stability (Figure 5.2(a)) [5]. In the NaTFSI-based electrolyte, the side reaction at 3.6 V is driven by the corrosion of aluminum foil. However, when using the same sodium salt and different organic solvents, there is a large difference in the ESW of the electrolyte (Figure 5.2(a)), which means that the solvent component is the main factor affecting the electrolyte ESW. The solvents of DEC and dimethyl carbonate (DMC) have the widest and narrowest ESW, respectively. The ESW is usually enhanced by using mixed solvents. As shown in Figure 5.2(a), the stability of EC: propylene carbonate (PC) exhibits better performance than single solvent during the voltage of 0 V to 5 V. Another feasible way to increase ESW is to replace organic solvents with ionic liquid, polymers or inorganic solid [6, 7]. These new electrolytes can provide higher ESW (>5 V). As shown in Figure 5.2(b)–(d), the ESW of solid electrolyte, in particular, can reach 6 V. Although the increase of viscosity will lead to the decrease of ionic conductivity, all-solid-state battery is still one of the most promising research directions to achieve high working voltage and high energy density.

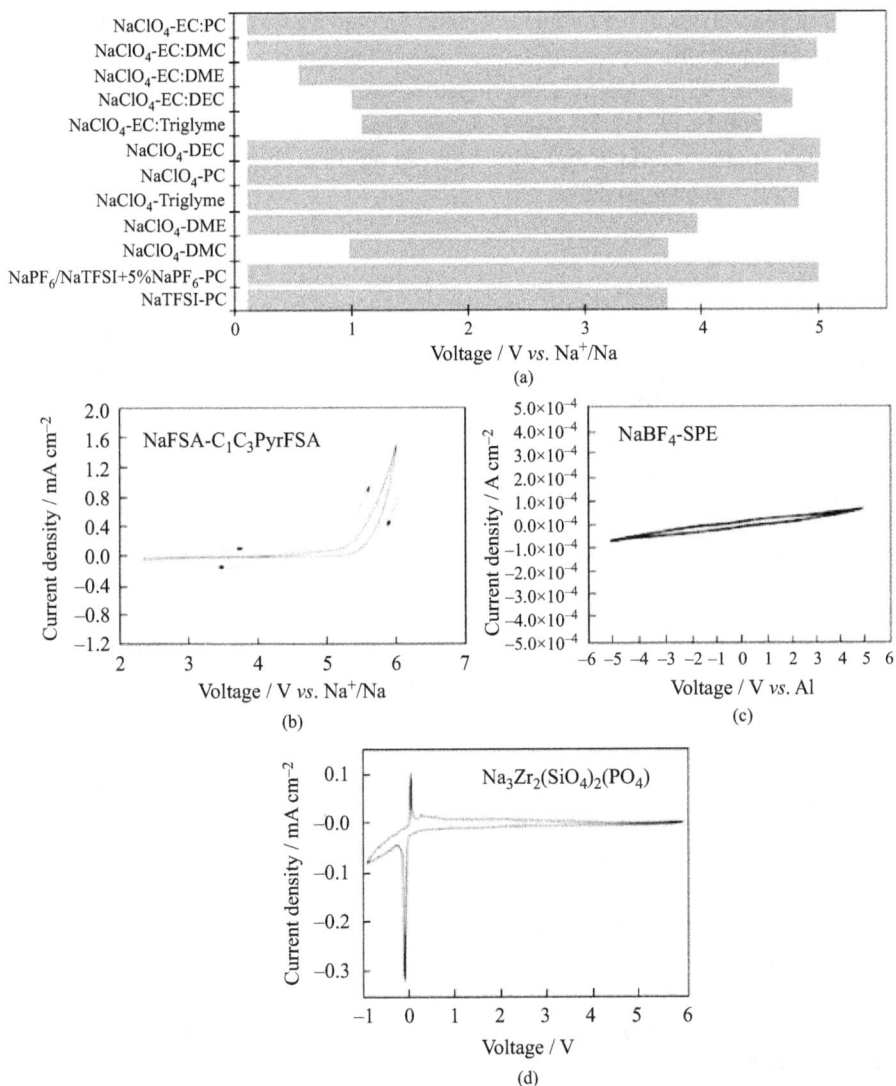

**Figure 5.2:** Electrochemical windows of different electrolytes: (a) different sodium salts in different solvents; (b) Sodium bis(fluorosulfonyl)amide (NaFSA) dissolved in $C_1C_3$PyrFSA [6]; (c) NaBF$_4$ dissolved in Solid polymer electrolytes (SPE) [7]; and (d) Na$_3$Zr$_2$(SiO$_4$)$_2$(PO$_4$) solid electrolyte [8].

At present, many methods have been proposed to expand ESW. For example, the ESW of liquid electrolyte is increased by improving interface stability and salt concentration. The stable interface between the electrode and the electrolyte is beneficial to inhibit the continuous decomposition of electrolytes and protect the activity of electrodes. In liquid electrolyte, a passivation layer forms in the interface between the electrode and the electrolyte. This unique layer formed by decomposing and depositing electrolytes in the initial cycle plays an important role in the stability of the interface. The mechanical properties of the passivation layer are improved and the impedance is reduced by using film-forming additives, such as fluorinated ethylene carbonate (FEC) and EC [9, 10]. Therefore, surface modification is an effective method to obtain high-quality interface and wide ESW. In addition, high-concentration liquid electrolyte can also increase ESW.

## 5.1.2 Thermal stability

Thermal stability is one of the most important properties of sodium-ion batteries in practical application. The freezing point and boiling point of the electrolyte must exceed the working temperature range of the battery. The ionic conductivity and safety of electrolytes are highly dependent on temperature, so electrolytes with high thermal stability are critical for the practical application of batteries. Compared with Li-based electrolytes using the same solvent, the melting point of Na-based electrolytes is much lower indicating that sodium-ion batteries possess a wider operating temperature range in applications.

Due to the inherent characteristics of aqueous electrolytes and inorganic solid electrolytes, these two electrolytes have nonflammability and high safety. In future research, in order to assemble high-safety batteries, the combination of those two electrolytes should be considered. Ionic liquids, the liquid with low freezing point and high boiling point, shows excellent electrochemical performance in a wide temperature range. In addition, the study of thermal stability is mainly focused on the organic liquid electrolyte of sodium-ion batteries. As shown in Figure 5.3(a), the thermal stability of sodium salt is better than that of lithium salt. Moreover, it can also be found from the figure that the type of solvent has a great influence on the thermal stability of electrolytes for sodium-ion batteries [4]. The initial decomposition temperature of a single carbonate-based solvent is about 80 °C, indicating that it has high flammability and low heat capacity (Figure 5.3(b)) [5]. The addition of linear carbonates with high reactivity (such as DMC and DEC) will lead to poor thermal stability [11]. The thermal stability of mixed carbonate solvents partially replaced by EC or PC is better than that of single solvent (Figure 5.3(b)). Developing different kinds of solvents and additives has been proved to be a feasible method for improving the thermal stability of sodium-ion batteries. Adding 10 vol% FEC as a film-forming additive to 0.8 M NaPF$_6$ and trimethyl phosphate (TMP) can obtain a new type of nonflammable organophosphate electrolyte, which will not be completely ignited even in fire (Figure 5.3(c)) [12]. The high-concentration electrolyte that uses only TMP solvent without any additives is also safe. The electrolyte takes hard carbon or graphite as negative electrode, the cycle life is more than 1,000 cycles, and the decomposition of the electrolyte is nonnegligible [13]. Using layered Transition metal oxides (TMO) as the cathode and alloy as the anode, the sodium-ion battery assembled with pure TMP solvent electrolyte showed high-specific capacity, cycle stability, and nonflammability. Addition of noncombustible additives is able to maintain the high electrochemical performance of carbonate-based electrolytes. Adding a small amount of flame retardant additive ethoxy (pentafluoro) cyclotriphosphazene (EFPN) to 1 M NaPF$_6$ and EC:DEC (1:1, volume ratio) electrolytes, obtained electrolyte displays incombustibility (Figure 5.3(c)) and electrochemical stability (Figure 5.4(d)) [14].Using ILs as solvents is another way to improve thermal stability. It broadens the lower limit of the working temperature of electrolytes. For example, using 1 M NaTFSI and butylmethyl pyrrolidine (BMP)-TFSIIL-based electrolytes, NaFePO$_4$ semicells exhibit high thermal stability and nonflammability at 400 °C, the electrochemical property of which is similar to those of 1 M NaClO$_4$'s EC: DEC solution electrolytes (Figure 5.3(e)) [15]. And ILs maintain high ionic conductivity at low temperatures (even as low as –40 °C) [16]. Inorganic solid electrolyte is not easy to burn at high temperature and has excellent stability, which is conducive to the preparation of sodium-ion batteries with high safety. As shown in Figure 5.3(f), the coefficient of thermal expansion of sulfur-based inorganic solid electrolytes is small at an extremely high temperature of 250 ℃, and the rising temperature has made little effect. Therefore, designing of reasonable inorganic solid electrolytes is one of the ways to improve the safety of batteries as well.

### 5.1.3 Ion transport performance

When conducting current, ions are charge carriers in the electrolyte rather than electrons, which indicates high ionic conductivity and low electron conductivity are regarded as characteristics of good electrolytes. The ionic conductivity of liquid electrolyte depends on the degree of dissociation of salt, the viscosity of solvent, and the migration number of sodium ion and corresponding anion [17]. The solubility of sodium salt in solvent is limited, and the safe working temperature also has a threshold, so it is difficult to improve the ionic conductivity of electrolytes from these two aspects. The ionic conductivity of aqueous electrolytes is mainly determined by the properties of sodium salts and solvents. The ionic conductivities of various types of electrolytes are listed in Table 5.1. In the same solvent, due to the low polarization of PF$_6^-$ anions, it is beneficial to salt hydrolysis and ion migration, and NaPF$_6$ has the highest ionic conductivity among the three sodium salts. When different solvents are used to dissolve the same sodium salt, the conductivity of binary electrolyte with EC as cosolvent is higher than that of other samples, because EC can reduce the viscosity and band gap width of other solvents and increase ion mobility. Among types of electrolytes, the ionic conductivity of aqueous electrolytes is higher than that of others. Without no doubt, the ionic conductivity of most inorganic solid electrolytes is very low ($1 \times 10^{-4}$ mS · cm$^{-1}$), which is only 1/10 of that of aqueous electrolytes. However, (NASICON), a superionic conductor containing sodium oxides and sodium, enables providing sufficient ionic conductivity at high temperatures [18, 19].

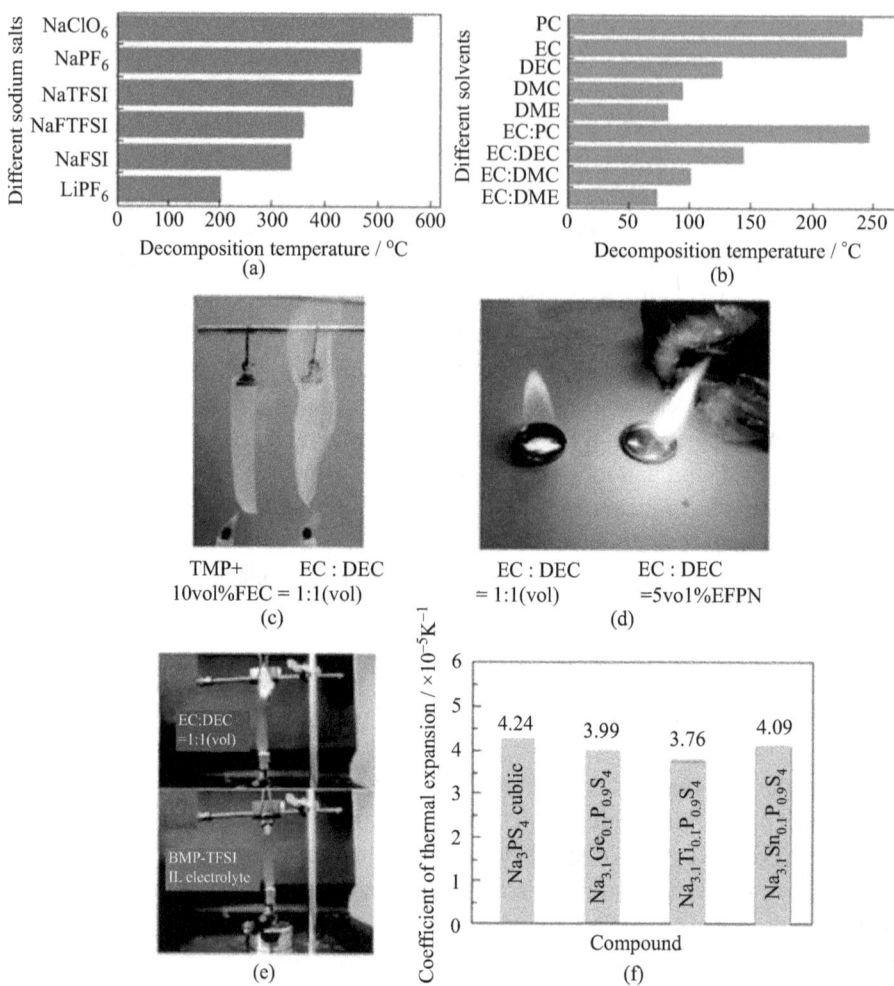

**Figure 5.3:** Thermal stability of electrolytes: (a) thermal decomposition temperature of different sodium salts; (b) thermal decomposition temperature of different solvents; Combustion tests of (c) TMP electrolytes; (d) carbonate electrolytes with EFPN addition and (e) BMP-TFSI IL electrolytes compared to conventional organic electrolytes, respectively; and (f) thermal expansion coefficient of $Na_3PS_4$ and its derivatives.

In order to further understand the factors that affect the ionic conductivity, the inherent characteristics and corresponding influencing factors of liquid sodium-ion batteries electrolytes are summarized, as well as the advantages and disadvantages of liquid sodium-ion batteries electrolytes compared with lithium-ion battery electrolytes.

First of all, the solvation between sodium ions and organic solvents is an important parameter of ionic conductivity and electrode reaction rate. The desolvation process is the decisive step of ion migration and electrode reaction rate, which can

**Table 5.1:** Ionic conductivity of various types of electrolytes.

| Electrolyte | $\sigma$/(mS/cm) | Electrolyte | $\sigma$/(mS/cm) |
|---|---|---|---|
| 1 M NaTFSI-PC | 6.2 | 1 M NaClO$_4$-EC: DEC | 6.35 |
| 1 M NaClO$_4$-PC | 6.4 | 1 M NaClO$_4$-EC: PC | 8.1 |
| 1 M NaPF$_6$-PC | 7.98 | 1 M NaClO$_4$-BMP-TFSI | 1.0 |

Note: $\sigma$ is the ionic conductivity of the electrolyte at room temperature.

be confirmed by the positive correlation between the activation barrier of the final solvent molecule and the desolvation energy. The desolvation energy of cations increases in the order of $Na^+ < Li^+ < Mg^{2+}$, and decreases with the increase of ion radius (Figure 5.4(a)) [20]. Compared with lithium-ion battery and magnesium-ion battery, sodium-ion batteries have lower desolvation energy (usually 40–70 kJ $\cdot$ mol$^{-1}$), which indicates that sodium-ion batteries have faster ion transfer rate and higher electrochemical reaction activity, which is also an important advantage of sodium-ion batteries. Therefore, sodium ions can act as weaker Lewis acids, providing lower desolvation energy and rapid electrode interface reaction. Moreover, the effect of solvent on desolvation energy is very small [21]. As there are a large number of free ions in the solvent, the ionic conductivity first tends to increase with the increase of salt concentration, and then decreases due to electrostatic interference (Figure 5.4(b)). In addition, the type of solvent has a significant effect on the energy of desolvation. As shown in Figure 5.4(c), there are significant differences in molar conductivity in solvents due to differences in the size, structure, and electronic configuration of solvated molecules. The exact solvation structures of $Li^+$, $Na^+$, and $K^+$ were revealed by first-principle molecular dynamics (MD) simulations [22]. It can be seen from Figure 5.4(d) that the coordination number of alkali ions increases gradually with the increase of ion radius. At the same time, compared with $Li^+$, the disordered and flexible solvation structures of $Na^+$ and $K^+$ ions lead to larger diffusion coefficients.

Second, the ionic conductivity is also affected by the cation–anion interaction between lithium salt and sodium salt. Due to the different characteristics of $Li^+$ and $Na^+$, the determinants of the number of transferable charges and carrier concentration in the electrolyte are different. When $Na^+$ is used as a cation instead of $Li^+$, the strength of the cation–anion interaction decreases by about 20% [20], which can be attributed to the small charge/radius of $Na^+$. Compared with quasi-$Li^+$ complexes, this leads to a decrease in the total bond energy between cations and anions. Therefore, the transmission speed of sodium ions in the electrolyte must be fast enough to obtain high-energy power. A similar situation can be observed by $K^+$. Theoretically, the solvation shell of anions is weaker than that of cations in any solvent, which is beneficial to the rapid movement of anions in electrolytes. Therefore,

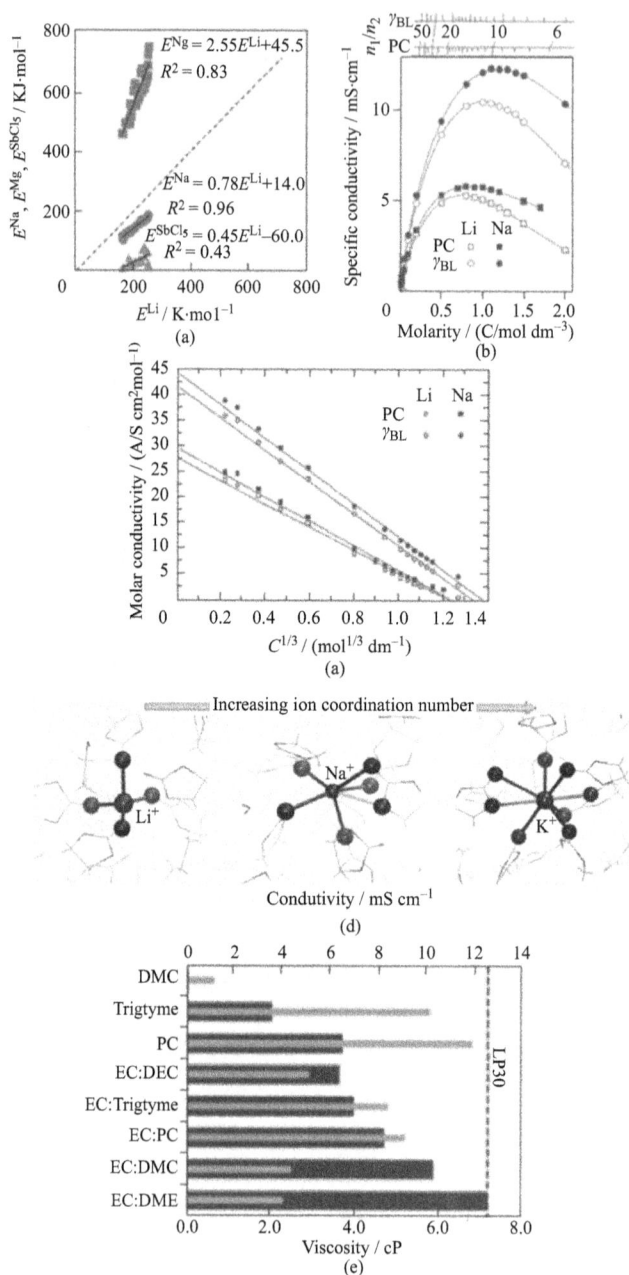

**Figure 5.4:** Factors affecting the conductivity of ions in electrolytes: (a) the desolvation energy barrier of Na⁺ versus the desolvation energy of other cations; (b) the relationship between salt concentration and conductivity; (c) the relationship between solvent type and molar conductivity; (d) solvation structure and coordination number of Li⁺, Na⁺, and K⁺ in EC solvents; and (e) viscosity of various organic electrolytes.

effective anion anchoring can be achieved in some single-ion conductive polymers, so that the migration number of $Na^+$ is close to 1.

Finally, the viscosity of the electrolyte is directly related to the ionic conductivity, which is obviously different for different electrolytes. The experimental results show that the number of ion migration is positively related to the fluidity of electrolyte and $Na^+$, but inversely proportional to the viscosity. In organic electrolytes with the same amount of $NaClO_4$, the viscosity order is DMC < EC: DMC < EC: DEC < EC: PC < PC (Figure 5.4(e)), which indicates that the presence of cosolvent can improve the dissociation of sodium salt and reduce the viscosity of the final electrolyte [5]. In the same solvent, $NaClO_4$, $NaPF_6$, and NaTFSI salts have similar viscosity, which indicates that the effect of anions on viscosity is negligible. The effect of salt concentration on viscosity cannot be ignored. Although high-concentration electrolyte shows a large $Na^+$ transfer number, high viscosity will not be conducive to ionic conductivity.

### 5.1.4 Other performance

Other properties of electrolytes cannot be ignored, such as environmental protection, low toxicity, rich resources, low cost, and so on, which are closely related to practical applications. Due to the sustainable development and environmental friendliness of sodium-ion batteries, the above properties should also be considered. Therefore, emphasis will be placed on the development of aqueous electrolytes combined with electrode materials in order to realize the above advantages in the future. At the same time, the existing organic electrolytes and electrolyte solutions should be improved to obtain higher energy density and longer cycle life.

## 5.2 Sodium salt

Sodium salt is mainly composed of sodium ions and anionic groups. It is the main component of liquid sodium-ion electrolytes and plays a decisive role in the electrochemical properties of the electrolyte. In order to avoid unnecessary side reactions and improve ion conductivity, sodium salt should meet the following requirements:
(1) Solubility determines the concentration of charge carriers in electrolytes [23];
(2) The redox potential limits the ESW of sodium salts [5];
(3) The chemical inertness of anions and cations affects the stability of the diaphragm, solvent, electrode, and current collector;
(4) Thermal stability affects the safety of batteries [24]; and
(5) Low price and nontoxic.

At present, researchers have obtained more and more sodium salts that meet the requirements via various modification methods.

Since the ionic radius of $Na^+$ is larger than that of $Li^+$, there are slightly more sodium salts than lithium salts that meet the minimum solubility requirements in low dielectric constant solvents [23]. Moreover, the structure that the anion in the sodium salt contains a stable central atom and an electronegative peripheral ligand connected by a weak coordination bond is conducive to ion transport [25]. The nature of the anion has a significant impact on the electrolyte. Therefore, the development of sodium salts is mainly focused on the structural optimization of anionic groups. Table 5.2 compares the physical and chemical properties of various sodium salts in the current literature in detail. It is seen from the table, due to the rapid ion migration, good compatibility, and low cost, $NaClO_4$ has been widely used in the performance test of electrodes [26], but its high water content, explosiveness, and high toxicity seriously obstruct the practical application of $NaClO_4$. Another common sodium salt is $NaPF_6$, which has the highest conductivity in PC-based electrolytes. However, many defects of $NaPF_6$ cannot be ignored, such as its high price, slight toxicity, and low decomposition temperature, which affect its practical application as well. In a single solvent of EC, PC, DEC, and DMC, the solubility of $NaPF_6$ is 1.4 M, 1.0 M, 0.8 M, and 0.6 M, respectively. As far as $NaPF_6$ is concerned, consequently, EC solvent is a necessary choice for preparing its multicomponent solvent. Generally speaking, ionic liquid-based anions including tetrafluoroborate ($BF_4^-$), TFSI, FSI, and trifluoromethanesulfonate (TF) exhibit excellent chemical properties such as stability and nontoxicity [27]. Normally, higher ionic conductivity is presented in sodium salts with larger anionic groups. It is noted that serious aluminum foil corrosion occurs in carbonate-based electrolytes with NaFSI or NaTFSI as the salt, which can be avoided by using ionic liquid-based electrolytes.

**Table 5.2:** The physical and chemical properties of various sodium salts in current literature.

| Sodium salt | Chemical structure | Relative molecular mass/ (g mol$^{-1}$) | Decomposition temperature/°C | Conductivity/ (mS cm$^{-1}$)* | Toxicity |
|---|---|---|---|---|---|
| $NaClO_4$ | | 122.4 | 480 | 6.4 | High |
| $NaPF_6$ | | 167.9 | 300 | 7.98 | Low |
| $NaBF_4$ | | 109.8 | 384 | – | High |

**Table 5.2** (continued)

| Sodium salt | Chemical structure | Relative molecular mass/ (g mol⁻¹) | Decomposition temperature/°C | Conductivity/ (mS cm⁻¹)* | Toxicity |
|---|---|---|---|---|---|
| NaFSI | | 203.3 | 118 | – | None |
| NaTFSI | | 303.1 | 257 | 6.2 | None |
| NaTf | | 172.1 | 248 | – | None |
| NaDFOB | | 159.8 | – | About 7 | None |
| NaTDI | | 208 | – | 4.47 | None |
| NaPDI | | 258 | – | 4.65 | None |
| NaBOB | | 209.8 | 345 | 0.256 (0.025 M) | None |
| NaBSB | | 263 | 353 | 0.239 (0.025 M) | None |

**Table 5.2** (continued)

| Sodium salt | Chemical structure | Relative molecular mass/ (g mol$^{-1}$) | Decomposition temperature/°C | Conductivity/ (mS cm$^{-1}$)* | Toxicity |
|---|---|---|---|---|---|
| NaBDSB | | 247 | 304 | 0.071 (0.025 M) | None |

Note: * presents ionic conductivity measured at room temperature in PC solvent with different sodium salts (1 M).

Although containing halogen atoms increases the risk and toxicity, they are still widely used in the electrolyte field due to the strong electronegativity and induction effects of F and Cl. At present, in order to obtain an electrolyte with higher conductivity and safety, a series of sodium salts containing F have been developed. The sodium-difluoro(oxalate) borate (NaDFOB) has been proved to own better electrical conductivity and contains more delocalized charges and anions that weakly interact with Na$^+$. Compared with the salt relying on the solvent (NaPF$_6$ and NaClO$_4$), NaDFOB possesses good compatibility with a variety of different solvents and enables achieving better cycle stability and higher specific capacity. Using the electrolyte of 1 M NaDFOB dissolved in EC:DEC solvent in the test of Na/Na$_{0.44}$MnO$_2$ half-cell, high reversible specific capacity and capacity retention rate are obtained at different charge and discharge rates. There are similar advantages in different salts with F atom and heterocyclic structure, such as 4,5-dicyano-2-(trifluoromethyl)imidazole sodium salt (NaTDI) and 4,5-dicyano-2- (Pentafluoroethyl) imidazole sodium salt (NaPDI) [28]. These chemicals with a wide ESW up to 4.5 V (versus Na$^+$/Na) show good thermal stability above 300 ° C and form a passivation layer on the aluminum foil to prevent further corrosion reactions. At the same time, good compatibility is shown in the ester electrolyte based on NaTDI and NaPDI salt, indicating promising application prospect in sodium-ion batteries. In order to further improve the chemical and thermal stability, researchers prepared halogen-free sodium salt by doping them with organic ligands [29]. Among these halogen-free sodium salts, sodium bis[oxalate]borate (NaBOB), sodium bis[salicylate]borate (NaBSB) and sodium [salicylic acid benzenediol]borate (NaBDSB) are easily soluble in common organic solvents, which display excellent electrical conductivity ($>1 \times 10^{-3}$ S cm$^{-1}$) at room temperature. However, the limited solubility of these salts is the obvious disadvantage, which hinders further improvement of conductivity.

In recent years, some researchers have proposed dual-ion batteries with excellent energy density based on the new concept of sodium-ion electrolytes [30]. In this system, the cathode, anode, and electrolyte are all active materials. During the charging process, Na$^+$ inserts anode and anions embed cathode at the same time.

Therefore, the difference in the concentration and type of sodium salt in the electro-lyte will exhibit different electrochemical behavior determining the energy density of the battery. When the salt concentration increases to 5 M, the specific capacity may exceed 100 W h kg$^{-1}$. In comparison, NaPF$_6$ has a higher energy density due to its suitable anion size and better solubility in mixed solvents. Furthermore, NaBF$_4$ and other sodium salts are also used as the transmission medium in dual-ion batter-ies. In the future, advanced sodium salts can not only optimize the electrochemical performance of sodium-ion batteries but also meet the application requirements of dual-ion systems or additional battery systems.

## 5.3 Organic liquid electrolytes

Generally, organic electrolytes consist of a solute dissolved in two or more solvents. The solute is composed of sodium salt and additives, and organic solvents can be divided into ester-based and ether-based solvents. As shown in Figure 5.5, based on the molecule structure, ester electrolytes are divided into cyclic ester solvents and chain ester solvents, which exhibit differences in physical and chemical properties, respectively. One of the important features of cyclic solvents is the high dielectric constant and this makes it essential for organic solvent. Compared with the former, chain solvents possess lower viscosity and better stability and usually function as cosolvents for ester electrolytes. In addition, in order to further improve compatibil-ity on electrolyte–electrode interface, cyclic ester additives such as EC and FEC are usually added to the ester electrolyte. Meanwhile, a fact revealed by theoretical sim-ulation analysis is that the coordination number of Na$^+$ in cyclic ester solvents is small, which is beneficial to the rapid migration and desolvation of Na$^+$. Ether sol-vents, including diglyme (DG), 1,2-dimethoxyethane (DEM), diethylene glycol di-methyl ether (DEGDME) and tetraethylene glycol dimethyl ether (TEGDME), and so on, have a chain structure, which display higher partial sodium-ion concentration improving the stability of the electrolyte. Owe to the high decomposition tempera-ture and nonflammability, organic phosphate ester solvents can also be used as high-safety electrolyte solvents. In addition, the similar Na$^+$ coordination structure in organic phosphate ester electrolyte and ester electrolyte means that it is feasible to form a better performance sodium-ion battery electrolyte by rational combination of organic phosphate solvent and other organic solvents.

### 5.3.1 Carbonate-based electrolytes

Carbonate-based electrolytes constituted by carbonate solvent, sodium salt, and addi-tives are the most commonly used and mature electrolytes for commercial lithium-ion batteries, owe to which largely satisfy the basic requirements for practical electrolytes

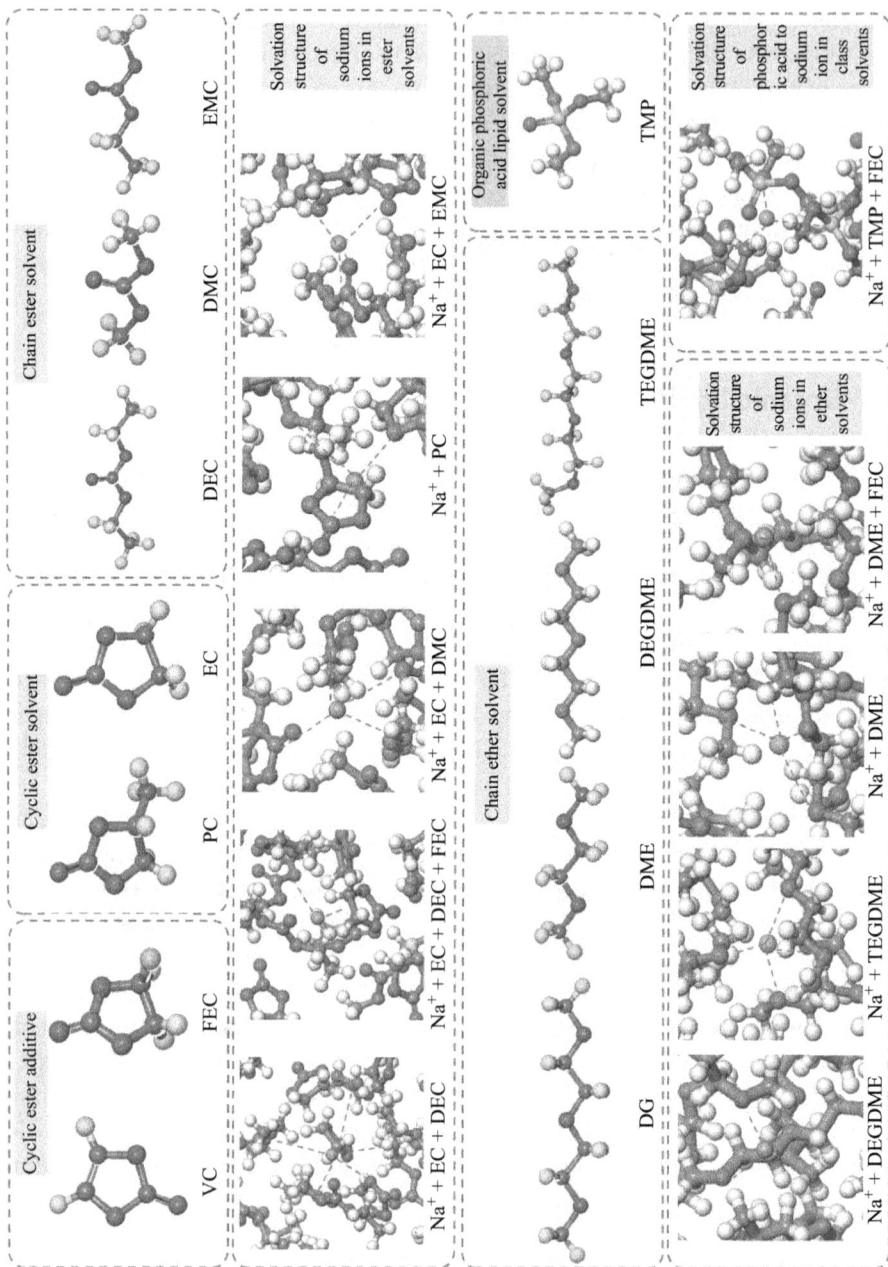

**Figure 5.5:** Schematic diagram of molecular structure of various organic solvent electrolytes (esters and ethers) and solvation structure of sodium ions in organic solvents.

in both property and price. In the current study, therefore, carbonate-based electrolytes also work as electrolytes for sodium-ion batteries.

### 5.3.1.1 Carbonate solvent

The most commonly used carbonate solvents are mainly based on two types of solvents: cyclic carbonates and linear carbonates. Cyclic carbonates mainly include PC and EC; linear carbonates mainly include ethyl methyl carbonate (EMC), DMC and DEC, etc. [1]. The relevant properties are shown in Table 5.3.

Table 5.3: Physical and chemical properties of ester solvents commonly used in SIB electrolyte.

| Solvent | Dielectric constant (25 °C) | Viscosity (25 °C) | Heat of vaporization/ (kJ mol$^{-1}$) | HOMO | LUMO |
|---------|------------------------------|--------------------|----------------------------------------|---------|---------|
| EC | 89.78 | 1.90 | 51.68 | −0.2585 | −0.0177 |
| PC | 64.92 | 2.53 | 45.73 | −0.2547 | −0.0149 |
| DMC | 3.107 | 0.59 | 36.53~38.56 | −0.2488 | −0.0091 |
| EMC | 2.958 | 0.65 | 34.63 | −0.2457 | −0.0062 |
| DEC | 2.805 | 0.75 | 41.96~44.7 | −0.2426 | −0.0036 |

Among these solvents, one of the commonly used and important solvents in the organic liquid electrolyte for sodium-ion batteries is EC, which can easily dissolve almost all sodium salt in research. The main reason for that is the higher dielectric constant (89.78) of EC than the other organic solvents. Now the dielectric constant of a solvent has been regarded as a measure of the chemical polarity. For polar molecules, EC and sodium salt are connected by a strong dipole–dipole intermolecular force to achieve an ideal dissolution state. Especially, the interaction between EC and the anion of the sodium salt effectively promotes the formation of a protective layer on the surface of various electrodes [5, 31, 32]. However, it is impossible for EC to work as a single solvent at room temperature owing to the melting point (36 °C) [33]. Although the dielectric constant (64.92) of PC is lower than that of EC, its liquid temperature range is wide, so it is also used as a solvent for the organic liquid electrolyte of sodium-ion batteries. PC is very compatible with hard carbon (HC). But some studies have shown that the decline of sodium-ion battery performance is related to the continuous decomposition of PC. In addition to the abovementioned cyclic carbonate solvents, linear carbonates such as DMC, DEC, and EMC are often used as cosolvents for EC or PC as well. This synergistic effect of the cosolvent system significantly improves the ionic conductivity, viscosity, electrochemical window of the electrolyte, and the safety of sodium-ion batteries.

At present, various researches on electrolytes are focused on the optimization of electrolyte solvents. Early in the research, Alcántara et al. tested the electrochemical performance of amorphous carbon electrodes in different electrolyte solutions containing NaClO$_4$ in EC:DMC, dimethoxyethane (DME), tetrahydrofuran (THF) and EC:THF [34]. After using THF, especially in combination with EC, the capacity retention rate of the battery can be improved (Figure 5.6(a)). However, the electrochemical window of the THF-based electrolyte is narrow (only 2 V for Na$^+$/Na) and its thermal stability is poor, which also limits the practical application of the THF-based electrolyte (Figure 5.6(b)). Ponrouch et al. systematically studied the basic characteristics of a series of electrolytes, such as viscosity, ionic conductivity, electrochemical stability,

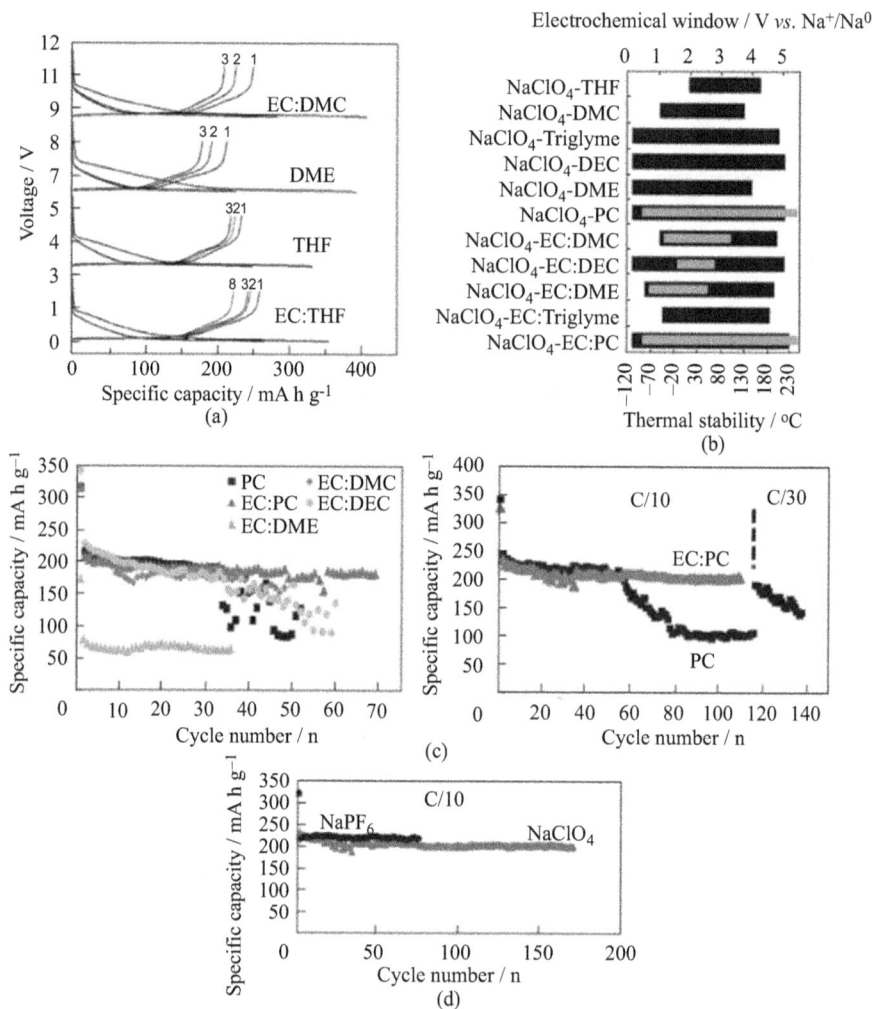

Figure 5.6: Performance of hard carbon anode in ester electrolytes: (a) charge and discharge curves of Na/HC half-cells in different electrolytes; (b) electrochemical potential windows of various electrolytes; (c) HC electrodes cycling stability in different solvents containing 1 M NaClO$_4$ [5]; and (d) cycling stability of HC electrode in EC:PC-based electrolyte containing 1 M NaPF$_6$ and NaClO$_4$ [5].

and thermal stability [35]. The study found that EC and PC are the best solvent combination, which make Na/HC have better performance in electrolytes containing $NaClO_4$ and $NaPF_6$ (Figure 5.6(c), (d)).

Adding a certain amount of DMC to the EC:PC cosolvent electrolyte can improve the cycle performance of the HC electrode (Figure 5.7(a)). The improvement of performance is not the modification of SEI, but the decrease of viscosity and the increase of ionic conductivity of the electrolyte after adding DMC. But adding DMC to the electrolyte will also produce some adverse effects (Figure 5.7(b)). For example, Komaba et al. used HC electrodes to test EC:DMC, EC:DEC, EC:EMC binary solvent electrolytes [36]. Due to the decomposition of DMC, in the EC:DMC electrolyte containing 1 M $NaClO_4$, the HC electrode appeared about 0.3 V ($Na^+$/Na) once during the initial discharge. This also shows that the DMC solvent has poor compatibility with HC electrodes.

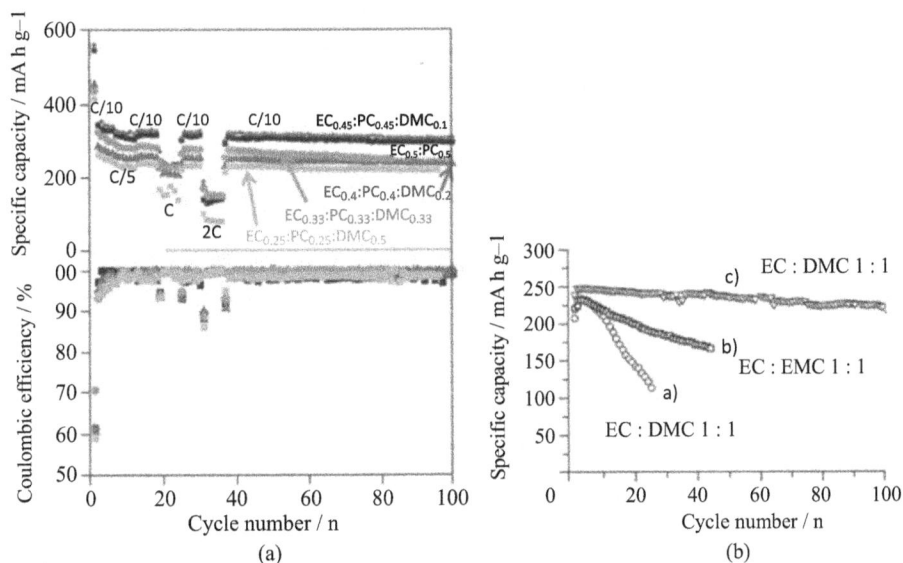

Figure 5.7: The performance of hard carbon anode in ester electrolytes: (a) the influence of different electrolytes on the cycle performance of Na/HC half-cell; and (b) the capacity retention rate of Na/HC half-cell cycling in different electrolytes.

## 5.3.1.2 Sodium salt in ester-based electrolyte

Sodium salt is one of the most indispensable components in the liquid electrolyte and exerts great influence on the performance of the battery. The ideal sodium salt can be completely dissolved in the electrolyte to form a stable complex where the dissolved ions, especially sodium ions, migrate without a kinetic energy barrier. As mentioned above, the core evaluation criteria of sodium salt in ester-based electrolyte is also the electrochemical stability, which reflects the stability of the electrode and the formation of the protective interface. Furthermore, another property for the sodium salt is the chemical inertia for other components of the battery (such as separator and current collector). Under these strict requirements, numbers of salts have been excluded, and the kind of sodium salt suitable for ester electrolytes is quite limited. Table 5.4 lists the physical and chemical properties of commonly used salts of SIB electrolyte.

The most commonly used sodium salt in sodium-ion batteries is $NaClO_4$. The salt shows better thermal stability than other sodium salts, but the disadvantage is that it is easy to explode and more difficult to remove moisture is obvious. Therefore, in the actual application, the use of sodium perchlorate is often avoided. $NaPF_6$ is another popular sodium salt in sodium-ion batteries. This salt is very sensitive to water and easily produces highly corrosive HF. The generated gas is easily reactive with the alkaline components of SEI and produces harmful gases that weaken the rigidity of the SEI film [37, 38]. When it comes to the electrochemical stability, however, $NaPF_6$ performs better than $LiPF_6$. Both Sodium bis(fluorosulfonyl)imide (NaFSI) and Sodium bis(trifluoroMethylsulfonyl)imide (NaTFSI) show high thermal stability and low nontoxicity, but they are extremely corrosive to aluminum, making it little possibility to be used as a single sodium salt in sodium-ion batteries. Nowadays, some reports have proved that high-concentration salt electrolytes are a feasible method to reduce the corrosion of aluminum [39].

In addition to abovementioned salts, some new salts are also developed, such as sodium borate (NaDFOB) [40], 5,5-dioxomethyl-2-(trifluoromethyl)imidazoline (NaTDI) and 4-5-dicyano-2-(pentafluoroethyl) sodium imidazole (NaPDI) [41], and so on. At present, however, studies focus on the practical application of these sodium salts is relatively less.

**Table 5.4:** Physical and chemical properties of common salts in SIB electrolytes.

| Sodium salts | Decomposition temperature/°C (lithium salt) | TGA/°C (mass loss) | Ionic conductivity (mS cm$^{-1}$) (lithium salt) |
| --- | --- | --- | --- |
| $NaPF_6$ | 300 (200) | 400 (8.14%) | 7.98 (5.8) |
| $NaClO_4$ | 472 (236) | 500 (0.09%) | 6.4 (5.6) |
| NaTFSI | 263 (234) | 400 (3.21%) | 6.2 (5.1) |
| NaFTFSI | 160 (94.5) | 300 (2.75%) | – |
| NaFSI | 122 (130) | 300 (16.15%) | – |

### 5.3.1.3 Additives in ester-based electrolytes

Electrolyte additives are the addition of a small amount (usually less than 10% by weight or volume) of non-solvent components in the electrolyte. It can effectively improve the electrochemical performance of electrolytes without changing the main composition of the electrolyte. Under normal circumstances, the additive will be consumed and contribute to the formation of an interfacial phase between the electrode and the electrolyte during the initial activation cycle. Electrolyte additives can be divided into the following categories according to their functions: ① film-forming additives; ② overcharge protection additives; and ③ other additives, such as bulk performance enhancers. In principle, the Fermi energy of an ideal film-forming electrolyte additive has to locate in the gap between the highest occupied molecular orbital and the lowest occupied molecular orbital of the salt and solvent, so that they can be preferentially reduced or oxidized to effectively adjust the intermediate phase. In addition to film-forming additives, there are flame retardant additives, anticoagulation additives, and high-pressure additives [38, 42]. The introduction of different additives will greatly enhance the functions of the ester electrolyte and meet the different application directions of the electrolyte. For example, adding 10% of the organic phosphate ester solvent to the electrolyte can realize the high-temperature noncombustibility of the electrolyte and optimize the passivation layer on the electrode surface. Moreover, it is also noticed that the environmental and negative effects of additives on the overall electrolyte system should be as low as possible.

In summary, the most researched and used additives are film-forming additives. Film-forming additives mainly improve the cycle performance and safety of sodium-ion batteries by chemically modifying the electrode–electrolyte interface. Regardless of whether it is a lithium-ion battery or a sodium-ion battery, the most used film-forming additive is FEC [10, 43, 44]. Komaba et al. studied the additives in sodium-ion batteries using $NaNi_{1/2}Mn_{1/2}O_2$ positive and HC negative electrodes [10]. FEC can effectively improve the cycle performance of sodium-ion batteries (Figure 5.8(a), (b)), and other additives used in lithium-ion batteries, such as sulfate, EC, difluoroethylene carbonate (DFEC), and so on, used in sodium-ion batteries do not have the corresponding effect (Figure 5.8(c)) [45].

FEC additives can be used not only in HC anodes but also in antimony-based, tin-based, and phosphorus-based anode materials with extremely high theoretical capacities [46–48]. Cao et al. compared the cycle performance of SiC-Sb-C electrodes in electrolytes with or without FEC [49]. The results show that the presence of FEC can prevent the occurrence of parasitic reactions, keep the electrode structure stable, and improve the electrode capacity retention rate (Figure 5.9(a), (b)). The combination of FEC and tris(trimethylsilyl) phosphite (TMSP) can further improve the cycle stability of $Sn_4P_3$ (Figure 5.9(c)). TMSP can remove HF generated by FEC decomposition. The FEC is not only helpful for building a protective film on the surface of $Sn_4P_3$, which prevents the electrolyte from decomposing and causing harmful side effects, but also prevents the formation of $Na_{15}Sn_4$ that brings great expansion of electrode volume. But the alternative product $Na_{15}Sn_4$, $Na_9Sn_4$, should also take responsibility for loss of reversible specific capacity (Figure 5.9(d)).

Each coin owning two sides, FEC will also give a negative impact on sodium-ion batteries as the addictive. It is reported that the use of 2% FEC additives significantly reduces the reversible specific capacity, Coulombic efficiency, and cyclability of sodium-HC batteries [11], which is blamed for the increase of overpotential caused by FEC (Figure 5.10). Besides, the main reason of sudden death of sodium-ion batteries is the exhaustion of FEC. In a similar failure mechanism of silicon-based lithium-ion batteries, FEC is continuously consumed [50, 51]. The electrolyte containing FEC produces a lot of gas after decomposition, especially when the battery is cycled at high temperature, the decomposition and gas production are more obvious, which definitely destroy the safety of the battery [52, 53].

Some novel additives can also be applied in sodium-ion batteries. Song et al. screened out adiponitrile as an electrolyte additive for sodium-ion batteries by calculating the frontier molecular orbitals to improve the cycle performance of sodium-ion batteries [54]. The physical characterization results reveal that ADN is able to form a uniform and dense SEI film on cathodes. Besides, adding ILs into the organic electrolyte is also proved to promote the ion mobility and interfacial properties of sodium ions. Manohar et al. compared electrochemical performance of $Na_3V_2(PO_4)_3/$ C cathode after introduction of different content of $N$-propyl $N$-methylpyrrolidone bis(trifluoromethylsulfonyl)imide (0, 3 wt%, 5 wt%, 7 wt%, 9 wt%, and 12 wt%, respectively) in carbonate-based electrolytes, and it is showed that cycle performance is the best when the additive content reach 5 wt% [55].

### 5.3.2 Ether-based electrolyte

In lithium-ion batteries, ether solvents are rarely used in electrolytes, mainly due to their poor film-forming ability on the surface of anodes and easy decomposition on the surface of cathodes. But in sodium-ion batteries, ether-based electrolytes are frequently applied. The core reason is that the ether-based electrolyte often has strong resistance to reduction, which makes the SEI on the surface of the negative electrode thinner, leading to a higher initial Coulombic efficiency than that of the ester-based electrolyte [56].

In various negative electrode materials, the storage of sodium ions has been significantly improved after the use of ether-based electrolytes. Cohn et al. found that in the $NaPF_6$-diglyme electrolyte, graphene is able to work as an ultrafast sodium storage negative electrode, and the diglyme solvation layer can encapsulate sodium ions as a "non-stick" coating to help sodium rapid ion insertion (Figure 5.11(a), (b)) [57]. Xu et al. developed a new method that combines the advantages of ester-based electrolytes and ether-based electrolytes to enable HC exhibiting superior cycle performance (Figure 5.11(c)) [58]. A simple pretreatment that put the HC anode in an ester-based electrolyte to form an SEI film in advance before assembling with the ether-based electrolyte has also been proved greatly improving performance of the electrode (Figure 5.11(d)). Kajita et al. studied electrochemical behavior of the layered FeSe anode in the ether-based electrolyte, and its reversible capacity in the first cycle was up to 294 mA h $g^{-1}$, and the Coulomb efficiency was increased to 87.2% [59]. As

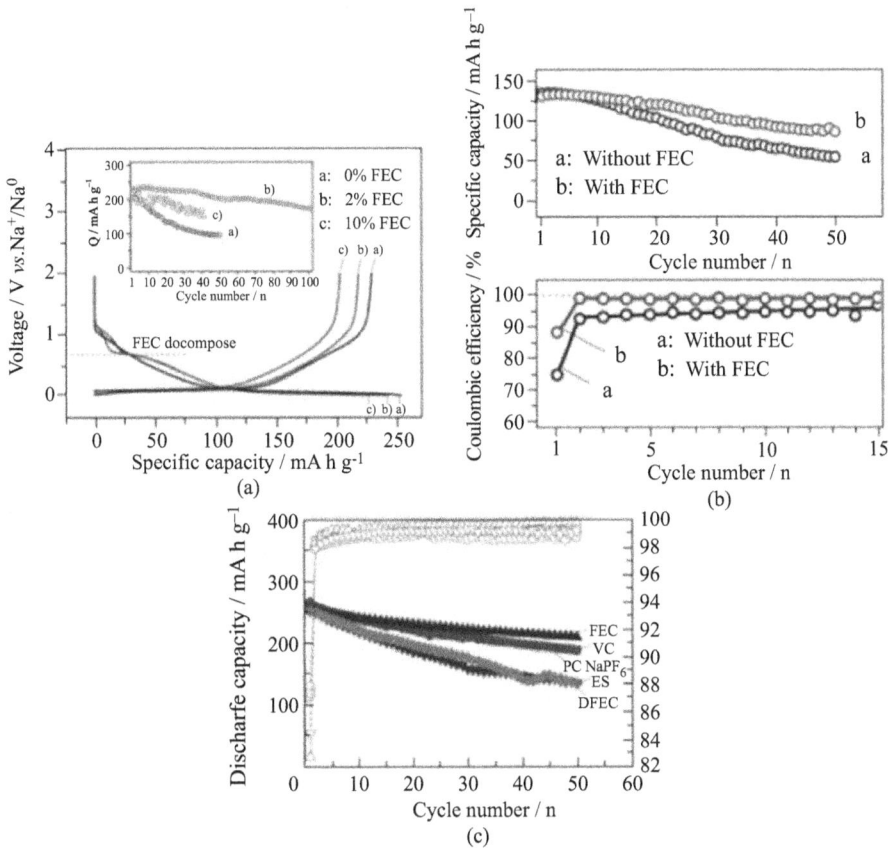

**Figure 5.8:** Analysis of the influence of ester electrolyte additives on the electrode performance of sodium-ion batteries: (a) cyclic performance of HC electrode in electrolyte containing FEC additive [10]; (b) cycle performance of NaNi$_{1/2}$Mn$_{1/2}$O$_2$ electrode in the electrolyte with or without FEC [10]; and (c) the influence of different additives on HC electrodes [45].

for reduced graphene oxide electrode, Yang et al. studied the influence of ester-based and ether-based electrolytes on the formation of SEI layer [60]. The results show that the reversibility and cycle life of batteries using ether-based electrolytes (NaOTF/diglyme) are significantly better than batteries using ester-based electrolytes (NaOTF/EC-DEC), which is the consequence of a thicker SEI film driven by the ester-based electrolytes. This team also studied the phase evolution of the interface between ether-based electrolyte and TiO$_2$ anode. When the diglyme electrolyte is applied in the TiO$_2$ electrode, its reversible capacity has increased to a new height compared with previous studies at different current densities (Figure 5.12(a), (b)) [61]. In situ X-ray diffraction (XRD) characterization reveals that the sodiumization process of TiO$_2$ in diglyme electrolyte is more complete than that in ester-based electrolytes (Figure 5.12(c), (d)). Furthermore, after replacing the ester-based electrolyte with the

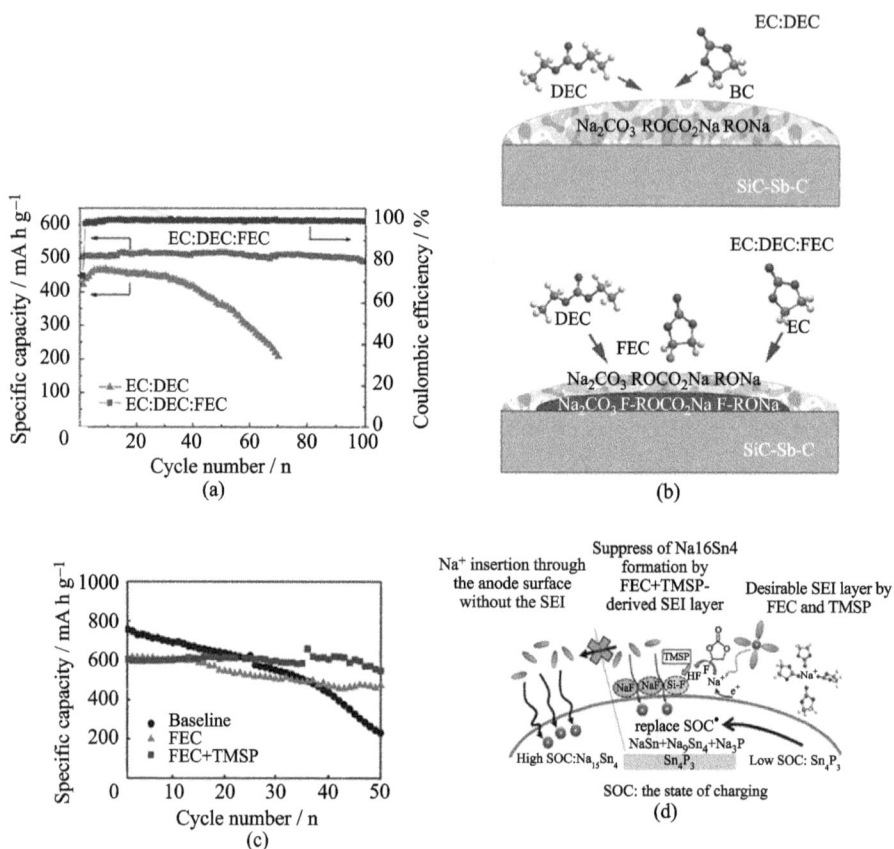

**Figure 5.9:** Analysis of the effect of ester electrolyte additives on the electrode performance of sodium-ion batteries: (a) cycling performance of SiC-Sb-C electrode in electrolytes with or without FEC additives [49]; (b) diagram of the mechanism of FEC action on the surface of SiC-Sb-C electrode [49]; (c) cycling performance of $Sn_4P_3$ electrode in electrolytes with or without FEC and FEC + TMSP additives [48]; and (d) schematic diagram of the sodiuming process of NaF-based SEI layer [48].

ether-based electrolyte, a lower sodium-ion interface transport barrier and charge transfer resistance can be obtained, which significantly improves the rate performance of $TiO_2$ (Figure 5.12(e)). These advantages are brought by the thinner and more uniform SEI film-forming in the ether-based electrolyte. What's more, the outside of the SEI layer in the ether-based electrolyte is composed of a dense layer of sodium alkyd, which reduces the diffusion distance of sodium ions and reduces the transmission barrier of $Na^+$. Therefore, an ideal interface with lower potential barrier and faster charge transfer kinetics can be achieved. In a word, faster kinetics further increases the adsorption and storage capacity of sodium ions on the electrode surface (Figure 5.12(e)), leading to a rapid charge storage and increased structure stability of $TiO_2$ electrode.

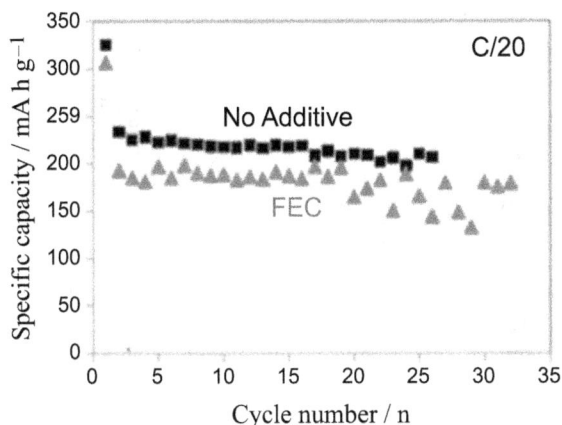

**Figure 5.10:** The electrochemical performance of HC electrode in the electrolyte with or without FEC. [11].

The attempt of applying commercial graphite to sodium-ion batteries with ether-based electrolytes is not new. Traditionally, it is difficult for sodium ions to get embedded in graphite electrodes because of the limitation of ionic radius and thermodynamics [61, 62]. Unlike ester-based electrolytes, however, solvated sodium ions can be intercalated between graphite layers to form a co-intercalation process when ether-based electrolytes are used. The size of the sodium-ion solvation structure formed in ester-based electrolyte badly hinders the success of a co-intercalation process, so the sodium ions cannot be transported to the sodium storage sites in the HC. By contrast, the solvated structure formed in ether-based solvents, such as dimethyl ether and so on can intercalate into the graphite electrode forming a ternary graphite intercalation compound (Figure 5.13(a)). This reaction process is reversible and thermodynamically beneficial. The van der Waals interaction between dimethyl ether and graphite enhances the coupling strength between layers, thereby stabilizing the mechanical integrity of graphite. Therefore, graphite shows good performance in sodium-ion batteries with ether-based electrolytes. After 1,000 cycles, the capacity retention rate is higher than 90%, and the Coulombic efficiency is up to 99.87% (Figure 5.13(b), (c)).

After graphite was successfully used in sodium-ion batteries for the first time, researchers conducted a lot of research to maximize the solvent co-intercalation mechanism and revealed the factors affecting its effect. Kang et al. studied the storage mechanism of sodium in natural graphite and compared the performance of graphite in different ether-based electrolytes (including diglyme, tetraglyme, and dimethyl ether) (Figure 5.13(d)) [63, 64]. First, the intercalation reaction of sodium ions in natural graphite was further revealed through transmission electron microscopy combined with infrared spectroscopy (Figure 5.13(e)). Second, by means of the combination of experiments and density functional theory calculations, researchers found that the intercalation reaction of sodium ions occurred through a multistage

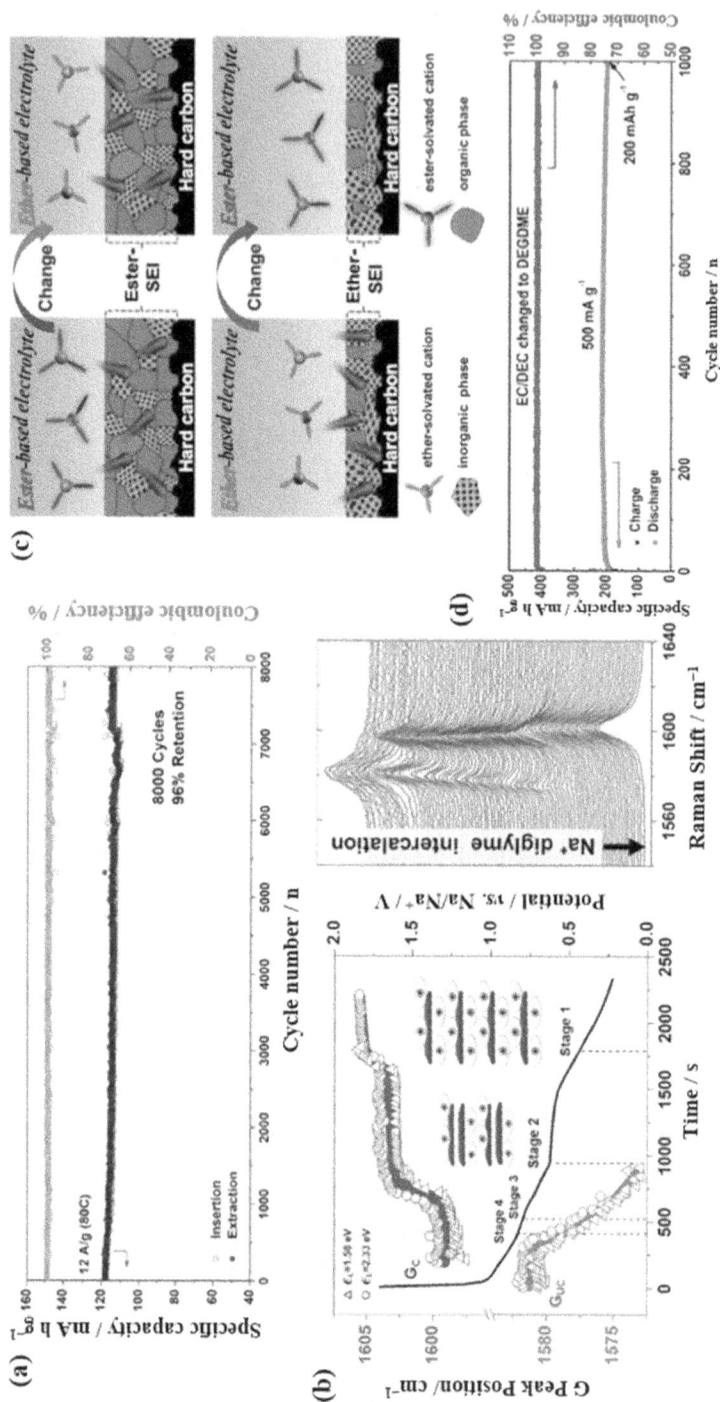

**Figure 5.11:** Performance of HC anode in ether-based electrolyte: (a) cycle performance of monolayer graphene electrode in diethylene glycol electrolyte at high current density; (b) in situ Raman of monolayer graphene electrode during discharge; (c) schematic representation of sodium ions entering SEI layer in different electrolytes; and (d) long cycle performance of HC electrode with pre-formed SEI.

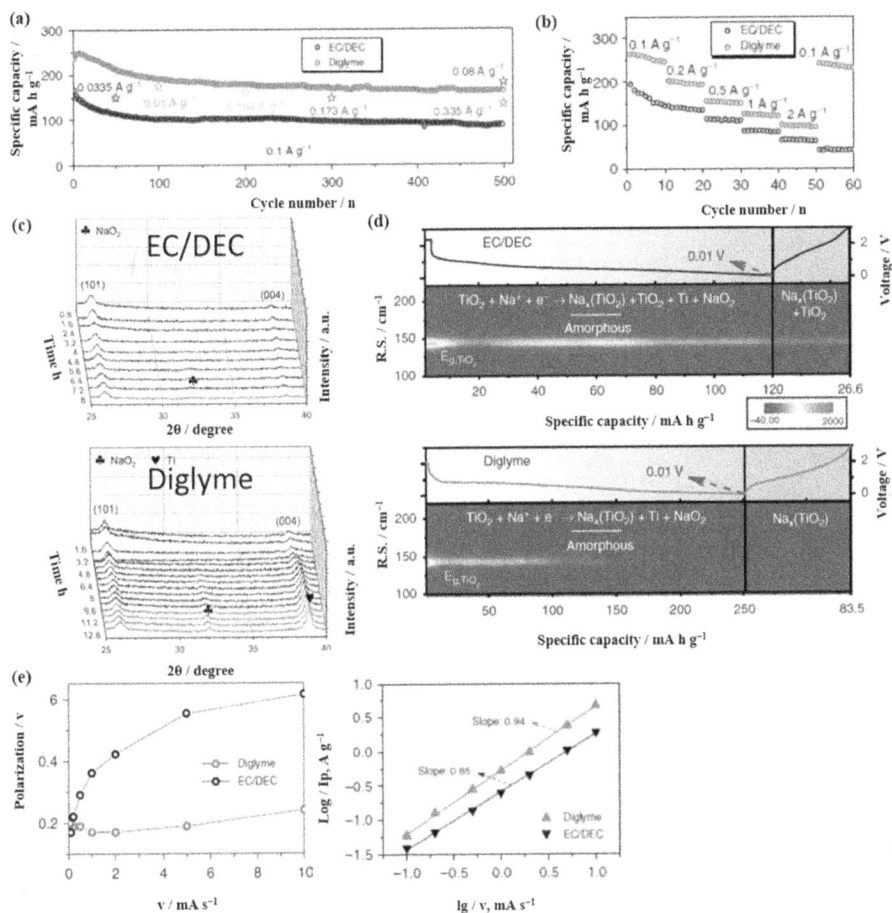

**Figure 5.12:** Performance of TiO₂ anode in ether-based electrolyte: (a) cycle performance of TiO₂ in ester-based and ether-based electrolytes; (b) rate performance of TiO₂ in ester-based and ether-based electrolytes; (c) in EC:DEC-based electrolyte and the diethylene glycol-based electrolyte, the in situ XRD pattern of the TiO₂ anode in the first charge and discharge cycle; (d) in the EC: DEC-based electrolyte and the diethylene glycol-based electrolyte, the TiO₂ anode is in the first charge and discharge in situ Raman in the cycle; and (e) in EC: DEC-based electrolyte and diethylene glycol-based electrolyte, the potential polarization of the TiO₂ anode between the cathode and anode peaks in the cyclic voltammetry (CV) curve, and the current corresponding b-value (slope) obtained from peak Ip = avb.

process, and the [Na-ether⁺] complex in the graphite channel overlapped the graphite layer in parallel (Figure 5.13(f)), leading to a significant increase in the interlayer spacing of graphite (about 346%). Although the volume of graphite has undergone tremendous changes, it is still stable after 2,500 cycles (Figure 5.13(g)), and there is no obvious structural change.

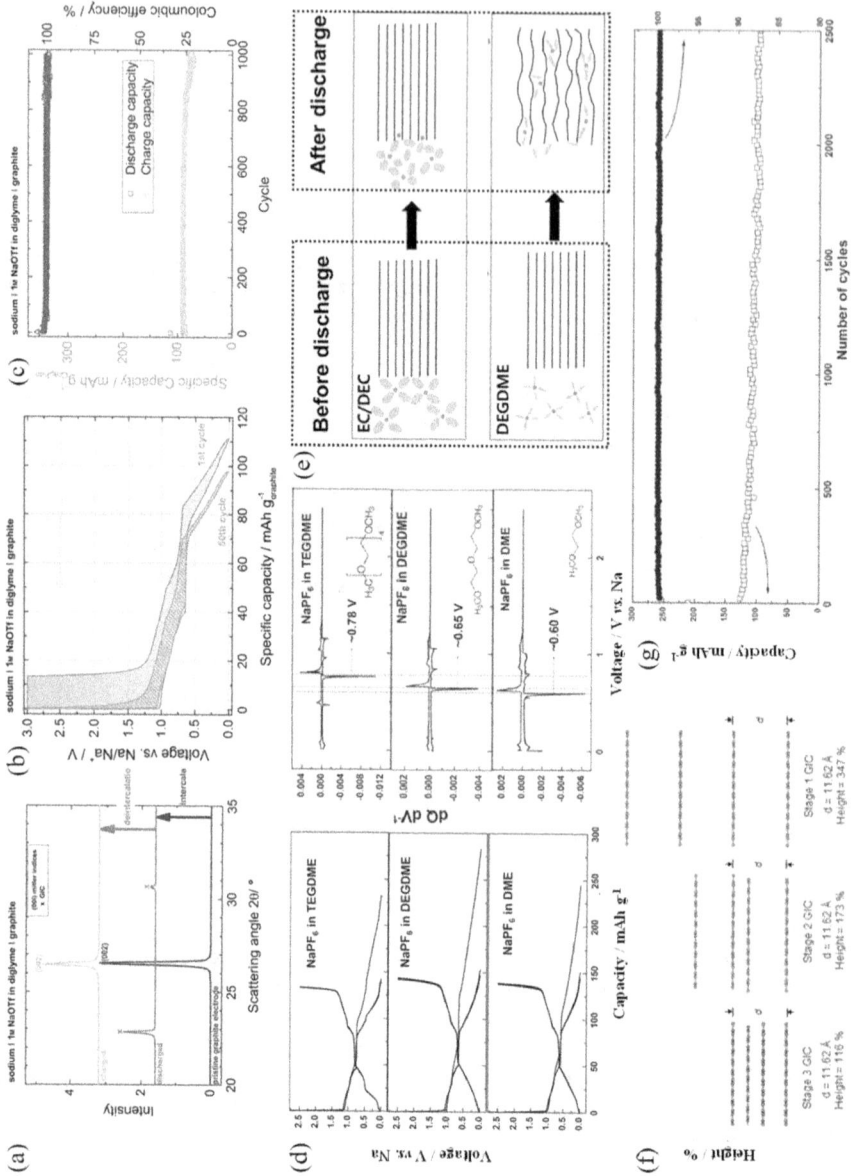

**Figure 5.13:** Performance of graphite anode in ether-based electrolytes: (a) diffractogram of graphite electrode in diethylene glycol-based electrolyte after three half cycles; (b) charge/discharge curves of graphite in diethylene glycol-based electrolyte; (c) cycle stability and Coulombic efficiency of sodium/graphite battery in diethylene glycol-based electrolyte; (d) electrochemical performance of graphite in different ether-based solvent electrolytes; (e) carbonate-based and ether-based electrolytes into natural graphite by the mechanism of Na⁺ embedding; (f) comparison of the height and c-dotting parameters of graphite in different graphite interlayer compounds; and (g) long-term cycling performance of natural graphite in ether-based electrolytes.

## 5.4 Ionic liquid electrolyte

Ionic liquids are materials that are liquid at normal temperature composed of anions and cations. There are many kinds of cations and anions, so there are many possibilities for the combination of cations and anions that make up ILs [65–69]. Common cations include imidazole salt cations, pyrrolidine cations, and quaternary ammonium salt cations. Common anions include FSI⁻, TFSI⁻, trifluoromethesulfonate ($CF_3SO_3^-$), and tetrafluoroborate ($BF_4^-$). The ionic liquid electrolytes were compared according to the three basic properties of viscosity, ionic conductivity, and decomposition temperature. The results show that the imidazolium salt electrolyte has lower viscosity and higher ionic conductivity and is close to the organic electrolyte in electrochemical properties. What's more, the stability is better than that of organic electrolyte. However, pyrrolidine ILs and quaternary ammonium salt ILs are superior to imidazolyl electrolytes in low-temperature and high-temperature performance, respectively. Therefore, these ionic liquid electrolytes have certain research significance. By expounding the physical, chemical, and electrochemical properties of ILs, this section summarizes the related properties and performance of ILs as electrolytes for sodium secondary batteries, as well as the treatment of ILs, especially the treatment of ILs applied in electrochemistry, and compares ILs for sodium-ion batteries with organic electrolytes and lithium battery systems.

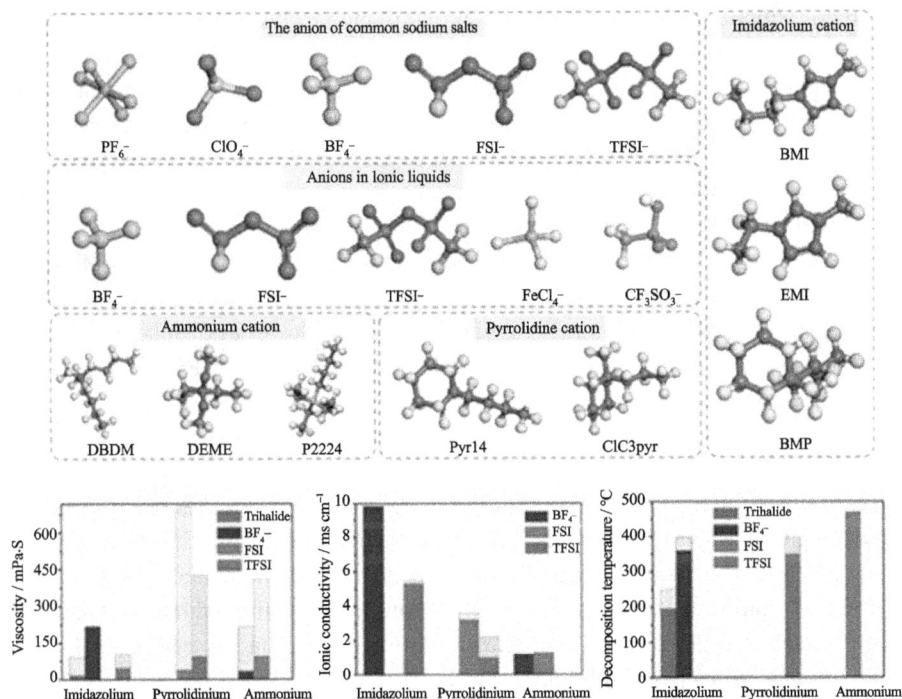

**Figure 5.14:** Composition and properties of common ionic liquid electrolytes.

### 5.4.1 Ion types of electrolytes

Figure 5.14 outlines the types of ions used in ILs, especially those commonly used in sodium-ion batteries and their abbreviations. Although some ILs can be used in secondary batteries in theory, the amount of ILs that can actually play a role is very limited. In the early study of ILs, fluorine complex anions ($PF_6^-$ and $BF_4^-$) are often used in the preparation of low-melting salts [70, 71]. $PF_6^-$-based ILs are also used in organic synthesis, but their low anti-hydrolysis stability and high viscosity limit their application in battery electrolytes. $BF_4^-$ has high anti-hydrolysis stability, but because of its hydrophilicity, the synthesis of high-purity $BF_4^-$-based ILs is expensive and time-consuming, so it is not suitable for practical applications. Sulfonamide-based (also known as sulfonimide) ILs are easy to synthesize (hydrophobic in most cases) and have relatively high ionic conductivity (e.g., [$C_2C_1$im] bis (triuoromethanesulfonyl) amide ([TFSA]-) is 8.8 mS cm$^{-1}$ [72], [$C_2C_1$im][FSA] is 15.4 mS cm$^{-1}$ [73, 74]), so they are widely used in electrolytes.

Cations in ILs can obtain more species changes by prolonging their organic structure. Therefore, there are usually more kinds of cations than anions. The introduction of large or long substituents can reduce the melting point, but the viscosity will increase and the ionic conductivity will decrease. Therefore, asymmetric cations with short alkyl chains, such as $C_3C_1Pyrr^+$ and $C_2C_1im^+$ are preferred in ILs. The cations most widely used in secondary batteries are nonaromatic pyrrolidone cations, and their reduction stability is higher than that of imidazolyl cations [75]. In a recent work, the physicochemical properties of a series of FSA- and TFSA$^-$-based ion structures (ammonium, pyrrole, piperidine, and morphine) have been systematically studied, and the properties of ether functional cations have been revealed [76].

There are two types of ILs that need to be discussed separately: inorganic ILs and solvent ILs. Inorganic ILs are also electrochemical attraction media at high temperature. In a series of Na[TFSA]-M[TFSA] binary systems, Na[TFSA]-Cs[TFSA] has the lowest eutectic temperature of 110 °C when the mole ratio of Na[TFSA]/M[TFSA] is 0.0715/0.93 [77]. After the introduction of the third component into the system to form a ternary system, the eutectic temperature can be lower and closer to the ambient temperature. The lowest eutectic temperature of Na[FSA]/K[FSA]/Cs[FSA] system is 36 °C, and the molar ratio is 40:25:35.144 [78].

In sodium-ion batteries, there are few studies on solvent ILs. Solvent ILs are composed of Lewis-based salts (for example, metal salts can be obtained by sintering at the same molar concentration). The representative is [Li(G3)][TFSA] (G3 = triethylene glycol dimethyl ether, also known as triethylene glycol), the melting point is 23 °C [79]. These metal ions coordinate with the metal center, and their behavior is different from that of pure metal or dilute solution, and the oxidation resistance and thermal stability of the complexes are higher than that of free radicals. Sodium-ion analogues [Na(G3)][TFSA], [Na(G4)][TFSA] and [Na(G5)][TFSA] (G4 = tetra ethylene glycol and G5 = melamine) are not room temperature ILs [80], but their behaviors in sodium-ion batteries have been studied and proved to be feasible [81].

### 5.4.2 Thermal stability

The thermal properties of ionic liquid electrolytes were studied by thermogravimetric (TG) and differential scanning calorimetry. Compared with organic electrolytes, ILs have higher thermal stability [75, 82]. However, special attention should be paid to the determination of thermal decomposition temperature of ILs by TG, which is affected by gas atmosphere, battery materials, scanning rate, sample quantity, impurities, and methods, and the thermal decomposition temperature of ILs is usually controlled by kinetics. Therefore, it is meaningless to compare the thermal decomposition temperatures determined under different conditions. Ionic liquids can increase the maximum operating temperature of the battery. However, due to the presence of oxygen and water in the air, the stability of ILs in the air is not as good as that in the inert environment [82].

The thermal decomposition temperature of ILs can be changed by adding sodium salt to ILs. The addition of $Na[BF_4]$, $Na[ClO_4]$, $Na[N (CN)_2]$, and $Na[TFSA]$ salts can reduce the decomposition temperature of $[C_4C_1Pyrr][TFSA]$ ILs from 400 °C to 300 °C [83, 84]. The ILs with $Na[PF_6]$ begin to decompose at lower temperatures (the mass decreases from 100 °C). The $Na[BF_4]^- [C_2C_1im][BF_4]$ ionic liquid system will not catch fire when it comes in contact with the flame, and its thermal stability can reach 380 °C [85]. FSA salt is extremely sensitive to water, and its thermal stability is controversial. Even if the melting point is higher than the decomposition temperature (such as $T_d = 70$ °C, $T_m = 130$ °C, Li[FSA]), the melting phenomenon can be observed during slow decomposition [86]. The thermal decomposition temperature of alkali metal FSA salt will increase with the increase of alkali metal cation size. The $[N_{4411}][FSA]$, $[As(4.5)] [FSA]$ and $[N_{6111}][FSA]$ electrolytes containing $Na[FSA]$ are dominated by the thermal decomposition temperature of Na [FSA] (130 °C) [87]. The thermal decomposition temperatures of these ionic liquid solvents are higher and tend to be the same: pure [N2 (202O1)3] [FSA] is 493 °C, while Na [FSA]-[N2 (202O1)3] [FSA] with 55 mol% Na [FSA] is 271 °C.

From the point of view of battery application, the decomposition temperature in the presence of electrode materials is more important. The electrode materials in the charged state are high reduction (negative electrode) or oxidation (positive) phase, and their contact with the electrolyte at high temperature will lead to thermal runaway. For example, 1.0 mol $dm^{-3}$ Li $[PF_6]$-EC/DMC will decompose exothermically in $Li/Li_{1-x}CoO_2$ system, while 0.32 mol $kg^{-1}$ Li [TFSA]-$[C_3C_1pip]$ [TFSA] will decompose to a small extent under the same conditions, indicating that ionic liquid electrolytes have more advantages in terms of thermal stability [88]. Although there are relatively few studies on the thermal properties of ILs in sodium-ion batteries, the studies on the organic electrolytes of different electrode materials (HC [5, 89–91], $Na_xSn$ [92], $Na_{0.5}Ni_{0.5}Mn_{0.5}O_2$ [93], $Na_{0.35}CrO_2$ [94], $Na_{0.58}FeO_2$ [95]) show that the charged electrode has higher reaction activity to the organic electrolyte (the presence of sodium salt affects the decomposition temperature). In $Na_3V_2(PO_4)_3$ symmetrical battery, the charged thermal stability of [$C_2C_1$im] [$BF_4$] ionic liquid electrolyte is compared with that of 1 M $NaClO_4$-PC electrolyte. The results show that ionic liquid has higher stability [96]. The thermal stability of electrolytes is also related to the stability of SEI membranes. In ILs containing Li, it is also confirmed that SEI films with high stability will be formed at high temperature. These factors should also be taken into account in the future research on the application of sodium-ion batteries [97].

Mixing sodium salt with organic salt forms a specific type of eutectic system (or appears to be eutectic). Such as Na[$C_2C_1$im][FSA], Na[FSA]-[$N_{4411}$] [FSA], Na[FSA]-[$N_{6111}$] [FSA], Na[FSA]-[AS(4.5)][FSA], Na[TFSA]-[$C_3C_1$Pyrr] [FSA], Na[TFSA]-[$C_2C_1$im] [TFSA], Na [TFSA]-[$C_4C_1$im] [TFSA], and Na [TFSA]-[$C_4C_1$Pyrr] [TFSA] [87, 98–102]. The liquid phase range of Na [FSA]-[$C_2C_1$im] [FSA] ($0 \le$ Na[FSA] $\le 50$ mol%) is wider than that of Na [TFSA]-[$C_2C_1$im] [TFSA] ($0 \le$ Na [TFSA] $\le$ about 20 mol%) at room temperature, indicating that FSA⁻ has the ability to provide a low melting point. In the chart, the decline of the liquid phase is usually steep, partly due to the existence of crystal gaps, only glass transition can be observed and the melting behavior disappears. However, the source of the crystallinity gap cannot be clearly explained at present, which may be caused by thermodynamics or kinetics. Long-term low-temperature aging (in the liquid range) does not lead to the crystallization of ILs [103]. At the glass transition temperature, the structure of the ionic liquid will be frozen into the glassy state. The glass transition temperature is related to the fluidity of ILs, so ILs have lower glass transition temperature [104, 105]. With the increase of sodium content, the glass transition temperature of ILs containing sodium salts increases, which is related to the low fluidity observed at high Na[FSA] molar ratio. With the increase of the size of organic cations, the glass transition temperature tends to increase.

### 5.4.3 Physical and chemical properties

Because of the strong Coulomb interaction between anions and cations of ILs, the viscosity of ILs is higher than that of organic electrolytes at the same sodium content. However, the concentration of ILs can be reduced by solvent dilution [106–108]. In addition, the increase of ionic volume will also increase the viscosity of ILs, while the introduction of ether groups will reduce the viscosity [76, 109, 110]. High-viscosity ILs are difficult to be used in batteries because they are difficult to impregnate into electrodes and diaphragms. The viscosity ($\eta$) is related to the molar ionic conductivity ($\lambda$), which accords with the Walden formula (or fractional Walden formula). The parameter $\alpha$ is between 0 and unit 1, corresponding to the slope of the Walden diagram (eq. (5.1)) (lg ($\eta$–1): lg ($\lambda$)): [104, 111–113]

$$\eta \cdot \lambda = \text{const. (or } \eta^\alpha \cdot \lambda = \text{const.)} \tag{5.1}$$

The Walden rule is usually applicable to pure ILs and ILs containing alkali metal cations. The $\alpha$ value represents the ratio of the B parameter in the Vogel–Tammann–Fulcher equation of viscosity and molar ionic conductivity. This shows that the addition of sodium salt to organic ILs can reduce the ionic conductivity ($\sigma$) of inorganic–organic hybrid ILs, which is due to the increase of sodium viscosity due to the enhancement of Coulombic interaction. This is compared with the maximum ionic conductivity in organic solvents at a certain molar concentration of sodium ions. Although this does not directly affect the transport of sodium ions, there is always an increase in resistance with the addition of sodium salt during the operation of the battery. The ionic conductivity of ILs containing sodium ions is highly dependent on the ionic structure, but the conductivity of ordinary ILs is similar to that of organic solutions of similar concentration (for example, Na[$C_3C_1$Pyrr]-[$C_3C_1$Pyrr] [FSA] is 3.6 mS cm$^{-1}$ at 0.98 mol dm$^{-3}$, Na[TFSA]-[$C_2C_1$im][TFSA] is 3.9 mS cm$^{-1}$ at 0.70 mol dm$^{-3}$). The ionic conductivity of ILs containing sodium ions is highly dependent on the ionic structure, but the conductivity of ordinary ILs is similar to that of organic solutions of similar concentration. Na [FSA]-[$C_2C_1$im] [FSA] is 8.5 mS cm$^{-1}$ under 1.1 mol dm$^{-3}$, Na [BF$_4$]-[$C_2C_1$im] [BF$_4$] is 411.8 mS cm$^{-1}$ under 0.75 mol dm$^{-3}$, and Na [PF$_6$]-PC is 7.98 mS cm$^{-1}$ under 1.0 mol dm$^{-3}$. 6.2 mM$^{-1}$/PC is 6.2 mS cm$^{-1}$/ PC (6.2 mm/TFSA) under 1.0 mol dm$^{-3}$). In the steady state, the sodium-ion transport is controlled by the sodium-ion conductivity ($\sigma$ Na$^+$).

$$\sigma_{Na^+} = \sigma \cdot t_{Na^+} \tag{5.2}$$

The migration number of sodium ion s in $t_{Na^+}$ eq. (5.3) was determined by AC-DC method [114, 115].

$$t_i = \frac{|z_i| c_i \mu_i}{\sum_i |z_i| c_i \mu_i} \left( \sum_i t_i = 1 \right) \tag{5.3}$$

$t_i$ is not simply derived from mobility. This simple form can be applied to limited situations, such as a single salt at 1:1 in the solution (such as NaPF$_6$ in PC). The ionic liquid electrolyte of sodium-ion battery containing organic cations is usually composed of more than three types of ions, so the ion concentration must be taken into consideration. Because ILs are highly concentrated systems, the ion transport number in ILs cannot be expressed by the simple ratio of component ion diffusion coefficient [113].

The relationship between sodium salt concentration and electrical conductivity is further discussed. $t_i$ increases with the increase of sodium concentration in Na [FSA]-[$C_3C_1$Pyrr] [FSA] and Na[FSA]-[$C_2C_1$im][FSA] systems (in the range of 0.13) (10 mol% Na [FSA])–0.35 (50 mol% Na[FSA][$C_2C_1$im][FSA][FSA]) [116, 117]. The migration number of sodium ions increases obviously with the increase of temperature. Under a certain Na[FSA] fraction, the ionic conductivity of sodium ions can reach the maximum, such as Na[FSAs]-[$C_3C_1$Pyrr][FSA] of 20 mol% Na[FSA] and Na [FSAs]-[$C_2C_1$im][FSA] of 30 mol% Na[FSAs]. Molecular dynamics simulations show that sodium ions and FSA⁻ can aggregate at high concentrations, which can promote position exchange or structural diffusion of sodium ions, thus increasing the ion migration number. Through Raman spectroscopy, nuclear magnetic resonance spectroscopy, and atomic force microscopy, we found the rapid exchange of sodium ions and FSA⁻ in different coordination environments and similar ionic liquid systems [118–120].

The fluidity of ILs ($\eta^{-1}$) deviates from the Arrhenius behavior when approaching the glass transition temperature, and the rate at which the transport properties of ILs change close to the glass transition temperature is called brittleness [104]. The relationship between viscosity and ionic conductivity of ILs with temperature can be fitted by Vogel-Fulcher-Tamman equation (VFT) (formulas (5.4) and (5.5)) [119, 121]. Inorganic–organic hybrid ionic liquid systems containing sodium ions also follow this formula:

$$\eta(T) = A_\eta \sqrt{T} \exp\left(\frac{B_\eta}{T - T_{0\eta}}\right) \tag{5.4}$$

$$\sigma(T) = \frac{A_\sigma}{\sqrt{T}} \exp\left(- \frac{B_\sigma}{T - T_{0\sigma}}\right) \tag{5.5}$$

The fitting parameters of $A_\eta$, $B_\eta$, $T_{0\eta}$, $A_\sigma$, $B_\sigma$, and $T_{0\sigma}$ are determined by mathematical fitting.

The glass transition temperature observed in the experiment is always higher than the ideal glass transition temperature ($T_g > T_0$, where $T_0$ is the ideal glass transition temperature in VTF equation).

The polarity of the solvent is considered to be an indicator of its ability to dissolve and stabilize ions. Polarity is usually discussed according to physical parameters such as relative dielectric constant, dipole moment, and refractive index. Dielectric spectra show that the relative permittivity of ILs containing $BF_4^-$, $PF_6^-$, $SO_3CF_3^-$, and TFSA⁻ ranges from 11 to 15, but these macroscopic parameters may not be effective in evaluating the polarity of ILs due to the great difference in solubility of inorganic salts (although the relative dielectric constants of ILs are similar) [122–125].

Ionic liquids are not homogeneous media, and local interactions play a more important role in electrochemical performance. The π–π* absorption bands of $E_T$ (30) and $\lambda_{Cu}$ parameters and the d–d absorption bands of $Cu^{2+}$ in [Cu (acac) (tmen)] (acac = acetylacetone and tmen = $N,N,N',N'$- tetramethylethylenediamine) were determined with solvent chromogenic dyes [126–131]. There is a good correlation between $E_T$(30) parameters and the acceptor number of solvents. In ILs, with the increase of alkyl chain length on cations, the morphology of cations changes from aromatic (imidazolium and pyridine) to nonaromatic (pyrrolidone and tetraalkylammonium), $E_T$(30) [132–137]. The anionic structure has little effect on the acceptor number of cations, so the $E_T$(30) parameters have little correlation with the anionic structure. On the contrary, the $\lambda_{Cu}$ parameter has a strong correlation with the number of solvent donors, so it is greatly affected by the anionic structure in ILs. The donor capacity shows the change rule of $[CF_3CO_2]^- > [SO_3CF_3]^- > [FSA]^- \approx [TFSA]^- > [BF_4]^- > [PF_6]^-$ [138, 139]. Other indicators called kamlettaft parameters ($\pi^*$, $\alpha$, and $\beta$) are widely used to discuss the polarity of ILs [140–143]. The $\pi^*$ parameters are related to the bipolarity/polarization of solvents. The parameters of ILs are generally higher than those of typical proton-free organic solvents and increase from nonaromatic to ILs based on aromatic cations. Compared with other ILs, TFSA-based ILs have smaller $\pi^*$ parameter values. The $\alpha$ and $\beta$ parameters are related to the hydrogen bond donor capacity and acceptor capacity, respectively, but they are not very important in the secondary battery electrolyte. In addition, the correlation between polarity and molar conductivity ratio ($A_{IMP}/A_{NMR}$) is also discussed [139, 144]. The molar conductivities measured by AC impedance method and nuclear magnetic resonance spectroscopy are $A_{IMP}$ and $A_{NMR}$, respectively. This correlation provides a basic insight for considering the ionic structure in the ionic liquid electrolyte of secondary batteries, because the $A_{IMP}/A_{NMR}$ ratio refers to the proportion of ions (charged species) that contribute to the ion conduction of all diffused species on the measured timescale.

The $\lambda_{Cu}$ parameter is mainly affected by the anion structure and has a strong correlation with the $A_{IMP}/A_{NMR}$ ratio. The increase of anion donor capacity will reduce the contribution of other ions to sodium-ion conduction. The relationship between the value of $E_T(30)$ and the ratio of $A_{IMP}/A_{NMR}$ is more complicated. The $A_{IMP}/A_{NMR}$ ratio increases with the decrease of $E_T(30)$, that is, the number of acceptors of cations decreases. Sodium ions in ILs are surrounded by other ions, so based on the polarity of anions, such as $\lambda_{Cu}$, it reflects the conduction behavior of ILs. However, the direct evidence of the correlation between these parameters and the properties of sodium ions has not been fully studied from the static and dynamic point of view. Therefore, further research is needed in this field, including computational research.

### 5.4.4 Electrochemical properties

Electrochemical window is the potential range of electrolytes from reduction limit to oxidation limit, and it is one of the most important properties in ion batteries. Theoretically, the electrochemical window is limited by the highest occupied molecular orbital (HOMO) and the lowest unoccupied molecular orbital (LUMO) [145]. However, in practice, it is often controlled by dynamics, which can be proved by the importance of SEI film [146, 147]. In ion batteries, the electrochemical window from $Na^+/Na$ redox potential to 5 V is an ideal choice for high energy density. Many ILs containing $BF_4^-$, $TFSA^-$, or $FSA^-$ can meet this requirement.

The study of the combination of MD and density functional theory (DET) shows that under a certain structure model, the stability of ion-counterion binding is the basis for studying the electrochemical stability of ILs. Therefore, when discussing the electrochemical stability of ILs, it is not enough to consider the stability of a single ion in vacuum [148]. Compared with $C_3C_1Pyrr^+$, the reduction of TFSA$^-$ has lower stability [149]. The thermodynamic evaluation of DFT data shows that the electrochemical window of 1-alkyl-3-methylimidazole cations has nothing to do with the length of alkyl chains, and the electrochemical stability of ILs with trifl-fluoromethanesulfonate ([TfO]$^-$) or TFSI$^-$ anions [150].

In ILs, the corrosion of aluminum electrode at high potential is very small. Unlike fluorine-containing passivation films formed instantly on aluminum foil by PF$^{6-}$ based electrolytes, TFSA$^-$ and FSA$^-$ oxide corrode aluminum foil in organic solvents, while aluminum is usually used as a fluid collector, which is disadvantageous to practical applications [74, 151–155]. However, the corrosion of aluminum is the least in lithium-based ionic liquid electrolytes and has a similar performance in sodium system. This is because in ILs, aluminum electrodes can form a stable passivation film at high potential to prevent further etching.

The dissolution–deposition of metal sodium is also one of the factors to be considered when evaluating the electrolyte of sodium-ion batteries. Transparent and colorless [$C_2C_1im$][TFSA] ILs browned after soaking in metallic sodium for four weeks, while [$C_2C_1im$][FSA] had no obvious change. FSA⁻ has high resistance to reduction and can effectively inhibit the decomposition of ILs with only 10 mol% FSA⁻. In the absence of FSA$^-$, the metal sodium deposited in alkyl imidazolyl ILs will not be dissolved. However, it can be observed in alkylpyrrolidone, tetraalkylamine, and tetraalkylphosphine ionic liquids, even in ILs without FSA⁻, indicating that the cationic structure has a strong effect on the deposition–dissolution behavior of metallic sodium. In the presence of FSA⁻, the Coulomb efficiency is higher in most cases.

The concentration of sodium salt in ILs also has an effect on the deposition–dissolution behavior of metals. The effect of the interface layer formed in the high-concentration electrolyte on the deposition–dissolution behavior is more important than the ionic conductivity of the electrolyte itself [106, 107]. In FSA⁻-based ILs, the addition of about 500 ppm water helps to form an ideal SEI film with smooth surface and low interfacial resistance, and a certain amount of ppm water additives can reduce the interfacial impedance.

Ionic liquid electrolyte has good cycle performance and excellent rate performance, which can form stable SEI film, make sodium-ion diffusion easier, and form a stable passivation layer on the surface of aluminum collector to prevent aluminum corrosion. Furthermore, ionic liquid electrolytes can work at high temperatures to promote the diffusion of sodium ions in electrodes and electrolytes. However, the cost of ILs is high, the processing process is high, and the viscosity of ILs at low temperature is high, some ILs decomposition may produce harmful products, these shortcomings will also limit the application of ionic liquid electrolytes.

## 5.5 Solid electrolyte

### 5.5.1 Polymer electrolyte

Polymer electrolyte is composed of polymer matrix and additive salt, which has the advantages of good flexibility, low interface resistance, and easy processing. The commonly used polymer substrates are Poly(ethylene oxide) (PEO), polyvinylidene fluoride (PVDF), polyvinylidene fluoride-hexafluoropropene (PVDF-HFP), polyacrylonitrile (PAN), and polymethyl methacrylate (PMMA) [156–160]. The main additive salts are $NaPF_6$, $NaClO_4$, $CF_3SO_3Na$, $NaN (SO_2CF_3)_2$, and $NaN (SO_2F)_2$ (NaFSI). Most types of sodium-ion polymer solid electrolytes are derived from the corresponding lithium-ion counterparts. The main problems of polymer solid electrolytes are low room temperature ionic conductivity ($< 10^{-4}$ S cm$^{-1}$) and low ion transport number ($< 0.5$) [33, 161].

According to the ion transport mechanism of polymer SSE, the migration of sodium ions depends on the movement of polymer short chains [161, 164]. The ionic conductivity can be increased by increasing the amorphous region, and the amorphous region can be increased by copolymerization, blending, cross-linking, and the addition of plasticizer [161, 163–165]. For example, quasi-solid electrolytes of PVDF-HFP matrix were prepared using $SiO_2$ and $NaClO_4$ TEGDME additives (Figure 5.15(a)) [156]. The room temperature ionic conductivity can reach $10^{-3}$ S $\cdot$ cm$^{-1}$, and the ESW can reach more than 4.5 V (Figure 5.15(b)). At present, the ESW of most polymer electrolytes is wide (>4 V Na$^+$/Na) [159, 163, 166], but there are few reports on the application of high-voltage cathode (>4 V) in polymer solid-state sodium-ion batteries. By adding ethylene carbonate-propylene carbonate mixed solvent, the room temperature ionic conductivity of perfluorosulfonic acid electrolysate is $2.8 \times 10^{-4}$ S cm$^{-1}$ [162], but the mechanical strength of its electrolyte is poor (modulus of elasticity: 34.78 MPa, Figure 5.15(c). It is worth noting that liquid additives in polymer electrolytes may also leak and burn.

Many inorganic plasticizers can also improve the ionic conductivity of polymer electrolytes. Commonly used inorganic plasticizers can be divided into active materials (NASICON) and inactive materials ($SiO_2$, $TiO_2$, $ZrO_2$) [156, 157, 163]. The addition of inorganic plasticizer can increase the specific amorphous region of the polymer, improve the ionic conductivity of the polymer, and improve the mechanical properties [167]. For example, adding 5% mass percentage of nano-$TiO_2$, to the $NaClO_4$/PEO-based electrolyte increases the ionic conductivity from $1.35 \times 10^{-4}$ to $2.62 \times 10^{-4}$ S cm$^{-1}$ [157] at 60 ℃. The addition of NASICON materials (such as $Na_3Zr_2Si_2PO_{12}$ and $Na_{3.4}Zr_{1.8}Mg_{0.2}Si_2PO_{12}$) can also improve the ionic conductivity of the polymer [165]. Compared with the unmodified PEO-NaFSI electrolyte, the electrolyte modified by NASICON did not break at 40–50 ℃, which indicated that the addition of NASICON reduced the crystallinity of PEO (Figure 5.15(d). Although the addition of inorganic fillers can improve the ionic conductivity of polymer electrolytes to some extent, the ionic conductivity can still not meet the requirements of all-solid-state sodium batteries at room temperature (Figure 5.16).

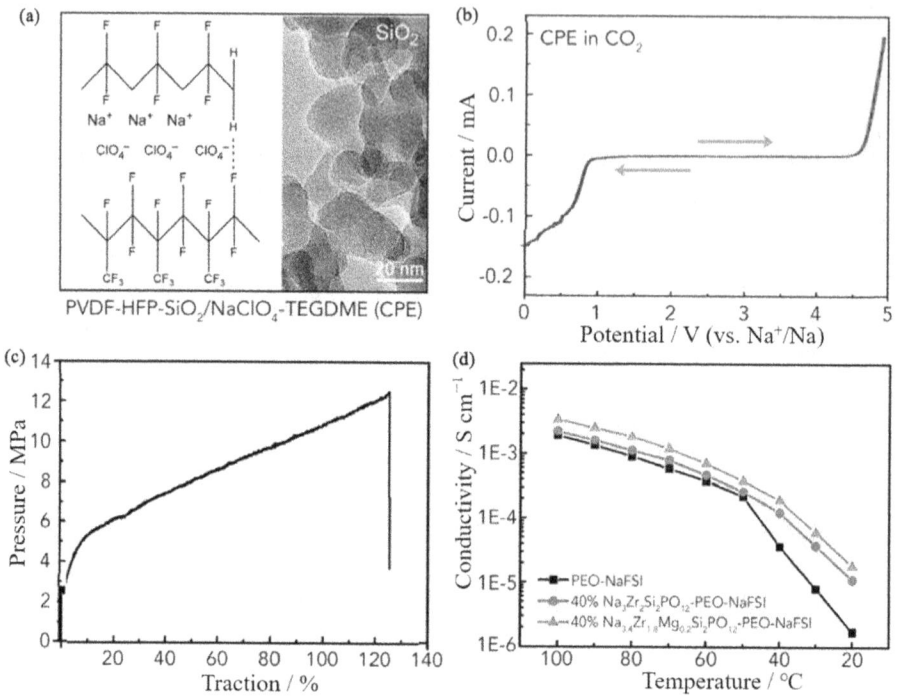

**Figure 5.15:** Electrochemical properties of polymeric solid electrolytes: (a) composition of composite polymer electrolyte (CPE), inset: transmission electron microscopy (TEM) image of $SiO_2$ [158]; (b) LSV curves in $CO_2$; (c) typical stress–strain curves of perfluorosulfonic acid-based electrolytes [164]; and (d) PEO-NaFSI, 40 mass percent $Na_3Zr_2Si_2PO_{12}$-PEO-NaFSI and 40 mass percent $Na_{3.4}Zr_{1.8}Mg_{0.2}Si_2PO_{12}$-PEO-NaFSI electrolytes with ionic conductivity versus temperature [165].

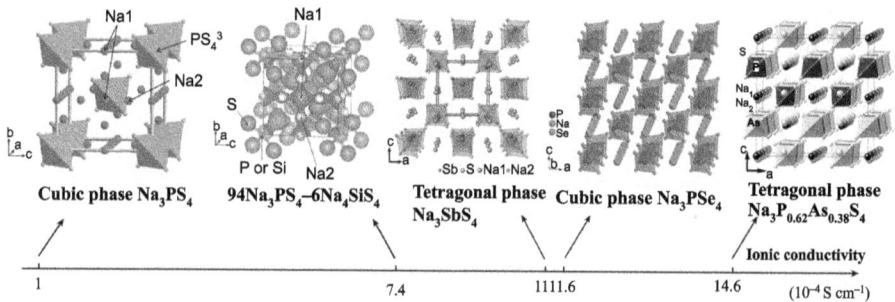

**Figure 5.16:** The crystal structure and ionic conductivity of representative sulfide-based inorganic solid-state electrolytes.

The reason for the low migration number of sodium ions in polymer electrolytes lies in the simultaneous migration of sodium ions and anions. Although inorganic additives can increase the migration number of sodium ions to some extent [169], the ultimate way to achieve the migration of high sodium ions is to fix anions. Using silane and sodium phosphonate groups to attach anions to the inorganic part of solid polymer electrolytes, the sodium-ion transfer number of solid polymer electrolytes can reach 0.9 [168]. In the next step, we attempt to study single-ion polymer electrolytes.

### 5.5.2 Inorganic solid electrolyte

#### 5.5.2.1 Solid electrolyte of $Al_2O_3$ and its analogues

In 1967, $\beta$-$Al_2O_3$ electrolyte was first discovered, which has high ionic conductivity and good thermal stability, and has been successfully used in large-scale energy storage systems, such as high-temperature sodium-sulfur battery and ZEBRA battery [170]. There are two kinds of crystal structures of $\beta$-$Al_2O_3$: $\beta$-$Al_2O_3$ (hexahedron; $P6_3/mmc$; $a_0 = b_0 = 5.58$ Å, $c_0 = 22.45$ Å), and $\beta$"-$Al_2O_3$ (rhombohedral; $R = 3$ m; $a_0 = b_0 = 5.61$ Å, $c_0 = 33.85$ Å) [171]; The chemical formula of $\beta$-$Al_2O_3$ is $Na_2O_{(8-11)}$ $Al_2O_3$, and the chemical formula of $\beta$ "- $Al_2O_3$ is $Na_2O_{(5-7)}$ $Al_2O_3$, But there are more sodium ions on the conduction surface, so the ionic conductivity of $\beta$"-$Al_2O_3$ is higher than that of $\beta$-$Al_2O_3$.

At 300 °C, the ionic conductivity of $\beta$"-$Al_2O_3$ is as high as 1 S cm$^{-1}$ [172]. The ionic conductivity of polycrystalline $\beta$"-$Al_2O_3$ is $2.0 \times 10^{-3}$ S cm$^{-1}$ and 0.2–0.4 S cm$^{-1}$ at room temperature and 300 °C, respectively [173, 174]. The low ionic conductivity of polycrystalline $\beta$"-$Al_2O_3$ is mainly caused by grain boundary resistance. Although its ionic conductivity is very high, pure $\beta$"-$Al_2O_3$ is very easy to react with water [175, 176] and has poor thermal stability, so it is difficult to prepare pure $\beta$"-$Al_2O_3$. $\beta$"-$Al_2O_3$, synthesized by traditional solid-state reaction will contain impurities such as $\beta$-$Al_2O_3$ and $NaAlO_2$, and the ionic conductivity of $\beta$-$Al_2O_3$, obtained by high-temperature evaporation will be reduced [177]. Although the abnormal growth of $\beta$"-$Al_2O_3$ grains has higher ionic conductivity, its mechanical properties are poor [178], and the $NaAlO_2$ impurities in $\beta$"-$Al_2O_3$ will react with $H_2O$ and $CO_2$ in the air. $\beta$"-$Al_2O_3$, with high purity can be prepared by adding doping stabilizers such as Li$^+$, Mg$^{2+}$, Ni$^{2+}$, and Ti$^{4+}$ in the reaction, but the mechanical properties of pure $\beta$"-$Al_2O_3$ are low (fracture strength is 200 MPa) [179].

#### 5.5.2.2 Solid electrolyte with NASICON frame structure

The open three-dimensional channel of sodium super ion conductor (NASICON) material enables sodium ion to transport quickly. In 1976, Goodenough et al. first discovered the NASICON-type material $Na_{1+x}Zr_2Si_xP_{3-x}O_{12}$ ($0 \leq x \leq 3$), which was obtained by using Si$^{4+}$ instead of P$^{5+}$ in $NaZr_2P_3O_{12}$ solid solution [180, 181]. When $x = 2$, the ionic conductivity of $Na_3Zr_2Si_2PO_{12}$ is the highest, which is $10^{-4} S \cdot cm^{-1}$ and $10^{-1} S \cdot cm^{-1}$ at room temperature and 300 ℃, respectively. Therefore, most of the studies of NASICON materials are based on $Na_3Zr_2Si_2PO_{12}$. $Na_3Zr_2Si_2PO_{12}$ has two kinds of crystal structure: rhombohedral phase and monoclinic phase. The rhombohedral crystal structure has two different sodium-ion sites, and the monoclinic crystal structure has three different sodium-ion sites (Figure 5.17(a)).

In NASICON solid electrolyte, the transport of sodium ion needs to be realized by vacancy jump. There are four different sodium vacancy regions in the electrolyte of monoclinic NASICON, including two Na1–Na2 and two Na1–Na3 channels (Figure 5.17(b)) [182]. The size of the vacancy region is directly related to the migration energy barrier and will have an impact on the ionic conductivity, so the size of sodium vacancy is generally expanded by introducing substituents to improve its ionic conductivity. Song et al. studied the crystal structure and ionic conductivity of NASICONs substituted by alkaline earth metal ions ($Na_{3.1}Zr_{1.95}M_{0.05}Si_2PO_{12}$, M = Mg, Ca, Sr, Ba). The crystal structure refinement parameters of NASICON in Figure 5.17(c) show that the sodium vacancy region increases as $Mg^{2+}$ plasma replaces $Zr^{4+}$. Instead of NASICON, $Mg^{2+}$ has the largest T1 region (6.522 $Å^2$), and its unreplaced area is only 5.223 $Å^2$. At room temperature, the ionic conductivity of $Na_{3.1}Zr_{1.95}Mg_{0.05}Si_2PO_{12}$ is $3.5 \times 10^{-3}$ S $cm^{-1}$. However, the sodium vacancy region in the larger $Ba^{2+}$ substituted NASICON is narrower than that in the unsubstituted materials, indicating that the ion radius of the substitutes has an important effect on the size of the sodium vacancy region in NASICON electrolytes. The ionic conductivity of $Sc^{3+}$ (74.5/72 pm), which is similar to that of $Zr^{4+}$, is $4.0 \times 10^{-3}$ S $cm^{-1}$ at room temperature, which reaches the highest value reported in the literature. However, the cost of Sc is high, which will limit the large-scale application of NASICON. Low-cost substitutes such as $Nd^{3+}$ and $Y^{3+}$, which can replace Sc, are also used to improve the ionic conductivity of NASICON, but its ionic conductivity is not as high as that of Sc [185].

In addition to increasing the ion transport rate by changing the size of the sodium vacancy region, the migration of sodium ions at the grain boundary will also affect the total ionic conductivity of inorganic solid electrolytes. By adjusting the grain size and the chemical composition of the grain boundary, the grain boundary resistance can be reduced [184, 186, 187]. Ihefeld et al. have studied the effect of size on the grain boundary resistance of $Na_{1+x}Zr_2Si_xP_{3-x}O_{12}$ (0.25 ≤ x ≤ 1.0). The results show that with the decrease of annealing temperature and the increase of Si content, the grain size becomes smaller and the ionic conductivity of grain boundary decreases. The researchers also synthesized $La^{3+}$-doped $Na_3Zr_2Si_2PO_{12}$ by self-assembly [186], and found that new phases of $Na_3La(PO_4)_2$, $La_2O_3$, and $LaPO_4$ appeared at the grain boundaries (Figure 5.17(d)). The emergence of new phases adjusts the chemical composition of the grain boundaries and promotes the rapid migration of sodium ions. Other studies have also shown that the new $Na_3PO_4$ at the grain boundary can promote the improvement of the ionic conductivity of the grain boundary [187].

The traditional method for the synthesis of $Na_{1+x}Zr_2P_{3-x}O_{12}$ $(0 \leq x \leq 3)$ is the high-temperature solid-state reaction (about 1,000 °C). The high-temperature reaction will lead to the evaporation of Na and P elements, forming impurities such as $ZrO_2$ [188]. Excess sodium and phosphorus are usually added to alleviate this problem [189]. If large-scale applications are to be realized, the sintering temperature must be reduced. By using the sol–gel method, the reactants can be mixed evenly at the molecular level, the mass transfer path can be reduced, and the processing temperature can be reduced to a certain extent [190]. In addition, inspired by the production of garnet ceramics, Suzuki et al. used liquid phase sintering method at a lower temperature of 700 °C to prepare $Na_3Zr_2Si_2PO_{12}$ using 9.1 wt% $Na_3BO_3$ (melting point is 680 °C) [191]. However, the preparation of the precursor system of NASICON still needs to be carried out at 1,100 °C.

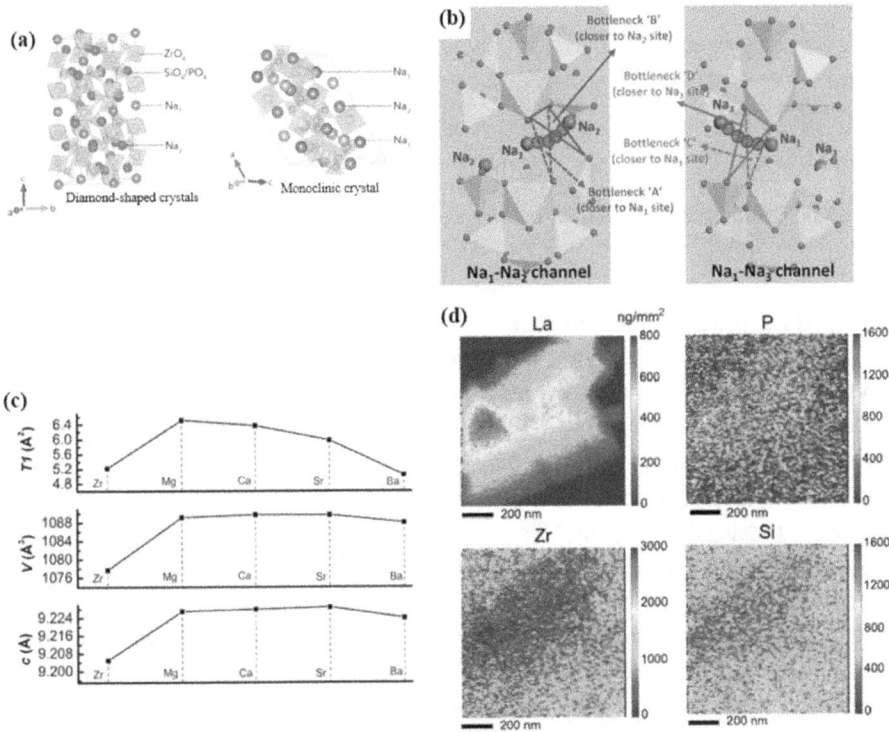

**Figure 5.17:** Electrochemical properties of NASICON-type solid-state electrolytes: (a) two crystal structures of $Na_3Zr_2Si_2PO_{12}$; (b) sodium-ion transport channels; (c) lattice parameters, unit cell volume, and bottleneck of $Na_3Zr_2Si_2PO_{12}$ and $Na_{3.1}Zr_{1.95}M_{0.05}Si_2PO_{12}$ (M = Mg, Ca, Sr, Ba) area of the region (embedding point T1) [183]; and (d) $Na_{3.3}Zr_{1.7}La_{0.3}Si_2PO_{12}$ electrolyte in La, P, Zr and Si high-resolution elemental distribution images (simultaneous X-ray nanoprobe analysis) [184].

### 5.5.2.3 Sulfide solid electrolyte

Sulfide-based solid-state electrolyte is a very promising sodium-ion solid-state electrolyte because of its high ionic conductivity, low grain boundary resistance, good plasticity, and mild synthesis conditions [192–195]. Compared with oxides, sulfide solid electrolytes have stronger sodium-ion migration ability, larger ion radius, and higher polarizability of sulfur atoms, which can weaken the electrostatic interaction with sodium ions. The most common sulfur-based SSE is $Na_3PS_4$, $Na_3PS_4$ with two crystal structures: cubic phase and tetragonal phase (Figure 5.18) [196]. $Na_3PS_4$ usually exists in the form of a tetragonal phase, but it will be transformed into a cubic phase at 530 K.

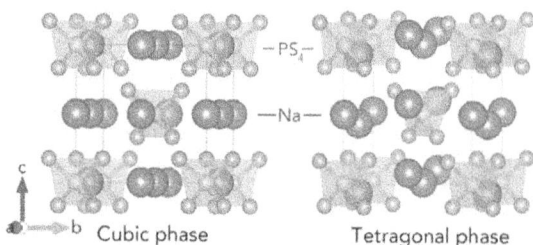

**Figure 5.18:** Crystal structure of $Na_3PS_4$ in cubic and tetragonal phases.

In 1992, Jansen and Henseler confirmed the existence of tetragonal phase by single crystal XRD. However, the ionic conductivity of tetragonal $Na_3PS_4$ at 50 °C is only $4.17 \times 10^{-6}$ S cm$^{-1}$, which cannot meet the needs of practical applications. In 2012, Hayashi et al. reported a glass-ceramic sulfide solid electrolyte with a room temperature ionic conductivity of $2 \times 10^{-4}$ S cm$^{-1}$ [195]. The stable existence of the cubic phase $Na_3PS_4$ at room temperature makes the ionic conductivity higher. By increasing the purity of the synthetic raw materials, the room temperature ionic conductivity can be increased to $4.6 \times 10^{-4}$ S cm$^{-1}$ [197].

The common method to improve the ionic conductivity of $Na_3PS_4$ is to introduce defects (sodium-ion gap or vacancy) into the lattice. Through the first-principle calculation, it can be proved that the sodium-ion gap is beneficial to improve the ionic conductivity of $Na_3PS_4$. Figure 5.19(a) shows the stable structure of cubic phase $Na_3PS_4$ with sodium-ion gap. The covalent cation doping of $M^{4+}$ and $P^{5+}$ can produce sodium gap, and a large number of experimental results have confirmed the positive effect of sodium gap. Some researchers have synthesized Si-doped sulfide solid electrolytes ($94Na_3PS_4 \cdot 6Na_4SiS_4$), which exhibit high ionic conductivity ($7.4 \times 10^{-4}$ S cm$^{-1}$) [198]. In addition to the gap, the sodium-ion vacancy also has a great influence on the ionic conductivity of sulfide-based inorganic solid electrolytes. DeKlerk et al. studied the effect of sodium vacancy on the ionic conductivity of $Na_3PS_4$ by MD simulation. Compared with the original cubic $Na_3PS_4$, the mobility of sodium ion in sulfide ($Na_{2.94}PS_4$) containing sodium vacancy increases sharply (Figure 5.19(c)),

300 K, the ionic conductivity can reach 0.17 S cm$^{-1}$. Bo et al. show that the perfect stoichiometric Na$_3$PSe$_4$ is not sensitive to the change of temperature, and the change of sodium-ion diffusion coefficient is negligible even at 900 K [200], but the sodium-ion conductivity increases significantly after introducing 2.1% sodium vacancy, which is close to the theoretical value of $2.89 \times 10^{-2}$ S cm$^{-1}$ at 300 K, and at 0 K, the diffusion barrier in Na$_3$PSe$_4$ with sodium-ion vacancy is only 5 kJ mol$^{-1}$, which means that sodium ion can be transported well (Figure 5.19(b)). According to the charge balance theory, sodium-ion vacancies can be generated by equivalent anions (such as Cl$^-$ and Br$^-$) instead of S$^{2-}$. The experimental results show that the ionic conductivity of Cl$^-$doped sulfide (Na$_{2.9375}$PS$_{3.9375}$Cl$_{0.0625}$) is $1.14 \times 10^{-3}$ S cm$^{-1}$ at 303 K, which is two orders of magnitude higher than that of the traditional tetragonal Na$_3$PS$_4$ ($5 \times 10^{-5}$ S cm$^{-1}$) [201].

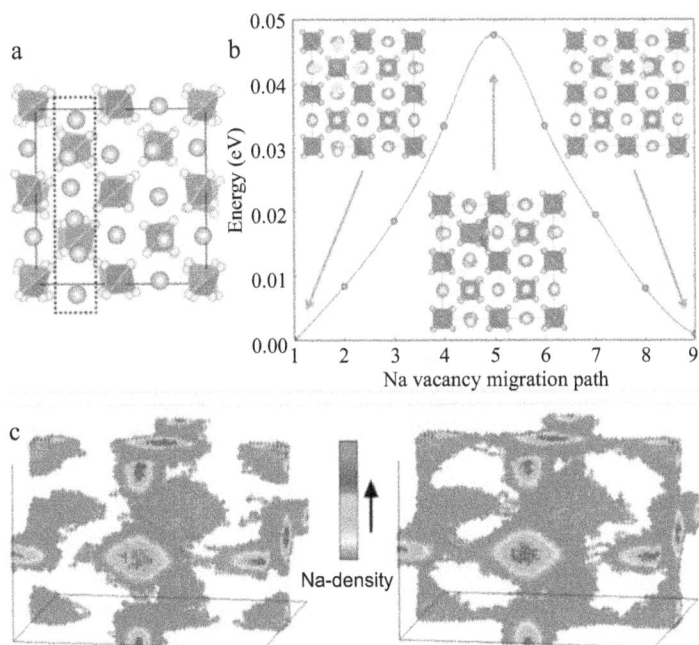

**Figure 5.19:** Theoretical simulation studies of sulfur-based inorganic solid electrolytes: (a) MD simulations of Na distribution at 100 ps for cubic Na$_3$PS$_4$ (left) and cubic Na$_{2.94}$PS$_4$ (right) at 525 K [170, 199]; (b) Na$^+$ vacancy-diffusion potential in cubic Na$_3$PSe$_4$; embedding diagram: the structures of the initial, lepton, and final states are shown in the upper left, lower middle, and upper right [171, 200]; and (c) lowest energy sequence of cubic Na$_3$PS$_4$ with sodium-ion gap [165].

Moreover, the interaction between sodium-ion and anion skeleton structure and the size of unit cell/channel also have a significant effect on the ionic conductivity. The theoretical calculation shows that Se doping can increase the polarizability of anion skeleton, soften the lattice, and reduce the activation energy barrier [202, 204]. At the same time, related experiments also show that all-Se substituted electrolytes ($Na_3PSe_4$) have higher ionic conductivity ($1.16 \times 10^{-3}$ S cm$^{-1}$). Sb-doped P-site can also be used to modify the structure of $Na_3PSe_4$ to obtain higher ionic conductivity. With the increase of Sb content, the ionic conductivity increases with the increase of cell and channel (Figure 5.20) [203]. The room temperature ionic conductivity of all-Sb-doped $Na_3SbSe_4$ can reach $3.7 \times 10^{-3}$ S cm$^{-1}$.

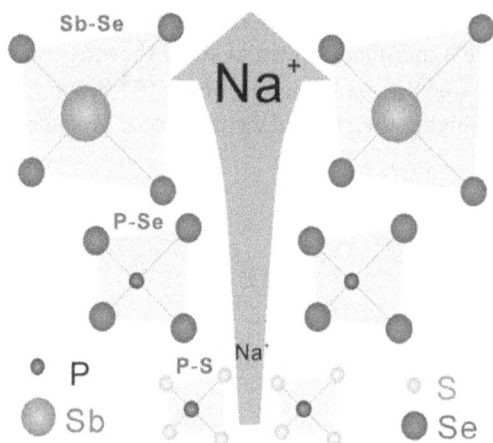

**Figure 5.20:** Schematic representation of sodium-ion diffusion channels in $Na_3PS_4$, $Na_3PSe_4$, and $Na_3SbSe_4$ [203].

With the discovery of lithium-sulfide inorganic solid electrolytes with high ionic conductivity such as $Li_{10}MP_2S_{12}$ and $Li_7P_3S_{11}$, many researchers have begun to develop similar inorganic solid electrolytes for sodium-ion transport [204, 205]. Theoretical studies show that the ionic conductivities of $Na_{10}GeP_2S_{12}$, $Na_7P_3S_{11}$, and $Na_7P_3Se_{11}$ can be as high as $4.7 \times 10^{-3}$ S cm$^{-1}$, $1.1 \times 10^{-2}$ S cm$^{-1}$, and $1.26 \times 10^{-2}$ S cm$^{-1}$ [206, 207]. Furthermore, through density functional theory simulation, it is found that the ionic conductivity of different atom-substituted ionic conductors $Na_{10}MP_2S_{12}$ is Si > Ge > Sn [208]. They further synthesized $Na_{10}SnP_2S_{12}$, with ionic conductivity of $4 \times 10^{-4}$ S cm$^{-1}$, but found that there are a large number of impurity phases in the samples, such as $P_2S_5$, $Na_2S$, and $Na_3PS_4$ (Figure 5.21(a). The pure sulfide of Na was synthesized by adjusting the ratio of $Na_{11}Sn_2PS_{12}$ to Sn, and its structure was verified by single crystal XRD and ab initio MD simulation [209]. At the same time, the same inorganic solid-state electrolyte with high ionic conductivity (($3.7 \pm 0.3$) $\times 10^{-3}$ S cm$^{-1}$) was synthesized by solid-state method, which is the highest value at present [210]. Due to the existence of vacancies in the lattice, it shows excellent ion conductivity (Figure 5.21(b)), and the three-dimensional conduction vacancy structure forms a good ion transport channel, and this view is also proved by theoretical calculation. In addition, the application of $Na_{11}M_2PS_{12}$ as an inorganic solid electrolyte is studied theoretically and

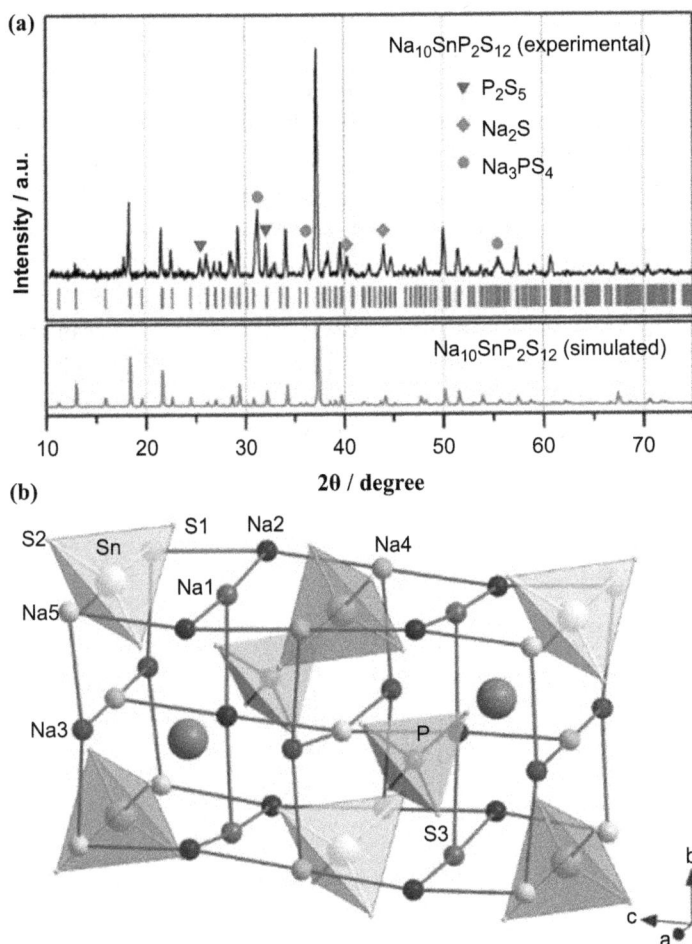

**Figure 5.21:** (a) Simulated and experimental XRD patterns of $Na_{10}SnP_2S_{12}$, indicating the presence of recrystallized $P_2S_5$, $Na_2S$, and $Na_3PS_4$ [208]; (b) crystal structure of $Na_{11}Sn_2PS_{12}$, with large spheres representing vacancies [210].

experimentally. It is found that the ionic conductivity of $Na_7P_3S_{11}$ and $Na_7P_3Se_{11}$ electrolytes imitating $Li_7P_3S_{11}$ analogues may be higher than that of $Na_{11}Sn_2PS_{12}$.

For possessing high ionic conductivity, sulfide-based inorganic solid electrolytes meet the basic need of practical applications. Unfortunately, however, most of sulfide-based inorganic solid electrolytes are unstable in air [211]. Methods like chemical coating or doping have been applied to overcome this challenge. For example, in $Na_2Se–Ga_2Se_3–GeSe_2$ ternary system, the crystals wrapped in inert glass matrix are stable relative to moisture [212]. However, due to the relatively weak P–S bond, most sulfide-based inorganic solid electrolytes can react with water and oxygen to form toxic hydrogen sulfide gas [33, 213]. Doping to replace P–S bond is an effective way to improve the chemical stability of sulfide-based inorganic solid electrolytes. When $P^{5+}$ is partially replaced by $As^{5+}$, sulfides ($Na_3P_{0.62}As_{0.38}S_4$) show stronger water molecular stability because As–S has a stronger affinity than P–S. When Sb is completely replaced, the stability of sulfide-based inorganic solid electrolyte ($Na_3SbS_4$) can be further improved [214, 215]. $Na_3SbS_4$ can exist stably in $O_2$ and $H_2O$, which can be explained by hard and soft acid-base theory. In $Na_3PS_4$, because of the strong interaction between $O^{2-}$ (hard base) and $P^{5+}$ (hard acid), $S^{2-}$ (soft base) is more easily replaced by $O^{2-}$, while in $Na_3SbS_4$, the bond between $Sb^{5+}$ (soft acid) and $S^{2-}$ is much stronger and is not easily destroyed by $O^{2-}$, and under appropriate conditions, $Na_3SbS_4$ can be synthesized by solvothermal method. When exposed to air, pure $Na_3SbS_4$ can easily absorb water to form $Na_3SbS_4 x H_2O$. After heating at 150 ℃ for 1 h, $H_2O$ can be removed (Figure 5.22). Although $Na_3SbS_4$ has high stability and ionic conductivity (about $10^{-3}$ S cm$^{-1}$), the toxicity of Sb limits its application.

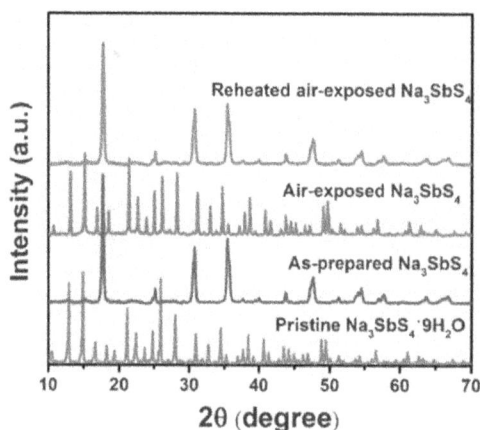

**Figure 5.22:** XRD patterns of $Na_3SbS_4 \cdot 9H_2O$, $Na_3SbS_4$, $Na_3SbS_4$ exposed to air, and $Na_3SbS_4$ heated after air exposure (150 °C for 1 h) [215].

In addition to chemical stability, the electrochemical stability of sulfide-based inorganic solid electrolytes is also a very important factor. The electrochemical window of sulfide-based inorganic solid electrolyte was tested by closed working electrode, and the electrochemical window of $Na_3PS_4$ and $Na_3SbS_4$ reached 5 V ($Na^+/Na$) [190, 216]. When a mixture of $Na_3PS_4$ and conductive carbon is used as the working electrode, the theoretical and experimental decomposition potentials of $Na_3PS_4$ are 2.3 V and 2.7 V, respectively, (relative to Figure 5.23 of $Na^+/Na$) [217]. When assembling the all-solid-state battery for testing, the sulfide inorganic solid-state electrolyte can be mixed into the cathode material to reduce the interface resistance. This test method will be more accurate, and the evaluation of the ESW should be carried out under conditions similar to those of the actual electrode. The ESW of many types of sulfur-based inorganic solid electrolytes is limited, so further research is needed to improve their electrochemical stability. Conductive carbon additives can promote the electrochemical decomposition of sulfide-based SSE ($Li_{10}GeP_2S_{12}$) during the cycle of all-solid-state lithium batteries, resulting in large interfacial impedance and capacity attenuation [218, 219]. It is of great significance to study whether conductive carbon will have an effect on sulfide-based inorganic solid electrolytes for sodium batteries.

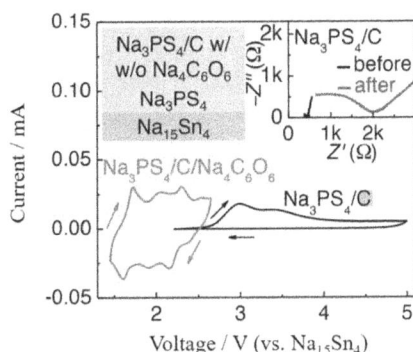

Figure 5.23: CV curves of $Na_3PS_4$ and $Na_4C_6O_6$ (0.05 mV s$^{-1}$); the left inset shows the cell composition for CV testing, and the right inset shows the EIS of the $Na_3PS_4$/C battery before and after CV testing [191].

### 5.5.2.4 Composite hydride solid electrolyte

In 2012, Orimo et al. proposed for the first time to use composite hydride as sodium-ion solid electrolyte [220]. The sodium-ion migration number of $NaAlH_4$ and $Na_3AlH_6$ is close to 1. Although the ionic conductivities of $NaAlH_4$ and $Na_3AlH_6$ are only $2.1 \times 10^{-10}$ S cm$^{-1}$ and $6.4 \times 10^{-7}$ S cm$^{-1}$ at room temperature, this study lays a foundation for the future study of sodium hydride solid electrolytes. In subsequent studies, they mixed $NaBH_4$ and $NaNH_2$ at a molar ratio of 1:1 to form $Na_2 (BH_4) (NH_2)$ with an ionic conductivity of $3 \times 10^{-6}$ S cm$^{-1}$ [221]. The perovskite structure of sodium vacancy makes $Na_2 (BH_4) (NH_2)$ have higher ionic conductivity.

Compared with small anions (such as $BH_4^-$ and $NH_2^-$), composite hydrides with large anions (such as $B_{12}H_{12}^-$ and $B_{10}H_{10}^-$) exhibit higher ionic conductivity above the order–disorder phase transition temperature [222–224]. For example, at 573 K, due to the existence of high-temperature disordered body-centered cubic phase (cation-rich vacancy structure), $Na_2B_{12}H_{12}$ exhibits higher ionic conductivity (>0.1 S cm$^{-1}$). Because the phase transition temperature of composite hydride is too high, in order to meet the needs of practical application, it is necessary to reduce or eliminate the phase transition temperature of composite hydride. There are three main methods to reduce or eliminate the phase transition temperature: anion chemical modification, anion mixing, and grain nano crystallization/disordering. For example, after introducing C for anionic modification of $Na_2B_{12}H_{12}$ and $Na_2B_{10}H_{10}$ (Figure 5.24(b)), their phase transition temperatures will be greatly reduced ($NaCB_{11}H_{12}$ and $Na_2B_{12}H_{12}$: 380 K and 529 K; $NaCB_9H_{10}$ and $Na_2B_{10}H_{10}$, 290 K and 380 K) [225]. The correction of local static cation–anion interaction, orientation deviation, and anion rotation reduce the phase transition temperature after C substitution [226]. When the hydrogen atom is completely replaced by halogen to form $Na_2B_{12}X_{12}$, the phase transition temperature is higher than the original $Na_2B_{12}H_{12}$ due to the introduction of halogen atom (Figure 5.24(c)) and the increase of anion size/weight and anisotropic electron density. In addition, by mixing different anions and introducing geometric dislocations, the transition temperature can be reduced or eliminated [227]. The ionic conductivity of $Na_2 (B_{12}H_{12})_{0.5} (B_{10}H_{10})_{0.5}$ obtained by mixing two different anions is $9 \times 10^{-4}$ S cm$^{-1}$ at 20 °C, and there is no obvious structural phase transition at –70–280 °C. When the grain size is reduced and the degree of disorder is increased by ball milling, the transition temperature can be reduced and the ionic conductivity at room temperature can be higher.

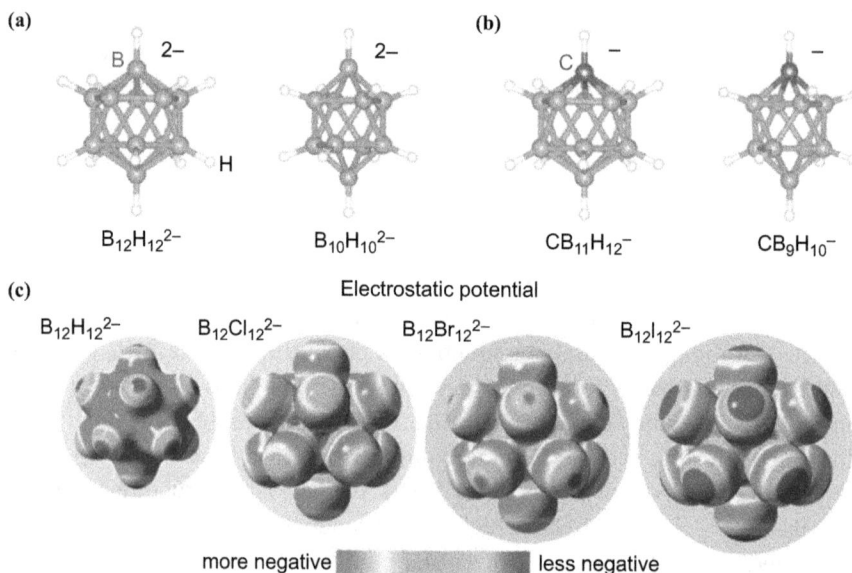

Figure 5.24: Theoretical simulation studies of composite hydride solid-state electrolytes: (a) relative geometries of $B_{12}H_{12}^{2-}$ and $B_{10}H_{10}^{2-}$ anions; (b) relative geometries of $CB_{11}H_{12}^{-}$ and $CB_9H_{10}^{-}$ anions shapes; and (c) Electrostatic potential surface diagrams of $B_{12}H_{12}^{2-}$, $B_{12}Cl_{12}^{2-}$, $B_{12}Br_{12}^{2-}$, and $B_{12}I_{12}^{2-}$ anions (calculated using HF/3-21G theory).

In the application of sodium hydride solid electrolyte, the chemical and electrochemical stability of composite hydride should be considered in addition to ionic conductivity. Although $Na_2B_{10}H_{10}$ is easy to absorb water, it is stable in air at room temperature. Through the study of the anodic decomposition process of electrolyte by CV, it was found that the ESW of $Na_2(B_{12}H_{12})_{0.5}(B_{10}H_{10})_{0.5}$ was about 3 V (Na +/Na). Then, based on this study, a stable 3 V solid-state sodium-ion battery was prepared. However, $Na_2$ $(B_{12}H_{12})_{0.5}(B_{10}H_{10})_{0.5}$ electrolyte is not suitable for positive electrode with high working voltage (> 3 V Na$^+$/Na). Future research should pay attention to the electrochemical stability of composite hydride and apply it to different solid-state batteries.

### 5.5.3 Mixed electrolyte

The new mixed electrolytes containing two different types of Na$^+$ transfer media combine their advantages and show excellent physical and electrochemical properties. Among them, the most representative mixed electrolytes are organic and IL-mixed electrolytes, which can provide high Na$^+$ mobility and excellent safety performance [228]. In addition, the new polymer electrolytes modified by organic or IL electrolytes also show good mechanical properties and cycle stability. However, solubility, ESW, compatibility, and other properties also need to be considered when designing mixed electrolytes. Furthermore, the performance of mixed electrolytes is usually affected by the synergism or inhibition between various precursors, rather than a simple superposition. Therefore, researchers should conduct further experimental and theoretical studies to obtain better-mixed electrolytes.

Most commercial full-cell electrolytes are generally based on organic solvents and have high ionic conductivity, but they have poor safety and mechanical properties [5]. Remedial safety measures alone are not enough to solve the problems of high flammability and volatility [229]. Coincidentally, IL solvents have high safety despite their high viscosity and low ionic conductivity. Therefore, the researchers studied the mixed electrolytes of organic solvents with ILs as cosolvent or additive to achieve the balance between high electrochemical performance and high thermal stability. For example, in half-cell tests using HC electrodes, organic/inorganic electrolyte mixtures consisting of mixed organic solvents (EC:PC), different IL (EMImTFSI, BMImTFSI, and Pyr13TFSI) as cosolvents, and NaTFSI salts showed unexpected Na$^+$ migration numbers [228].

The proper ratio of organic solvent to IL solvent is the key factor to optimize the performance. The experimental results show that the mixed electrolyte with 10–50% ILs content has higher ionic conductivity (Figure 5.25(a)), and its safety has been enhanced to some extent (Figure 5.25(b)). At the same time, the mechanical stability and electrochemical stability of SEI film can also be improved by adding organic solvent to IL solvent. In addition, the combustion test showed that the safety of the mixed electrolyte was also enhanced by the addition of IL cosolvent. It should be noted that verified by Raman spectroscopy (Figure 5.25(c)), Na$^+$ the first solvated shell is modified by different electrolyte components and exhibits an increased area ratio. Na$^+$-cosolvent and NaTFSI salt showed high Na$^+$ transfer number. In addition to HC negative electrode, the interface optimization effect of IL/organic electrolyte on NVP cathode was also studied [230]. A more stable passivation layer containing NaTFSI compounds can be formed in the mixed electrolyte, which can effectively improve the stability of the NVP electrode.

SPEs have good dimensional stability, toughness, and transparency. However, the ion migration rate and amount of SPE are limited, so organic solvents or ILs have been used to improve their electrochemical properties. $NaCF_3SO_3$ and EMITf were immobilized in PVdF-HFP to form a new composite electrolyte, as shown in Figure 5.25(d). It can be observed that the conformational change of PVDF-HFP is caused by the incorporation of EMITf or EMITf + NaTf solution (Figure 5.25(e) and (f)), which indicates that there is an interaction between the liquid and the gel polymer. The ionic conductivity of the mixed electrolyte at room temperature is as high as $5.74 \times 10^{-3}$ S cm$^{-1}$ and has a high Na$^+$ transport number (0.23), which can be attributed to the transport of anions with the conduction of ions in the IL component. At the same time, as shown in Figure 5.25(g), a width of –3–3 V ESW can be measured in the battery SS/EMITf$^+$NaTf/SS (SS: stainless steel).

The combination of inorganic solid electrolytes (ISEs) and IL further refreshes the concept of mixed electrolytes [232], both of which have nonflammability and high electrochemical stability, indicating that high safety is an outstanding advantage of mixed electrolytes composed of solid electrolytes and ILs. For example, at high temperature, the $Na_{0.66}Ni_{0.33}Mn_{0.67}O_2$ electrode using IL-modified Na- β"-$Al_2O_3$ electrolyte can achieve high cycle life and high Coulomb efficiency at the same time, at the same time, a small amount of ILs acts as a cosolvent on the positive side. (Figure 5.26(a)), reduces the grain boundary resistance of the electrode and SEI [184]. So, the electrode/IL- solid electrolyte/Na with sandwich structure solves the problem of low interface dynamics. In addition, the ILs used as modification layers make up for the interfacial compatibility defects of solid-state electrolytes without harming the safety and stability of all-solid-state electrolytes. A similar combination can be made between highly stable organic electrolytes and flexible polymer electrolytes [234]. A flexible full battery was prepared from a mixed solid membrane electrolyte containing organic electrolyte, polymer electrolyte, and solid electrolyte, and a breakthrough in safety and electrochemical properties was achieved through the coordination of three ion channels (Figure 5.26(b)).

As shown in Figure 5.27(a), people try to use organic electrolytes, solid ceramic electrolytes, and aqueous electrolytes as ternary electrolytes in SIB. The mixed electrolyte battery with replaceable positive electrode structure uses seawater as positive electrode and nonaqueous electrolyte as negative electrode. NASICON ceramics prevent direct contact between the two liquid electrolytes, but allow Na$^+$ to migrate quickly. In this system, the lower limit of cutoff voltage is broadened by using organic negative electrode–electrolyte, and the specific capacity of battery is increased by using metal Na negative electrode. In addition, aqueous electrolyte has high ionic conductivity and low cost. Therefore, the mixed electrolyte battery with high energy density, long cycle life, and low price has a cost-effective ratio compared with lead-acid battery, and the electrochemical performance of mixed electrolyte battery is better than that of general water system SIBs. Similar devices are assembled using PBA positive electrodes, water electrolytes, ceramic diaphragms, organic electrolytes, and

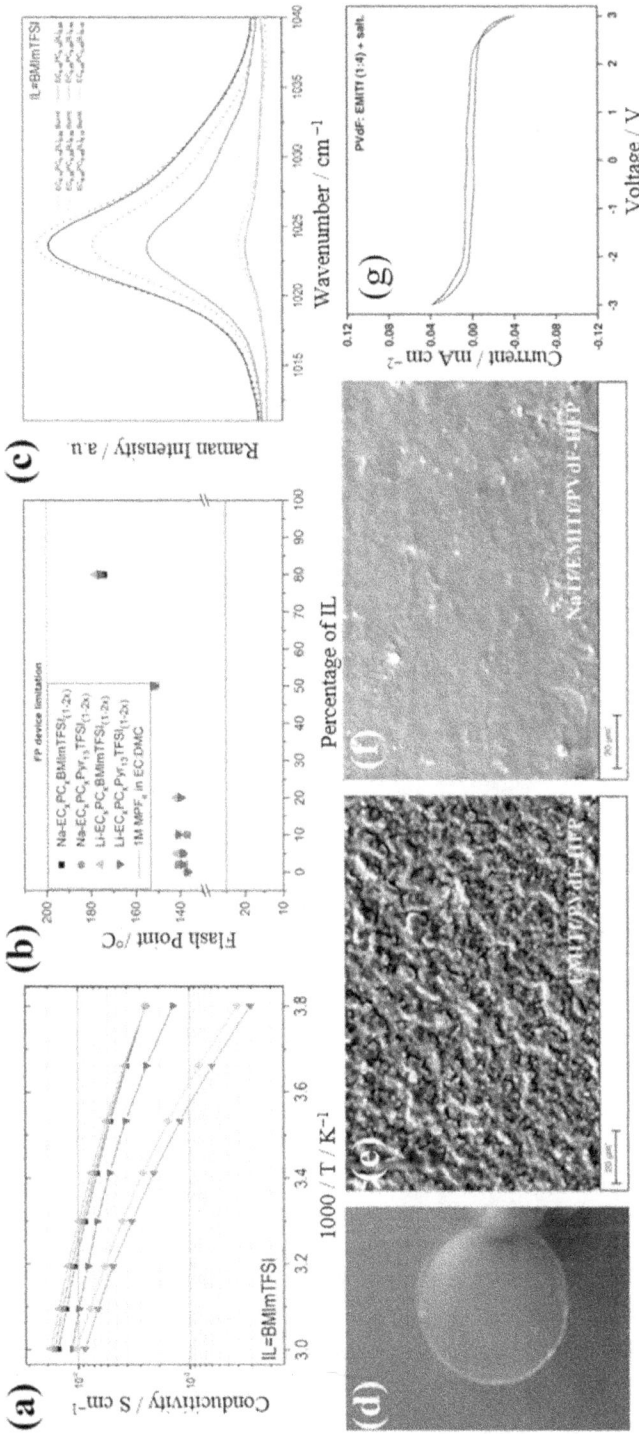

**Figure 5.25:** Electrochemical properties of composite electrolytes: (a) ionic conductivity of 0.8 M NaTFSI in EC:PC:IL; (b) flash point diagram; (c) Raman spectrum [228]; (d) photographs of NaTf/EMITf/PVDF-HFP gel polymer; (e) EMITf/PVDF-HFP; (f) NaTf/EMITf/PVDF-HFP gel polymer SEM images; and (g) NaTf/EMITf/PVDF-HFP gel polymer electrolyte CV curves [231].

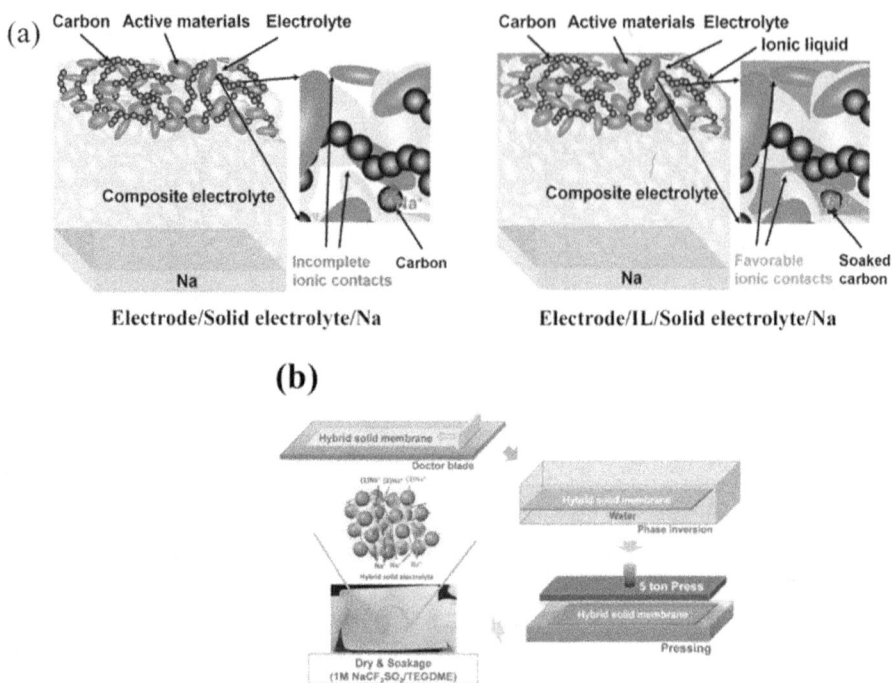

**Figure 5.26:** Schematic diagram of solid-state composite electrolytes: (a) IL–ISE hybrid electrolyte [184]; and (b) organic polymer–ISE hybrid electrolyte [234].

metal Na negative electrodes, as shown in Figure 5.27(b). Due to the redox reaction of $[Fe^{II}(CN)_6]^{4-}/Fe^{III}(CN)_6]^{3-}$, the use of sodium hexacyanoferrate soluble in aqueous solution as a redox active electrolyte instead of traditional $Na_2SO_4$ aqueous solution can provide additional capacity [236]. In addition, semiliquid $NaC_{12}H_{10}$ dissolved in TEGDME has been used as a fluid negative electrode to replace the Na metal negative electrode (Figure 5.27(c)), this system has higher safety and lower interface impedance, so the whole cell exhibits high energy density and excellent rate performance. In addition, some researchers have innovatively mixed organic electrolytes (8 M sodium trifluoromethanesulfonate (NaOTf) in PC) and aqueous electrolytes (7MNaTOf in water) into NVP//NTP whole cells, which show wider ESW and high conductivity [237]. Among them, the super-concentrated "water-salt" and "solvent salts" systems play an important role in the synergistic effect of mixed electrolytes [238]. Similarly, the researchers prepared a novel organic-water composite electrolyte in which 1 M $NaClO_4$ is dissolved in a cosolvent containing 90% acetonitrile and 10% water to solve the solubility problem of Mn-based PBA positive electrodes. In addition to mixtures of different solvents, composite water batteries can also be composed of two salts, such as a mixture of sodium and potassium salts. Due to the existence of selective cation channels in the PBA electrode, the battery can provide sufficient capacity and voltage.

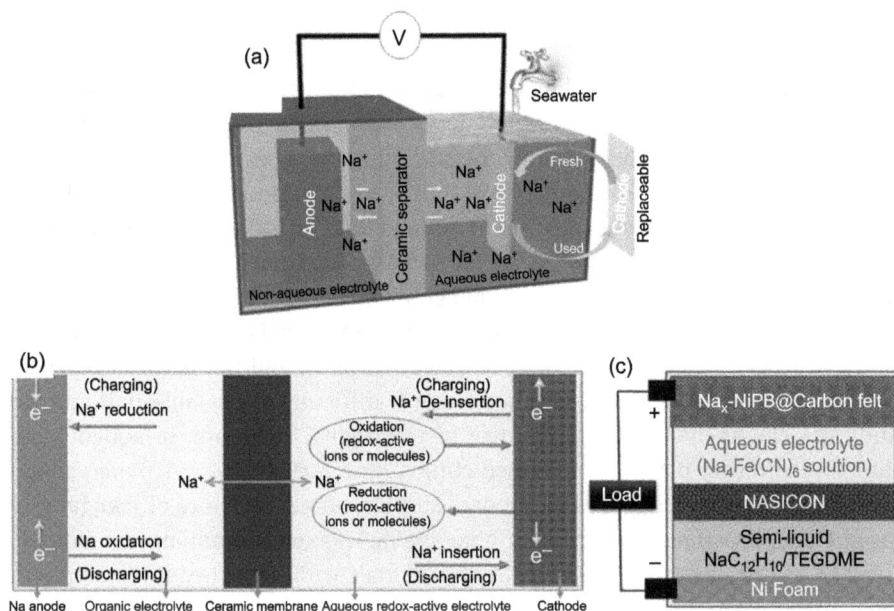

**Figure 5.27:** Schematic diagram of water complex electrolyte: organic–aqueous–Ises complex electrolyte [235, 236].

Recent studies have shown that composite solid electrolytes composed of ceramics and polymers can be used as electrolytes for SIBs with high safety and high electrochemical stability [33]. Due to the filling of NASICON solid ceramics, the combination of solid–solid electrolytes can also exhibit excellent ionic conductivity at 80 ℃ (2.4 mS cm$^{-1}$). At the same time, the electrolyte exhibits excellent flexibility due to the loading of organic polymers. Surprisingly, the interface resistance between electrolyte and electrode is reduced due to good contact and rapid ion transfer because polymer electrolytes with good mechanical properties can form a good transport path in the gap between active materials and solid electrolytes.

Generally speaking, because the mixed electrolyte can combine the advantages of different electrolytes, its design should follow the following requirements: ① the organic solvent as the main component increases the viscosity and ionic conductivity; ② the IL solvent as an additive improves the safety or modifies the solid–solid interface; ③ SPE as the substrate can improve the mechanical properties or as a modification to optimize the interface compatibility; ④ ISE as the main component improves the safety and Na$^+$ transport quantity, and ⑤ the aqueous electrolyte is used to improve the ionic conductivity of the electrolyte and reduce the cost. In the future research, we should not only pay attention to the composition and content of mixed electrolytes, but also try a variety of combination modes between different types of electrolytes.

## 5.6 Aqueous electrolyte

Low cost and high safety are the characteristics of aqueous sodium-ion batteries, which makes them suitable for future energy storage applications. At present, the study of aqueous electrolytes mainly focuses on two important aspects, including ion storage mechanism and interface stability in sodium-ion batteries [239]. The general aqueous electrolyte uses 1 M sodium sulfate or sodium nitrate as the sodium salt and deionized water (DW) as the solvent. The composition of the aqueous electrolyte has the advantages of high ion conductivity and nonflammability [240], but for aqueous electrolytes, the choice of electrode material is limited by the potential of hydrogen evolution and oxygen evolution [241], and the reaction between metallic sodium and DW is violent, making it difficult to assemble half-cells to study the electrochemical performance of electrodes. Therefore, in aqueous sodium-ion batteries, platinum, activated carbon, and $NaTi_2(PO_4)_3$ (NTP) are usually used as the counter electrodes or anodes, and the tunnel structure of manganese-based oxide, Prussian blue, Prussian blue analogues, and polyanionic compounds are used as cathodes [246–250]. Electrode materials with high chemical stability in water-based electrolytes can also be used as positive or negative electrodes [247], and organic electrode materials with high chemical stability is applied to partially replace the above materials in aqueous electrolytes. Due to kinetic reasons, the actual stability window of aqueous electrolytes is slightly larger than the thermodynamic limit, so more materials can be used in these systems [248].

The difference between aqueous and nonaqueous electrolytes lies in solvent molecules. In order to improve the electrochemical performance of aqueous sodium-ion batteries, it is necessary to understand the migration and storage process of sodium ions in aqueous electrolytes. The uniform film of $Na_2Ni[Fe(CN)_6]$ model electrode, studied the influence of water molecules on the intercalation process of sodium ions [249]. The insertion of sodium ions in the aqueous system is carried in three steps: first, an oxidation process of rapid electron transfer occurs on the electrode material; second, the charge-induced effect causes the anions to be trapped in the aqueous solution, thus balancing the instantaneous charge on the electrode. When the charge is transferred, the sodium ion is removed from the electrode; finally, the sodium ions and anions are removed from the electrode in an electrochemically inert step. Therefore, the anions separated from the sodium salt play an important role in the electrode reaction kinetics. In addition, the ion hydration effect and the size of cations are the main parameters that control the insertion potential of alkali metals (Figure 5.28). Therefore, more accurate models need to be established in the future to provide guidance for the design of aqueous batteries.

In order to determine the formation mechanism of the SEI film of the electrode in the aqueous electrolyte, some researchers conducted a series of studies on the solid–liquid interface evolution of the NTP electrode in a 1 M sodium sulfate aqueous

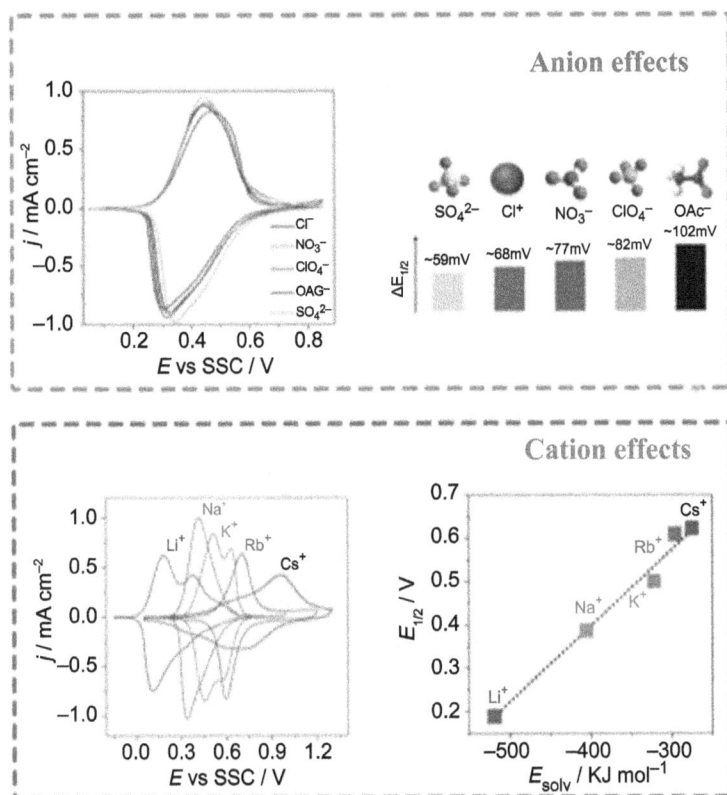

**Figure 5.28:** Influence of different cations and anions on electrochemical behavior in aqueous sodium-ion batteries.

solution [250]. Through the analysis of the Nyquist and Bode diagrams of the NTP electrode, there is no conventional SEI layer formed on the surface of the NTP electrode, but in the absence of electrochemical storage of sodium ions, the presence of insoluble components is found [251]. According to X-ray photoelectron spectroscopy (XPS) analysis (Figure 5.29(a)), these insoluble components are composed of amorphous transition metal phosphate layer and titanium sulfate. The slow kinetics and capacity decay are mainly due to the dissolution of $Na^+$ and $Ti^{4+}$ and the hydrolysis of surface groups. It is caused by the formed insulating phosphate layer (Figure 5.29(b)), in which sulfate ions will generate insoluble titanium sulfate on the electrode surface, which hinders the migration of ions and electrons.

However, titanium sulfate as a passivation layer also prevents continuous hydrolysis of the active electrode, but these large-sized precipitates also block the pores through which the electrolyte penetrates. After that, a series of work was carried out to study the interface stability under different conditions, revealing the influence of the anions in the sodium salt on the aqueous electrolyte. For example, by

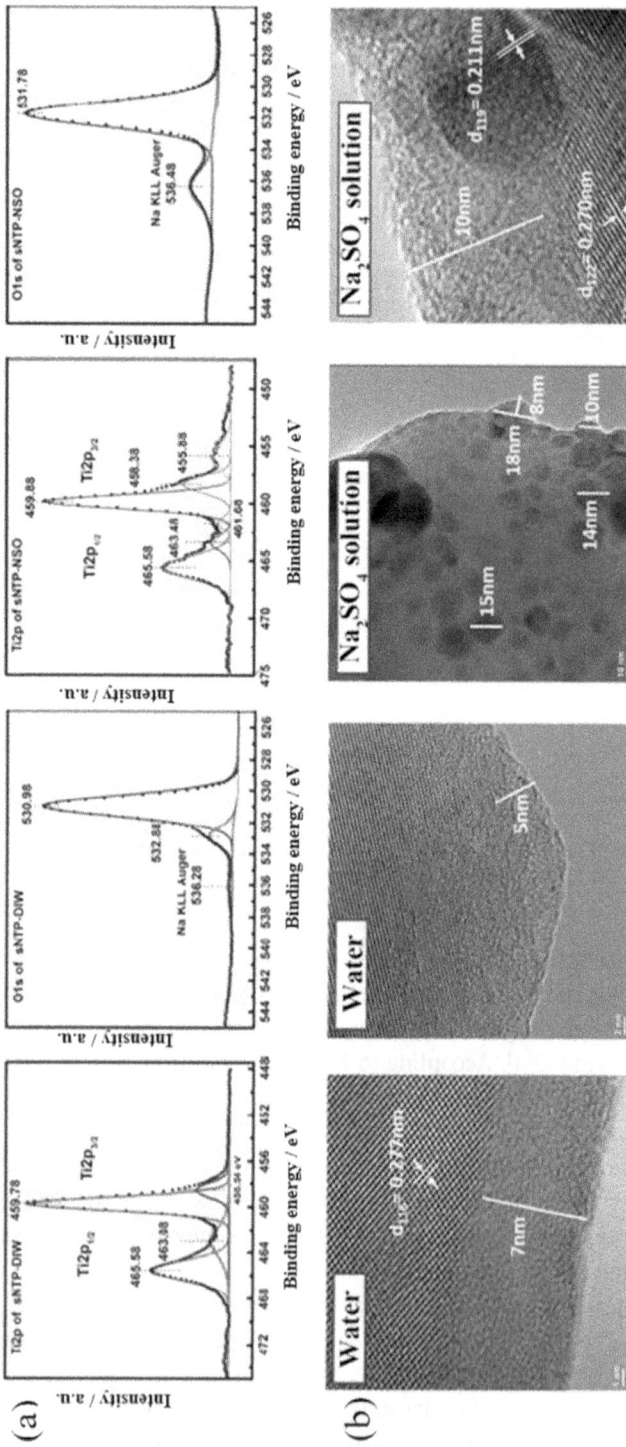

**Figure 5.29:** Study of interfacial evolution process of NTP electrode in 1 M aqueous Na$_2$SO$_4$ electrolyte: (a) XPS of NTP electrode surface; and (b) TEM pattern of NTP electrode surface in pure water and 1 M aqueous Na$_2$SO$_4$ solution.

matching 2 M $Na_2SO_4$, 4 M $NaNO_3$, and 4 M $NaClO_4$ with $Na_2FeP_2O_7$/NTP to assemble a full battery, compared with nonaqueous electrolytes, the cycle and rate performance of the battery are improved [252]. However, in this full battery, the hydrogen evolution and corrosion side reactions of the 4 M $NaNO_3$ electrolyte will cause a lot of irreversible capacity loss, which is mainly due to the decomposition of nitrate ions at high potentials to generate $HNO_2$ and $HNO_3$. Therefore, the anion of sodium salt plays an important role in interface stability and electrode side reactions.

In addition to the effects of anions and cations, the concentration of the electrolyte impacts great influence on ion conductivity and rate performance. Improving the concentration of the electrolyte is also an important method to improve the performance of aqueous electrolyte [253]. In the range of 1 M to 5 M, the ion conductivity increases and the rate performance of the NTP/$Na_{0.44}MnO_2$ battery is improved as the salt concentration increases (Figure 5.30(a), (b)). The high molar concentration of the electrolyte reduces the solubility of oxygen in the electrolyte and improves the stability of the electrolyte. The self-discharge caused by the diffusion of reactants determines the application potential of the battery. The dissolved oxygen in the aqueous electrolyte will produce self-discharge. Increasing the salt concentration can effectively inhibit the self-discharge phenomenon related to oxygen. This can be verified by the extended Open-circuit potential (OCP) curve of the charged anode (Figure 5.30(c)). However, high-concentration electrolytes have some other problems, such as corrosion, especially at extremely high electrochemical potentials. This problem can be solved by optimization in electrode protection or composition. Applying the concept of "water in salt" to aqueous sodium-ion batteries can also expand the ESW. The high concentration of $NaCF_3SO_3$ is dissolved in DW and assembled with $Na_{0.66}[Mn_{0.66}Ti_{0.34}]O_2$/NTP to assemble a full battery, which can form a stable sodium-ion-containing SEI film on the electrode surface, which can provide 2.5 V electrochemistry. The stability window is mainly through the strong interaction between cations and anions (Figure 5.30(d)) and significant ion aggregation, thereby increasing the reduction potential of anions and preventing water decomposition. At the same time, the high concentration of $NaCF_3SO_3$ aqueous solution can effectively inhibit the dissolution of NaF in the passivation layer.

Adding an appropriate dosage of additives to the aqueous electrolyte effectively improves the chemical and electrochemical stability of aqueous sodium-ion batteries [254]. Kumar et al. studied the effects of additives on the performance of aqueous electrolytes [255], and the results showed that adding 2 vol% VC to 10 M $NaClO_4$ aqueous electrolyte can increase the electrochemical window by 0 to 0.9 V (Figure 5.30 (e)), (f)). In addition to adjusting the pH of the electrolyte and continuously injecting nitrogen, VC additives can also be used to form a protective layer on the electrode surface to isolate the electrode from the reaction of $O_2$. This protective layer composed of VC degradation products can also prevent occurrence in high-concentration electrolytes. Therefore, the development of water-based electrolyte additives can promote the application of aqueous batteries.

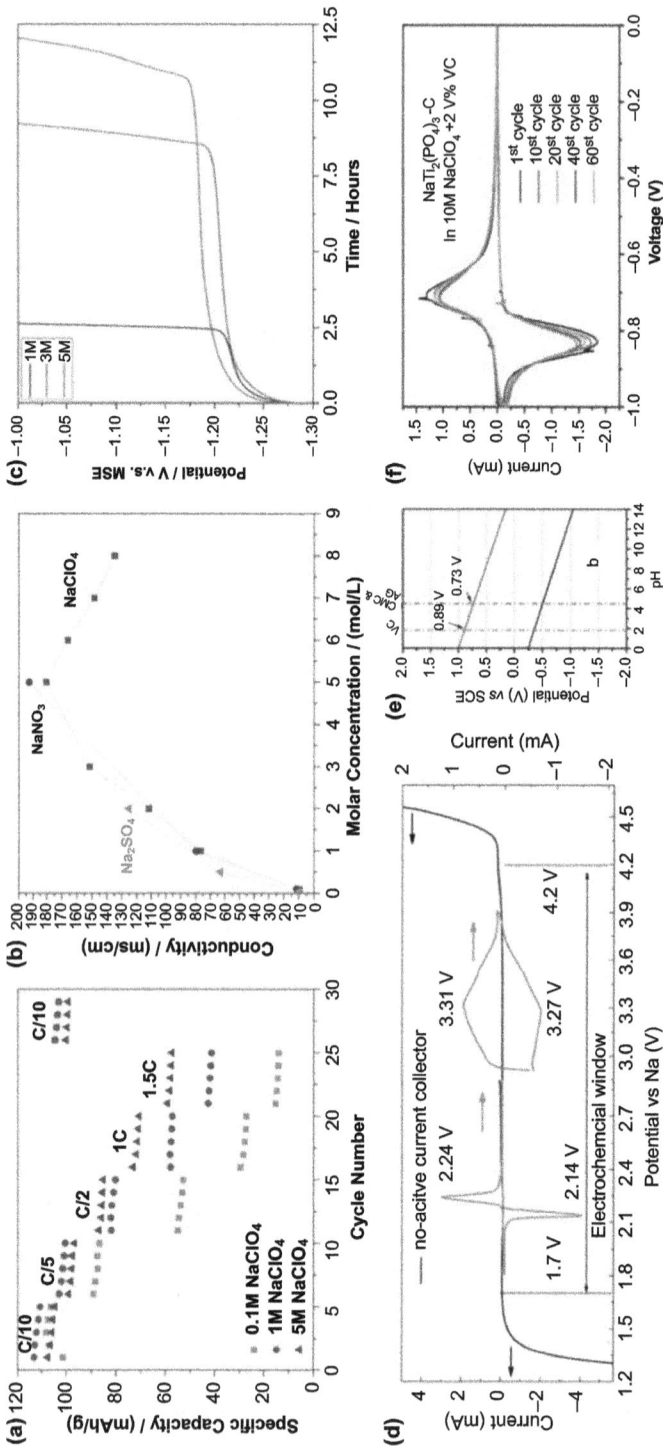

**Figure 5.30:** Study on the influence of electrolyte salt concentration and additives on the electrochemical performance of water-based batteries: the influence of electrolytes with different concentrations of sodium salt on (a) rate performance; (b) ionic conductivity; (c) OCP curve; (d) the electrochemical window of "water-in-salt" electrolyte; and (e) the influence of VC additive on the pH and ESW of aqueous electrolyte; and (f) CV curve of NTP electrode in 10 M NaClO$_4$ + 2 vol% VC electrolyte.

Compared with other electrolytes, the research on aqueous electrolytes is very limited, especially related reports on the solid—liquid interface. In fact, the passivation layer on the electrode surface in the aqueous medium not only improves the cycle stability, but also broadens the ESW. Factors such as the type of anion in the electrolyte, electrode material, and the solvent determine the ion transport and mechanical properties of the passivation layer. Therefore, in order to improve the aqueous electrolyte performance of sodium-ion batteries, the following aspects should be considered: ① strict control of solvent properties, including pH balance and removal of dissolved oxygen; ② select appropriate sodium salt and anions to form a stable low resistance on the electrode surface protective layer; and ③ add additives to improve interface stability and inhibit side reactions of high-concentration electrolytes.

## 5.7 Electrode—electrolyte interface

### 5.7.1 Basic characteristics of electrode—electrolyte interface

The process of embedding $Na^+$ in the electrolyte into the electrode can be divided into four steps: first, the solvated $Na^+$ is transported near the electrode surface; second, the $Na^+$ goes through the desolvation process to form free ions; third, the irreversible process that part of the $Na^+$ is captured by the dense inorganic layer to form $Na_2CO_3$ or NaF, covering the electrode surface is the reason for the low Coulomb efficiency in the first week; finally, $Na^+$ is embedded in the main structure of the electrode material, which is a reversible sodium storage process. Therefore, $Na^+$ can be embedded in the electrode material through SEI film, and the $Na^+$ diffusion barrier of this process may be the key factor determining the reaction kinetics of sodium storage process.

As shown in Figure 5.31, in the whole sodium-ion batteries, the formation of SEI on the electrode surface mainly comes from the decomposition of binder, the dissolution of electrode material, and the decomposition of electrolyte. The composition of these three parts directly affects the nature of SEI. SEI is mainly composed of two components, including organic components (polyester and polyether) and inorganic components (NaF). SEI needs to pay attention not only to its composition but also to its structure. The thickness, smoothness, and ordered/disordered structure of SEI will affect its stability and the transmission rate of $Na^+$ in it. In order to build a highly stable electrode, an artificial SEI layer can be designed to improve the Coulomb efficiency, reaction rate, and reversibility of the battery. Compared with the positive electrode, many special properties of the metal Na interface in the negative electrode of sodium-ion batteries are different from those of ordinary porous electrode, so it needs to be studied emphatically.

Due to the high reactivity of metal Na, the construction of stable SEI films is the key to the preparation of high-performance sodium-ion batteries [256]. The physical and chemical properties of $Na^+$ are different from Li + in ion radius, solvent

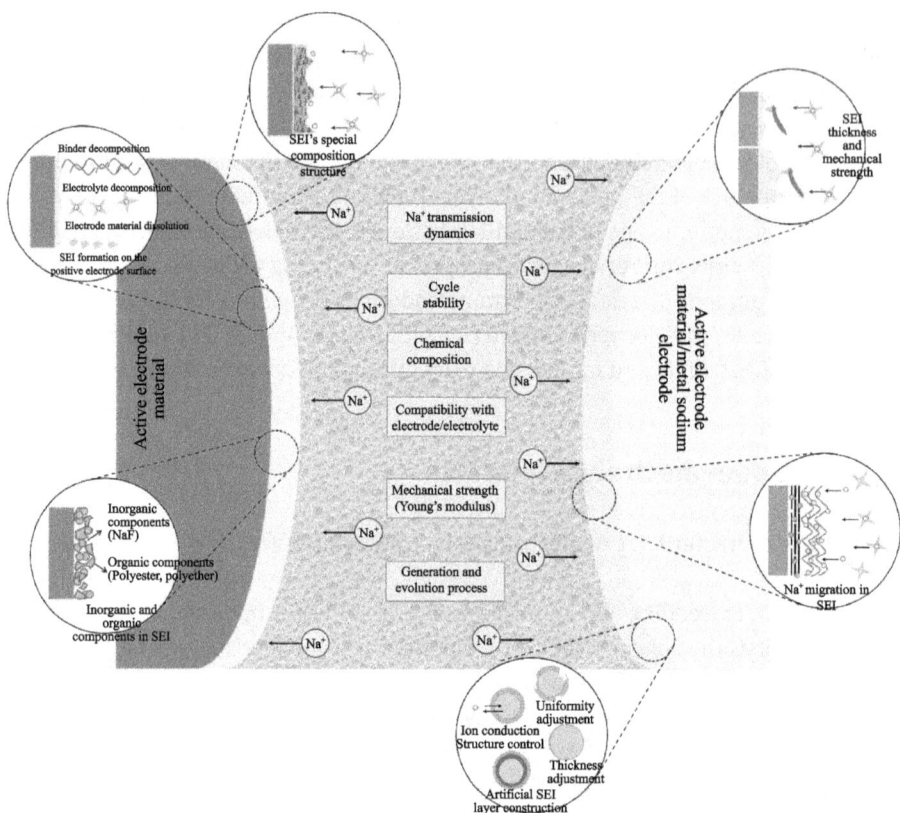

**Figure 5.31:** Schematic diagram of vital property and process of electrode–electrolyte interface in sodium-ion batteries.

properties, and redox potential. These differences make the SEI film formed in lithium-ion batteries and sodium-ion batteries have significant differences. The SEI film formed by 1 M $LiClO_4$ + PC electrolyte and electrode in lithium-ion battery is more stable than that formed by 1 M $NaClO_4$ + PC electrolyte and electrode in sodium-ion batteries [257]. Some experiments have confirmed that the redox potential of electrolytes using EC as solvent in sodium-ion batteries is higher than that in lithium-ion batteries, which indicates that electrolytes in sodium-ion batteries are more likely to deteriorate [258]. Studies on the transport mechanism of $Li^+/Na^+$ inorganic components (such as LiF, NaF, $Li_2CO_3$, and $Na_2CO_3$) in SEI membranes show that the $Li^+$ transport in Na-based SEI components follows the "knock-off" or "direct hopping" mechanism (knock or direct transmission mechanism), while the $Na^+$ transport in the Li-based SEI layer follows the "knock-off" or "vacancy-diffusion" mechanism (knock or vacancy diffusion mechanism). Therefore, it can be inferred that the ion transport in SEI membranes containing $Li^+$ and $Na^+$ components shows different processes, which is due to the kinetic differences caused by different ion sizes. In addition, the results show that the SEI film formed in sodium-ion batteries is mainly composed of inorganic compounds such as $Na_2O$, NaCl, NaF, or $Na_2CO_3$ [259]. On the contrary, the SEI film in lithium-ion battery is mainly composed of organic compounds. From the SEM images, it can be seen that the SEI films formed by sodium-ion batteries and lithium-ion batteries show different morphologies. The deposition layer of the former is rough and uneven, while the electrode surface of the latter is a uniform SEI film. Although the crystals LiF and NaF are stable components in the SEI films of the two kinds of batteries, their effects on defect thermodynamics, diffusion carrier concentration, and diffusion barrier are different. For NaF, the ionic conductivity of either positive or negative electrode is several orders of magnitude lower than that of LiF. Therefore, the inorganic components (such as NaF) play a decisive role in the electrochemical properties of SEI films in sodium-ion batteries.

In fact, in addition to the influence of cations, the choice of solvents also has an important influence on the optimization of SEI membranes. The thickness, composition, and properties of SEI films formed by organic electrolytes, ester-based, and ether-based solvents are also different. This phenomenon can be observed on both the carbon negative electrode and the sulfide electrode [260]. For the commonly used ester-based electrolytes, the structure of solvent molecules is also one of the main factors affecting the composition of SEI. Some researchers have conducted in-depth studies on the reduction mechanism of cyclic carbonates such as EC, PC, and FEC in sodium-ion batteries by simulation calculation. However, there is still a lack of research on chain carbonates. The purity of the solvent is also very important for the preparation of high-performance sodium-ion batteries. The impurities in the solvent decompose on the electrode surface to produce components that reduce the performance of SEI films [261].

The introduction of film-forming additives such as VC and FEC, can improve the performance of SEI membrane in sodium-ion batteries. When VC is added to 1 M $NaClO_4$ in PC electrolyte as an additive, the oxidized HC electrode will form a stable SEI film [10]. On the contrary, the same study found that FEC additive is beneficial to the formation of stable SEI film on the same electrode. In addition, FEC additive has been used in various anode materials and has good adaptability. The addition of FEC additive helps to produce excellent sodium carbonate and NaF passivation layer, which can reduce the decomposition rate of organic carbonate solvents [9]. F atom has a strong inducing effect on EC molecules. The reaction with $Na^+$ and ring-opening polymerization of FEC can form cross-linking networks of organic and inorganic salts and limit the dissolution of these compounds [10]. In addition, the FEC additive protects the PVDF binder from decomposition by replacing the consumption of F atoms, which ensures higher efficiency in the early cycle. However, VC additive can only reduce the formation energy of SEI film in sodium-ion batteries but not improve the cycle stability of the electrode. The role of additives in sodium-ion batteries has a second meaning, which can replace the decomposition of solvent molecules and change the decomposition products [262]. According to the quantum chemical simulation, the additive molecules with obviously high decomposition voltage will be decomposed before the solvent molecules, which can also improve the stability of the electrolyte. Because the reduction potential of FEC additive is higher than that of VC additive, FEC additive has a more significant inhibitory effect on electrolyte decomposition. At the same time, solvent molecules and additive molecules form dimers, resulting in new decomposition pathways corresponding to the formation of stable reduction products. In addition, NaF produced by FEC additive can further enhance the durability of SEI layer. Therefore, on the whole, FEC additive is more suitable for sodium-ion batteries.

### 5.7.2 Electrode–electrolyte interface model

The study of SEI films mainly focuses on the electrochemical window, crystal structure, surface properties, and chemical composition of different electrode–electrolyte interfaces. Therefore, the SEI films of different electrodes should be studied in detail by combining experiments and theories, so as to optimize the performance of the battery.

#### 5.7.2.1 SEI in hard carbon

As the most potential negative electrode, HC has high capacity and low voltage platform. The properties of SEI films play a key role in the reversibility and stability of the electrochemical reaction of the electrode [259]. Nowadays, various cyclic carbonate solvents have been widely used in HC negative electrode, among which EC and PC solvents show good electrochemical properties [263, 264]. In their early studies, Ponrouch et al. proposed that EC and PC electrolytes without FEC can produce uniform SEI films with higher conductivity on the electrode surface than those containing FEC [11]. The electrolyte containing FEC forms a dense passivation film on the electrode surface, which also increases the overpotential of the reaction, resulting in a larger irreversible specific capacity in the initial cycle (Figure 5.32(a)). However, the effects of additives in different electrolyte solvents tend to be different, for example, because FEC decomposes before the PC solvent participates in the formation of the passivation layer [10], PC electrolytes containing FEC additives can achieve a more stable cycle (Figure 5.32(b)). From the CV curve of the Al foil electrode, it can be seen that the oxidation and the reduction peak corresponding to the peak current are very large (Figure 5.32(c)), which indicates that the addition of FEC in the electrolyte promotes the reversible intercalation/detachment of metal Na.

Through the in-depth study of electrolyte stability and SEI formation, the researchers believe that the different effects of FEC on electrode performance in the

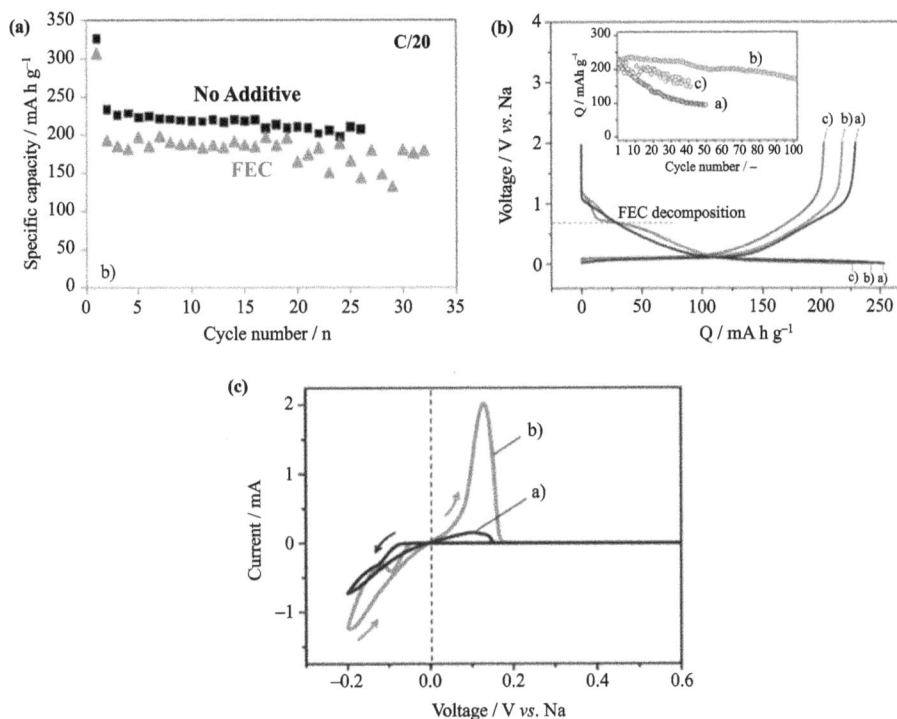

**Figure 5.32:** Different performance of hard carbon with and without electrolyte additives:
(a) cycling performance of HC electrode measured in EC:PC electrolyte solution with 1 M NaClO$_4$
with/without FEC addition; (b) charge/discharge curve and cycling of HC electrode measured in PC
electrolyte solution with 1 M NaClO$_4$ (with/without FEC addition) performance (inset); and (c) CV
curve of the second cycle of Al foil electrode with electrolyte of FEC added/not added 1 M NaClO$_4$
in PC solution with a scan rate of 3 mV min$^{-1}$.

above two experiments are mainly due to the following three reasons: ① the decomposition voltage of 1 PC is different from that of the mixture of PC and EC; ②
the quality of SEI mainly depends on the purity of electrolyte; and ③ the roughness and composition of HC electrode are different. Some researchers have proposed that the effect of FEC additive on the film-forming property of SEI is also
related to the binder used. Because FEC can inhibit the decomposition of PVDF
binder, it shows effective film-forming performance for HC electrode with PVDF
binder [265]. When Sodium Carboxymethyl Cellulose (CMC) is used as a binder, the
effect of FEC additives is less obvious in comparison (Figure 5.33). Studies on sodium
salts and FEC additives using hard and soft XPS) show that F-ions can help to form a
thin passivation layer, so the electrolytes containing NaPF$_6$ and FEC show the best reversibility and capacity retention. To sum up, electrolyte composition, solvent purity,
electrode performance, and additives are all important factors that determine the quality of SEI layer on the surface of HC.

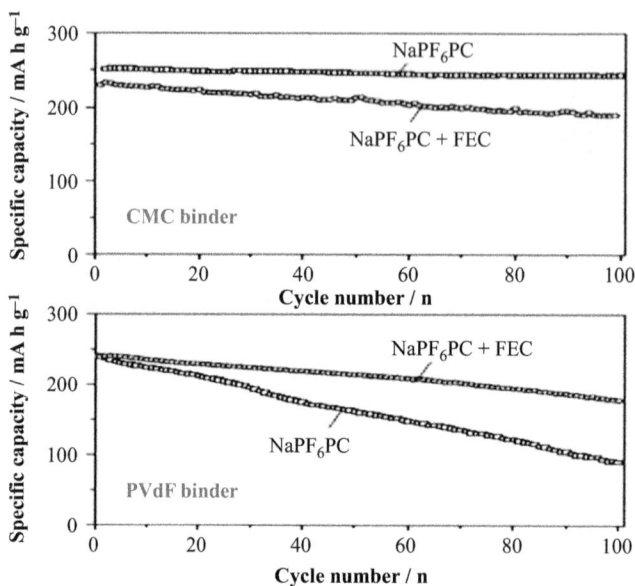

**Figure 5.33:** Cycling performance of HC electrodes with or without FEC additives, using $NaPF_6$ and $NaClO_4$ as sodium salts, respectively.

The researchers also analyzed the composition and structural evolution of SEI films on the surface of HC electrodes in 1 M $NaClO_4$ EC:DEC electrolytes by XPS [266]. The characteristic peaks in XPS spectra belong to sp2 carbon and-$CH_2$-, ester bond. RO-$CO_2Na$, $Na_2CO_3$, and-$CF_2$-, show that there is HC in the graphene phase, SEI layer, and PVDF binder at the interface. Among them, alkali metal carbonate, alkyl carbonate, and polymer are considered to be important components of SEI membrane. In addition to the carbonate electrolyte, the irreversible voltage platform corresponding to the formation of SEI film on the HC electrode can also be observed in the IL electrolyte. Then the study on the film-forming ability of SEI on the negative electrode of HC shows that the excellent film-forming ability of superconcentrated salt electrolyte is produced by the reduction of anions, and the electrolyte can effectively improve the capacity and cycle stability of the electrode. Therefore, it can be inferred that the SEI on the HC electrode is not only related to the kinds of solvents and additives, but also affected by the sodium salt concentrate. The latest research shows that the main cause of capacity attenuation is the continuous formation of SEI rather than the degradation of HC materials. The gradual thickening of SEI layer leads to the slow reaction kinetics of HC negative electrode, so the construction of highly stable thin SEI film is the key to realize high-performance HC negative electrode in the future.

### 5.7.2.2 SEI in other carbon materials

In addition to HC electrode, HSSAC materials such as reduced graphene oxide (rGO), activated carbon (AC), and ordered mesoporous carbon (CMK-3) can store sodium reversibly in thermodynamics. However, the high-specific surface area of these materials will lead to serious interface problems, thus reducing the cycle life of the battery. Therefore, compared with HC electrodes, the performance of SEI films on these high-specific-surface-area carbon (HSSAC) electrodes is more important. The results show that the composition of SEI membrane on HSSAC electrode can be effectively improved by using ether-based electrolyte, and the first-week Coulomb efficiency and cycle stability can be effectively improved. The SEI membrane formed by ether-based electrolyte shows ultrafast electron and ion transfer kinetics. The XPS analysis of rGO electrodes in different electrolytes shows that due to the low content of polyether uniformly deposited outside the SEI layer, the SEI films observed in ether-based electrolytes are thin and dense (Figure 5.34). On the contrary, there are a large number of polyester outside the SEI film of the ester-based electrolyte, and the whole SEI film is thick and loose. In addition, thermodynamically unstable $Na_2CO_3/Na_2CO_2R$ [267] was detected in the ester system, while F-C (sp2), which can improve the reversible electrochemical activity of surface defects, was detected in the ether system.

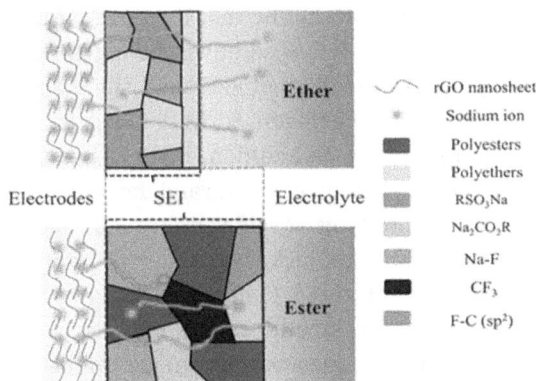

**Figure 5.34:** Diagrammatic representation of the different components of the SEI layer formed on the surface of HSSAC electrodes in ether- and ester-based electrolytes.

### 5.7.2.3 SEI on Ti-based anodes

SEI film on the titanium-based negative electrode also has a significant effect on the electrochemical performance of the Ti-based negative electrode. Because of the poor electronic conductivity of titanium-based negative electrodes, the structures of Ti-based negative electrodes, including $TiO_2$, $Na_2Ti_3O_7$ (NTO), and $NaTi_2(PO_4)_3$ (NTP), are usually combined with different carbon sources to solve the problem of low conductivity. Anatase $TiO_2$ (A-$TiO_2$), which is used as sodium storage negative electrode, shows low first-week Coulomb efficiency due to obvious irreversible film-formation reaction. Subsequently, the Coulomb efficiency increased gradually, indicating that there was an activation process in both SEI membrane and A-TiO$_2$ electrode, and the rapid increase in efficiency and capacity further reflected that the activation rate of SEI membrane was relatively faster [268]. The activation process of A-TiO$_2$ is the capture of Na$^+$ and the formation of sodium-containing compounds. The stability of the SEI film formed during the cycle is related to the type of electrolyte selected. When the nano-sized A-TiO$_2$ electrode is matched with the ether-based electrolyte, a thin and strong SEI film can be formed on the amorphous TiO$_2$ negative electrode [269]. Moreover, in the process of charge and discharge, uniform SEI film can cause pseudo-capacitive sodium storage effect and improve the kinetics of electrode reaction.

At present, researchers have constructed different kinds of electrolytes by using different kinds of sodium salts, solvents, and additives and matched them with A-TiO$_2$ electrodes, thus forming SEI films with different properties [270]. The electrolyte with NaPF$_6$ as sodium salt shows the most stable cycle performance and higher Coulomb efficiency (Figure 5.35(a)), but the specific capacity is only 120 mA h g$^{-1}$, which is much lower than that of electrolyte with NaTFSI and NaClO$_4$ as sodium salt. Comparatively speaking, the electrolyte with NaClO$_4$ as sodium salt has higher specific capacity and rapid activation process, so it can be considered as the most suitable sodium salt for A-TiO$_2$ electrode. It is worth noting that the three electrolytes using PC, EC:PC and EC:DMC as solvents and FEC as additives differ greatly in the Coulomb efficiency in the first week, and the addition of FEC has a slight negative effect on the improvement of Coulomb efficiency (Figure 5.35(b)), which is consistent with the previous research conclusions [11]. In addition, there is a certain correlation between the morphology of the TiO$_2$/carbon electrode and the properties of the SEI layer. The TiO$_2$/carbon electrode with large specific surface area shows high SEI content, but the cycle stability is poor; the thickness of the SEI layer increases with the carbon content in the electrode (Figure 5.35(c)), thick SEI layer limits the diffusion of Na$^+$ and reduces the reaction kinetics of the TiO$_2$ electrode. According to the above conclusions, the atomic arrangement technique can be applied to the surface chemical improvement and interface modification of titanium dioxide electrode.

The performance of NTO electrode is closely related to the properties of electrolyte and electrode–electrolyte interface. From the matching of different sodium salts (NaClO$_4$, NaFSI) and solvents (EC:DMC, EC:DEC, PC) with NTO electrodes (Figure 5.36), it can be seen that sodium NaClO$_4$ exhibits a very low Coulomb efficiency and capacity retention in all three electrolytes. In 1 M NaFSI $\cdot$ PC electrolyte, the initial Coulomb efficiency and cycle performance of NTO electrode were improved. However, since the stability of the SEI layer is directly related to the capacity attenuation of the battery, the film-forming additive can further improve the interface impedance and cycle stability of the NTO electrode with large surface area [271, 272]. FEC additives contribute to the formation of stable oxide-rich compounds and inert

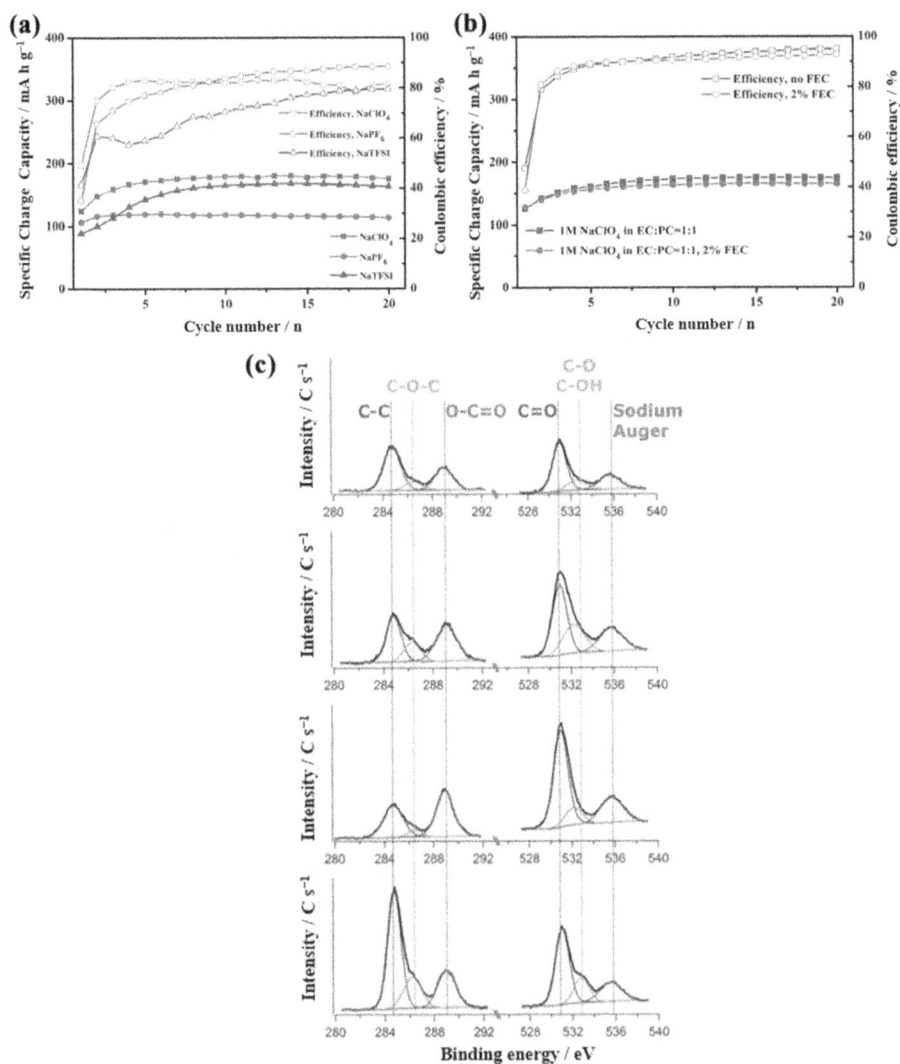

**Figure 5.35:** Performance effects of $TiO_2$ cathode with and without electrolyte additives: (a) electrochemical properties of A-$TiO_2$ electrode measured in PC electrolyte using different sodium salts; (b) electrochemical properties of A-$TiO_2$ electrode measured in electrolyte with/without FEC addition; and (c) C 1s (left) and O 1s (right) of A-$TiO_2$/C cathode with different carbon contents after 30 cycles.

components of NaF in the SEI layer. The SEI film on the interface is mainly composed of carbonate, semi-polycarbonate, NaF, and polyethylene oxide. Among them, oxygen-rich compounds play an important role in the stability of binders and SEI films. It is certain that the optimization of the interface layer of the NTO electrode may reduce the catalytic activity of NTO and improve the stability of the electrode and electrolyte. At present, some researchers have obtained the ideal sodium storage characteristics by modifying the surface of NTO electrode. The NTP electrode is tested in a high-voltage range of 1 V to 3 V without forming a SEI film, thus achieving ultra-stable cycle performance and ultra-high Coulomb efficiency [273]. However, the energy density of NTP is limited by low theoretical capacity and high operating voltage. Therefore, the ESW of NTP is expanded from 0.01 V to 3 V, which can achieve a high specific capacity of about 210 mA h g$^{-1}$, which corresponds to the two-step electron transfer process between Ti$^{2+}$, Ti$^{3+}$, and Ti$^{4+}$. When the EC:PC electrolyte is applied to the system, a new short platform is observed near 2.0 V, which is related to the formation of a passivation layer on the surface of the negative electrode. On the contrary, no polarization step was observed in pure PC solvent, which indicated that the SEI film was not formed.

### 5.7.2.4 SEI in organic electrodes

Due to the characteristics of low cost, simple synthesis and high redox activity, organic electrode materials have attracted more and more attention of researchers [274]. However, because these materials are easily soluble in organic electrolytes, the cycle stability of electrodes is poor. To solve this problem, the relationship between organic electrodes and organic electrolytes was analyzed by XPS [275]. The results show that the substances in the SEI film formed in the battery pack cycle are mainly inorganic compounds, and their properties are determined by these inorganic compounds. In the process of charge and discharge, the continuous degradation of these salts leads to the decrease of cycle stability. On the contrary, the SEI film in lithium-ion battery is gradually formed in the process of cycle, and the main component of SEI film is organic matter. But similarly, with the increase of cycle number, the solvent will be degraded to produce inorganic substances, while the battery capacity will decline, which has been confirmed by the researchers through XPS (Figure 5.37). Therefore, adjusting the composition of SEI layer is the key to limit the dissolution of electrodes, and the problem of dissolution of organic electrodes can also be fundamentally solved by using ILs or solid electrolytes.

### 5.7.2.5 SEI in alloy anodes

The alloy negative electrode has been widely concerned by the research community because of its high theoretical specific capacity and low redox potential. However, the significant volume expansion during the alloying reaction will lead to the crushing of the electrode material and the rupture of the SEI film. Therefore, the systematic study of the passivation layer on the alloy negative electrode is very important to achieve excellent cycle stability. XPS measurements on the surface of Sn electrode show that the SEI layer is mainly composed of carbonates (Na$_2$CO$_3$ and Na$_2$CO$_3$R), and the decomposition of electrolyte can be attributed to Sn$^{4+}$ covered nano-Sn particles rather than Sn$^{2+}$ [276]. The researchers used the quantitative data of XPS to

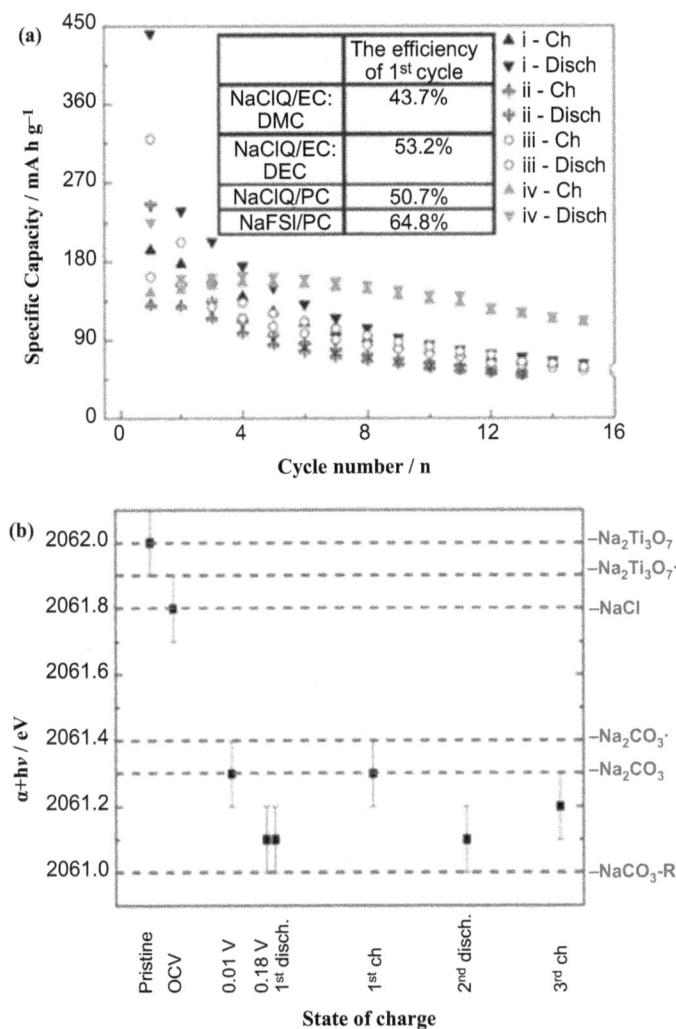

(a)

| | The efficiency of 1st cycle |
|---|---|
| NaClO/EC: DMC | 43.7% |
| NaClO/EC: DEC | 53.2% |
| NaClO/PC | 50.7% |
| NaFSI/PC | 64.8% |

▲ i - Ch
▼ i - Disch
✦ ii - Ch
✦ ii - Disch
○ iii - Ch
○ iii - Disch
▲ iv - Ch
▼ iv - Disch

(b)

Figure 5.36: Effect of electrolyte additives in NTO anode on performance: (a) effect of electrolyte on cycling performance of NTO electrode; and (b) Auger parameters applicable to NTO electrode with different charging/discharging states.

study the main components of SEI film and its evolution during Sb/Na cell cycle. It is proved that the main components of SEI layer are organic carbonate, sodium carbonate, and alkyl carbonate [277].

In addition to the above studies, the XPS results of the recycled black phosphorus electrode show that the organic and inorganic substances produced by the decomposition of solvents and additives coexist in the SEI membrane [267]. The experimental results show that FEC and VC additives can form a stable and uniform

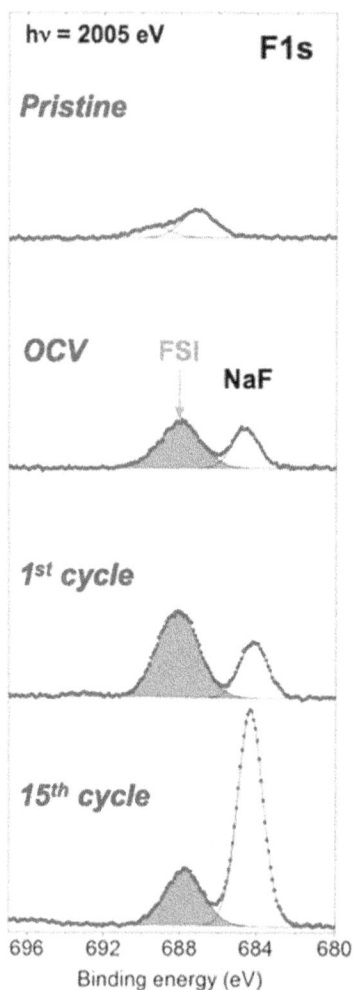

**Figure 5.37:** F 1s of metal Na electrodes measured at different charge/discharge states.

SEI film on the surface of black phosphorus electrode, while the electrolyte without additives forms a thicker SEI layer, in which the passivation layer formed by VC does not have the effect of electrode protection. In fact, the composition of the SEI film formed by the electrolyte with FEC and VC as additives is different. The surface layer of the SEI film of the former is mainly composed of inorganic compounds, while the latter is composed of organic and inorganic compounds. Among the electrolytes containing FEC, NaF, $Na_2CO_3$, and olefins are the main components of SEI membrane. At the same time, the oxygen-containing polymer with high molecular weight in VC electrolyte plays a good role in optimizing the interface and improving the stability of the electrode. The different results of the two additives can be attributed to the strong reducibility and multielectron reaction process of black phosphorus electrode.

However, it is certain that the SEI membranes modified by FEC and VC additives can achieve longer cycle life and higher reaction reversibility.

### 5.7.2.6 SEI in cathodes

Because the electrolyte will decompose under the condition of high voltage, the passivation layer will also be formed on the surface of the positive electrode. Although little attention has been paid to the passivation layer on the positive surface, (CEI) has a great influence on the performance of sodium-ion batteries. Take the PBAs cathode as an example, when it is used in the whole battery, the interfacial compatibility between PBAs and electrolyte should be considered first [278]. However, a relatively stable CEI can be formed by adding FEC to the organic solution. Sulfone-based electrolytes can also be used in PBA positive electrodes and are more stable at high voltage, and the electrode–electrolyte interface is optimized [279]. In addition to external factors, the trace amount of residual water in electrodes and electrolytes is an internal factor that cannot be ignored. Because the side reaction between interstitial water and organic electrolyte produces $Na_2CO_3$, which not only protects the electrode from electrolyte corrosion, but also promotes charge transfer at the interface [280].

Recently, the solid–liquid interface of TMO electrode–electrolyte has received a lot of attention [281]. It is found that the Mn on the oxide surface of the electrolyte with mixed organic solvent is reduced and an artificial CEI layer is formed. The protective layer consists of reduced transition metal cations and metal-organic compounds, which not only maintains the chemical stability of the particles in the atmospheric environment, but also improves the electrochemical stability of the electrode during charge-discharge. Similarly, the polyanion cathode can form a CEI layer at high potential [230]. When IL is introduced into conventional organic electrolytes, a thin and highly conductive CEI layer is formed on the electrode surface to reduce the charge transfer resistance. However, the introduction of IL may increase the impedance of metal Na surface, indicating that there is an equilibrium point in the use of mixed electrolytes. In the future, the importance of CEI should be further studied in order to manufacture high-performance sodium batteries.

### 5.7.3 Study on the modification of electrode–electrolyte

Artificial SEI film or preformed passivation layer, such as $Al_2O_3$ coating, can effectively improve the interfacial properties. The coating with appropriate thickness can reduce the thickness of the SEI film and the continuous consumption of the electrolyte during the subsequent cycle. $Al_2O_3$ coating is an excellent inorganic passivation film, which is suitable for the modification of positive and negative electrodes of sodium-ion batteries [282]. As shown in Figure 5.38(a), the theoretical relative energy level calculation reveals the action mechanism of the $Al_2O_3$ coating. Because the redox potential ($\mu A$) of the negative electrode is higher than the LUMO value of each component in the electrolyte, the solvent will decompose on the surface of the negative electrode. However, this process can be prevented by forming a passivated SEI layer on the surface of the electrode. Similarly, the positive electrochemical potential ($\mu C$) below the HOMO value will lead to electrolyte oxidation. According to the DFT calculation results, the $Al_2O_3$ coating has appropriate band gap and $Na^+$ conductivity, which can effectively prevent electrolyte decomposition, thus reducing the consumption of $Na^+$. At the same time, the $Al_2O_3$ coating has a similar effect to the SEI film, transporting $Na^+$ and preventing electron transfer into the electrolyte (Figure 5.38(b)).

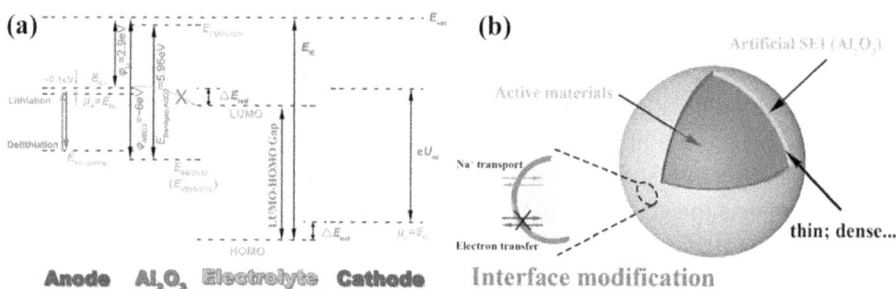

**Figure 5.38:** Mechanism of SEI formation on the electrode surface: (a) energy-level diagram in a typical Lithium-ion batteries (LIBs) using $Al_2O_3$ coating; and (b) diagram of artificial SEI ($Al_2O_3$) modified on the active material as a dense protective layer.

The preparation methods of $Al_2O_3$ coating include wet chemical method, sol–gel method and atomic layer deposition (ALD) method. The ALD method can control the thickness of the $Al_2O_3$ coating, thus shortening the sodium-ion diffusion distance. However, the ALD method is difficult to guarantee the quality of the coating, and it is even more difficult to realize industrial application. Therefore, the simple and cheap sol–gel method and chemical precipitation method are considered to be effective alternatives to the ALD method. At present, the method of coating $Al_2O_3$ on the surface of negative and positive electrodes has achieved good experimental results. As shown in Figure 5.39(a), the $Al_2O_3$ protective layer was uniformly coated on the surface of $Na_{2/3}[Ni_{1/3}Mn_{2/3}]O_2$ by chemical method. A high-quality CEI film was formed on the surface of $Al_2O_3$ during charge and discharge, which suppressed the side reaction at high voltage, so the capacity retention of the electrode was significantly improved after 300 cycles. In addition, the ultrathin $Al_2O_3$ coating (1–2 nm) coated on the organic negative electrode by ALD method plays the role of SEI substitution, which significantly reduces the consumption of $Na^+$ in the initial cycle and improves the Coulombic efficiency in the first week. This dense passivation layer acts like a high-quality SEI film to prevent pulverization and dissolution of the material. The $Al_2O_3$ coating also shows good mechanical properties, which is used to alleviate the volume change caused by the expansion and shrinkage of Sn nanoparticles (SnNPs) in the negative electrode of Sn, which can not only avoid the continuous growth of SEI film on the electrode surface (Figure 5.39(b)), but also prevent the pulverization of active materials. Due to the protective effect of $Al_2O_3$ coating on the electrode structure, the specific capacity of Sn electrode remained as high as 650 mA h $g^{-1}$ after 40 cycles. To sum up, $Al_2O_3$ coating can be used as a substitute for SEI and CEI films in sodium-ion batteries.

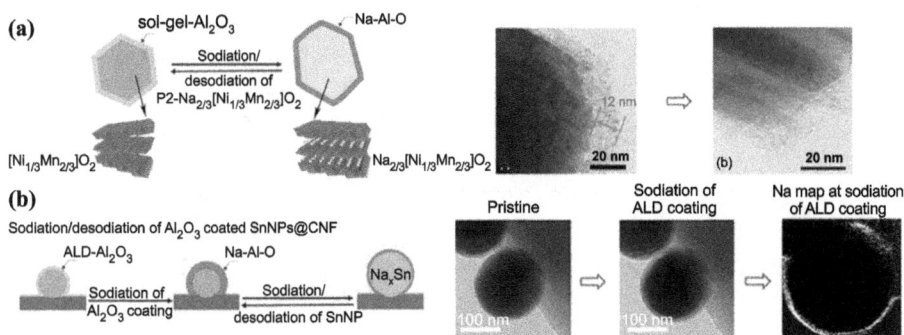

**Figure 5.39:** Artificial SEI film design: (a) illustration of sodium storage in $Al_2O_3$-coated P2-$Na_{2/3}$[$Ni_{1/3}Mn_{2/3}$]$O_2$ and corresponding TEM images; and (b) illustration of sodium storage in $Al_2O_3$-coated SnNP and corresponding TEM images.

Although $Al_2O_3$ has been successfully applied to most of the negative and positive electrodes of SIBs, other artificial SEI films still need to be studied to meet various needs. For example, in lithium-ion batteries, coating $Li_3PO_4$ on the positive electrode of high-voltage layered TMO can significantly limit the dissolution of transition metal ions and reduce the consumption of $Li^+$ in irreversible reactions [283, 284]. Similarly, sodium phosphate ($NaPO_3$) nano-layer (about 10 nm thickness) coated on P2 type $Na_{2/3}$[$Ni_{1/3}Mn_{2/3}$]$O_2$ can also provide a good ion transport pathway [285]. As shown in Figure 5.40(a), the thin $NaPO_3$ nano-layer can effectively block the corrosion of HF and $H_2O$ on the electrode surface and inhibit the formation of $Mn_3O_4$ on the electrode surface. The $\beta$-$NaCaPO_4$ coating shows the same effect at high voltage (Figure 5.40(b)) [286], which can effectively inhibit the dissolution of lattice oxygen and make the electrode exhibit a highly reversible sodium storage process at high rates.

In addition to coating the active material with inert components, the inorganic solid electrolyte material can also be coated on the surface of the electrode as a modification layer. The $Na_{2.9}PS_{3.95}Se_{0.05}$ electrolyte with $1.21 \times 10^{-4}$ S $cm^{-1}$ excellent ionic conductivity modified Fe1-xS electrode by in situ liquid phase method (Figure 5.41). The good interfacial compatibility between the solid electrode and the solid electrolyte reduces the interfacial resistance and reduces the polarization of

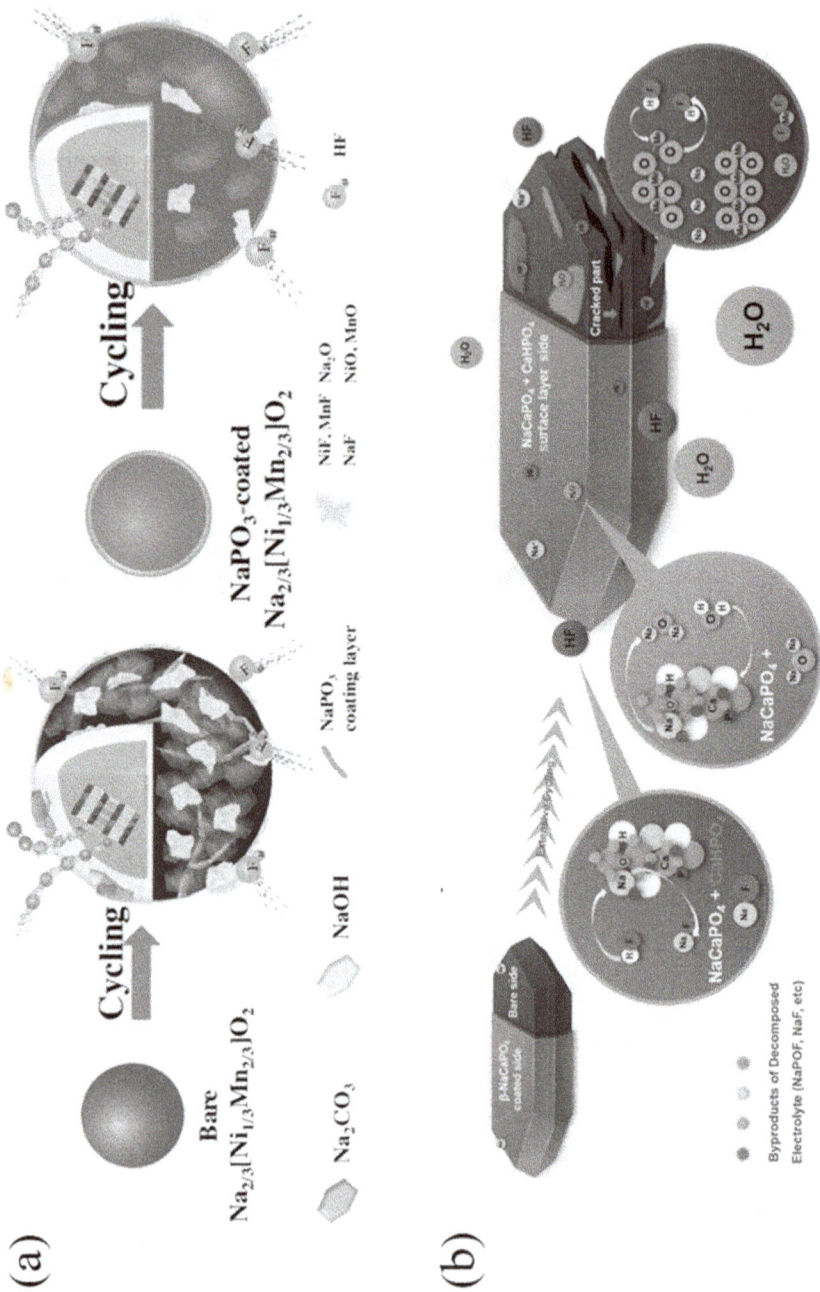

**Figure 5.40:** Design of artificial SEI film:(a) NaPO$_3$ layer coated on Na$_{2/3}$[Ni$_{1/3}$Mn$_{2/3}$]O$_2$; (b) Schematic diagram of the structure and protective mechanism of β-NaCaPO$_4$ coating on Na$_{2/3}$[Ni$_{1/3}$Mn$_{2/3}$]O$_2$.

the whole cell [287]. Recently, researchers have prepared a new type of toothpaste electrode attached to the surface of solid electrolyte, which shows good interfacial compatibility [233]. In the working process, the interface between electrolyte and electrode can be kept moist by IL smearing, and a stable passivation layer can be obtained at the same time. This experiment provides us with innovative ideas to improve the electrode–electrolyte interface in solid electrolytes. In the future, more attention will be paid to the interface properties of sodium-ion batteries and the commercialization of electrode materials. The improvement of initial Coulomb efficiency and cycle stability will be the key to the practical application of sodium-ion batteries. In the future, it is expected to improve the artificial SEI membrane technology to solve the above problems and promote the development of sodium-ion batteries.

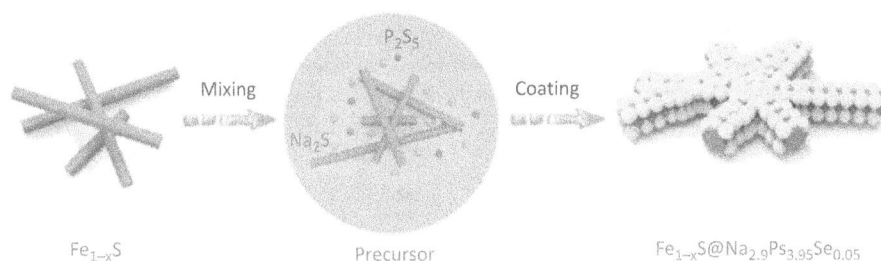

**Figure 5.41:** Artificial SEI film design: ISE layer coated on $Fe_{1-x}S$ ($Na_{2.9}PS_{3.95}Se_{0.05}$).
Note: The inset shows the relevant images and membrane performance of the bare and coated active material.

# References

[1]   Xu K. Electrolytes and Interphases in Li-Ion Batteries and Beyond[J]. Chemical Reviews, 2014, 114(23): 11503–11618.
[2]   Eshetu G G, Grugeon S, Kim H. et al., Comprehensive Insights into the Reactivity of Electrolytes Based on Sodium Ions[J]. ChemSusChem, 2016, 9(5): 462–471.
[3]   Kühnel R, Lübke M, Winter M. et al., Suppression of Aluminum Current Collector Corrosion in Ionic Liquid Containing Electrolytes[J]. Journal of Power Sources, 2012, 214: 178–184.
[4]   Yamada Y, Chiang C H, Sodeyama K. et al., Corrosion Prevention Mechanism of Aluminum Metal in Superconcentrated Electrolytes[J]. ChemElectroChem, 2015, 2(11): 1627.
[5]   Ponrouch A, Marchante E, Courty M. et al., In Search of an Optimized Electrolyte for Na-ion Batteries[J]. Energy & Environmental Science, 2012, 5(9): 8572–8583.
[6]   Ding C, Nohira T, Hagiwara R. et al., Na[FSA]-[$C_3C_1$Pyrr][FSA]Ionic Liquids as Electrolytes for Sodium Secondary Batteries: Effects of Na Ion Concentration and Operation Temperature[J]. Journal of Power Sources, 2014, 269: 124–128.
[7]   Xue Y, Quesnel D J. Synthesis and Electrochemical Study of Sodium Ion Transport Polymer Gel Electrolytes[J]. RSC Advances, 2016, 6(9): 7504–7510.

[8]   Ma Q, Guin M, Naqash S. et al., Scandium-Substituted $Na_3Zr_2(SiO_4)_2(PO_4)$ Prepared by a Solution-Assisted Solid-State Reaction Method as Sodium Ion Conductors[J]. Chemistry of Materials, 2016, 28(13): 4821–4828.

[9]   Webb S A, Baggetto L, Bridges C A. et al., The Electrochemical Reactions of Pure Indium with Li and Na: Anomalous Electrolyte Decomposition, Benefits of FEC Additive, Phase Transitions and Electrode Performance[J]. Journal of Power Sources, 2014, 248: 1105–1117.

[10]  Komaba S, Ishikawa T, Yabuuchi N. et al., Fluorinated Ethylene Carbonate as Electrolyte Additive for Rechargeable Na Batteries[J]. ACS Applied Materials & Interfaces, 2011, 3(11): 4165–4168.

[11]  Ponrouch A, Goñi A R, Palacín M R. High Capacity Hard Carbon Anodes for Sodium-ion Batteries in Additive Free Electrolyte[J]. Electrochemistry Communications, 2013, 27: 85–88.

[12]  Zeng Z, Jiang X, Li R. et al., A Safer Sodium Ion Battery Based on Nonflammable Organic Phosphate Electrolyte[J]. Advanced Science, 2016, 3(9): 1600066.

[13]  Wang J, Yamada Y, Sodeyama K. et al., Fire-extinguishing Organic Electrolytes for Safe Batteries[J]. Nature Energy, 2018, 3(1): 22–29.

[14]  Feng J, An Y, Ci L. et al., Nonflammable Electrolyte for Safer Non-Aqueous Sodium Batteries [J]. Journal of Materials Chemistry A, 2015, 3(28): 14539–14544.

[15]  Wongittharom N, Lee T, Wang C. et al., Electrochemical Performance of $Na/NaFePO_4$ Sodium-ion Batteries with Ionic Liquid Electrolytes[J]. Journal of Materials Chemistry A, 2014, 2(16): 5655.

[16]  You Y, Yao H, Xin S. et al., Subzero-Temperature Cathode for a Sodium Ion Battery[J]. Advanced Materials, 2016, 28(33): 7243–7248.

[17]  Ding M S, Richard J T. How Conductivities and Viscosities of PC-DEC and PC-EC Solutions of $LiBF_4$, $LiPF_6$, LiBOB, $Et_4NBF_4$, and $Et_4NPF_6$ Differ and Why[J]. Journal of the Electrochemical Society, 2004, 151(12): A2007.

[18]  Lalère F, Leriche J B, Courty M. et al., An All-solid State NASICON Sodium Battery Operating at 200 °C[J]. Journal of Power Sources, 2014, 247: 975–980.

[19]  Wei T, Gong Y, Zhao X. et al., An All-Ceramic Solid-State Rechargeable $Na^+$-Battery Operated at Intermediate Temperatures[J]. Advanced Functional Materials, 2014, 24(34): 5380–5384.

[20]  Qu X, Jain A, Rajput N N. et al., The Electrolyte Genome Project: A Big Data Approach in Battery Materials Discovery[J]. Computational Materials Science, 2015, 103: 56–67.

[21]  Kuratani K, Uemura N, Senoh H. et al., Conductivity, Viscosity and Density of $MClO_4$ (M = Li and Na) Dissolved in Propylene Carbonate and γ-Butyrolactone at High Concentrations[J]. Journal of Power Sources, 2013, 223: 175–182.

[22]  Pham T A, Kweon K E, Samanta A. et al., Solvation and Dynamics of Sodium and Potassium in Ethylene Carbonate from Ab Initio Molecular Dynamics Simulations[J]. The Journal of Physical Chemistry C, 2017, 121(40): 21913–21920.

[23]  Chen S, Ishii J, Horiuchi S. et al., Difference in Chemical Bonding between Lithium and Sodium Salts: Influence of Covalency on Their Solubility Dagger[J]. Physical Chemistry Chemical Physics, 2017, 19(26): 17366–17372.

[24]  Tanaka T, Doi T, Okada S. et al., Effects of Salts in Methyl Difluoroacetate-based Electrolytes on Their Thermal Stability in Lithium-ion Batteries[J]. Fuel Cells, 2009, 9(3): 269–272.

[25]  Jnsson E, Armand M, Johansson P. Novel Pseudo-Delocalized Anions for Lithium Battery Electrolytes[J]. Physical Chemistry Chemical Physics: PCCP, 2012, 14(17): 621–625.

[26]  Li H, Peng L, Zhu Y. et al., An Advanced High-Energy Sodium Ion Full Battery Based on Nanostructured $Na_2Ti_3O_7VOPO_4$ Layered Materials[J]. Energy & Environmental Science, 2016, 9(11): 3399–3405.

[27] Hashmi S A, Bhat M Y, Singh M K. et al., Ionic Liquid-Based Sodium Ion -conducting Composite Gel Polymer Electrolytes: Effect of Active and Passive Fillers[J]. Journal of Solid State Electrochemistry, 2016, 20(10): 2817–2826.

[28] Bitner-Michalska A, Krztoń-Maziopa A, Żukowska G. et al., Liquid Electrolytes Containing New Tailored Salts for Sodium-ion batteries[J]. Electrochimica Acta, 2016, 222: 108–115.

[29] Ge C, Wang L, Xue L. et al., Synthesis of Novel Organic-Ligand-Doped Sodium Bis(Oxalate)-Borate Complexes with Tailored Thermal Stability and Enhanced Ion Conductivity for Sodium-ion batteries[J]. Journal of Power Sources, 2014, 248: 77–82.

[30] Bordet F, Ahlbrecht K, Tübke J. et al., Anion Intercalation into Graphite from a Sodium-Containing Electrolyte[J]. Electrochimica Acta, 2015, 174: 1317–1323.

[31] Shakourian-Fard M, Kamath G, Smith K. et al., Trends in Na-ion Solvation with Alkylcarbonate Electrolytes for Sodium-ion Batteries: Insights from Firstprinciples Calculations[J]. Journal of Physical Chemistry C, 2015, 119(40): 22747–22759.

[32] Xing L D, Zheng X W, Schroeder M. et al., Deciphering the Ethylene Carbonate–propylene Carbonate Mystery in Li-ion Batteries[J]. Accounts of Chemical Research, 2018, 51(2): 282–289.

[33] Zhao C L, Liu L L, Qi X G. et al., Solid-state Sodium Batteries[J]. Advanced Energy Materials, 2018, 8(17): 1703012.

[34] Alcántara R, Lavela P, Ortiz G F. et al., Carbon Microspheres Obtained from Resorcinol-formaldehyde as High-capacity Electrodes for Sodium-ion batteries[J]. Electrochemical and Solid State Letters, 2005, 8(4): A222–A225.

[35] Ponrouch A, Dedryvère R, Monti D. et al., Towards High Energy Density Sodium-ion Batteries through Electrolyte Optimization[J]. Energy & Environmental Science, 2013, 6(8): 2361–2369.

[36] Komaba S, Murata W, Ishikawa T. et al., Electrochemical Na Insertion and Solid Electrolyte Interphase for Hard-Carbon Electrodes and Application to Na-Ion Batteries[J]. Advanced Functional Materials, 2011, 21(20): 3859–3867.

[37] Zhu Y M, Luo X Y, Zhi H Z. et al., Diethyl(Thiophen-2-Ylmethyl) Phosphonate: A Novel Multifunctional Electrolyte Additive For High Voltage Batteries[J]. Journal of Materials Chemistry A, 2018, 6(23): 10990–11004.

[38] Zhu Y M, Luo X Y, Zhi H Z. et al., Structural Exfoliation of Layered Cathode under High Voltage and Its Suppression by Interface Film Derived from Electrolyte Additive[J]. ACS Applied Materials & Interfaces, 2017, 9(13): 12021–12034.

[39] Geng C X, Buchholz D, Kim G T. et al., Influence of Salt Concentration on the Properties of Sodium-Based Electrolytes[J]. Small Methods, 2018, 3: 1800208.

[40] Chen J E, Huang Z G, Wang C Y. et al., Sodium-difluoro(oxalato) Borate (Nadfob): A New Electrolyte Salt for Na-ion Batteries[J]. Chemical Communications, 2015, 51(48): 9809–9812.

[41] Plewa-Marczewska A, Trzeciak T, Bitner A. et al., New Tailored Sodium Salts for Battery Applications[J]. Chemistry of Materials, 2014, 26(17): 4908–4914.

[42] Goodenough J B, Kim Y S. Challenges for Rechargeable Li Batteries[J].Chemistry of Materials, 2010, 22(3): 587–603.

[43] Markevich E, Salitra G, Aurbach D. Fluoroethylene Carbonate as an Important Component for the Formation of an Effective Solid Electrolyte Interphase on Anodes and Cathodes for Advanced Li-ion Batteries[J]. ACS Energy Letters, 2017, 2(6): 1337–1345.

[44] Huang Y X, Xie M, Zhang J T. et al., A Novel Border-Rich Prussian Blue Synthetized by Inhibitor Control as Cathode for Sodium-ion batteries[J]. Nano Energy, 2017, 39: 273–283.

[45] Dahbi M, Nakano T, Yabuuchi N. et al., Effect of Hexafluorophosphate and Fluoroethylene Carbonate on Electrochemical Performance and the Surface Layer of Hard Carbon for Sodium-ion batteries[J]. ChemElectroChem, 2016, 3(11): 1856–1867.

[46]  Qian J F, Chen Y, Wu L. et al., High Capacity Na-storage and Superior Cyclability of Nanocomposite Sb/C Anode for Na-ion Batteries[J]. Chemical Communications, 2012, 48(56): 7070–7072.

[47]  Sadan M K, Choi S H, Kim H H. et al., Effect of Sodium Salts on the Cycling Performance of Tin Anode in Sodium-ion batteries[J]. Ionics, 2018, 24(3): 753–761.

[48]  Jang J Y, Lee Y, Kim Y. et al., Interfacial Architectures Based on a Binary Additive Combination for High-Performance $Sn_4P_3$ Anodes in Sodium-ion batteries[J]. Journal of Materials Chemistry A, 2015, 3(16): 8332–8338.

[49]  Lu H Y, Wu L F, Xiao L F. et al., Investigation of the Effect of Fluoroethylene Carbonate Additive on Electrochemical Performance of Sb-based Anode for Sodium-ion batteries[J]. Electrochimica Acta, 2016, 190: 402–408.

[50]  Dugas R, Ponrouch A, Gachot G. et al., Na Reactivity toward Carbonate-Based Electrolytes: The Effect of FEC as Additive[J]. Journal of the Electrochemical Society, 2016, 163(10): A2333–A2339.

[51]  Jung R, Metzger M, Haering D. et al., Consumption of Fluoroethylene Carbonate (FEC) on Si-C Composite Electrodes for Li-ion Batteries[J]. Journal of the Electrochemical Society, 2016, 163 (8): A1705–A1716.

[52]  Schiele A, Breitung B, Hatsukade T. et al., The Critical Role of Fluoroethylene Carbonate in the Gassing of Silicon Anodes for Lithium-Ion Batteries[J]. ACS Energy Letters, 2017, 2(10): 2228–2233.

[53]  Wang Y H, Sinha N N, Burns J C. et al., A Comparative Study of Vinylene Carbonate and Fluoroethylene Carbonate Additives for $LiCoO_2$/Graphite Pouch Cells[J]. Journal of the Electrochemical Society, 2014, 161(4): A467–A472.

[54]  Song X N, Meng T, Deng Y M. et al., The Effects of the Functional Electrolyte Additive on the Cathode Material $Na_{0.76}Ni_{0.3}Fe_{0.4}Mn_{0.3}O_2$ for Sodium-ion batteries[J]. Electrochimica Acta, 2018, 281: 370–377.

[55]  Manohar C V, Forsyth M, MacFarlane D R. et al., Role of N-propyl-N-methyl Pyrrolidinium Bis (trifluoromethanesulfonyl)imide as an Electrolyte Additive in Sodium Battery Electrochemistry[J]. Energy Technology, 2018, 6(11): 2232–2237.

[56]  Lee M, Hong J, Lopez J. et al., High-performance Sodium–organic Battery by Realizing Four-Sodium Storage in Disodium Rhodizonate[J]. Nature Energy, 2017, 2(11): 861–868.

[57]  Cohn A P, Share K, Carter R. et al., Ultrafast Solventassisted Sodium Ion Intercalation into Highly Crystalline Fewlayered Graphene[J]. Nano Letters, 2015, 16(1): 543–548.

[58]  Bai P, He Y, Xiong P. et al., Long Cycle Life and High Rate Sodium Ion Chemistry for Hard Carbon Anodes[J]. Energy Storage Materials, 2018, 13: 274–282.

[59]  Kajita T, Noji T, Imai Y. et al., Electrochemical Performance of Layered FeSe for Sodium-ion Batteries Using Ether-Based Solvents[J]. Journal of the Electrochemical Society, 2018, 165 (14): A3582–A3585.

[60]  Zhang J, Wang D W, Lv W. et al., Achieving Superb Sodium Storage Performance on Carbon Anodes through an Ether-Derived Solid Electrolyte Interphase[J]. Energy & Environmental Science, 2017, 10(1): 370–376.

[61]  Li K K, Zhang J, Lin D M. et al., Evolution of the Electrochemical Interface in Sodium-ion Batteries with Ether Electrolytes[J]. Nature Communications, 2019, 10(1): 725.

[62]  Jache B, Adelhelm P. Use of Graphite as a Highly Reversible Electrode with Superior Cycle Life for Sodium-ion Batteries by Making Use of Co-Intercalation Phenomena[J]. Angewandte Chemie International Edition, 2014, 53(38): 10169–10173.

[63]  Kim H, Hong J, Park Y. et al., Sodium Storage Behavior in Natural Graphite Using Ether-based Electrolyte Systems[J]. Advanced Functional Materials, 2015, 25(4): 534–541.

[64]  Kim H, Hong J, Yoon G. et al., Sodium Intercalation Chemistry in Graphite[J]. Energy & Environmental Science, 2015, 8(10): 2963–2969.

[65]  Hallett J P, Welton T. Room-Temperature Ionic Liquids: Solventsfor Synthesis and Catalysis[J]. Chemical Reviews, 2011, 111(5): 3508–3576.

[66]  Hagiwara R, Ito Y. Room Temperature Ionic Liquids of Alkylimidazolium Cations and Fluoroanions[J]. Journal of Fluorine Chemistry. 2000, 105(2): 221–227.

[67]  Xue H, Verma R, Shreeve J M. Review of Ionic Liquids with Fluorine-Containing Anions[J]. Cheminform, 2006, 127(2): 159–176.

[68]  Ohno H, Yoshizawa M, Mizumo T. Electrochemical Aspects of Ionic Liquids[M], Hoboken NJ: John Wiley & Sons Inc., 2011, 199–203.

[69]  Petkovic M, Seddon K, Rebelo L. et al., Ionic Liquids: Apathway to Environmental Acceptability[J]. Chemical Society Reviews, 2011, 40(3): 1383–1403.

[70]  Fuller J, Carlin R T, Long H C D. et al., Structure of 1-Ethyl-3- Methylimidazolium Hexafluorophosphate: Modelfor Room Temperature Molten Salts[J]. Journal of the Chemical Society. Chemical Communications, 1994: 299–300.

[71]  Wilkes J S, Zaworotko M J. Air and Water Stable 1-Ethyl- 3-Methylimidazolium Based Ionic Liquids[J]. Journal of the Chemical Society, 1992: 965–967.

[72]  Pierre B, Dias A P, Papageorgiou N. et al., Hydrophobic, Highly Conductive Ambient-Temperature Molten Salts[J]. Inorganic Chemistry, 1996, 35(5): 1168–1178.

[73]  Matsumoto H, Sakaebe H, Tatsumi K. Fast Cycling of Li/LiCoO$_2$ Cell with Low-Viscosity Ionic Liquids Based on Bis(Fluorosulfonyl)imide [FSI][J]. Journal of Power Sources, 2006, 160(2): 1308–1313.

[74]  Zhang H, Feng W, Nie J. et al., Recent Progresses on Electrolytes of Fluorosulfonimide Anions for Improving the Performances of Rechargeable Li and Li-ion Battery[J]. Journal of Fluorine Chemistry, 2015, 174: 49–61.

[75]  Xue Z, Qin L, Jiang J. et al., Thermal, Electrochemical and Radiolytic Stabilities of Ionic Liquids[J]. Physical Chemistry Chemical Physics, 2018, 20: 8382–8402.

[76]  Fischer P J, Phuong M. et al., Synthesis and Physicochemical Characterization of Room Temperature Ionic Liquids and Their Application in Sodium-ion batteries[J]. Physical Chemistry Chemical Physics, 2018, 20: 29412–29422.

[77]  Hagiwara R, Tamaki K, Kubota K. et al., Thermal Properties of Mixed Alkali Bis (trifluoromethylsulfonyl)amides[J]. Chemical and Engineering Data, 2008, 53(2): 355–358.

[78]  Keigo K, Toshiyuki N, Rika H. New Inorganic Ionic Liquids Possessing Low Melting Temperatures and Wide Electrochemical Windows: TernaryMixtures of Alkali bis (fluorosulfonyl)amides[J]. Electrochimica Acta, 2012, 66: 320–324.

[79]  Henderson W A, Mckenna F, Khan M A. et al., Glyme-lithium Bis (Trifluoromethanesulfonyl) imide and Glyme-lithium Bis(perfluoroet- Hanesulfonyl) Imide Phase Behavior and Solvate structures[J]. Chemistry of Materials, 2005, 17(9): 2284–2289.

[80]  Mandai T, Yoshida K, Tsuzuki S. et al., Effect of Ionic Size on Solvate Stability of Glyme-Based Solvate Ionic Liquids[J]. Journal of Physical Chemistry B, 2015, 119(4): 1523–1534.

[81]  Terada S, Mandai T, Nozawa R. et al., Physicochemical Properties of Pentaglyme-sodium Bis (Trifluoromethanesulfonyl)amide Solvate Ionic Liquid[J]. Physical Chemistry Chemical Physics Cambridge Royal Society of Chemistry, 2014 16(23): 11737–11746.

[82]  Maton C, De Vos N, Stevens C V. Ionic Liquid Thermal Stabilities: Decomposition Mechanisms and Analysis Tools[J]. Chemical Society Reviews. 2013, 42(13): 5963–5977.

[83]  Wongittharom N, Wang C H, Wang Y C. et al., Ionic Liquid Electrolytes with Various Sodium Solutes for Rechargeable Na/NaFePO$_4$ Batteries Operated at Elevated Temperatures[J]. ACS Applied Materials & Interfaces, 2014, 6(20): 17564–17570.

[84]  Wang C H, Yeh Y W, Wongittharom N. et al., Rechargeable Na/Na$_{0.44}$MnO$_2$ Cells with Ionic Liquid Electrolytes Containing Various Sodium Solutes[J. Journal of Power Sources, 2015, 274: 1016–1023.

[85] Wu F, Zhu N, Bai Y. et al., Highly Safe Ionic Liquid Electrolytes for Sodium Ion Battery: WideElectrochemical Window and Good Thermal Stability[J]. ACS Applied Materials & Interfaces, 2016, 8(33): 21381–21386.

[86] Kubota K, Nohira T, Hagiwara R. Thermal Properties of Alkali Bis(fluorosul- Fonyl)amides and Their Binary Mixtures[J]. Chemical and Engineering Data, 2010, 55(9): 2546–2549.

[87] Matsumoto K, Taniki R, Nohira T. Inorganic-Organic Hybrid Ionic Liquid Electrolytes for Na Secondary Batteries[J]. Electrochemical Society, 2015, 162(7): A1409–A1414.

[88] Sakaebe H, Matsumoto H, Tatsumi K. Application of Room Temperature Ionic Liquids to Li Batteries[J]. Electrochimica Acta, 2008, 53(3): 1048–1054.

[89] Zhao J, Zhao L, Dimov N. et al., Electrochemical and Thermal Properties of Hard Carbon-Type Anodes for Na-Ion Batteries[J]. Power Sources, 2013, 244(5): 752–757.

[90] Xia X, Obrovac M N, Dahn J R. Comparison of the Reactivity of NaxC$_6$ and LixC$_6$ with Non-Aqueous Solvents and Electrolytes[J]. Electrochemical and Solid-State Letters, 2011, 14(9): A130–A133.

[91] Xia X, Dahn J R. Study of the Reactivity of Na/Hard Carbon with Different Solvents and Electrolytes[J]. Electrochemical Society, 2012, 159(5): A515–A519.

[92] Lee Y, Lim H, Kim S O. et al., Thermal Stability of Sn Anode Material with Non-Aqueous Electrolytes in Sodium-ion batteries[J]. Materials Chemistry A, 2018, 6(41): 20383–20392.

[93] Xin X, Dahn J R. A Study of the Reactivity of De-intercalated NaNi$_{0.5}$Mn$_{0.5}$O$_2$ with Non-aqueous Solvent and Electrolyte by Accelerating Rate calorimetry[J]. Electrochemical Society, 2012, 159(7): A1048–A1051.

[94] Xia X, Dahn J R. A Study of the Reactivity of De-Intercalated P2-NaxCoO$_2$ with Non-Aqueous Solvent and Electrolyte by Accelerating Rate Calorimetry[J]. Electrochemical Society, 2012, 159(5): A1048–A1051.

[95] Zhao J, Zhao L, Dimov N. et al., Electrochemical and Thermal Properties of -nafeo$_2$ Cathode for Na-Ion Batteries[J]. Electrochemical Society, 2013, 160(5): A3077–A3081.

[96] Plashnitsa L S, Kobayashi E, Noguchi Y. et al., Performance of NASICON Symmetric Cell with Ionic Liquid Electrolyte[J]. Electrochemical Society, 2007, 157(4): A536–A543.

[97] Rodrigues M T F, Sayed F N, Gullapalli H. et al., High-temperature Solid Electrolyte Interphases (SEI) in Graphite electrodes[J]. Power Sources, 2018, 381: 107–115.

[98] Kazuhiko M, Takafumi H, Toshiyuki N. et al., The Na[FSA]–[C$_2$C$_1$im][FSA] (C$_2$c$_1$im$^+$:1-ethyl-3-methylimidazolium and FSA$^-$: Bis(fluorosulfonyl)amide) Ionic Liquid Electrolytes for Sodium Secondary Batteries[J]. Power Sources, 2014, 265: 36–39.

[99] Yoon H, Zhu H, Hervault A. et al., Physicochemical Properties of N-propyl- N-Methylpyrrolidinium Bis(fluorosulfonyl)imide for Sodium Metal Battery Applications[J]. Physical Chemistry Chemical Physics, 2014, 16: 12350–12355.

[100] Monti D, Jonsson E, Palacin M R. et al., Ionic Liquid Based Electrolytes for Sodium-ion Batteries: Na$^+$Solvation and Ionic Conductivity[J]. Power Sources, 2014, 245: 630–636.

[101] Mohd Noor S A, Howlett P C, MacFarlane D R. Properties of Sodium-Based Ionic Liquid Electrolytes for Sodium Secondary Battery Applications[J]. Electrochimica Acta, 2013, 114: 766–771.

[102] Serra Moreno J, Maresca G, Panero S. et al., Sodium-Conducting Ionic Liquid-Based Electrolytes[J]. Electrochemistry Communications, 2014, 43: 1–4.

[103] Hayes R, Warr G G, Atkin R. Structure and Nanostructure in Ionic Liquids[J]. Chemical Reviews, 2015, 115(13): 6357.

[104] Xu W, Cooper E I, Angell C A. Ionic Liquids: IonMobilities, Glass Temperatures, and Fragilities [J]. Physical Chemistry B, 2003, 107(25): 6170–6178.

[105] Moynihan C T, Macedo P B, Montrose C J. et al., Structure Relaxation in Vitreous Materials[J]. Annals of the New York Academy Sciences, 1976, 279(1): 15–35.

[106] Ferdousi S A, Hilder M, Basile A. et al., Water as an Effective Additive for High-Energy-Density Na Metal Batteries? Studies in a Superconcentrated Ionic Liquid Electrolyte[J]. Chemistry Sustainability Energy Materials, 2019, 12(8): 1700–1711.

[107] Basile A, Ferdousi S A, Makhlooghiazad F. et al., Beneficial Effect of Added Water on Sodium Metal Cycling in Super Concentrated Ionic Liquid Sodium Electrolytes[J]. Power Sources, 2018, 379: 344–349.

[108] Seddon K R, Stark A, Torres M-J. Influence of Chloride, Water, and Organic Solvents on the Physical Properties of Ionic Liquids[J]. Pure & Applied Chemistry, 2000, 72(12): 2275–2287.

[109] John M, Slattery C. et al., How to Predict the Physical Properties of Ionic Liquids: AVolume-Based Approach[J]. Angewandte Chemie International Edition, 2007, 119(28): 5480–5484.

[110] Yuyama K, Masuda G, Yoshida H. Ionic Liquids Containing the Tetrafluoroborate Anion Have the Best Performance and Stability for Electric Double Layer Capacitor Applications[J]. Power Sources, 2006, 162(2): 1401–1408.

[111] Macfarlane D R, Forsyth M, Izgorodina E I. et al., On the Concept of Ionicity in Ionic Liquids[J]. Physical Chemistry Chemical Physics Pccp, 2009, 11(25): 4962–4967.

[112] Masahiro Y, Wu X, Angell C A. Ionic Liquids by Proton Transfer: Vapor Pressure, Conductivity, and the Relevance of Δpka from Aqueous Solutions[J]. American Chemical Society, 2003, 125(50): 15411–15419.

[113] Harris K R. Relations between the Fractional Stokes–Einstein and Nernst–Einstein Equations and Velocity Correlation Coefficients in Ionic Liquids and Molten Salts[J]. Physical Chemistry B, 2010, 114(29): 9572–9577.

[114] Evans J, Vincent C A, Bruce P G. Electrochemical Measurement of Transference Numbers in Polymer Electrolytes[J]. Polymer, 1987, 28(13): 2324–2328.

[115] Abraham K M, Jiang Z, Carroll B. Highly Conductive PEO-Like Polymer Electrolytes[J]. Chemistry of Materials, 1997, 9(9): 1978–1988.

[116] Forsyth M, Yoon H, Chen F. et al., Novel Na$^+$ Ion Diffusion Mechanism in Mixed Organic-Inorganic Ionic Liquid Electrolyte Leading to High Na$^+$ Transference Number and Stable, High Rate Electrochemical Cycling of Sodium Cells[J]. Physical Chemistry C, 2016, 120(8): 4276–4286.

[117] Chen C Y, Kiko T, Hosokawa T. et al., Ionic Liquid Electrolytes with High Sodium Ion Fraction for High-Rate and Long-Life Sodium Secondary Batteries[J]. Power Sources, 2016, 332(15): 51–59.

[118] Carstens T, Lahiri A, Borisenko N. et al., [Py1,4]fsi-nafsi-based Ionic Liquid Electrolyte for Sodium Batteries: Na+Solvation and Interfacial Nanostructure on Au(111)[J]. The Journal of Physical Chemistry C, 2016, 120(27): 14736–14741.

[119] Vogel H. The Temperature Dependence Law of the Viscosity of Fluids[J]. Phys. Z, 1921, 22: 645–646.

[120] Fulcher G S. Analysis of Recent Measurements of the Viscosity of Glasses[J]. American Ceramic Society, 1925, 8(6): 339–355.

[121] Tokuda H, Hayamizu K, Ishii K. et al., Physicochemical Properties and Structures of Room Temperature Ionic Liquids. 1. Variation of Anionic Species[J]. Physical Chemistry. B, 2004, 108(42): 16593–16600.

[122] Huang M M, Jiang Y, Sasisanker P. et al., Static Relative Dielectric Permittivities of Ionic Liquids at 25°C[J]. Chemical and Engineering Data, 2011, 56(4): 1494.

[123] Wakai C, Oleinikova A, Ott M. et al., How Polar are Ionic Liquids? Determination of the Static Dielectric Constant of an Imidazolium- Based Ionic Liquid by Microwave Dielectric Spectroscopy[J]. Physical Chemistry B, 2005, 109(36): 17028–17030.

[124] Weingrtner H. Understanding Ionic Liquids at the Molecular Level: Facts, Problems, and Controversies[J]. Angewandte Chemie International Edition. 2008, 47(4): 654–670.

[125] Krossing I, Slattery J, Daguenet C. et al., Why are Ionic Liquids Liquid? A Simple Explanation Based on Lattice and Solvation Energies[J]. American Chemical Society, 2006, 128(41): 13427–13434.

[126] Reichardt C. Polarity of Ionic Liquids Determined Empirically by Means of Solvatochromic Pyridinium N-Phenolate Betaine Dyes[J]. Green Chemistry, 2005, 7(5): 339–351.

[127] Caricato M, Mennucci B, Tomasi J. et al., Formation and Relaxation of Excited States in Solution: A New Time Dependent Polarizable Continuum Model Based on Time Dependent Density Functional Theory[J]. Chemical Physics, 2006, 124(12): 875–887.

[128] Fletcher K A, Storey I A, Hendricks A E. et al., Behavior of the Solvatochromic Probes Reichardt's Dye, Pyrene, Dansylamide, Nile Red and 1-pyrenecarbaldehyde within the Room-Temperature Ionic Liquid bmimPF6[J]. Green Chemistry, 2001, 3: 210–215.

[129] Baker S N, Baker G A, Kane M A. et al., The Cybotactic Region Surrounding Fluorescent Probes Dissolved in 1Butyl3-methylimidazolium Hexafluorophosphate: Effects of Temperature and Added Carbon Dioxide[J]. Physical Chemistry B, 2001, 105(39): 9663–9668.

[130] Jin H, Baker G A, Arzhantsev S. et al., Solvation and Rotational Dynamics of Coumarin 153 in Ionic Liquids: Comparisonsto Conventional Solvents.[J]. Physical Chemistry B, 2007, 111(25): 7291–7302.

[131] Ab Rani M A, Brant A, Crowhurst L. et al., Understanding the Polarity of Ionic liquids[J]. Physical Chemistry Chemical Physics, 2011, 13(37): 16831–16840.

[132] Reichardt C. Solvatochromic Dyes as Solvent Polarity Indicators[J]. Chemical Reviews, 1994, 94(8): 416–431.

[133] Werner S R, Kozo S. (Acetylacetonato)(n,n,n′,n′- Tetramethy- Lethylenediamine) Copper(II) Tetraphenylborate as a Solvent Basicity Indicator [J]. Bulletin of the Chemical Society of Japan, 1987, 60(6): 2286–2288.

[134] Muldoon M J, Gordon C M, Dunkin I R. Investigations of Solvent–Solute Interactions in Room Temperature Ionic Liquids Using Solvatochromic Dyes[J]. Chemical Society, Perkin Transactions, 2001, 2(4): 433–435.

[135] Dzyuba S V, Bartsch R A. Expanding the Polarity Range of Ionic Liquids[J]. Tetrahedron Letters. 2002, 43(26): 4657–4659.

[136] Crowhurst L, Mawdsley P R, Perez-Arlandis J M. et al., Solvent–Solute Interactions in Ionic Liquids[J]. Physical Chemistry Chemical Physics, 2003, 5(13): 2790–2794.

[137] Kaar J L, Jesionowski A M, Berberich J A. et al., Impact of Ionic Liquid Physical Properties on Lipase Activity and Stability[J]. American Chemical Society, 2003, 125(14): 4125–4131.

[138] Ueno K, Yoshida K, Tsuchiya M. et al., Glyme–Lithium Salt Equimolar Molten Mixtures: Concentrated Solutions or Solvate Ionic Liquids?[J]. Physical Chemistry B, 2012, 116(36): 11323–11331.

[139] Tokuda H, Tsuzuki S, Susan M A B H. et al., How Ionic are Room-Temperature Ionic Liquids? An Indicator of the Physicochemical Properties[J]. Physical Chemistry B, 2006, 110(39): 19593–19600.

[140] Taft R W, Kamlet M J. The Solvatochromic Comparison Method. 2. The. Beta-Scale of Solvent Hydrogen-Bond Acceptor (HBA) Basicities[J]. American Chemical Society, 1976, 98(10): 2886–2894.

[141] Kamlet M J, Taft R W. The Solvatochromic Comparison Method. I. The. Beta-Scale of Solvent Hydrogen-Bond Acceptor (HBA) Basicities[J]. American Chemical Society, 1976, 98(2): 377–383.

[142] Tai Y, Taft R W, Mortimer J K. The Solvatochromic Comparison Method. 3. Hydrogen Bonding by Some 2-nitroaniline Derivatives[J]. American Chemical Society, 1976, 98(11): 3233–3237.

[143] Kamlet M J, Abboud Z J L, Taft R W. The Solvatochromic Comparison Method. 6. The T* Scale of Solvent Polarities1[J]. American Chemical Society, 1977, 99(18): 6027–6038.

[144] Ueno K, Tokuda H, Watanabe M. Ionicity in Ionic Liquids: Correlation with Ionic Structure and Physicochemical Properties[J]. Physical Chemistry Chemical Physics, 2010, 12(8): 1649–1658.

[145] Pekka P, Hubert G. Electrochemical Potential Window of Battery Electrolytes: The HOMO–LUMO Misconception[J]. Energy & Environmental Science, 2018, 11(9): 2306–2309.

[146] Peled E, Menkin S. Review–SEI: Past, Present and Future[J]. Electrochemical Society, 2017, 164(7): A1703–A1719.

[147] Bommier C, Ji X. Electrolytes, SEI Formation, and Binders: A Review of Nonelectrode Factors for Sodium Ion Battery Anodes[J]. Small, 2018, 14(16): 1703576.

[148] Ong S P, Andreussi O, Wu Y. et al., Electrochemical Windows of Room- Temperature Ionic Liquids from Molecular Dynamics and Density Functional Theory Calculations[J]. Chemistry of Materials, 2011, 23(11): 2979–2986.

[149] Howlett P C, Izgorodina E I, Forsyth M. et al., Electrochemistry at Negative Potentials in Bis (trifluoromethanesulfonyl)amide Ionic Liquids[J]. Ztschrift Für Physikalische Chemie, 2006, 220(10): 1483–1498.

[150] Kazemiabnavi S, Thornton K. et al, Electrochemical Stability Window of Imidazolium-Based Ionic Liquids as Electrolytes for Lithium Batteries[J]. Physical Chemistry B Condensed Matter Materials Surfaces Interfaces & Biophysical, 2016, 120(25): 5691–5702.

[151] Ma T, Xu G L, Li Y. et al., Revisiting the Corrosion of the Aluminum Current Collector in Lithium-Ion Batteries[J]. Physical Chemistry Letters, 2017, 8(5): 1072–1077.

[152] Krause L J, Lamanna W, Summerfield J. Corrosion of Aluminum at High Voltages in Non-Aqueous Electrolytes Containing Perfluoroalkylsulfonyl Imides; New Lithium Salts for Lithium-Ion Cells[J]. Power Sources, 1997, 68(2): 320–325.

[153] Dahbi M, Ghamouss F, Tran-Van F. et al., Comparative Study of EC/DMC LiTFSI and LiPF$_6$ Electrolytes for Electrochemical Storage[J]. Power Sources, 2011, 196(22): 9743–9750.

[154] Han H B, Si-Si Z, Dai-Jun Z. et al., Lithium Bis(fluorosulfonyl)imide (Lifsi) as Conducting Salt for Nonaqueous Liquid Electrolytes for Lithium-Ion Batteries: Physicochemicaland Electrochemical Properties[J]. Power Sources, 2011, 196(7): 3623–3632.

[155] Abouimrane A, Ding J, Davidson I J. Liquid Electrolyte Based on Lithium Bis-Fluorosulfonyl Imide Salt: AluminumCorrosion Studies and Lithium Ion Battery Investigations[J]. Power Sources, 2009, 189(1): 693–696.

[156] Hu X, Li Z, Zhao Y. et al., Quasi-Solid State Rechargeable Na-CO$_2$ Batteries with Reduced Graphene Oxide Na Anodes[J]. Science Advances, 2017, 3(2): e1602396.

[157] Ni'mah Y L, Cheng M, Cheng J H. et al., Solid-State Polymer Nanocomposite Electrolyte of TiO$_2$/PEO/NaClO$_4$ for Sodium-ion batteries[J]. Power Sources, 2015, 278: 375–381.

[158] Hu X, Li Z, Chen J. Flexible Li-CO$_2$ Batteries with Liquid-Free Electrolyte[J]. Angewandte Chemie International Edition, 2017, 56(21): 5785–5789.

[159] Zhou D, Liu R, Zhang J. et al., In Situ Synthesis of Hierarchical Poly(Ionic Liquid)-Based Solid Electrolytes for High-Safety Lithium-Ion and Sodium-ion batteries[J]. Nano Energy, 2017, 33: 45–54.

[160] Khurana R, Schaefer J L, Archer L A. et al., Suppression of Lithium Dendrite Growth Using Cross-Linked Polyethylene/Poly(ethylene Oxide) Electrolytes: A New Approach for Practical Lithium-Metal Polymer Batteries[J]. American Chemical Society, 2014, 136(20): 7395–7402.

[161] Zhang Q, Liu K, Ding F. et al, Recent Advances in Solid Polymer Electrolytes for Lithium Batteries[J]. Nano Research, 2017, 10(12): 4139–4174.

[162] Hou H, Xu Q, Pang Y. et al., Efficient Storing Energy Harvested by Triboelectric Nanogenerators Using a Safe and Durable All-Solid-State Sodium Ion Battery[J]. Advanced Science, 2017, 4(8): 1700072.

[163] Zhang Z, Zhang Q, Ren C. et al., A Ceramic/Polymer Composite Solid Electrolyte for Sodium Batteries[J]. Materials Chemistry A, 2016, 4(41): 15823–15828.

[164] Ratner M A, Johansson P, Shriver D F. Polymer Electrolytes: Ionic Transport Mechanisms and Relaxation Coupling[J]. MRS Bulletin. 2000, 25(3): 31–37.

[165] Zhang Z, Zhang Q, Ren C. et al., A Ceramic/Polymer Composite Solid Electrolyte for Sodium Batteries[J]. Materials Chemistry A, 2016, 4: 15823–15828.

[166] Zhou D, Liu R, Zhang J. et al., In Situ Synthesis of Hierarchical Poly(Ionic Liquid)-Based Solid Electrolytes for High-Safety Lithium-Ion and Sodium-ion batteries[J]. Nano Energy, 2017, 33: 45–54.

[167] Hu X, Dawut G, Wang J. et al., Room-temperature Rechargeable Na-SO$_2$ Batteries with Gel-Polymer Electrolyte[J]. Chemical Communications, 2018, 54: 5315–5318.

[168] Bronstein L M, Karlinsey R L, Stein B. et al., Solid Polymer Single-Ion Conductors: Synthesis and Properties[J]. Chemistry of Materials, 2006, 18(3): 708–715.

[169] Yung F, Yao Y, Kummer J T. Ion Exchange Properties of and Rates of Ionic Diffusion in Beta-Alumina[J]. Inorganic and Nuclear Chemistry, 1967, 29(9): 2453–2475.

[170] Hueso K B, Palomares V, Armand M. et al., Challenges and Perspectives on High and Intermediate-Temperature Sodium Batteries[J]. Nano Research, 2017, 10(12): 4082–4114.

[171] Lee S, Lee D, Lee S. et al., Effects of Calcium Impurity on Phase Relationship, Ionic Conductivity and Microstructure of Na$^+$-β/β″- Alumina Solid Electrolyte[J]. Bulletin of Materials Science, 2016, 39(3): 729–735.

[172] Bates J B, Engstrom H, Wang J C. et al., Composition, Ion-ion Correlations and Conductivity of Beta″-Alumina[J]. Solid State Ionics, 1981, 5: 159–162.

[173] Lu X, Xia G, Lemmon J P. et al., Advanced Materials for Sodium-Beta Alumina Batteries: Status, Challenges and Perspectives[J]. Power Sources, 2010, 195(9): 2431–2442.

[174] Lu Y, Li L, Zhang Q. et al., Electrolyte and Interface Engineering for Solid-State Sodium Batteries[J]. Joule, 2018, 2(9): 1747–1770.

[175] Will F G. Effect of Water on Beta Alumina Conductivity[J]. Electrochemical Society, 1976, 123(6): 834.

[176] Flor G, Marini A, Massarotti V. et al., Reactivity of B-aluminas with Water[J]. Solid State Ionics, 1981, 2(3): 195–204.

[177] Baffier N, Badot J C, Colomban P. Conductivity of Ion Rich β and β″ Alumina: Sodium and Potassium Compounds[J]. Materials Research Bulletin, 1981, 16(3): 259–265.

[178] Green D J. Transformation Toughening and Grain Size Control in Beta″-Al$_2$O$_3$/ZrO$_2$ Composites[J]. Materials Science, 1985, 20(7): 2639–2646.

[179] Virkar A V, Gordon R S. Fracture Properties of Polycrystalline Lithia-Stabilizedβ″- Alumina[J]. American Ceramic Society, 2006.

[180] Goodenough J B, Hong H, Kafalas J A. Fast Na-Ion Transport in Skeleton Structures[J]. Materials Research Bulletin, 1976, 11(2): 203–220.

[181] Hong Y P, Kafalas J A, Goodenough J B. Crystal Chemistry in the System MSbO$_3$[J]. Solid State Chemistry, 1974, 9(4): 345–351.

[182] Park H, Jung K, Nezafati M. et al., Sodium Ion Diffusion in Nasicon (Na$_3$zr$_2$si$_2$po$_{12}$) Solid Electrolytes: Effects of Excess Sodium[J]. ACS Applied Materials & Interfaces, 2016, 8(41): 27814–27824.

[184] Song S, Duong H M, Korsunsky A M. et al., A Na$^+$ Superionic Conductor for Room-Temperature Sodium Batteries[J]. Scientific Reports, 2016, 6(1): 32330.

[185] Zhang Z, Zhang Q, Shi J. et al., A Self-Forming Composite Electrolyte for Solid-State Sodium Battery with Ultralong Cycle Life[J]. Advanced Energy Materials, 2017, 7(4): 1601196.

[186] Ruan Y, Song S, Liu J. et al., Improved Structural Stability and Ionic Conductivity of Na$_3$Zr$_2$Si$_2$PO$_{12}$ Solid Electrolyte by Rare Earth Metal Substitutions[J]. Ceramics International, 2017, 43(10): 7810–7815.

[187] Ihlefeld J F, Gurniak E, Jones B H. et al., Scaling Effects in Sodium Zirconium Silicate Phosphate ($Na_{1+x}Zr_2Si_xP_{3-x}O_{12}$) Ion-Conducting Thin Films[J]. American Ceramic Society, 2016, 99(8): 2729–2736.

[188] Samiee M, Radhakrishnan B, Rice Z. et al., Divalent-Doped $Na_3Zr_2Si_2PO_{12}$ Natrium Superionic Conductor: Improving the Ionic Conductivity via Simultaneously Optimizing the Phase and Chemistry of the Primary and Secondary Phases[J]. Power Sources, 2017, 347: 229–237.

[189] Ahmad A, Wheat T A, Kuriakose A K. et al., Dependence of the Properties of Nasicons on Their Composition and Processing[J]. Solid State Ionics, 1987, 24(1): 89–97.

[190] Bell N S, Edney C, Wheeler J S. et al., The Influences of Excess Sodium on Low-Temperature NaSICON Synthesis[J]. American Ceramic Society, 2014, 97(12): 3744–3748.

[191] 章志珍, 施思齐, 胡勇胜, 等. 溶胶凝胶法制备钠离子固态电解质$Na_3Zr_2Si_2PO_{12}$及其电导性能研究[J]. 无机材料学报, 2013, 28(11): 1255–1260.

[192] Suzuki K, Noi K, Hayashi A. et al., Low Temperature Sintering of $Na_{1+x}Zr_2SiP_{3-x}O_{12}$ by the Addition of $Na_3BO_3$[J]. Scripta Materialia, 2018, 145: 67–70.

[193] Tian Y, Shi T, Richards W D. et al., Compatibility Issues between Electrodes and Electrolytes in Solid-State Batteries[J]. Energy & Environmental Science, 2017, 10(5): 1150–1166.

[194] Banerjee A, Park K H, Heo J W. et al., $Na_3SbS_4$: A Solution Processable Sodium Superionic Conductor for All-Solid-State Sodium-ion batteries[J]. Angewandte Chemie, 2016, 128(33): 9786–9790.

[195] Zhang Z, Ramos E, Lalère F. et al., $Na_{11}Sn_2PS_{12}$: A New Solid State Sodium Superionic Conductor[J]. Energy & Environmental Science, 2018, 11(1): 87–93.

[196] Hayashl A, Noi K, Sakuda A. et al., Superionic Glass-Ceramic Electrolytes for Room-Temperature Rechargeable Sodium Batteries[J]. Nature Communications, 2012, 3(1): 1–5.

[197] Nishimura S, Tanibata N, Hayashi A. et al., The Crystal Structure and Sodium Disorder of High-Temperature Polymorph β-$Na_3PS_4$[J]. Materials Chemistry A, 2017, 5(47): 25025–25030.

[198] Hayashi A, Noi K, Tanibata N. et al., High Sodium Ion Conductivity of Glass–Ceramic Electrolytes with Cubic $Na_3PS_4$[J]. Power Sources, 2014, 258: 420–423.

[199] Tanibata N, Noi K, Hayashi A. et al., Preparation and Characterization of Highly Sodium Ion Conducting $Na_3PS_4$–$Na_4SiS_4$ Solid Electrolytes[J]. RSC Advances, 2014, 4(33): 17120–17123.

[200] de Klerk N J J, Wagemaker M. Diffusion Mechanism of the Sodium Ion Solid Electrolyte $Na_3PS_4$ and Potential Improvements of Halogen Doping[J]. Chemistry of Materials, 2016, 28(9): 3122–3130.

[201] Bo S, Wang Y, Kim J C. et al., Computational and Experimental Investigations of Na-Ion Conduction in Cubic $Na_3PSe_4$[J]. Chemistry of Materials, 2016, 28(1): 252–258.

[202] Chu I, Kompella C S, Nguyen H. et al., Room-Temperature All-solid-state Rechargeable Sodium-ion Batteries with a Cl-doped $Na_3PS_4$ Superionic Conductor[J]. Scientific Reports, 2016, 6(3): 33733.

[203] Krauskopf T, Pompe C, Kraft M A. et al., Influence of Lattice Dynamics on Na+Transport in the Solid Electrolyte $Na_3PS_4$–$xSe_x$[J]. Chemistry of Materials, 2017, 29(20): 8859–8869.

[204] Wang N, Yang K, Zhang L. et al., Improvement in Ion Transport in $Na_3PSe_4$–$Na_3SbSe_4$ by Sb Substitution[J]. Materials Science, 2018, 53(3): 1987–1994.

[205] Mo Y, Ong S P, Ceder G. First Principles Study of the $Li_{10}GeP_2S_{12}$ Lithium Super Ionic Conductor Material[J]. Chemistry of Materials, 2011, 24(1): 15–17.

[206] Wang Y, LU D, Bowden M. et al., Mechanism of Formation of Li7P3S11 Solid Electrolytes through Liquid Phase Synthesis[J]. Chemistry of Materials, 2018, 30(3): 990–997.

[207] Kandagal V S, Bharadwaj M D, Waghmare U V. Theoretical Prediction of a Highly Conducting Solid Electrolyte for Sodium Batteries: $Na_{10}GeP_2S_{12}$[J]. Materials Chemistry A, 2015, 3(24): 12992–12999.

[208] Wang Y, Richards W D, Bo S. et al., Computational Prediction and Evaluation of Solid-State Sodium Superionic Conductors $Na_7P_3X_{11}$(X = O, S, Se)[J]. Chemistry of Materials, 2017, 29(17): 7475–7482.

[209] Richards W D, Tsujimura T, Miara L J. et al., Design and Synthesis of the Superionic Conductor $Na_{10}SnP_2S_{12}$[J]. Nature Communications, 2016, 7(1): 11009.

[210] Zhang Z, Ramos E, Lalere F. et al., $Na_{11}Sn_2PS_{12}$: A New Solid State Sodium Superionic Conductor[J]. Energy & Environmental Science, 2018, 11(1): 87–93.

[211] Benedetto M D, Lidozzi A, Solero L. et al., Small-Signal Model of the Five-Level Unidirectional T-Rectifier[J]. IEEE Transactions on Power Electronics, 2017, 32(7): 5741–5751.

[212] Zhang L, Yang K, Mi J. et al., $Na_3PSe_4$: A Novel Chalcogenide Solid Electrolyte with High Ionic Conductivity[J]. Advanced Energy Materials, 2015, 5(24): 1501294.

[213] Kim S K, Mao A, Sen S. et al., Fast Na-Ion Conduction in a Chalcogenide Glass–Ceramic in the Ternary System $Na_2Se$–$Ga_2Se_3$– $GeSe_2$[J]. Chemistry of Materials, 2014, 26(19): 5695–5699.

[214] Kim J, Yoon K, Park I. et al., Progress in the Development of Sodium Ion Solid Electrolytes[J]. Small Methods, 2017, 1(10): 1700219.

[215] Zhang L, Zhang D, Yang K. et al., Vacancy-Contained Tetragonal $Na_3SbS_4$ Superionic Conductor[J]. Advanced Science, 2016, 3(10): 1600089.

[216] Wang H, Chen Y, Hood Z D. et al., An Air-Stable $Na_3SbS_4$ Superionic Conductor Prepared by a Rapid and Economic Synthetic Procedure[J]. Angewandte Chemie International Edition, 2016, 55(30): 8551–8555.

[217] Yoon K, Kim J, Seong W M. et al., Investigation on the Interface between $Li_{10}GeP_2S_{12}$ Electrolyte and Carbon Conductive Agents in All-Solid-State Lithium Battery[J]. Scientific Reports, 2018, 8(1): 1–7.

[218] Zhang W, Leichtweiß T, Culver S P. et al., The Detrimental Effects of Carbon Additives in $Li_{10}GeP_2S_{12}$-Based Solid-State Batteries[J]. ACS Applied Materials & Interfaces, 2017, 9(41): 35888–35896.

[219] Oguchi H, Matsuo M, Kuromoto S. et al., Sodium Ion Conduction in Complex Hydrides $NaAlH_4$ and $Na_3AlH_6$[J]. Applied Physics, 2012, 111(3): 36102.

[220] Matsuo M, Oguchi H, Sato T. et al., Sodium and Magnesium Ionic Conduction in Complex Hydrides[J]. Journal of Alloys and Compounds, 2013, 580: S98–S101.

[221] Arado O D, Luft M, Moenig H. et al,, Understanding Molecular Self-Assembly of A Diol Compound by considering Competitive Interactions[J]. Physical Chemistry Chemical Physics, 2016, 18(39): 27390–27395.

[222] Udovic T J, Matsuo M, Tang W S. et al., Exceptional Superionic Conductivity in Disordered Sodium Decahydro-closo-decaborate[J]. Advanced Materials, 2014, 26(45): 7622–7626.

[223] Tang W S, Matsuo M, Wu H. et al., Liquid-Like Ionic Conduction in Solid Lithium and Sodium Monocarba-closo-Decaborates near or at Room Temperature[J]. Advanced Energy Materials, 2016, 6(8): 1502237.

[224] Dimitrievska M, Shea P, Kweon K E. et al., Carbon Incorporation and Anion Dynamics as Synergistic Drivers for Ultrafast Diffusion in Superionic $LiCB_{11}H_{12}$ and $NaCB_{11}H_{12}$[J]. Advanced Energy Materials, 2018, 8(15): 1703422.

[224] Tang W S, Yoshida K, Soloninin A V. et al., Stabilizing Superionic-Conducting Structures via Mixed-Anion Solid Solutions of Monocarba-closo-borate Salts[J]. ACS Energy Letters, 2016, 1(4): 659–664.

[225] Duchene L, Kuhnel R S, Rentsch D. et al., A Highly Stable Sodium Solid-State Electrolyte Based on A Dodeca/Deca-Borate Equimolar Mixture[J]. Chemical Communications, 2017, 53(30): 4195–4198.

[226] Hansen B R S, Paskevicius M, Jørgensen M. et al., Halogenated Sodium-closo-Dodecaboranes as Solid-State Ion Conductors[J]. Chemistry of Materials, 2017, 29(8): 3423–3430.

[227] Duchêne L, Kühnel R S, Stilp E. et al., A Stable 3 V All-Solid-State Sodium Ion Battery Based on A Closo-Borate Electrolyte[J]. Energy & Environmental Science, 2017, 1(12): 269–2615.

[228] Monti D, Ponrouch A, Palacín M R. et al., Towards Safer Sodium-ion Batteries via Organic Solvent/Ionic Liquid Based Hybrid Electrolytes [J]. Power Sources, 2016, 324: 712–721.

[229] Mauger A, Julien C M. Critical Review on lithium-Ion Batteries: Are They Safe? Sustainable? [J]. Ionic, 2017(23): 1933–1947.

[230] Manohar C V, Raj K A, Kar M. et al., Stability Enhancing Ionic Liquid Hybrid Electrolyte for NVP@C Cathode Based Sodium Batteries [J]. Sustainable Energy & Fuels, 2018, 2: 566–576.

[231] Kumar D, Hashmi S A. Ionic Liquid Based Sodium Ion Conducting Gel Polymer Electrolytes [J]. Solid State Ionics, 2010, 181: 416–423.

[232] Song S, Kotobuki M, Zheng F. et al., A Hybrid Polymer/ Oxide/Ionic-Liquid Solid Electrolyte for Na-Metal Batteries [J]. Materials Chemistry A, 2017, 5: 6424–6431.

[233] Liu L, Qi X, Ma Q. et al., Toothpaste-like Electrode: A Novel Approach to Optimize the Interface for Solid-State Sodium-ion Batteries with Ultralong Cycle Life [J]. ACS Applied Materials & Interfaces, 2016, 8: 32631–32636.

[234] Kim J-K, Lim Y J, Kim H. et al., A Hybrid Solid Electrolyte For Flexible Solid-State Sodium Batteries [J]. Energy & Environmental Science, 2015, 8: 3589–3596.

[235] Senthilkumar S T, Abirami M, Kim J. et al., Sodium Ion Hybrid Electrolyte Battery for Sustainable Energy Storage Applications [J]. Power Sources, 2017, 341: 404–410.

[236] Senthilkumar S T, Bae H, Han J. et al., Enhancing Capacity Performance by Utilizing the Redox Chemistry of the Electrolyte in a Dual-Electrolyte Sodium Ion Battery [J]. Angewandte Chemie International Edtion. In English, 2018, 57: 5335–5339.

[237] Zhang H, Qin B, Han J. et al., Aqueous/Nonaqueous Hybrid Electrolyte for Sodium-ion Batteries [J]. ACS Energy Letters, 2018(3): 1769–1770.

[238] Suo L, Borodin O, Gao T. et al., "Water-in-salt" Electrolyte Enables High-Voltage Aqueous Lithium-Ion Chemistries [J]. Science, 2015, 350: 938–943.

[239] Wessells C D, Peddada S V, Huggins R A. et al., Nickel Hexacyanoferrate Nanoparticle Electrodes for Aqueous Sodium and Potassium Ion Batteries [J]. Nano Letters, 2011, 11: 5421–5425.

[240] Wang G, Fu L, Zhao N. et al., An Aqueous Rechargeable Lithium Battery with Good Cycling Performance [J]. Angewandte Chemie International Edtion. In English, 2007, 46: 295–297.

[241] Wang Y, Liu J, Lee B. et al., Ti-substituted Tunnel-type $Na_{0.44}MnO_2$ Oxide as a Negative Electrode for Aqueous Sodium-ion Batteries [J]. Nature Communications, 2015, 6: 6401.

[242] Wu X, Cao Y, Ai X. et al., A Low-Cost and Environmentally Benign Aqueous Rechargeable Sodium Ion Battery Based on $NaTi_2(PO_4)_3$–$Na_2$ NiFe(CN)$_6$ Intercalation Chemistry [J]. Electrochemistry Communications, 2013, 31: 145–148.

[243] Wu X, Luo Y, Sun M. et al., Low-defect Prussian Blue Nanocubes as High Capacity and Long Life Cathodes for Aqueous Na-Ion Batteries [J]. Nano Energy, 2015, 13: 117–123.

[244] Kim D J, Ponraj R, Kannan A G. et al., Diffusion Behavior of Sodium Ion S in $Na_{0.44}MnO_2$ in Aqueous and Non-Aqueous Electrolytes [J]. Power Sources, 2013, 244: 758–763.

[245] Sun Y, Zhao L, Pan H. et al., Direct Atomic-scale Confirmation of Three-Phase Storage Mechanism in $Li_4Ti_5O_{12}$ Anodes for Room-Temperature Sodium-ion Batteries [J]. Nature Communications, 2013, 4: 1870.

[246] Deng W, Shen Y, Qian J. et al., A Polyimide Anode with High Capacity and Superior Cyclability for Aqueous Na-ion Batteries [J]. Chemical Communications (Camb), 2015, 51: 5097–5099.

[247] Luo J Y, Cui W J, He P. et al., Raising the Cycling Stability of Aqueous Lithium-Ion Batteries by Eliminating Oxygen in the Electrolyte [J]. Nature Chemistry, 2010, 2: 760–765.

[248] Yun J, Pfisterer J, Bandarenka A S. How Simple are the Models of Na Intercalation in Aqueous Media? [J]. Energy & Environmental Science, 2016, 9: 955–961.

[249] Zhan X, Shirpour M. Evolution of Solid/Aqueous Interface in Aqueous Sodium-ion Batteries [J]. Chemical Communications (Camb), 2016, 53: 204–207.

[250] Alias N, Mohamad A. Advances of Aqueous Rechargeable Lithium-ion Battery: A Review [J]. Power Sources, 2015, 274: 237–251.

[251] Nakamoto K, Kano Y, Kitajou A. et al., Electrolyte Dependence of the Performance of a $Na_2FeP_2O_7//NaTi_2(PO_4)_3$ Rechargeable Aqueous Sodium Ion Battery [J]. Power Sources, 2016, 327: 327–332.

[252] Wu W, Shabhag S, Chang J. et al., Relating Electrolyte Concentration to Performance and Stability for $NaTi_2(PO_4)_3/Na_{0.44}MnO_2$ Aqueous Sodium-ion Batteries [J]. Electrochemical Society, 2015, 162: A803–A808.

[253] Wessells C D, Huggins R A, Cui Y. Copper Hexacyanoferrate Battery Electrodes with Long Cycle Life and High Power [J]. Nature Communications, 2011, 2: 550.

[254] Kumar P R, Jung Y H, Moorthy B. et al., Effect of Electrolyte Additives on $NaTi_2(PO_4)_3$-C// $Na_3V_2O_2X(PO_4)2F_3$-2X-MWCNT Aqueous Rechargeable Sodium Ion Battery Performance [J]. Electrochemical Society, 2016, 163: A1484–A1492.

[255] Larcher D, Tarascon J. Towards Greener and More Sustainable Batteries for Electrical Energy Storage[J]. Nature Chemistry, 2015, 7(1): 19–29.

[256] Moshkovich M, Gofer Y, Aurbach D. Investigation of the Electrochemical Windows of Aprotic Alkali Metal (Li, Na, K) Salt Solutions[J]. Electrochemical Society, 2001, 148(4): E155.

[257] Vogdanis L, Martens B, Uchtmann H. et al., Synthetic and Thermodynamic Investigations in the Polymerization of Ethylene Carbonate[J]. Makromolekulare Chemie-Macromolecular Chemistry and Physics, 1990, 191(3): 465–472.

[258] Komaba S, Murata W, Ishikawa T. et al., Electrochemical Na Insertion and Solid Electrolyte Interphase for Hard-Carbon Electrodes and Application to Na-Ion Batteries[J]. Advanced Functional Materials, 2011, 21(20): 3859–3867.

[259] Li Q, Wei Q, Zuo W. et al., Greigite $Fe_3S_4$ as a New Anode Material for High-Performance Sodium-ion batteries[J]. Chemical Science, 2017, 8(1): 160–164.

[260] Bommier C, Leonard D, Jian Z. et al., New Paradigms on the Nature of Solid Electrolyte Interphase Formation and Capacity Fading of Hard Carbon Anodes in Na-Ion Batteries[J]. Advanced Materials Interfaces, 2016, 3(19): 1600449.

[261] Kumar H, Detsi E, Abraham D P. et al., Fundamental Mechanisms of Solvent Decomposition Involved in Solid-Electrolyte Interphase Formation in Sodium-ion batteries[J]. Chemistry of Materials, 2016, 28(24): 8930–8941.

[262] Zheng Y, Wang Y, Lu Y. et al., A High-Performance Sodium Ion Battery Enhanced by Macadamia Shell Derived Hard Carbon Anode[J]. Nano Energy, 2017, 39: 489–498.

[263] Xiao L, Cao Y, Henderson W A. et al., Hard Carbon Nanoparticles as High-Capacity, High-Stability Anodic Materials for Na-Ion Batteries[J]. Nano Energy, 2016, 19: 279–288.

[264] Dahbi M, Nakano T, Yabuuchi N. et al., Sodium Carboxymethyl Cellulose as a Potential Binder for Hard-Carbon Negative Electrodes in Sodium-ion batteries[J]. Electrochemistry Communications, 2014, 44: 66–69.

[265] Zhu Y S, Li L L, Li C Y. et al., $Na_{1+x}Al_xGe_{2-x}P_3O_{12}$ (X = 0.5) Glass– Ceramic as a Solid Ionic Conductor for Sodium Ion [J]. Solid State Ionics, 2016, 289: 113–117.

[266] Dahbi M, Yabuuchi N, Fukunishi M. et al., Black Phosphorus as a High-Capacity, High-Capability Negative Electrode for Sodium-ion Batteries: Investigation of the Electrode/ Electrolyte Interface[J]. Chemistry of Materials, 2016, 28(6): 1625–1635.

[267] Yang X, Wang C, Yang Y. et al., Anatase $TiO_2$ Nanocubes for Fast and Durable Sodium Ion Battery Anodes[J]. Materials Chemistry A, 2015, 3(16): 8800–8807.

[268] Xu Z, Lim K, Park K. et al., Engineering Solid Electrolyte Interphase for Pseudocapacitive Anatase $TiO_2$ Anodes in Sodium-ion batteries[J]. Advanced Functional Materials, 2018, 28(29): 1802099.

[269] Wu L, Buchholz D, Bresser D. et al., Anatase $TiO_2$ Nanoparticles for High Power Sodium Ion Anodes[J]. Power Sources, 2014, 251: 379–385.

[270] Shoaib Anwer Y H J L. Nature Inspired $Na_2Ti_3O_7$ Nanosheets Formed Three- Dimensional Micro-Flowers Architecture as a High-Performance Anode Material for Rechargeable Sodium-ion batteries[J]. ACS Applied Materials & Interfaces, 2017, 9(13): 11669–11677.

[271] Fu S, Ni J, Xu Y. et al., Hydrogenation Driven Conductive $Na_2Ti_3O_7$ Nanoarrays as Robust Binder-Free Anodes for Sodium-ion batteries[J]. Nano Letters, 2016, 16(7): 4544–4551.

[272] Pang G, Nie P, Yuan C. et al., Mesoporous $NaTi_2(PO4)_3$/CMK-3 Nanohybrid as Anode for Long-Life Na-Ion Batteries[J]. Materials Chemistry. A, 2014, 2(48): 20659–20666.

[273] Padhy H, Chen Y, Lüder J. et al., Charge and Discharge Processes and Sodium Storage in Disodium Pyridine-2,5-Dicarboxylate Anode-Insights from Experiments and Theory[J]. Advanced Energy Materials, 2018, 8(7): 1701572.

[274] Oltean V A, Philippe B, Renault S. et al., Investigating the Interfacial Chemistry of Organic Electrodes in Li and Na-Ion Batteries[J]. Chemistry of Materials, 2016, 28(23): 8742–8751.

[275] Baggetto L, Ganesh P, Meisner R P. et al., Characterization of Sodium Ion Electrochemical Reaction with Tin Anodes: Experiment and Theory[J]. Power Sources, 2013, 234: 48–59.

[276] Darwiche A, Bodenes L, Madec L. et al., Impact of the Salts and Solvents on the SEI Formation in Sb/Na Batteries: An XPS Analysis[J]. Electrochimica Acta, 2016, 207: 284–292.

[277] Piernas-Muñoz M J, Castillo-Martínez E, Gómez-Cámer J L. et al., Optimizing the Electrolyte and Binder Composition for Sodium Prussian Blue, $Na_{1-x}Fe_{x+(1/3)}(CN)_6·yH_2O$, as Cathode in Sodium-ion batteries[J]. Electrochimica Acta, 2016, 200: 123–130.

[278] Chen R, Huang Y, Xie M. et al., Chemical Inhibition Method to Synthesize Highly Crystalline Prussian Blue Analogs for Sodium Ion Battery Cathodes[J]. ACS Applied Materials & Interfaces, 2016, 8(46): 31669–31676.

[279] Fu H, Xia M, Qi R. et al., Improved Rate Performance of Prussian Blue Cathode Materials for Sodium-ion Batteries Induced by Ion-Conductive Solid-Electrolyte Interphase Layer[J]. Power Sources, 2018, 399: 42–48.

[280] Mu L, Rahman M M, Zhang Y. et al., Surface Transformation by a "Cocktail" Solvent Enables Stable Cathode Materials for Sodium-ion batteries[J]. Materials Chemistry A, 2018, 6(6): 2758–2766.

[281] Feng T, Xu Y, Zhang Z. et al., Low-Cost $Al_2O_3$ Coating Layer as a Preformed SEI on Natural Graphite Powder to Improve Coulombic Efficiency and High-Rate Cycling Stability of Lithium-Ion Batteries[J]. ACS Applied Materials & Interfaces, 2016, 8(10): 6512–6519.

[282] Yuan L, Wang Z, Zhang W. et al., Development and Challenges of $LiFePO_4$ Cathode Material for Lithium-Ion Batteries[J]. Energy Environ. Sci., 2011, 4(2): 269–284.

[283] Liu H, Chen C, Du C. et al., Lithium-Rich $Li_{1.2}Ni_{0.13}Co_{0.13}Mn_{0.54}O_2$ Oxide Coated by $Li_3PO_4$ and Carbon Nanocomposite Layers as High Performance Cathode Materials for Lithium Ion Batteries[J]. Materials Chemistry A, 2015, 3(6): 2634–2641.

[284] Jo J H, Choi J U, Konarov A. et al., Sodium-ion Batteries: Building Effective Layered Cathode Materials with Long-Term Cycling by Modifying the Surface via Sodium Phosphate[J]. Advanced Functional Materials, 2018, 28(14): 1705968.

[285] Jo C, Jo J, Yashiro H. et al., Bioinspired Surface Layer for the Cathode Material of High-Energy-Density Sodium-ion batteries[J]. Advanced Energy Materials, 2018, 8(13): 1702942.

[286] Wan H, Mwizerwa J P, Qi X. et al., Core–Shell $Fe_{1-x}S@Na_{2.9}PS_{3.95}Se_{0.05}$ Nanorods for Room Temperature All-Solid-State Sodium Batteries with High Energy Density[J]. ACS Nano, 2018, 12(3): 2809–2817.

# Chapter 6
# The commercialization of sodium-ion batteries

With the continuous development of the modern industry, the demand for electrical energy has been rising in our society every year, and the amount of primary energy, such as oil and natural gas, which are indispensable in the field of electricity supply, has been increasing every year. The dependence on primary energy has also been rising (Figure 6.1) [1]. In recent years, in order to solve the energy demand and security problems, as well as to alleviate the environmental pollution caused by the burning of fossil fuels, China has vigorously promoted the development of renewable energy technologies. The National Development and Reform Commission and the National Energy Administration have published documents such as the "Thirteenth Five-Year Plan for Renewable Energy Development" and the "2018 Energy Work Guidance Circular", which highlight the importance of improving clean energy utilization in building an efficient and clean modern energy system [2]. However, due to the intermittent nature of renewable energy sources, such as wind and solar energy, their power generation systems have an unstable and discontinuous supply of electricity to the grid, while the power consumption in industry, transportation sector, and households is cyclical [3, 4]. Therefore, we need to build large-scale energy storage systems to improve the utilization of renewable energy sources.

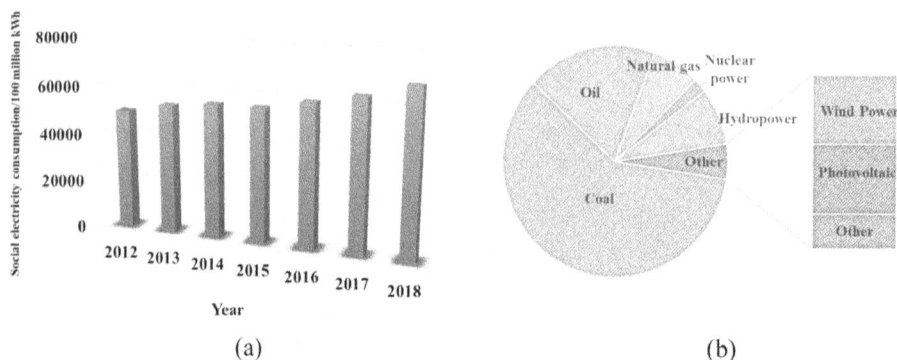

(a)                                                         (b)

**Figure 6.1:** Current situation of energy utilization in China: (a) total social electricity consumption in China during 2012–2018; and (b) China's energy consumption structure in 2018.

Among the current energy storage technologies, secondary batteries have attracted widespread attention due to their short construction cycle, long cycle life, and easy maintenance [5–8]. However, unlike mobile electronics (~4 W h) and electrified transportation (~40 kW h), large-scale energy storage must be able to store megawatt-hour capacities to meet the needs of renewable energy and smart grids [9]. Such large-scale power storage requires a large number of secondary batteries and hence, metrics such

https://doi.org/10.1515/9783110749069-006

as material abundance, battery cost, and safety become its main limiting factors [10]. Currently, lithium-ion batteries, with characteristics such as high energy density and light weight occupy a monopoly in electrochemical energy storage systems; however, the limited lithium resources and the concentration of their distribution limit their sustainability in this field [5]. Sodium-ion batteries, which have high similarity with lithium-ion batteries in terms of reaction mechanism and cell structure, not only have obvious advantages in terms of resources and costs, but also have excellent multiplicative characteristics and high recyclability, which are more suitable for the construction of next generation large-scale energy storage systems. Although sodium-ion battery research started almost at the same time as lithium-ion batteries, its substantial development started in the early twenty-first century, which makes it only 20 years old [11]. At present, some results have been achieved in the research of sodium-ion batteries in the fields of cathode and anode materials, electrolytes, and diaphragms, but the application of a battery system still depends on the development of full battery research and the growth of market demand. This chapter will explore the current market pain points of sodium-ion batteries commercialization through SWOT (strengths, weaknesses, opportunities, and challenges) analysis, as well as summarize the representative sodium-ion full batteries in research, analyze the technical difficulties in their industrial application, and then discuss the performance and structure design, and future development trends of sodium-ion batteries required to meet the requirements of large-scale energy storage.

## 6.1 SWOT analysis of sodium-ion batteries

Sodium-ion batteries will inevitably compete with lithium-ion batteries in the future commercialization process; so sodium-ion batteries must give full play to their advantages and build full batteries at low cost, and with fast response capability and ultralong cycle life. As shown in Table 6.1, in this chapter, we carried out a SWOT analysis on the feasibility of sodium-ion batteries in large-scale energy storage applications and summarized the unique advantages of this battery system in terms of cost resources, power characteristics, climate characteristics, and recyclability. At present, sodium-ion full batteries of organic system are relatively mature, and companies such as HiNa BATTERY and STAR SODIUM have the ability to produce industrial-grade sodium-ion full batteries. However, aqueous sodium-ion batteries are gradually gaining market attention as a kind of low-cost, high safety, and environment-friendly full batteries. The hybrid aqueous sodium-ion capacitor battery is simple in structure, costs less, is environmentally friendly, and effectively circumvents the challenge of selecting a suitable sodium storage cathode material. Aqueion USA has started the commercialization of aqueous sodium-ion batteries based on this system of full batteries. Solid state electrolyte has a very important role in improving the energy density, operating voltage, and safety performance of the battery system; so the research on all-solid-state sodium batteries has become increasingly in-depth. Currently, all-solid-state sodium batteries are still in the laboratory research stage and have a long way to go before commercial application, but the application of this system will not only have an impact on large-scale energy storage systems, but will also make the application of flexible electronic wearable devices possible. In conclusion, as an important energy storage component for grid storage and smart grid construction, the commercialization of sodium-ion batteries will be accelerated with the gradual adjustment of the China's energy structure, and will become an important component of the future energy storage market.

**Table 6.1:** SWOT analysis of sodium-ion battery.

| | **Advantage** | **Disadvantage** |
|---|---|---|
| Internal factor analysis | 1. Resources are plentiful and costs are low<br>2. Excellent power characteristics<br>3. High recyclability<br>4. Strong adaptability to climate<br>5. High safety | 1. Low energy density<br>2. Lack of suitable commercial electrode materials<br>3. Lack of low-cost commercial electrolytes |
| | **Opportunity** | **Threat** |
| External environmental analysis | 1. National energy storage policy support<br>2. Grid-side energy storage and renewable energy demand<br>3. Low speed electric vehicle market demand | 1. Lithium-ion batteries are monopolized in the industry<br>2. New energy storage technologies are developing rapidly |

### 6.1.1 Advantage

#### 6.1.1.1 Resources and costs

Lithium resources are limited in availability and are unevenly distributed; mostly located in remote or politically-sensitive areas such as Chile (52%) and Argentina (14%), thus increasing transportation risks and transportation costs [12–14]. In contrast, sodium resources are abundant and widely distributed, and its elemental abundance in the earth is more than 1,000 times that of lithium (Figure 6.2). In lithium-ion batteries, copper foil is often used as the collector for the negative electrode and aluminum foil for the positive electrode. For sodium-ion batteries, on the other hand, aluminum foil can be used as the collector for both positive and negative electrodes, and the cost of aluminum foil is lower than that of copper foil due to its abundance and wide distribution [5].

The cathode materials commonly used in commercial lithium-ion batteries include LCO, NCA, NCM, etc. All these cathode materials contain Li, Co, or Ni elements, all of

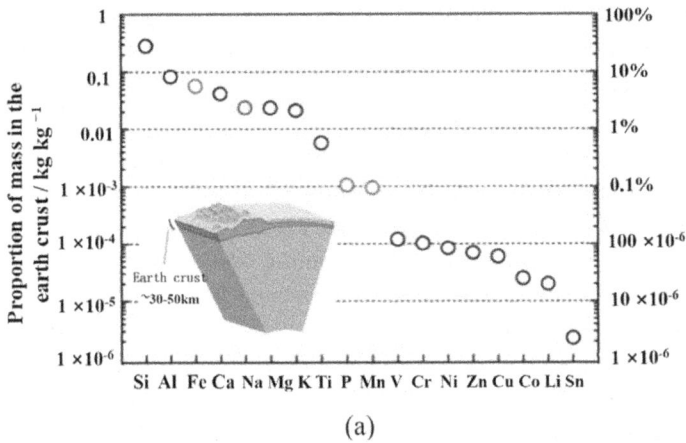

(a)

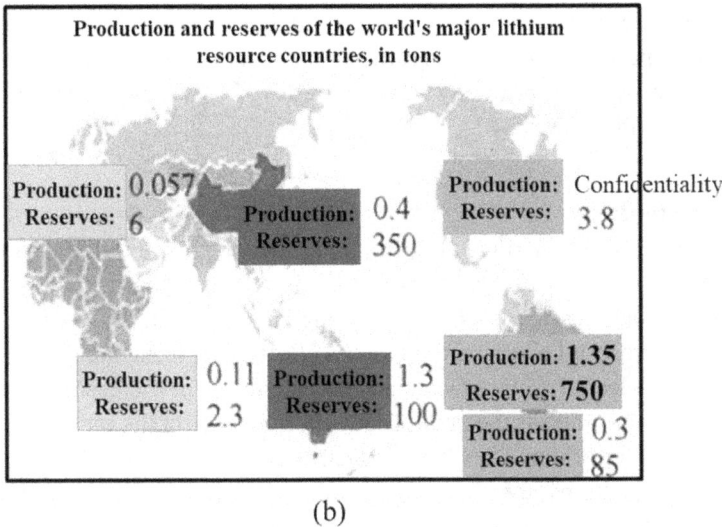

(b)

**Figure 6.2:** Reserves and distribution of lithium resources: (a) distribution of elements on the Earth's crust [11]; (b) global distribution of lithium resources [18].

which are relatively limited on the earth's crust. The cathode materials for sodium-ion batteries are mainly divided into metal oxides, polyanionic compounds, Prussian blue analogues, organic compounds, etc. The larger layer spacing of layered transition metal oxide materials enables the reversible deintercalating of sodium ions (1.02 Å), and the representative layered transition metal oxides are $NaFeO_2$, $Na_{0.44}MnO_2$, $Na_{7/9}Cu_{2/9}Fe_{1/9}Mn_{2/3}O_2$, etc. [15–17]. These materials have excellent electrochemical properties, and their raw materials are based on Fe and Mn, which are abundant on the earth's crust and are of low cost.

The electrolyte for lithium-ion/sodium-ion batteries consists of an electrolyte salt and a solvent, with the electrolyte salt accounting for 10% to 20% of the total electrolyte mass. Sodium salts are relatively easy to obtain and their processing costs are much lower than lithium analogs, and the organic solvents used in both electrolytes are nearly identical; therefore, the use of sodium salts, instead of lithium salts in electrolytes, can further reduce battery costs.

As shown in Figure 6.3, Vaalma et al. modeled the cost of sodium-ion and lithium-ion full batteries and compared them [19]. As shown in Figure 6.3, for a capacity of 11.5 kW h, the cost of Li-ion battery, with $LiMn_2O_2$ as the positive electrode and graphite as the negative electrode, is $ 1,022, and the cost of sodium-ion battery, with sodium-manganese-based material as the positive electrode and graphite as the negative electrode is $ 894; the cost of sodium-ion battery is relatively lower. Hu et al. calculated and analyzed the cell raw material cost for two specific lithium-ion and sodium-ion batteries. The cost ($ 0.37/W h) was somewhat lower for both metal oxide type sodium-ion batteries and Prussian blue type sodium-ion batteries, relative to lithium-ion phosphate batteries ($ 0.47/W h), for the same capacity and voltage [18]. With the specific capacity of the cathode material reaching 120 mA h $g^{-1}$, the raw material cost of sodium-ion battery cells can be as low as ¥ 0.31 /W h, which has a more prominent cost advantage.

### 6.1.1.2 Power characteristics

Compared to lithium metal, sodium metal has a larger ionic radius, and the larger ionic radius makes the sodium ion have a lower desolvation energy barrier in polar solvents, and thus the sodium salt has a higher solubility in the solvent [20]. The desolvation energy has a strong influence on the kinetics of the alkali metal-ion deintercalating process at the electrolyte interface, and the magnitude of the desolvation energy of the alkali metal ion affects the multiplicative performance of the cell [21, 22]. Compared to sodium ions, lithium ions have a small radius and a relatively high charge density around their ions; therefore, they require a relatively large energy for the desolvation process, compared to sodium ions. This conclusion is supported by results of first-principles calculations, where the activation energy for diffusion is relatively smaller for $NaCoO_2$ compared to $LiCoO_2$ [23]. In addition, the higher ionic conductivity of the electrolyte of sodium-ion batteries, compared to that of lithium-ion battery, helps to improve the electrochemical performance of the battery. A comparative study of the molar conductivity of $NaClO_4$ and $LiClO_4$ showed that $NaClO_4$ has a relatively lower viscosity and higher conductivity in the same nonprotonic solvent [24]. The relatively large ionic radius of sodium ions increases the flexibility of material design. The large size gap in the sodium-ion radius makes it easy to prepare layered oxides, with different stacking patterns from many different three-dimensional transition metals (SC-Ni) [25].

Currently, layered metal oxides, NASICON-type materials, carbon materials, and alloy-transformed materials have all shown excellent multiplicative performance in sodium-ion batteries studies. As shown in Figure 6.4(a), the carbon-coated $NaCrO_2$,

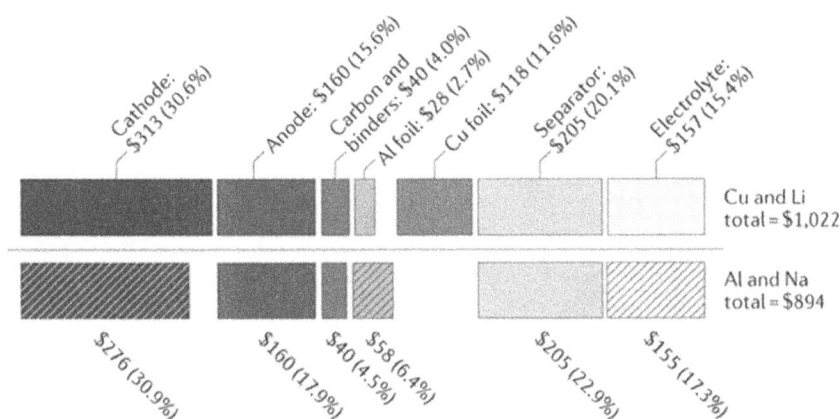

**Figure 6.3:** Cost analysis model of sodium-ion batteries and lithium-ion batteries [19].

prepared by Yu et al., has good capacity retention and up to 150 C charge/discharge multiplicity, which is equivalent to complete charging or discharging in 27 s [26]. $Na_3V_2(PO_4)_3$ is a representative NASICON-type material, as shown in Figure 6.4(c), and Ren et al. prepared a novel three-dimensional nano fiber network structure of $Na_3V_2(PO_4)_3$ material [27]. The three-dimensional nanofiber network structure provides a multi-channel ion diffusion pathway, enhances the electron conduction ability, and improves the structural stability. The material can still discharge a specific capacity of 94 mA h g$^{-1}$ at a multiplicity of 100 C, reflecting the excellent multiplicity performance. For carbon materials, the embedding of sodium ions in graphite interlayers can be achieved by selecting a suitable electrolyte solvent; for example, graphite anodes have achieved low overpotential sodium storage of approximately 100 mA h g$^{-1}$ in diethylene glycol dimethyl ether-based electrolytes and have been able to undergo 1,000 cycles [28–30]. The application of graphite materials in sodium-ion batteries can also be achieved by expanding the layer spacing. Expanded graphite, with a 4.3 Å layer spacing, can be prepared by first oxidizing and then partially reducing the graphite material, which can discharge a high reversible specific capacity of 284 mA h g$^{-1}$ at a current density of 20 mA g$^{-1}$, when used as a sodium storage anode [31]. Alloyed and converted electrode materials also exhibit excellent multiplicative performance, which facilitates the realization of high-power sodium-ion batteries. As shown in Figure 6.4(b), the Bi@graphite complex is able to provide a discharge specific capacity of 160 mA h g$^{-1}$ at a current density of 1 C and maintain 70% specific capacity (~110 mA h g$^{-1}$) at 30 C, which corresponds to a complete charge and discharge in 12 s [32]. In summary, many electrode materials in sodium-ion batteries exhibit excellent multiplicative performance and thus have a greater advantage over lithium-ion batteries in terms of this characteristic of fast response energy storage systems.

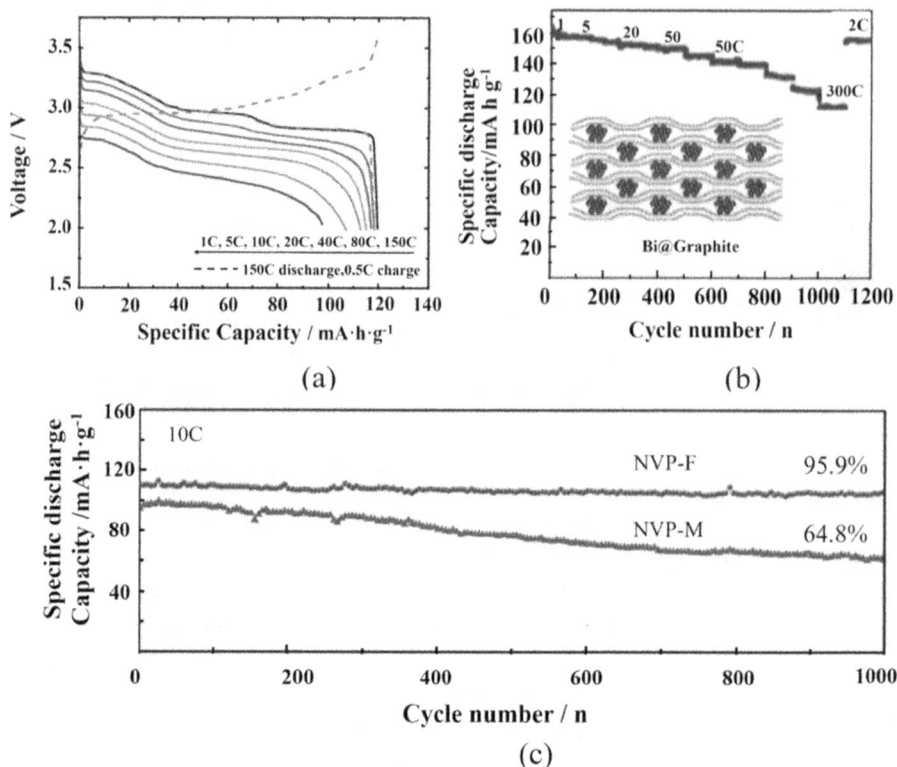

**Figure 6.4:** Sodium plate materials with typical high power characteristics: (a) specific capacity–voltage curves of C-NaCrO$_2$ with mass percentage of 3.4% at different magnification rates [26]; (b) cycling of bi@Graphite complexes at different charge and discharge rates – specific capacity curve [32]; and (c) cyclic specific capacity curve of NVP-F and NVP-M at 10 C charge/discharge rate [27].

### 6.1.1.3 Recyclability

Lithium-ion batteries have the same composition as sodium-ion batteries and consist of six main components: anode, cathode, electrolyte, diaphragm, collector, and shell [33]. In lithium-ion and sodium-ion batteries, the shell, electrolyte, and diaphragm can be effectively recycled, but the recycling of the positive and negative electrode materials and the collector fluid is still a major challenge [34]. For example, lithium-ion batteries with a single electrode structure, the positive and negative materials are coated on Al and Cu foils, respectively. During large-scale recycling of Li-ion batteries, the positive and negative electrode materials and the collector fluid inevitably mix together, making it difficult to separate them, increasing the difficulty and cost of recycling [35]. Thus, the valuable metals in lithium-ion batteries are usually recovered using wet and fire methods, and the recovered metals can be reused in the production of new batteries. However, the recycling process is costly and generates solid, liquid, and gaseous wastes that pollute the environment [34]. For batteries with large-scale energy storage and applications, this causes greater consumption of rare metal sources (e.g., lithium) and a more serious environmental impact, if there is no green and effective low-cost recycling method.

As shown in Figure 6.5, researchers designed sodium-ion batteries with a bipolar electrode structure: the positive electrode material, $Na_2V_2(PO_4)_3$ (NVP), and the negative electrode materials were coated on both sides of a common collector (Al foil) [36]. No alloying reaction occurs between Na and Al during charging and discharging; thus the electrode material can be easily collected from the spent sodium-ion batteries by a simple treatment. More than 98% of the solid fraction of spent sodium-ion batteries is recovered without releasing toxic waste. High-value NVP was separated and collected with 100% recovery efficiency. The NVP, treated by a simple regeneration process, is used in new sodium-ion batteries and exhibits excellent electrochemical performance. The use of this structure for sodium-ion batteries, with double electrodes, gives the NVP material a closed-loop sustainability.

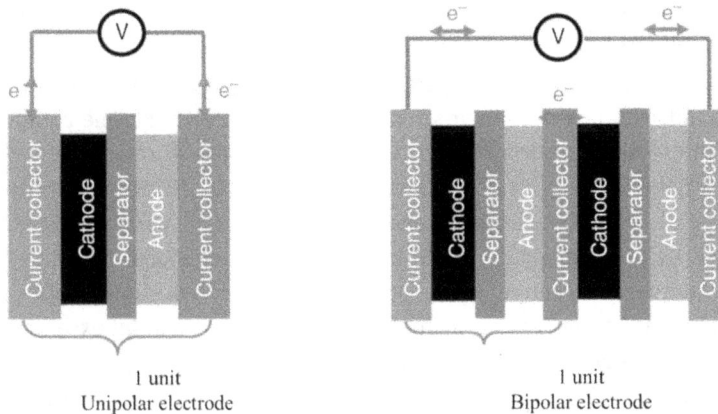

|  I unit | I unit |
| Unipolar electrode | Bipolar electrode |

**Figure 6.5:** Schematic diagram of single-stage battery system and novel bipolar battery system [36].

### 6.1.1.4 Climatic characteristics

The reaction kinetics of electrode materials in most secondary cells are temperature-dependent. Usually, the battery performance increases as the operating temperature increases. Since sodium-ion batteries use organic electrolytes, their operating temperature must be kept below the threshold of organic solvents to avoid the risk of fire or electrochemical degradation of the battery at low temperatures. Therefore, large-scale energy storage systems, using secondary cells as energy storage units, must be paired with temperature management systems for safe and stable energy storage and delivery [37]. Researchers have designed an internal self-heating nickel foil assembly with superior temperature management capabilities, compared to the conventional external temperature control assemblies. The internal heating system consumes only 3.8% of the battery capacity to rapidly warm up the battery system from a low temperature environment (–20 °C) to 20 °C. Nevertheless, the temperature component configuration has a negative impact on the energy density of the battery and the cost of the stationary electrochemical energy storage systems, and should be addressed before the energy storage method is widely used. The climate characteristics of sodium-ion batteries are summarized below, demonstrating that sodium-ion batteries can operate over an extremely wide temperature range (–30 to 55 °C) without additional temperature control to maintain the operating temperature. This climatic characteristic is crucial for large-scale energy storage, demonstrating their strong adaptability to various climatic environments and highlighting their high reliability and safety.

All-climate electrodes are essential for sodium-ion batteries to achieve an extremely wide operating temperature window, which requires electrodes with a stable phase structure and an electrode-electrolyte interface that can quickly deliver sufficient charge. Layered transition metal oxides are not thermally stable at high temperatures; so many researchers have worked on introducing inactive elements, such as $Li^+$, $Mg^{2+}$, $Al^{3+}$, and $Mn^{4+}$, into the transition metal layers of layered oxides to enhance their structural stability. Compared to layered transition metal oxides, phosphate compounds can maintain their structural stability under extreme conditions. Therefore, the seminal reports related to the all-climate performance of sodium-ion batteries have focused on phosphate compounds. NASICON type is the most representative all-climate electrode, as shown in Figure 6.6(a), where Liu et al. prepared a $Na_2V_2(PO_4)_3$@C composite, which exhibited electrochemical properties over an excellent temperature range, from –30 to 55 °C[38]. In addition to phosphate compounds, several other advanced electrode materials also exhibit excellent electrochemical properties at low temperatures. Ponrouch et al. investigated the sodium storage performance of hard carbon at low temperatures and experimentally demonstrated its ability to put out a specific capacity of 265 mA h $g^{-1}$ at 5 °C [39]. You et al. prepared a Prussian-blue iron-based composite electrode, which exhibited excellent electrochemical properties at –25 °C [40]. It is worth noting, however, that all the above sodium-ion batteries were evaluated using half-cells for the electrode materials, neglecting the high impedance of the sodium metal anode at low temperatures.

As the operating temperature decreases, the half-circle diameter in the high to mid-frequency range of the impedance curve of the sodium-ion battery becomes significantly larger, indicating an increase in the internal impedance of the battery, an increase that is caused by the negative sodium metal electrode in the sodium-ion battery and not by the working electrode [38]. The high impedance of the sodium metal electrode leads to a cell with a large overpotential during cycling, which narrows the actual voltage window for charging/discharging, and eventually leads to the failure of the NVP@C cathode. Therefore, in order to better evaluate the kinetic properties of the electrode material, full-cell or symmetrical electrode cells are applied, instead of half-cell tests, in order to provide more valid information about the material. As shown in Figure 6.6(b), the full cell assembled with $Na_2V_2(PO_4)_2O_2F$ cathode and $NaTi_2(PO_4)_3$ anode (NVPOF-NTP) [41], the specific capacity of the full cell still reached 96.1 mA h g$^{-1}$ when the temperature was lowered from 25 °C to −25 °C. and has two distinct plateaus with a capacity retention of 76.4% at 25 °C, which implies that NVPOF-NTP electrodes have good low-temperature kinetics.

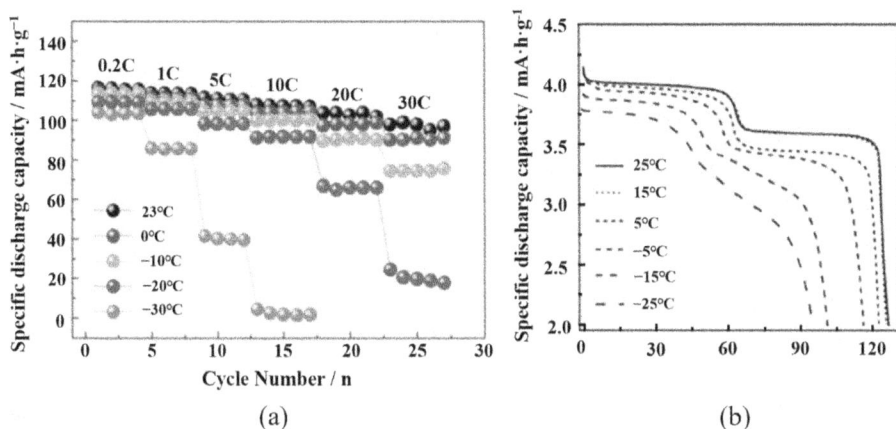

**Figure 6.6:** Electrochemical performance of an all-climate electrode: (a) rate performance of NVP@C at different temperatures [38]; (b) constant-current discharge curve of NVPF-NTP cycling in the temperature range of −25 ℃ to 25℃ [41].

Temperature significantly affects the ion diffusion at the interface between the electrolyte and the solid electrolyte. Performance tests [42] on various electrolytes containing different solvent mixtures (including cyclic/cyclic carbonate and glycol dimethyl ether) and sodium salts containing F groups or perchlorate anions (including electrolyte viscosity, ionic conductivity, thermal stability, and electrochemical stability) showed that binary solvents of ethylene carbonate (EC) and propylene carbonate (PC) were the best electrolyte solvents, and their combination with $NaClO_4$ or $NaPF_6$ salts constituting the electrolyte can be used to evaluate the electrochemical performance of Na/rigid carbon cells. Since the glass transition temperature of PC is about −95 °C, electrolyte solidification was not observed experimentally when PC was used as a co-solvent. This property is important for sodium-ion batteries used in low temperature conditions, allowing the liquid electrolyte to exhibit ionic liquid-like electrochemical properties at very low temperatures. In addition, high temperatures are a major challenge for organic electrolytes. Compared to nonaqueous organic electrolytes, solid-state electrolytes have better mechanical properties, higher thermal stability, and wider electrochemical windows. Therefore, solid-state sodium-ion batteries may be the ultimate choice for building battery systems with higher energy density and safer characteristics.

### 6.1.1.5 Security

For secondary batteries, safety is a basic requirement for wide application. Lithium-ion batteries are now widely used in mobile 3 C devices, electric vehicles, and large-scale energy storage. However, in recent years, reports of lithium-ion battery safety accidents have emerged one after another, including electric car fires and cell phone battery explosions, impacting people's confidence in the safety of lithium-ion batteries. For batteries in scale energy storage applications, the level of safety directly determines the possibility of their future applications. Although the reaction mechanism and material system of sodium-ion batteries are similar to those of lithium-ion batteries, sodium-ion batteries have a higher safety level. This is not only because of the relatively low energy density of sodium-ion batteries, but in this system, researchers can mitigate or even eliminate the safety problems caused by the electrolyte from the electrolyte design and electrode/electrolyte interface optimization.

Battery abuse often leads to thermal runaway, thus causing a rapid increase in the internal temperature of the battery, which in turn triggers a series of uncontrollable chemical reactions, such as the decomposition of the SEI film on the negative electrode, the reaction between the negative electrode and the electrolyte, and the decomposition reaction of the electrolyte [43]. Thermal runaway is often accompanied by fire and toxic smoke, which directly threatens the life of users. This runaway process starts with the disruption of the interface homeostasis and ends with the fire of the electrolyte; so the construction of a stable electrode/electrolyte interface is the key to solve the battery safety problem. For nonaqueous sodium-ion batteries, the following strategies can be used to improve their battery safety: (i) using stable Na salts or high concentration electrolytes; (ii) using nonflammable solvents or ionic liquids (ILs); and (iii) adding flame retardant additives and overcharge protection additives. Using stable Na salt can improve the thermal stability of electrolyte, while increasing salt concentration can reduce the flammability of electrolyte. The use of noncombustible solvents or ionic liquids can make the electrolyte noncombustible. Introduced flame retardants and overcharge protection additives can inhibit combustion reactions by trapping free radicals and suppress the voltage rise of the cell under overcharge conditions, respectively. However, understanding how to improve the safety of electrolytes without compromising the electrochemical performance of sodium-ion batteries still requires high attention. Since ether and ester electrolytes have been widely used and studied in Li-ion batteries, studies have shown that the thermal stability of pure solvents follows the order DME < DMC < DEC < EC < PC [43], with ether having lower thermal stability, compared to esters. Moreover, the thermal stability of solvents is closely related to their flash points. Therefore, the use of some other solvents with high flash points, such as sulfones, nitrile, silicon-based solvents, and fluorinated solvents, can also help to improve the safety of nonaqueous electrolytes for sodium-ion batteries.

Aqueous sodium-ion batteries have higher safety. In this system, 1 M $Na_2SO_4$ aqueous solution or 10 M $NaClO_4$ aqueous solution is used as the electrolyte, which has high flash point and ignition point, and with the low operating voltage of the battery, the probability of thermal runaway is greatly reduced and there is no harsh environmental impact. However, the disadvantage of aqueous electrolyte is also very obvious. If overcharged, the gas generated by the decomposition of electrolyte will make the battery bloated. Adding redox additives to the electrolyte can effectively inhibit the generation of oxygen and other gases. All in all, the safety of sodium-ion batteries proves that it can be used in large-scale energy storage system without serious explosion and fire accidents due to its own defects.

## 6.1.2 Disadvantage

### 6.1.2.1 Energy density
Although sodium-ion batteries have obvious advantages in terms of material cost and resource reserves, they have more obvious disadvantages in terms of specific mass energy density and specific volume energy density due to the larger relative atomic mass and ionic radius of sodium ions, compared to lithium-ion batteries. Taking $LiCoO_3$ and $NaCoO_3$ as examples, assuming that only one electron redox ($CO^{3+}/Co^{4+}$ redox) occurs for cobalt ions, the theoretical specific capacities of $LiCoO_3$ and $NaCoO_3$ can be calculated to be 274 mA h g$^{-1}$ and 235 mA h g$^{-1}$, respectively, and the reversible capacity decreased by 14% year-on-year [11]. As shown in (Figures 6.7), the voltage reduction of $NaCoO_3$ becomes more pronounced with the increase of sodium content in the structure. Therefore, the actual available energy density of the Na-ion battery system is much lower.

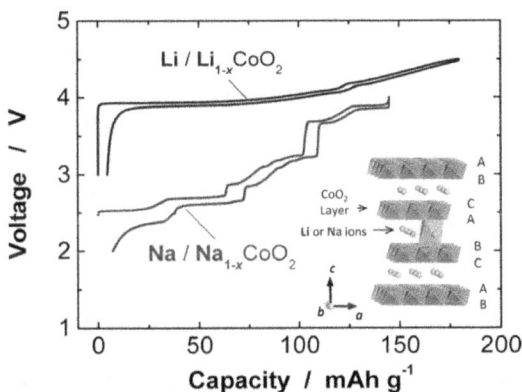

**Figure 6.7:** Charge/discharge curves of Li/LiCoO₂ and Na/NaCoO₂ batteries [44].

### 6.1.2.2 Electrode materials

Sodium-ion batteries are developed on the basis of lithium-ion batteries, and they have great similarity in terms of cell reaction mechanism, structure and components, but the radius of sodium ions is larger than that of lithium ions, and thus some materials are not suitable for application in sodium-ion batteries. For example, natural graphite, a widely used anode material in lithium-ion batteries, initially had almost no capacity when applied to sodium-ion batteries [23, 44, 45]. Therefore, the further development of sodium-ion batteries requires the exploration of cathode and anode materials suitable for this system. Currently, the closest to commercialization of anode materials for sodium-ion batteries is hard carbon, which has a practical capacity close to that of graphite, but requires extremely high temperature (>1,000 °C) in its preparation, and the higher cost and safety of the preparation process become the key factors limiting its commercial application. Although transition metal oxides (sulfides) themselves have a very high discharge capacity, the problems of material pulverization and electrode polarization during the reaction process make their cycle life poor. Therefore, in order to solve this problem, researchers often enhance the material performance by means of carbon coating and structural design, but the complex modification process and the use of high-cost modification materials (such as graphene) likewise limit their large-scale production and application.

Among the cathode materials, layered metal oxides have high reversible capacity and voltage plateau, but the presence of multiphase reactions during the charging and discharging process make their material structure poorly stable, resulting in poor cycling performance. Polyanionic compounds have good electrochemical and thermal stability, but their capacity is low. Prussian blue material not only has high capacity but also has high structural stability and excellent cycling performance. In particular, its good electrolyte compatibility makes it possible to be used in sodium-ion batteries with two different systems, aqueous and organic, and is a key cathode material for the future commercial application of sodium-ion batteries. However, these materials are still facing problems, such as structural defects and high crystalline water content. It is noteworthy that although researchers have made some breakthroughs in key materials for sodium-ion batteries, industrial applications require a material preparation process that is inexpensive, requires easy-to-use raw materials, and entails a simple synthesis process, while the preparation process of sodium-ion batteries electrode materials in laboratory studies is complicated and the product yield is low. Therefore, a difficult problem to be solved is producing electrodes of high quality production and low cost, with simple and efficient synthesis methods, before commercialization of sodium-ion batteries.

### 6.1.2.3 Electrolyte

The matching performance of electrolyte and electrode determines the quality of the battery and the fields in which it can be applied. In organic system electrolytes, to ensure good stability of the electrode/electrolyte interface, fluorinated sodium salts and fluorinated additives are usually used, which are relatively expensive; they directly affect the cost advantage of sodium-ion batteries. To improve the safety performance of this system, flame retardant additives or overcharge protection agents are added to the electrolyte, and although the percentage of such additives is small, they still reduce the electrochemical performance of the battery. The use of ionic liquids can balance the safety and electrochemical performance of the battery, but the price of ionic liquids is very high, making it difficult to promote the commercialization of - ion batteries. Currently, ether- and sulfur-based electrolytes have only shown good electrochemical performance in the matching of some electrode materials, and there is still a long way to go before their commercial application.

Aqueous sodium-ion batteries have good safety and environmental friendliness, but the electrochemical window of water is only 1.23 V. In order to avoid water electrolysis reactions, the voltage of aqueous sodium-ion batteries is usually around 1.5 V and cannot be higher than 1.8 V, even when kinetics are taken into account [46]. Moreover, the selection of electrode materials must also be considered to suppress the effects of water decomposition side reactions. The potential of the embedded sodium reaction of the positive electrode material should be lower than the oxygen precipitation overpotential of water, while the potential of the embedded sodium reaction of the negative electrode material should be higher than the hydrogen precipitation overpotential of water, which limits the application of many sodium storage positive and negative electrode materials that perform well in organic-based electrolytes. In addition, many sodium salt compounds are highly soluble in water or readily decompose in water, which further limits the choice of sodium storage materials. A variety of side reactions with electrode materials can occur in aqueous electrolytes, reducing battery reversibility. For example, the reaction of electrode materials with $H_2O$ and residual $O_2$ in the electrolyte leads to the dissolution of electrode materials and corrosion of the collector fluid. In this regard, researchers have proposed a series of improvement measures. Wang et al. proposed a new electrolyte, "Water in Salt", through which a dense solid electrolyte interface (SEI) is formed on the negative electrode and the electrochemical activity of water is reduced; thus expanding the electrochemical stability window and improving the energy density [47] [47]. However, the high concentration of electrolyte also causes corrosion problems of the electrode materials. Therefore, the application of water-based sodium-ion batteries remains a challenge.

## 6.1.3 Opportunity

### 6.1.3.1 Government policies

With the growing demand for energy storage, countries have formulated a series of policies to regulate the energy storage market in response to their own energy consumption characteristics. This book summarizes the energy storage policies of various countries, as shown in Table 6.2. The energy storage policies of the U.S. focuses on guaranteeing fairness in the energy storage market; Japan and Germany focus on government guidance and policy subsidies; while China, which has the most urgent demand for energy storage, has further emphasized the importance of energy storage in recent years through its active layout in four areas: energy storage industry, energy storage technology, renewable energy consumption, and peak-to-valley tariff adjustment, which are important to improve China's energy structure and energy security.

**Table 6.2:** Energy storage policies of various countries.

| Country | Purpose | Policy | Content |
|---|---|---|---|
| The United States | Ensure market fairness | 'Federal Energy Regulatory Commission Order No. 809' | Allow demand-side resources to participate in transmission planning. |
| | | 'Federal Energy Regulatory Commission Order No. 719' | Modify the wholesale electricity market. |
| | | 'Federal Energy Regulatory Commission Order No. 745' | Pricing demand response according to the wholesale market level. |
| | | 'Federal Energy Regulatory Commission Order No. 755' | Compensation for FM services in the electricity market. |
| | | 'Muti-Year Program Plan 2011–2015' | Systematic understanding and analysis of energy storage value chain. |
| Japan | Rely on subsidy policy to guide industry | 'New Sunshine Project' | Use energy storage to solve the problem of unstable power supply. |
| | | 'Strategic Energy Plan' and 'Electricity Business Act' | Make energy storage an important part of power reform. |
| Germany | Rely on subsidy policy to guide industry | Subsidy policy for photovoltaics | Expanded from only subsidized power generation units to subsidized energy storage units that guarantee photovoltaic power generation. |

**Table 6.2** (continued)

| Country | Purpose | Policy | Content |
|---|---|---|---|
| China | Develop the energy storage industry and adjust the power market structure | 'A Guideline On Emerging Sectors Of Strategic Importance During The 13th Five-Year Plan Period (2016–20)' | Emphasize the important role of energy storage and support investment in the construction of pure facilities. |
| | Energy storage technology equipment research and development support | '12th Five-Year Plan For Energy Technology' and 'Strategic Energy Action Plan (2014–2020)' | Develop large-scale grid-connected, large-capacity and fast energy storage and related devices to form an independent intellectual property system for energy storage technology. |
| | Improve the efficiency of renewable energy utilization | '13th Five-Year Plan Development Plan For Renewable Energy' | Promote research on large-capacity energy storage technology, install an appropriate proportion of equipment in the grid to improve the intermittent and volatility of power. |
| | Adjust peak and valley electric charges | 'Guiding Opinions On Promoting Energy Storage Technology And Industry Development' | Emphasize the use of energy storage to solve the problem of the mismatch of power demand in time and space. |

### 6.1.3.2 Market demand status quo

At present, the most widely used method in the field of energy storage is still physical energy storage (such as pumped storage), but this storage method has the disadvantages of long construction cycle, high maintenance costs, and strong environmental constraints, while electrochemical energy storage has greater application value in the development of clean energy consumption, grid-side energy storage, etc. In the first half of 2018, China's installed capacity of electrochemical energy storage increased by 127% year-on-year and a total of 300 MW h grid-side energy storage projects in Jiangsu and Henan have been started. The combined demand for "thermal power + energy storage" will also exceed 100 MW h. This huge demand for energy storage will definitely require large-scale, cost-effective, and environment-friendly energy storage system, and there is still a gap between lithium-ion batteries, which have limited lithium resources, poor safety and high cost, lead-acid batteries, which are less environment-friendly (lead is more toxic to human body), and super capacitors, which have low energy density and inadequate system construction.

Low-speed electric vehicles have a wide range of applications in China, and have a wide market demand in third- and fourth-tier cities, rural areas, and also in other developing countries. At present, the power source of low-speed electric vehicles is mainly lead-acid batteries, whose advantages are mature battery preparation and recycling technology, excellent high- and low-temperature performance, high safety, and outstanding cost advantages. However, lead-acid batteries run the risk of lead pollution in the production and processing of recycled lead, and poor management may cause harm to the environment and human health. Further, if the content is too high, it may cause lead poisoning. In addition, lead-acid batteries have low energy density (about 30 W h kg$^{-1}$) and short cycle life, which cannot meet the growing demand of low-speed electric vehicles. Although lithium-ion batteries have high energy density and cycle life, the cost and safety of the batteries limit their widespread use in this field. Sodium-ion batteries not only have high safety and cycle life, but also have obvious advantages over lead-acid batteries in terms of energy density, making them a good alternative to lead-acid batteries in the field of low-speed electric vehicles. In 2018, HiNa BATTERY showed the world's first sodium-ion batteries low-speed electric vehicle manufactured by the company to the public, demonstrating the good prospect of sodium-ion batteries application in the field of low-speed electric vehicles. At present, the energy density of the sodium-ion batteries products manufactured by the company has reached 120 Wh kg$^{-1}$, which is about three times that of lead-acid batteries, and is expected to be widely used in low-speed electric vehicles, electric ships, and other equipment in the future.

In addition, for the power grid, as the load density of load centers, such as cities, continues to rise, in addition to the gradual retirement of old thermal power units in load centers, the power supply capacity of the power grid will become increasingly tight. Creating new power supply capacity requires a longer cycle time and greater cost. It may even be difficult to achieve the needs of the power grid, which is not conducive to the commercial and industrial users. Hence, the use of energy storage systems, as a distributed power supply, to alleviate the power supply gap in the power grid has become a method to assure electrical energy and ensure the safe operation of the grid. In addition, the energy storage system should also be equipped to provide peak regulation, frequency regulation, and voltage regulation for the grid to ensure a stable and efficient operation of the auxiliary services. Therefore, the construction of grid-side energy storage system will become a key link in the revival of the grid system. 2018 China's grid-side energy storage ushered in a real explosion. Jiangsu, Henan, Hunan, Zhejiang, Jilin, and other places have begun to explore the grid-side energy storage project. The explosion of the grid-side energy storage market has stimulated the development of the entire energy storage market, and the national energy storage installation in 2018 exceeded the sum of the energy storage installations in the past few years. The energy storage market in China can reach 14.5 GW in 2020. The energy storage project in Zhenjiang, Jiangsu, is currently the world's largest comprehensive energy storage power plant and the world's first millisecond response source network load energy storage system. It is also the detonation point of the national energy storage market in 2018. The rapid development of grid-side energy storage in the future will also further enhance the demand for electrochemical energy storage devices; so the commercialization of low-cost sodium-ion batteries with high power characteristics, high safety, and good recyclability will also be further accelerated.

Wind, solar, tidal energy, and other renewable energy sources are discontinuous and show poor stability. Renewable energy can only be a part of the power and can be integrated into the grid. Thus, the renewable energy consumption problem, which profoundly reflects China's current power planning, operation, and institutional mechanisms model, has been increasingly unable to adapt to the development of renewable energy generation industry. This is a deep-rooted contradiction in the institutional mechanism of the electric power industry. According to statistics, in 2018, the abandoned power of renewable energy generation in China was as high as 120.39 billion kW h, and there is a high proportion of power limitation in some areas. The abandoned wind rate in Xinjiang, Gansu, Jilin, and other areas was more than 10% [48]. Without deep institutional and mechanical changes in the development of energy and electricity, consumption will become increasingly difficult, and there is a risk that the scope of power limitation will expand and the quantity of abandoned power will increase.

Energy storage system helps to solve the problem of renewable energy consumption. The introduction of energy storage system can provide a certain buffer for wind and photoelectric stations to connect to the grid; play the role of smoothening the wind and light output and energy dispatch; and can solve the problem of unstable power availability of new energy generation to a considerable extent; thus improving the power quality and utilization. On October 11, 2017, the first industry policy in the field of energy storage, "Guidance on Promoting Energy Storage Technology and Industry Development" was released. It aims to clearly put forward a mechanism to promote energy storage to enhance the level of renewable energy utilization, encourage renewable energy field stations to reasonably configure energy storage systems, promote the coordinated operation of energy storage systems and renewable energy, study the establishment of renewable energy field station side energy storage compensation mechanism, and support a variety of energy storage systems to promote renewable energy consumption.

A large number of wind and light storage power plant demonstration projects have been put into use in China. Taking the Zhangjiakou wind and light storage demonstration project as an example, through the combination of 6 combined power generation methods of wind, light, and storage with 4 functions of smooth processing, tracking, system frequency regulation, and peak shaving, obvious results have been achieved, which provides a good reference for the in-depth promotion of large-scale energy storage system in the field of new energy grid connection. As of the end of 2017, the installed capacity of domestic centralized photovoltaic power plants is about 101 GW, and the installed capacity of wind power is about

164 GW. Assuming that 10% of energy storage devices are matched, it will bring 26.5 GW of installed energy storage demand, and the market space will continue to increase as the installed capacity of new energy continues to rise.

### 6.1.3.3 Social and technical environment

Today, an awareness of green environment has gradually taken root. People are increasingly aware of the important role of new energy development in improving the environmental energy consumption structure and reducing environmental pollution; so the construction of related energy storage systems is also receiving increasing attention. National energy storage demonstration projects have been built in Hebei, Jiangsu, Dalian, etc. Gansu 720 MW h large-scale energy storage power plant project is the world's largest national energy storage demonstration project. Although lithium-ion battery is the most widely used energy storage device in the current energy storage system, in recent years, the numerous cell phone battery explosions and electric car fires have caused people to question the safety of lithium-ion batteries, and in the field of large-scale energy storage, the safety of energy storage equipment is an extremely important consideration in the construction of energy storage systems. High-safety sodium-ion batteries are more likely to be favored by people, which is more conducive to the market promotion of products with sodium-ion batteries.

Lithium-ion battery industry has developed over the last 20 years and the basic technology is relatively mature. Sodium-ion battery is very similar to lithium-ion battery in terms of reaction mechanism, material preparation, and battery testing; so experience of lithium-ion battery industrialization can be learnt and the production line construction model of lithium-ion battery can be applied to sodium-ion battery production. At present, STAR SODIUM has built the world's first sodium-ion batteries production line, taking an important step toward the industrialization of sodium-ion batteries. Meanwhile, China is not only the largest battery consumption market at present, but is also the largest battery production market in the world. Taking lithium-ion batteries as an example, China's lithium-ion battery production capacity accounts for 73% of the world, and has established a robust recycling mechanism, recycling 67,000 tons of lithium-ion batteries in 2018 alone, accounting for 69% of the global lithium-ion battery recycling market. The perfect industrial structure and the abundant talent pool are the unique advantages of China's battery market, which will contribute to the rapid industrialization of sodium-ion batteries in the future.

### 6.1.4 Threat

### 6.1.4.1 Market competition

As shown in Figure 6.8, among the current energy storage fields in China, mechanical energy storage is the most mature and commonly used. According to the data,

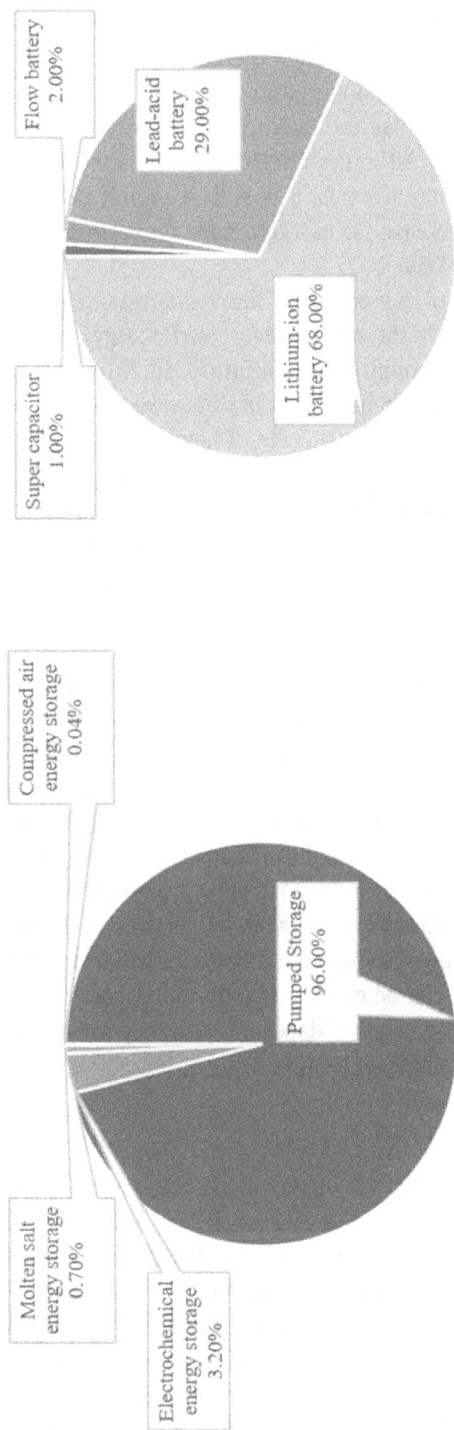

**Figure 6.8:** Installed scale distribution of energy storage projects and electrochemical energy storage projects in China in 2018.

as of the end of December 2018, the cumulative installed scale of commissioned energy storage projects in China was 31.2 GW, among which the cumulative installed scale of pumped energy storage was the largest, about 30.0 GW, accounting for as much as 96%. Pumped energy storage occupies the majority of the total, mainly due to its mature technology and low cost. Although physical energy storage has long construction cycles and high maintenance costs, it is still extremely difficult for electrochemical energy storage to increase its market share or even replace physical energy storage in a short period of time.

However, the easy control, low construction cost, and fast response of electrochemical energy storage determine that it is the most widely used energy storage technology with the greatest development potential in the future. In 2018, the installed scale of new energy storage projects commissioned in China was 2.3 GW, of which electrochemical energy storage had the largest scale of new commissioning at 0.6 GW, up 414% year-on-year.

In electrochemical energy storage, the dominant position is currently occupied by lithium-ion batteries. In the commissioned electrochemical energy storage projects, the cumulative installed capacity of lithium-ion batteries accounts for 68%; followed by lead-acid batteries, accounting for 29%; liquid-flow batteries, supercapacitors, etc. account for a limited share. The application of sodium-ion batteries in the field of energy storage is still in its infancy, with representative companies that include Aquion Energy, HiNa BATTERY, and Zhejiang Sodium Creation, among which Aquion Energy, a pioneer in the industrialization of water-based sodium-ion batteries, declared bankruptcy in 2017. The main reason for its failure is still the lithium battery that has been belittled. In 2018–2030, it is expected that the price of lithium-ion batteries will fall by 52%, and the cost of energy storage systems is expected to fall faster than expected. Therefore, it is very challenging for sodium-ion batteries to occupy a place in the energy storage market while establishing a profitable business model. As the cost of lithium-ion batteries decreases, the cost advantage of sodium-ion batteries will be further weakened. Only by giving full play to the characteristics of sodium-ion batteries in terms of fast response, material recycling, and climate adaptation, can they truly open their own market and realize the vision of commercialization.

### 6.1.4.2 Business model

The main application area of sodium-ion batteries is electrical energy storage, especially on the power generation side. However, up to now, a suitable business model for energy storage has not been established. At present, some of the energy storage pilot models are not even profitable. The open degree of auxiliary frequency modulation service market, and power peak and valley regulation market is relatively low, and the profit distribution between them and the power generation market is still not clear, which leads to the delay in establishing the profit model of the energy storage market. In grid-side energy storage, the construction of energy storage system, combined with secondary batteries, is very scarce, and the relevant industry standards have not yet been issued, which cannot effectively measure the safety and economy of energy storage system; thus inhibiting the advantages of the energy storage system.

## 6.2 Research progress of sodium-ion full battery

In the past seven years, the technology associated with sodium-ion batteries has advanced leaps and bounds, and both academic research and commercialization have taken off. In order to achieve the application objectives, key electrode materials and related scientific problems must be solved, while the electrolyte and its interface must be optimized. Currently, a large number of anode and cathode materials have demonstrated excellent electrochemical properties in commercial applications. There is, therefore, a need to reduce the cost of mass production of high-quality batteries and improve their safety. In the short term, the commercialization of organic sodium-ion batteries is more realistic. If the water-based sodium-ion batteries can solve the problem of narrow electrochemical window, it will show more practical value. In fact, these two systems have been applied in small-scale production and have achieved remarkable practical and economic benefits. In the long term, ionic liquids and solid electrolytes can not only provide higher weight and volume energy density, but also solve safety problems. However, the high cost may limit its development; so this problem can be solved by designing mixed electrolytes.

Although large-scale energy storage is still in the early stages of development, sodium-ion based full cells, button cells, and battery packs have been manufactured and used in various types of electronic devices. In particular, full cells represent a bridge between laboratory half-cells and commercial battery packs. The $MoO_2$/aqueous electrolyte/HC battery pack (25 A h), based on a 1 M $Na_2SO_4$ solution, is the system of choice for low-cost, high-safety sodium-ion batteries, with a bulk energy density of 60 Wh $L^{-1}$ and a first-cycle coulombic efficiency of 100%, which remains above 100% after 500 cycles. For the organic system, investigators prepared button cells (2 A h) with a low-cost carbon cathode and transition metal oxide cathode in a 0.8 M $NaPF_6$ – EC: DMC electrolyte, with an energy density of about 100 W h $kg^{-1}$. No explosions were observed in this cell under short-circuit, overcharge, and puncture tests. To obtain higher energy density, researchers assembled button cells with good interfacial stability using NVP as the positive electrode, $FeSe_2$ as the negative electrode, and 1 M $NaPF_6$ – DEGDME as the electrolyte, with a specific capacity of up to 400 mA h $g^{-1}$. Many reports have now focused on flexible batteries, based on solid-state electrolytes. These solid-state electrolytes have good mechanical strength and high electrical conductivity, which can be applied to flexible wearable devices in the future.

In addition to being economically efficient during production and use, sodium-ion batteries are also highly environmentally efficient. Researchers have carried out a systematic life cycle evaluation of commercial sodium-ion batteries. Relative to lithium-ion batteries, the environmental impact of sodium-ion batteries can be reduced to less than 20% while still maintaining 80% of their initial capacity after 5,000 cycles. These values are higher than most lithium-ion batteries. In future developments, the focus should be on achieving long life and high energy density to improve environmental efficiency. Many experimental reports have presented design principles applicable to commercial sodium-ion batteries and the compatibility of cathodes, anodes, and electrolytes in different systems.

### 6.2.1 Organic sodium-ion full battery

The electrolyte used in organic system sodium-ion batteries consists of three parts: sodium salt, organic solvent, and additives. The commonly used sodium salts in this system electrolyte include sodium hexafluorophosphate ($NaPF_6$), sodium perchlorate ($NaClO_4$), sodium bis(trifluoromethanesulfonyl)imide (NaTFSI), etc. The commonly used solvents include ethylene carbonate (EC), propylene carbonate (PC), dimethyl carbonate (DMC), diethyl carbonate (DEC), etc. The commonly used additives include fluorosubstituted ethylene carbonate (FEC), vinylidene carbonate (VC), ethylene glycol sulfite (ES), etc. [49]. In organic sodium-ion full cells, the nature of the electrolyte (including the type and concentration of the sodium salt, the ratio of the solvent, and the concentration of the additives) can have a large impact on the performance of the cell. Komaba et al. investigated the effect of three different sodium salts ($NaClO_4$, $NaPO_4F_6$, and NTFSI) in the same solvent on the performance of $HC//Na[Ni_{0.5}Mn_{0.5}]O_2$ full cells of the full cell [50]. The results showed that the full cell, with the electrolyte using $NaClO_4$ electrolyte salt, had the highest initial discharge specific capacity, but the capacity declined very rapidly, while the full cell, with the remaining two electrolytes, had higher cycling stability than the former, and still had 75% capacity retention after 50 cycles.

Since the nature of the electrolyte has a large impact on the energy density, cycle life, multiplier performance and safety of the battery, the ideal electrolyte should have high ionic conductivity, good electrochemical stability and thermal stability, should be nontoxic and cost low. Among liquid electrolytes, ionic liquid has high ionic conductivity, excellent electrochemical stability, and is not easy to burn, which is a kind of sodium-ion batteries electrolyte with excellent performance. However, its use in commercial sodium-ion batteries will reduce the cost advantage of sodium-ion batteries over lithium-ion batteries due to its very high price. Aqueous electrolytes are characterized by high ionic conductivity and environmental friendliness, but the narrow electrochemical window of their solvents results in a lower average operating voltage of aqueous sodium-ion full batteries and thus a lower energy density. Compared with the first two types of electrolytes, organic electrolytes are the best choice for commercial sodium-ion batteries, which are not only cheaper than ionic liquids, but also have higher operating voltages, and, therefore, have a natural advantage over aqueous sodium-ion batteries in terms of energy density. This book summarizes the organic system sodium-ion full batteries in Table 6.3.

**Table 6.3:** Sodium-ion full cell of organic system.

| Type | Full batteries | Voltage/V | Capacity/ mA h g$^{-1}$ | Capacity retention (cycles) | Energy density / W h kg$^{-1}$ |
|---|---|---|---|---|---|
| Symmetric cells | $Na_3V_2(PO_4)_3//$ $Na_3V_2(PO_4)_3$[a] | 1.75 | 70 (10 C) | 77% (100) | 150 |
| | $Na_{0.6}[Cr_{0.6}Ti_{0.4}]O_2//$ $Na_{0.6}[Cr_{0.6}Ti_{0.4}]O_2$[b] | 2.53 | 70 (1 C) | 87.5% (100) | 82 |
| | $Na_{0.8}Ni_{0.4}Ti_{0.6}O_2//$ $Na_{0.8}Ni_{0.4}Ti_{0.6}O_2$[b] | 2.8 | 85 (0.2 C) | ~75% (150) | 96 |
| Asymmetric cells | $NiSb//Na_{0.4}Mn_{0.54}Co_{0.46}$ $O_2$[b] | ~2.5 | 301 (300 mA·g−1) | 75% (20) | – |
| | $Na_2Ti_3O_7//VOPO_4$[b] | 2.9 | 114 (0.1 C) | 92% (100, at 1C) | 220 |
| | Graphite// $Na_{1.5}VPO_{4.8}F_{0.7}$[b] | 2.92 | 103 (100 mA·g$^{-1}$) | 70% (250, at 500 mA·g$^{-1}$) | 120 |
| | $Sb/C//Na_3Ni_2SbO_6$[a] | 2.4 | ~105 (0.1 C) | 70% (50) | – |
| | $Sb@TiO_{2-x}//Na_3V_2$ $(PO_4)_3-C$[b] | 2.5 | 90 (100 mA·g$^{-1}$) | ~72.2% (100) | 150 |
| | $Sb2S3@rGO//Na_{2/3}$ $Ni_{1/3}Mn_{2/3}O_2$ | 2.18 | 110 µA·h (0.2 C) | 97% (10) | 80 |
| | $FeSe_2//Na_3V_2(PO_4)_3$[b] | 1.7 | 298 (1 A·g$^{-1}$) | 81.4% (200) | – |
| | $HC// NaNi_{1/2}Mn_{1/2}O_2$[b] | 3 | >200 | ~60% (80, at 1 C) | ~240 |
| | $HC//Na_{7/9}Cu_{2/9}Fe_{1/9}$ $Mn_{2/3}O_2$[b] | 3.5 | 313 (0.2 C) | 89% (50) | 195 |

**Table 6.3** (continued)

| Type | Full batteries | Voltage/V | Capacity/ mA h g$^{-1}$ | Cycling (cycles) | Energy density / W h kg$^{-1}$ |
|------|----------------|-----------|-------------------------|------------------|--------------------------------|
| Asymmetric cells | HC//Na$_{0.9}$[Cu$_{0.22}$Fe$_{0.30}$ Mn$_{0.48}$]O$_2$[b)] | 3.2 | 301 (0.5 C) | 100% (100) | 210 |
| | HC//Na[Fe$_{0.5}$Mn$_{0.5}$]O$_2$ + 10 wt% Na$_3$P[a)] | 2.6 | 247 (0.1 C) | 84.5% (20) | 210 |
| | HC// R-Na$_{1.92}$Fe [Fe(CN)$_6$][a)] | 3 | 119.4 (10 mA·g$^{-1}$) | 94% (50) | – |

## 6.2.1.1 Symmetric sodium-ion full cell

Symmetrical sodium-ion batteries refer to full batteries assembled with the same active material for both positive and negative electrodes. Currently, the materials used as electrodes for symmetrical sodium-ion batteries in research are mainly vanadium-based compounds and titanium-based phosphate materials, whose reactions rely on two redox pairs, $M^{3+}/M^{2+}$ and $M^{4+}/M^{3+}$ (M = V, Ti, Cr or Ni). For example, as shown in Figure 6.9(a)–(c), a symmetric Na$_3$V$_2$(PO$_4$)$_3$//Na$_3$V$_2$(PO$_4$)$_3$ sodium-ion full cell was assembled by Zhu et al. under laboratory conditions, with an average cell operating voltage of 1.75 V. The specific weight energy density, after 100 cycles at a current density of 10 C, reached 150 W h kg$^{-1}$, demonstrating excellent multiplicative performance and cycling performance [51]. Figure 6.9(d)–(f) shows the P$_2$-type Na$_{0.6}$ [Cr$_{0.6}$Ti$_{0.4}$]O$_2$ symmetric sodium-ion full cell designed by Wang et al. The reaction is based on two redox pairs of Cr$^{4+}$/ Cr$^{3+}$ and Ti$^{4+}$/ Ti$^{3+}$, with an average cell operating voltage of 2.53 V. The cell energy density, at a discharge multiplicity of 0.2 C, reached 94 W h kg$^{-1}$ [52]. In addition, some researchers have also assembled symmetric full cells with O3-type Na$_{0.8}$Ni$_{0.4}$Ti$_{0.6}$O$_2$ as the electrode materials, with Ni$^{4+}$/Ni$^{2+}$ and Ti$^{4+}$/ Ti$^{3+}$ as the redox pairs for the positive and negative electrodes, respectively, and the average operating voltage of the cell was up to 2.8 V [53]. Symmetric sodium-ion all-cells have good multiplicative performance, but their low energy density and poor cycling stability have hindered their commercial application. Therefore, the key to the commercialization of symmetric sodium-ion batteries still lies in the development of active materials with high energy density and long cycle life.

## 6.2.1.2 Asymmetric sodium-ion full cell

Asymmetric sodium-ion batteries are full batteries assembled with different electrode materials for positive and negative electrodes, which have relatively high energy density. However, the large amount of irreversible capacity generated at the negative electrode during cycling and the lack of de-embedded sodium ions lead to the poor cycling stability of these batteries; thus limiting their commercialization. To solve the above problems, the most commonly used means include a pre-sodiuming process and an addition of compensating sodium salts, but the addition of compensating sodium salts (e.g., $Na_3P$, $Na_2CO_3$, etc.) can also lead to the generation of side reactions, which affect the battery performance [54, 55]. Therefore, the pre-sodiuming process is more suitable for the improvement of the performance of asymmetric sodium-ion full batteries. In this book, the batteries are divided into two major categories, according to whether the anode needs pre-sodiuming treatment, i.e., batteries with pre-sodiuming materials as the anode and batteries with raw materials as the anode, where the batteries with raw materials as the anode are further divided into batteries with carbon materials as the anode and batteries with non-carbon materials as the batteries. Raw materials as anode are divided into those with carbon material as anode and those with non-carbon material as anode, and are summarized and analyzed for different classifications of batteries to explore the possibilities and also understand the challenges of their commercialization.

Sb-based and Sn-based negative electrode materials have high theoretical capacities and relatively low redox potentials, but generate significant irreversible capacities due to the large volume expansion generated during cycling that can lead to chalking of the electrodes and shedding of active materials. However, pre-sodiuming of Sb-based and Sn-based materials, prior to the assembly of the full cell, can reduce the irreversible capacity of the full cell [56, 57]. As shown in Figure 6.10(a) and (b), the NiSb// $Na_{0.4}Mn_{0.54}Co_{0.46}O_2$ full cell assembled by Yu et al. exhibited a reversible discharge specific capacity of 301 mA h $g^{-1}$ after 20 cycles of cycling at a current density of 300 mA $g^{-1}$, and achieved a 75% capacity retention [58]. The transition metal oxide anode has relatively high cycling stability and superior multiplicity performance, compared to the alloy anode. The $Na_2Ti_3O_7$//$VOPO_4$ full cell has excellent multiplicity and excellent cycling stability, exhibiting an energy density of up to 220 W h $g^{-1}$, close to that of the existing Li-ion batteries [59]. Carbon materials, such as hard carbon and graphite, can also be used as sodium-ion battery anode materials after pre-sodiumization treatment, thereby enhancing the full cell performance. As shown in Figure 6.10(c) and (d), a full cell, assembled with natural graphite as the negative electrode and $Na_{1.5}VPO_{4.8}F_{0.7}$ as the positive electrode, has an average operating voltage of up to 2.95 V and was able to achieve an energy density of 120 W h $g^{-1}$ by weight [60]. In summary, sodium-ion full batteries, composed of pre-sodiumed cathode materials, exhibit high energy density and cycling stability, but pre-sodiuming is relatively unfavorable for the simplification of the battery process and the control of production costs; so using raw materials as sodium-ion battery cathodes may be a better choice for commercial sodium-ion batteries.

Among the non-carbon material anodes, alloy anodes have received much attention from researchers because of their high theoretical specific capacity. Cao et al. assembled a full cell with Sb/C as the negative electrode and $Na_3Ni_2SbO_6$ as the positive electrode, and the cell had an average discharge voltage of 2.4 V and still had 70%

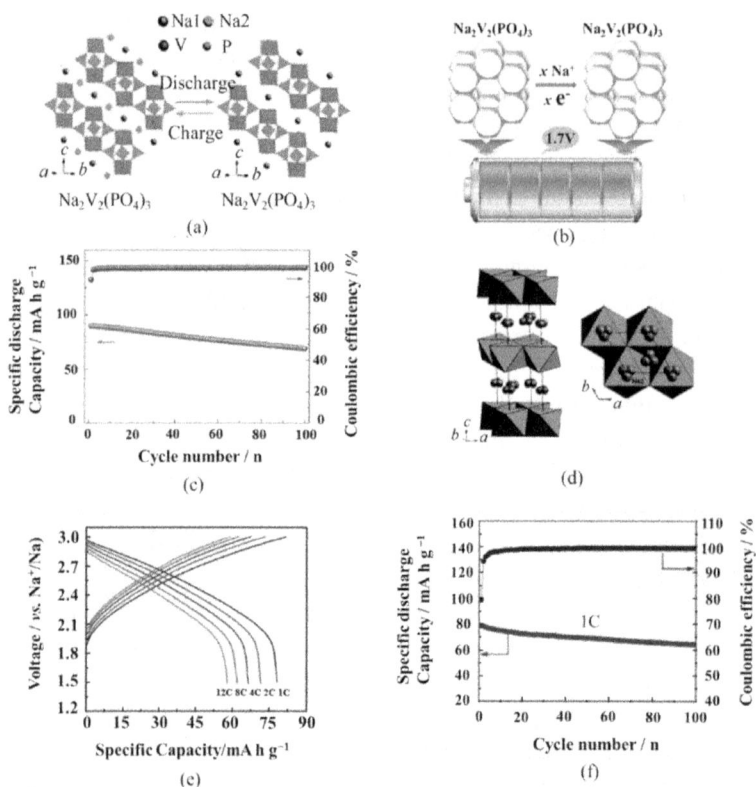

**Figure 6.9:** $Na_3V_2(PO_4)_3$ symmetrical battery [51]: (a), (b) schematic diagram of electrode material structure and charge/discharge mechanism; (c) cycle performance. $Na_{0.6}[Cr_{0.6}Ti_{0.4}]O_2//Na_{0.6}$ $[Cr_{0.6}Ti_{0.4}]O_2$ full battery [52]: (d) material structure; (e) rate performance; and (f) cycle performance.

capacity retention after 50 cycles of cycling at a discharge multiplicity of 0.1C [61]. A physical model of a $Sb@TiO_{2-x}//Na_3V_2(PO_4)_{3-x}$ full cell was assembled by Yang et al. Its electrochemical performance are shown in Figure 6.11(a)–(c), from which it can be seen that this cell has a weight specific energy density of 151 W h kg$^{-1}$ and exhibits excellent multiplicative performance [62]. Besides, the sulfur-containing alloy anode and selenium-containing alloy anode are well matched with the cathode material, when used as the anode of sodium-ion batteries, and show good potential for commercial application. As an example, $Sb_2S_3$, whose composite material formed with redox graphene (rGO) and $Na_{2/3}Ni_{1/3}Mn_{2/3}O_2$ as the positive electrode, can achieve a full battery, assembled with a weight specific energy density of 80 W h kg$^{-1}$ [63]. As shown in Figure 6.11(d)–(f), a sodium-ion full cell, assembled with excess $Na_3V_2(PO_4)_3$ as the positive electrode and $FeSe_2$ as the negative electrode, has excellent multiplicative performance, and a specific capacity of 88 mA h g$^{-1}$ can be obtained at a discharge multiplicity of 50 C [64]. Meanwhile, the battery excels in cycling stability, with

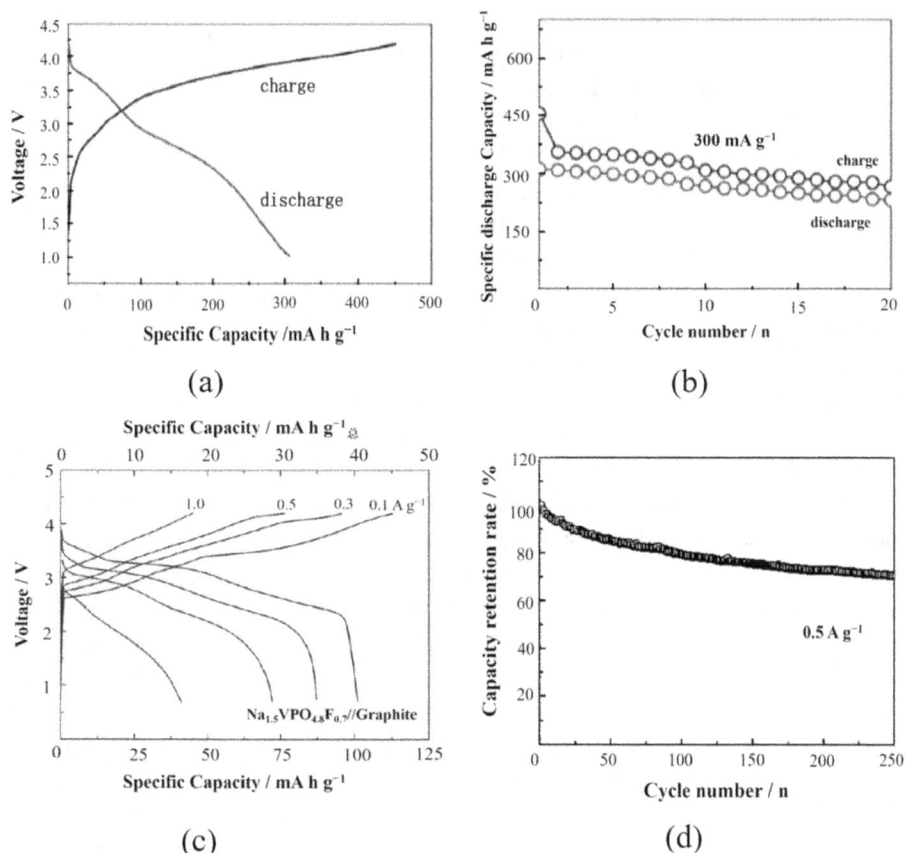

**Figure 6.10:** NiSb // $Na_{0.4}$ $Mn_{0.54}$ $Co_{0.46}$ $O_2$ full battery [58]: (a) capacity–voltage curve; (b) cycle performance. Natural graphite // $Na_{1.5}VPO_{4.8}F_{0.7}$ full battery [60]: (c) rate performance; and (d) long cycle stability.

80% capacity retention even after 1,000 cycles at a current density of 10 C. Sodium-ion full batteries, assembled with alloy materials as the anode, achieved high energy density and excellent multiplicative performance; some even demonstrated outstanding cycle life. However, the low first-cycle coulombic efficiency of the cathode material and the large volume change during charging and discharging remain key issues to be addressed. While the excess cathode material in the full cell not only increases the cost of the cell, it also exacerbates the risk of electrode material detachment from the collector during long cycling.

Sodium-ion full batteries, with carbon as the anode, have higher operating voltage and lower irreversible capacity than sodium-ion full batteries with non-carbon material as the anode, and thus have an advantage in energy density. The classical system in such batteries is a layered transition metal oxide cathode, matched with a hard carbon cathode. A hard carbon// $NaNi_{1/2}Mn_{1/2}O_2$ full cell, reported by Komaba et al., had an average cell operating voltage of 3 V and the calculated weight-for-weight energy density, at a discharge multiplier of 1 C, was about 60% of that of a graphite//$LiCoO_2$ cell [65]. Hu et al. assembled HC // $Na_{7/9}Cu_{2/9}Fe_{1/9}Mn_{2/3}O_2$ and HC//$Na_{0.9}[Cu_{0.22}Fe_{0.30}Mn_{0.48}]O_2$, two full cells with energy densities of 195 W h $kg^{-1}$ and 210 W h $kg^{-1}$, calculated with the total mass of positive and negative electrodes, respectively, where HC // $Na_{0.9}[Cu_{0.22}Fe_{0.30}Mn_{0.48}]O2$ cells showed no significant capacity decay after 100 cycles of cycling [66, 67]. The addition of compensating sodium salts to the cathode material can also improve cell cycling stability and thus the battery performance. Tarascon's team assembled a series of full cells using carbon materials as the negative electrode, with the C // $Na[Fe_{0.5}Mn_{0.5}]O_2 + 10$ wt% $Na_3P$ full cell exhibiting optimal electrochemical performance, discharging a reversible specific capacity of 155 mA h $g^{-1}$ in the first cycle and a stable discharge specific capacity of 131 mA h $g^{-1}$ after 20 cycles [68] (Figure 6.11(g)). Prussian blue analogs can also be used as sodium-ion battery cathodes, and their matching with the commercially available hard carbon cathodes can result in sodium-ion full cells with good cycling stability. The Goodenough team assembled a full cell [69]. The cell had a first-cycle discharge-specific capacity of 119.4 mA h $g^{-1}$, a first-cycle coulombic efficiency of 78%, and a capacity retention of 94% after 50 cycles of cycling. It was shown that the crystalline water affected the electrochemical performance of the material, so the group prepared a Prussian blue analogue material, $Na_2MnFe(CN)_6$, with less crystalline water, which was assembled with a hard carbon cathode to form a full cell and exhibited excellent electrochemical performance [70]. In addition to layered transition metal oxides and Prussian blue analogs, sodium-rich polyanionic compounds can also be matched with hard carbon cathodes to assemble sodium-ion full cells and have higher first-period coulombic efficiency and cycling stability, relative to general polyanionic compounds.

Hard carbon is currently the most commonly used anode material in sodium-ion full batteries research, but the preparation process of hard carbon requires high temperature and, thus, its relatively high requirements for production equipment preparation and the cost is difficult to compete with graphite. To solve this problem, organic anode materials have been prepared and matched with layered transition metal oxide cathodes, and the assembled sodium-ion full batteries have good electrochemical performance and show potential for commercial application [71, 72].

### 6.2.2 Aqueous sodium-ion full cell

The advantages of organic electrolytes in sodium-ion batteries are significant, but there are also some obvious shortcomings, such as sensitivity to traces of water in the system, poor solubility of some sodium salts, flammability of the organic solvents used, volatility, easy leaking, etc., which can pose serious safety hazards in the production

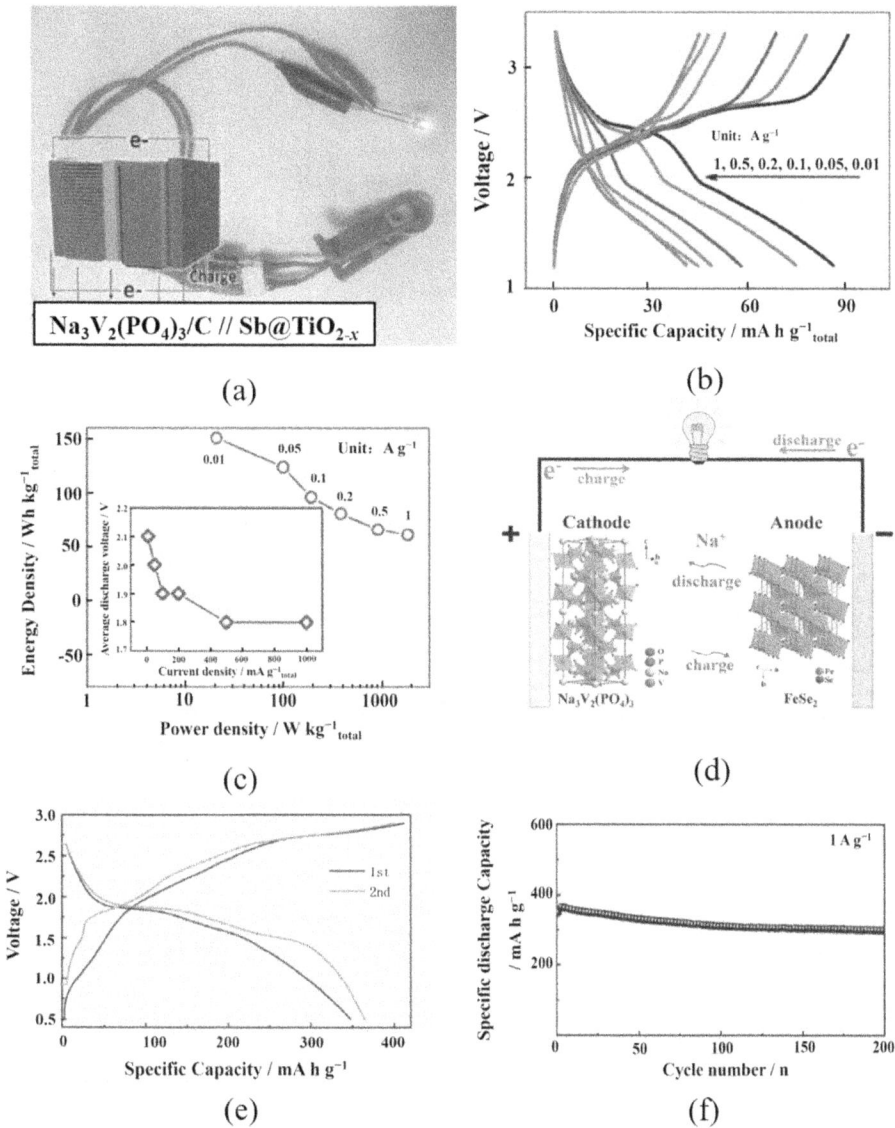

(a)

(b)

(c)

(d)

(e)

(f)

**Figure 6.11:** Sb@TiO$_{2-x}$//Na$_3$V$_2$(PO$_4$)$_3$-C full battery [62]: (a) physical model plot; (b) rate performance; (c) Ragone plot. FeSe$_2$//Na$_3$V$_2$(PO$_4$)$_3$ full battery [64]: (d) physical model plot; (e) specific capacity–voltage curve; and (f) cycle curve.

and use process. Water-based electrolyte has high safety, high ionic conductivity, low price and easy availability, and low production conditions, which not only reduces the production and technical costs of the battery, but also greatly reduces the environmental pollution during the production, use, and disposal of the battery, so the water-based sodium-ion battery has a very broad application prospect in the field

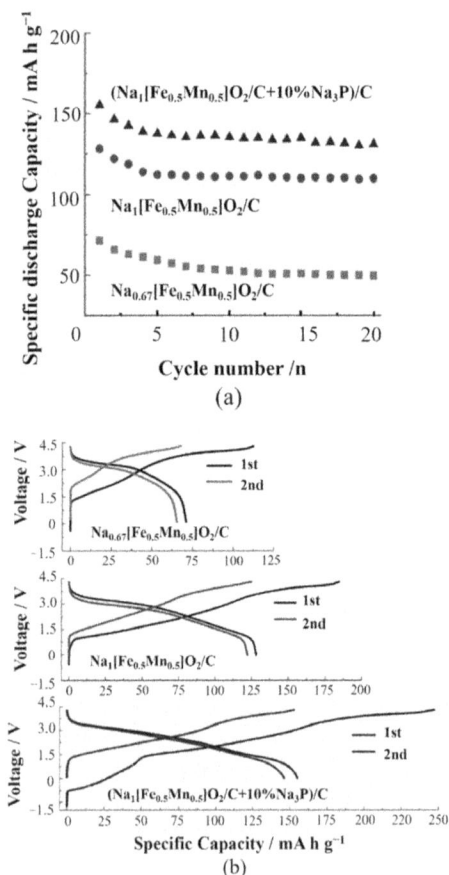

Figure 6.12: $C//P2$-$Na_{0.67}[Fe_{0.5}Mn_{0.5}]O_2$, $C//Na(Fe_{0.5}Mn_{0.5})O_2$ and $C//Na(Fe_{0.5}Mn_{0.5})O_2 + 10\%$ $Na_3P$ three batteries [68]: (a) cycle performance and (b) constant current charge/discharge curves.

of large-scale energy storage. Despite the incomparable advantages of aqueous sodium-ion batteries, there are several shortcomings: (i) the electrochemical window is narrow – the thermodynamic electrochemical window of water is only 1.23 V. In order to avoid the electrolysis reaction of water, the voltage of aqueous sodium-ion batteries should be around 1.5 V and not more than 2 V; (ii) the electrode materials cannot form an effective SEI film in water, which affects the stability and lifetime of the cycle. (iii) Many sodium salt compounds are more soluble in water, and some even decompose in water, limiting the choice of sodium storage materials.

The cathode materials used in the current study for the assembly of aqueous sodium-ion full batteries are mainly metal oxides, polyanionic compounds, and Prussian blue analogues, while the cathode materials are mainly carbon, oxide, and phosphate materials. The aqueous sodium-ion batteries can be classified into embedded negative-/embedded cathode-type full batteries, capacitive negative/

embedded cathode-type full batteries, and novel system full batteries according to the cell reaction mechanism. In this book, the three types of aqueous sodium-ion full cells are summarized and analyzed for the characteristics of different systems, and the typical aqueous sodium-ion full cells are summarized in Table 6.4.

**Table 6.4:** Typical aqueous sodium-ion full cell.

| Types | Full battery | Electrolytes | Operating voltage/V | Capacity retention rate (cycle cycles) | energy density/ W h kg$^{-1}$ |
|---|---|---|---|---|---|
| Reactive | NaTi$_2$(PO)$_4$@C// NaMnO$_2$b) | 2 M CH$_3$COONa | 1.3 | 75% (500, at 5 C) | 30 |
| | NaTi$_2$(PO$_4$)$_3$@C// Na$_{0.44}$MnO$_2$a) | 1 M Na$_2$SO$_4$ | 1.1 | 60 (700) | 65 |
| | NaTi$_2$(PO)$_4$@C// MnO$_2$@CNTb) | 1 M Na$_2$SO$_4$ + 2 M MgSO$_4$ | 1.4 | 72% (1,000) | 46.5 |
| | NaTi$_2$(PO$_4$)$_3$//Na$_2$FeP$_2$O$_7$ | 2 M Na$_2$SO$_4$ | 1.4 | 95% (30) | — |
| | NaTi$_2$(PO$_4$)3@MWCNT// Na$_4$MnV(PO$_4$)$_3$@rGO | 1 M Na$_2$SO$_4$ | 1.3 | Approx. 50% (100, at 10 C) | About 130 |

## 6.2.2.1 Embedded negative/embedded positive-type

The reaction mechanism of the embedded negative/embedded positive full cell is the same as that of the conventional organic-based sodium-ion battery, which relies on the reversible de-embedding of sodium ions in the embedded sodium compound at the positive and negative electrodes to achieve electrical energy storage and release. The energy density and operating voltage of these batteries are relatively higher, but the cycle stability of the batteries is a challenge due to the side reaction of water electrolysis. Currently, this system has been extensively studied for manganese-based oxides with NaTi$_2$(PO)$_4$ system. As shown in Figure 6.13, Hou et al. assembled a sodium-ion full cell with NaTi$_2$(PO)$_4$@C as the negative electrode, NaMnO$_2$ as the positive electrode, and 2 M CH$_3$COONa as the electrolyte, with an operating voltage window of 0.5 to 1.8 V. At 50 W kg$^{-1}$ and 1,000 W kg$^{-1}$ power density of 30 W h kg$^{-1}$ and 20 W h kg$^{-1}$, respectively, exhibiting excellent multiplicative performance [73]. Li et al. constructed an aqueous sodium-ion battery with NaTi$_2$(PO$_4$)$_3$@C water as the negative electrode and Na$_{0.44}$MnO$_2$ as the positive electrode [74]. The cell exhibited excellent multiplicative performance with 60% capacity retention after 700 cycles of cycling in the range of 5 to 50 C and a volumetric energy density of 65 W h L$^{-1}$ in terms of the whole cell mass. Liu et al. assembled an aqueous sodium-ion battery with MnO2@CNT as the positive electrode, NaTi$_2$(PO)$_4$@C as the negative electrode, 1 M Na$_2$SO$_4$–2 M MgSO$_4$ mixture solution as the electrolyte for a sodium-ion full cell

[75]. The cell exhibited a good cycling stability with an operating voltage of 1.4 V, a discharge specific capacity of 97 mA h g$^{-1}$ at a current density of 1 C, and a capacity retention of 72% after 1,000 cycles of cycling. Notably, this NaTi$_2$(PO)$_4$@C//MnO$_2$@ CNT achieves an energy density of 46.5 W h kg$^{-1}$ at a power density that is not only higher than the energy density of most aqueous sodium-ion batteries, but also exceeds the energy density of conventional lead-acid batteries (about 30 W h kg$^{-1}$), showing great prospects for commercial applications.

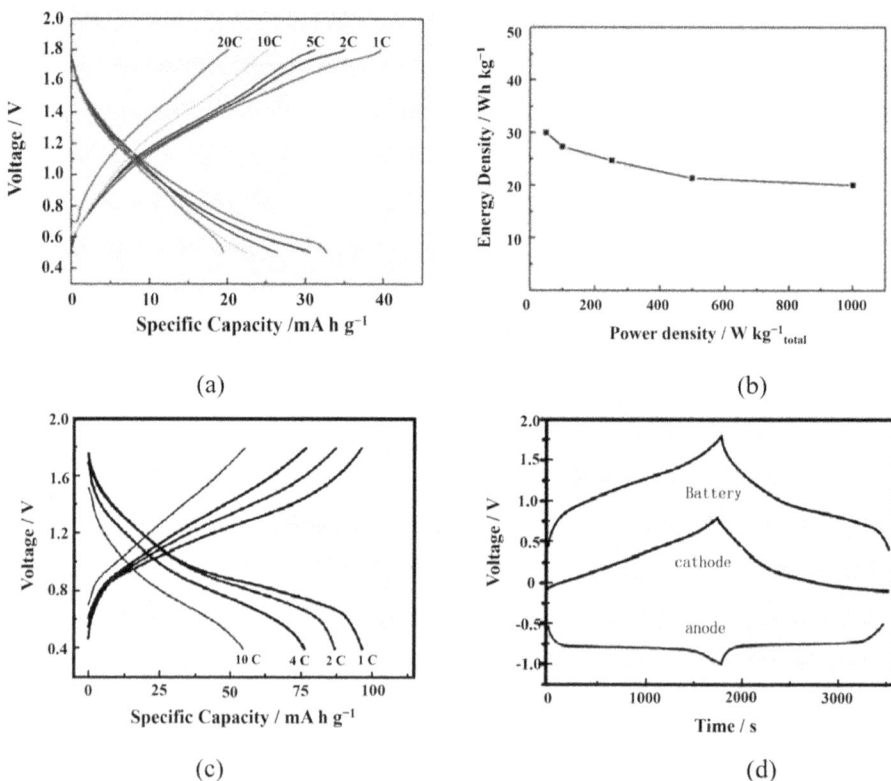

(a)

(b)

(c)

(d)

**Figure 6.13:** NaTi$_2$(PO$_4$)$_3$//NaMnO$_2$ full battery [73]: (a) rate performance; (b) Ragone diagram. NaTi$_2$ (PO$_4$)$_3$//MnO$_2$ full battery [75]: (c) rate performance; and (d) time–voltage curves.

Among the polyanionic compounds, sodium-ion fast conductor structure (NASICON) materials have attracted the attention of researchers because of their high ion diffusion rate and their nontoxic and abundant storage capacity. Researchers have used NASICON-type materials with representative types, such as Na$_2$V$_2$(PO$_4$)$_3$, NaVPO$_4$F, and Na$_2$FeP$_2$O$_7$, as the positive electrode and suitable negative electrode materials to assemble aqueous sodium-ion full batteries, and the batteries not only have good cycling stability but also exhibit excellent multiplicative performance. Kosuke et al. used 1 M NaClO$_4$-PC, 2 M Na$_2$SO$_4$-aq, 4 M NaNO$_3$-aq, and 4 M NaClO$_4$ as electrolytes to construct the cells [76]. Compared to the full cells using organic-based electrolytes, the full cells with 2 M $Na_2SO_4$-aq and 4 M $NaClO_4$ as the electrolyte showed improved performance in terms of multiplicity performance, cycling performance, and polarization during charging and discharging, which was related to the relatively high dielectric constant and low viscosity of the solvent water. Kumar et al. synthesized $Na_4MnV$ $(PO_4)_3$@rGO cathode material by a simple sol-gel method, which was matched with electrolytes of different systems (1 M $Na_2SO_4$, 5 M $NaClO_4$, 10 M $NaClO_4$ + 2 wt% VC) and assembled with $NaTi_2(PO_4)_3$@MWCNT negative electrode to form a full cell [77]. In the organic system, the cell can obtain a discharge-specific capacity of 68 mA h $g^{-1}$ by cycling for 100 cycles at a discharge multiplicity of 0.2 C. In contrast, in the aqueous electrolyte, the cell has a high discharge-specific capacity of 97 mA h $g^{-1}$ in the first cycle at a discharge multiplicity of 10 C and still has a specific capacity of about 50 mA h $g^{-1}$, after 100 cycles of cycling. The excellent multiplicity performance of the cell is mainly attributed to the fast ion transport ability in the NASICON structure and the high rGO electronic conductivity. The energy density of the $NaTi_2(PO_4)_3$@MWCNT// $Na_4MnV(PO_4)_3$@rGO cell is about 130 W h $kg^{-1}$ by weight of the whole cell, which is higher than the energy density of most aqueous sodium-ion full cells that have been reported. Due to the large size of sodium ions and their tendency to form stable structures after embedding in the oxide lattice, the number of sodium ions that can participate in reversible embedding/detachment reactions is limited, so the aqueous sodium-ion battery system has the disadvantage of low capacity utilization for the anode material of transition metal oxides. The Prussian blue compound has a cubic crystal shape in which the transition metal atoms are hexa-coordinated with $C \equiv N$ bonds to form a three-dimensional tunneling structure, forming a large number of void sites that can undergo rapid exchange with sodium ions in aqueous solutions and is a highly promising cathode material for aqueous sodium-ion batteries. As shown in Figure 6.14, Wu et al. assembled a sodium-ion full cell using $Na_2CuFe(CN)_6$ ($Na_2CuHCF$) as the positive electrode, $NaTi_2(PO_4)_3$ as the negative electrode, and 1 M $Na_2SO_4$-aq as the electrolyte [78]. The cell had a discharge voltage of 1.4 V, an energy density of 48 W h $kg^{-1}$, and 88% capacity retention after 1,000 cycles of cycling at a discharge multiplicity of 10 C. Pasta et al. constructed an aqueous sodium-ion full cell using Prussian blue material with an open crystal structure for both positive and negative electrodes [79]. As shown in Figure 6.14(d), the cell uses $Cu_{II}$-N $\equiv$ C-$Fe_{III}$ (CuHCF) as the positive electrode, $Mn_{II}$-N $\equiv$ C-$Mn_{III}$ (MnHCF) as the negative electrode, and 10 M $NaClO_4$-aq as the electrolyte. At a discharge multiplicity of 10 C, the capacity of the cell did not decay significantly after 1,000 cycles of cycling, and when the discharge multiplicity was increased to 50 C, the energy retention rate was still as high as 84.2%.

### 6.2.2.2 Capacitor negative/embedded positive-type battery system

The reaction mechanism of this type of battery system is different from that of conventional secondary batteries, which usually use carbon material with high specific

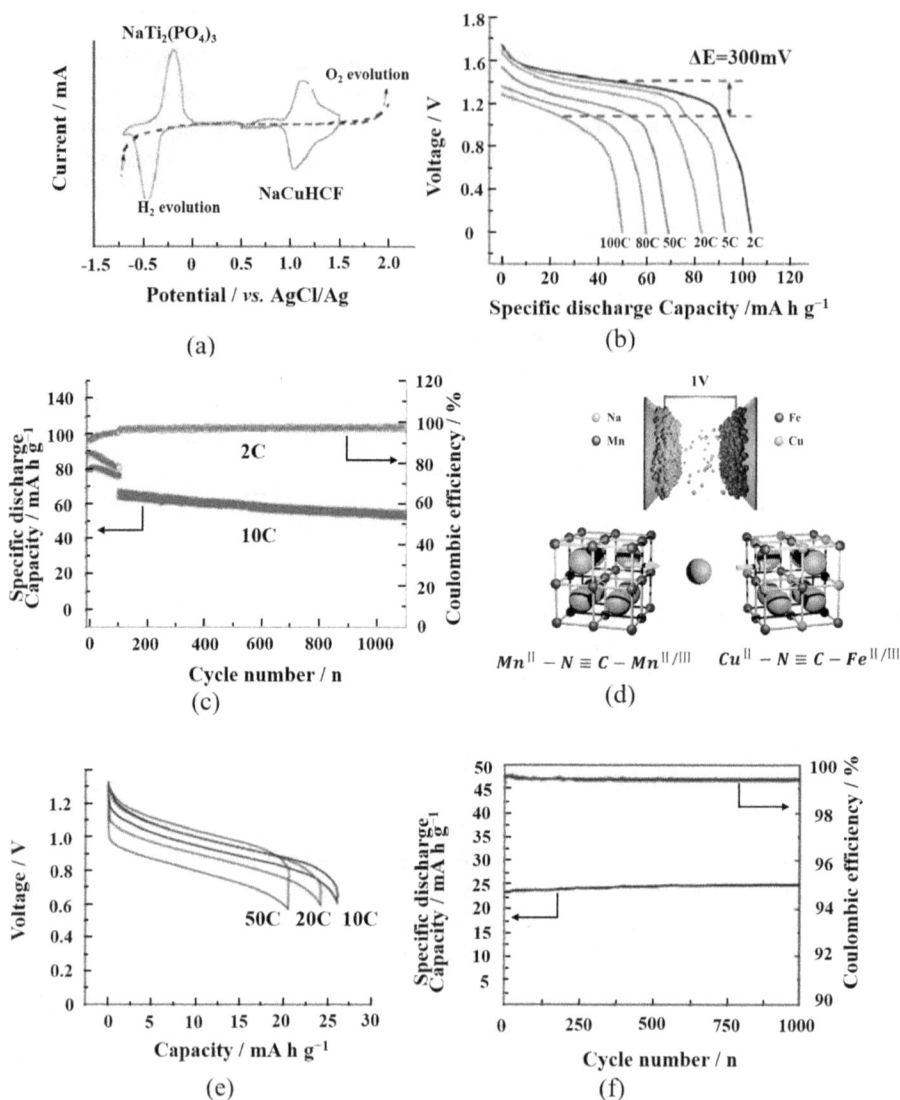

**Figure 6.14:** NaTi$_2$(PO$_4$)$_3$// Na$_2$CuHCF full battery [78]:(a) CV curve; (b) rate performance; (c) cycle performance. NaTi$_2$(PO$_4$)$_3$// Na$_2$CuFe(CN)$_6$ full battery [79]: (d) Schematic diagram of structure; (e) rate performance; and (f) cycle performance.

surface area as the negative electrode, and the reaction is the adsorption/desorption of sodium ions at the negative electrode, and no electrochemical reaction occurs. The positive electrode is an embedded sodium compound, and the process of charging and discharging is the de-embedding of sodium ions, and there is a change of elemental valence in the reaction, so the capacitive negative-/embedded positive-sodium-ion battery is also called a hybrid aqueous sodium-ion capacitor battery. Since the negative electrode is made of carbon material, its theoretical capacity and vibrational density are lower, and the operating voltage of the battery is limited by the electrochemical window of the electrolyte, so the energy density of these batteries is generally lower than that of the embedded negative-/embedded cathode-type sodium-ion battery. However, hybrid aqueous sodium-ion capacitor batteries are simple, inexpensive, and environment-friendly and effectively circumvent the challenge of selecting a suitable sodium storage anode material, making them a potential candidate for large-scale energy storage systems.

In 2010, Whitacre et al. constructed the first hybrid sodium-ion battery by using $Na_{0.44}MnO_2$ synthesized by solid-phase method as the positive electrode, activated carbon as the negative electrode, and 1 M $Na_2SO_4$ as the electrolyte [80]. Thereafter, the group assembled a hybrid aqueous sodium-ion battery using $Na_{0.44}MnO_2$ and $\lambda$-$MnO_2$ as the positive electrode and activated carbon as the negative electrode, respectively (Figure 6.15(a)–(c)) [81]. Electrochemical tests showed that the specific capacity of the full cell constructed with spinel-type $\lambda$-$MnO_2$ was higher compared to $Na_{0.44}MnO_2$, and the specific energy density of the full cell constructed with activated carbon was about three times higher than that of the hybrid aqueous sodium-ion battery with $Na_{0.44}MnO_2$ as the anode. Zhang et al. prepared $Na_{0.35}MnO_2$ in nanowire structure using a simple, low-energy hydrothermal method [82]. Despite the low sodium content of this material, when it was assembled with an activated carbon cathode to form a hybrid aqueous sodium electrode, the cell exhibited excellent cycling performance even without the removal of dissolved oxygen from the electrolyte, yielding a specific energy density of 42.6 W h $kg^{-1}$, much higher than that of a full cell assembled with a rod $Na_{0.35}MnO_2$ as the cathode (27.3 W h $kg^{-1}$). As shown in Figure 6.15(d), Feng et al. constructed a new hybrid aqueous sodium-ion battery using flake graphite as the positive electrode, activated carbon as the negative electrode, and a mixed sodium-manganese solution (1 M $Na_2SO_4$ + 1 M $MnSO_4$ + 0.1 M $H_2SO_4$) as the electrolyte [83]. The charging and discharging process occurs as a reversible conversion reaction of $Mn^{2+}$ to $MnO_2$ at the positive electrode, with $Mn^{2+}$ eventually forming $\gamma$-$MnO_2$ with oxygen vacancies, while the absorption/desorption process of $Na^+$ on activated carbon occurs at the negative electrode. The system combines the characteristics of both batteries and supercapacitors, combining the advantages of both electrode reactions to achieve improved cycling performance. As shown in Figure 6.15(e), the battery has almost no capacity degradation after cycling at a current density of 200 mA $cm^{-1}$ for 7,000 cycles and has a coulombic efficiency of 99.2%.

### 6.2.2.3 New systems

In order to further improve the performance indexes such as energy density, cycle life, and multiplicative performance of the battery system, researchers have proposed

(a)

(b)

(c)

(d)

(e)

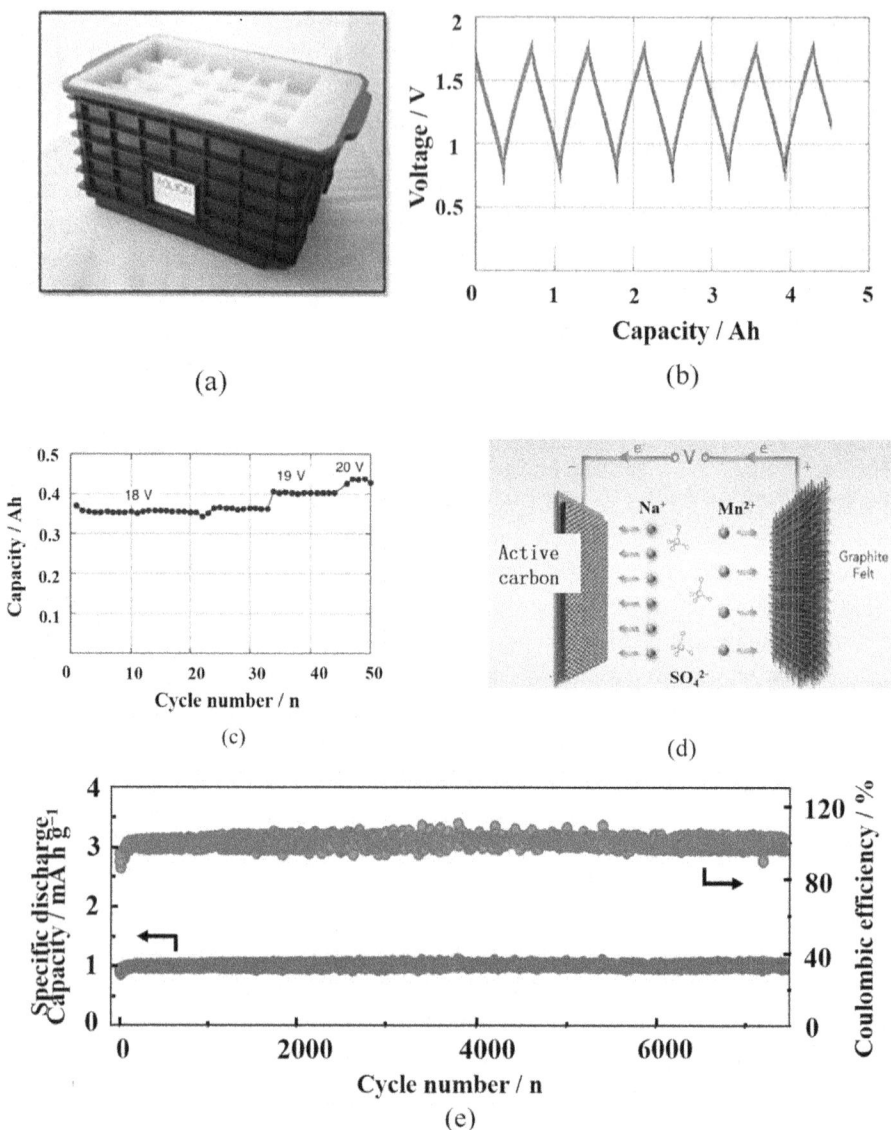

**Figure 6.15:** Hybrid aqueous sodium-ion batteries [80, 81]: (a) engineering sample; (b) capacity-voltage curve; (c) cyclic performance. Sodium–manganese hybrid full batteries [83]: (d) schematic diagram of working principle; and (e) long cycle test.

some new aqueous sodium-ion battery systems based on the above two full-cell reaction mechanisms. Zhang et al. assembled a new aqueous asymmetric sodium-ion battery for the first time by using the embedded sodium compound $NaTi_2(PO)_4$ as the negative electrode and activated carbon as the positive electrode [84]. As shown in

Figure 6.16(a), the positive and negative electrode structures of the battery are distinguished from the conventional electrodes by a self-supporting structure with a thickness greater than 1 mm. The operating voltage of the cell is 1.15 V, and the specific capacity is increased compared to the conventional capacitor negative-/embedded sodium cathode-type full cell with a discharge-specific capacity of 27.5 mA h g$^{-1}$ and a specific energy density of 31.6 W h kg$^{-1}$, based on the mass of active material in the whole cell. migrating aqueous secondary battery [85]. The cell uses NaTi$_2$(PO$_4$)$_3$@C with a layered structure prepared by coprecipitation–annealing method as the negative electrode, LiMn$_2$O$_4$ as the positive electrode, and a mixed lithium–sodium aqueous solution (0.5 M Li$_2$SO$_4$ + 0.5 M Na$_2$SO$_4$) as the electrolyte. The NaTi$_2$(PO$_4$)$_3$@C// LiMn$_2$O$_4$ cells exhibited a specific capacity of 75.7 mA h g$^{-1}$ and an energy density of 43 W h kg$^{-1}$ after 220 cycles of cycling at a charge/discharge multiplier of 1 C.

The performance of current sodium-ion full batteries is usually limited by the limited cycle life and slow reaction kinetics. In addition, the cell operating voltage depends on the high potential of the redox reaction of the metal element. Therefore, Xia, Y. Yao et al. proposed a two-phase aqueous sodium-ion battery system based on the redox electric pair of nonmetal ions [86]. As shown in Figure 6.16(b), the battery uses a mixed iodine–sodium solution as the positive electrode and a polyimide as the negative electrode, the electrolyte uses an aqueous solution of sodium nitrate or sodium sulfate, and a polymer ion-exchange membrane acts as a diaphragm to separate the liquid positive electrode from the solid negative electrode. The whole process is a redox reaction without the participation of metal elements, and the charge is transferred only by the migration of iodide ions and sodium ions between the positive and negative electrodes. The system combines the high energy density of batteries and the long cycle life and high power density of capacitors, and the batteries can even cycle for 50,000 cycles within the electrochemical window of 0 to 1.6 V. Such a system achieves a high safety and low cost design, and is one of the options for grid-scale energy storage.

## 6.2.3 All-solid-state sodium batteries

Although conventional liquid electrolytes with high ionic conductivity are now commonly used in sodium-ion full batteries, the safety issues associated with the volatility and flammability of liquid electrolytes continue to seriously hinder their application in large-scale energy storage. Therefore, solid electrolytes, which are nonflammable and have characteristics such as good thermal stability and mechanical properties, have received extensive attention from researchers. In general, electrolytes can be divided into polymeric solid-state electrolytes and inorganic solid-state electrolytes [87]. Among them, polymeric solid-state electrolytes, which are mainly based on polymer matrix, usually have good flexibility and low conductivity ($10^{-5}$ to $10^{-7}$ S cm$^{-1}$). Inorganic solid-state electrolytes are classified into four types: sulfide, sodium

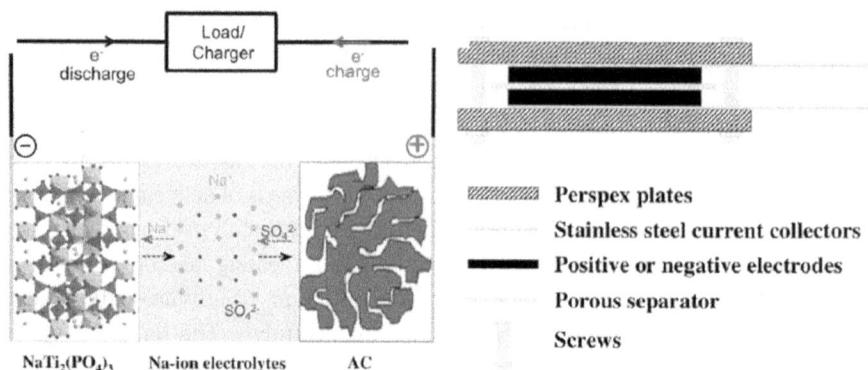

(a)

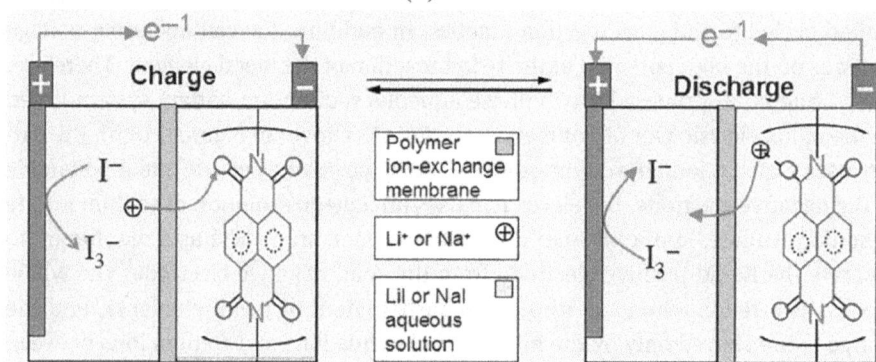

(b)

**Figure 6.16:** New system aqueous sodium-ion batteries: (a) schematic diagram of the charge/discharge process model of aqueous sodium-ion batteries and the structure of its assembled hybrid power battery [84]. (b) Schematic diagram of cell structure and electrode reaction of dual-phase sodium-ion batteries [86].

β alumina, NASICON type, and complexed hydrides, which have high ionic conductivity ($>10^{-4}$ S cm$^{-1}$) and Na$^+$ transfer number ($\approx 1$) and wide electrochemical window ($\sim 5$ V).

Although organic electrolytes have been successfully applied in small-scale production, there are still three pressing issues for eventual commercialization, namely high cost of NaPF$_6$ salt, narrow temperature window, and poor interfacial stability. Therefore, the development of novel additives is key to optimizing organic electrolytes. As another very promising battery system, aqueous sodium-ion batteries offer several advantages such as low cost, high conductivity, and nonflammability, but the pH gradient and dissolved oxygen problems in aqueous electrolytes must be overcome before high volume production. In contrast, ionic liquids have a wide electrochemical window and are well compatible with most developed electrodes. To achieve high energy densities in batteries, ionic liquids can be applied to layered transition metal oxide cathodes and hard carbon or conversion/alloy-type cathodes. Solid state electrolytes also offer broad electrochemistry and high stability, especially thermal stability and mechanical strength. Polyanionic cathodes and alloyed cathodes matched with solid-state electrolytes show that they are capable of good applications. It has been shown that both ionic liquids and solid-state electrolytes suppress the dissolution problems of organic electrodes and can be used to prepare flexible cells. However, low electrical conductivity, poor grain boundary compatibility, and environmentally hazardous preparation processes are also important challenges for practical applications. However, ionic liquids and solid-state electrolytes will receive more attention after reducing the cost and optimizing their compatibility.

### 6.2.3.1 Embedded negative-type

$Na_3V_2(PO_4)_3$ (NVP) is a representative NASICON-type material in which reversible de-embedding of two $Na^+$ can occur in the reaction with a small change in material volume during the process (8.26%), and this material shows great potential for application as both a sodium-ion battery cathode and a sodium-ion battery anode. Symmetrical NVP//NVP full cells have been shown to have good electrochemical performance in the current study; however, this practical system is limited by its low operating voltage (~1.8 V). In order to increase the operating voltage of vanadium-based NASICON structural materials, Sun et al. prepared $Na_3V_2(PO_4)_2O_2F$ (NVPOF) by introducing more electronegative $O^{2-}$ and $F^-$ instead of $PO_4^{3-}$ and used it as a sodium-ion battery cathode with $Na_3V_2(PO_4)_3$ negative electrode to construct an embedded negative-type all-solid-state sodium-ion battery (Figure 6.17(a)–(c)) [88]. Electric system replaces the flammable organic liquid electrolyte with a NASICON-structured solid-state electrolyte ($Na_5YSi_4O_{12}$, NYSO), which has a significantly higher operating voltage than the NVP // NVP full cell, in addition to a higher safety profile. Stretchable sodium-ion batteries could be used in flexible wearable devices in the future, and the application of solid-state electrolytes is making this idea possible. As shown in Figure 6.17(d), Li et al. developed a stretchable all-solid-state sodium-ion battery based on a modified graphene with poly(dimethylsiloxane) sponge electrode and an elastic gel electrolyte [89]. The cell has good electrochemical properties and excellent mechanical deformation properties that remain well maintained under many different stretching conditions and after hundreds of stretch-release cycles (Figure 6.17(e) and (f)). This novel design that integrates all stretchable components provides ideas and directions for the next generation of wearable energy devices in modern electronic devices.

### 6.2.3.2 Metal negative type

To meet the requirement of high energy density, researchers have replaced the hard carbon negative electrode of sodium-ion batteries with a metallic sodium negative electrode that can be reversibly deposited and stripped in the absence of sodium

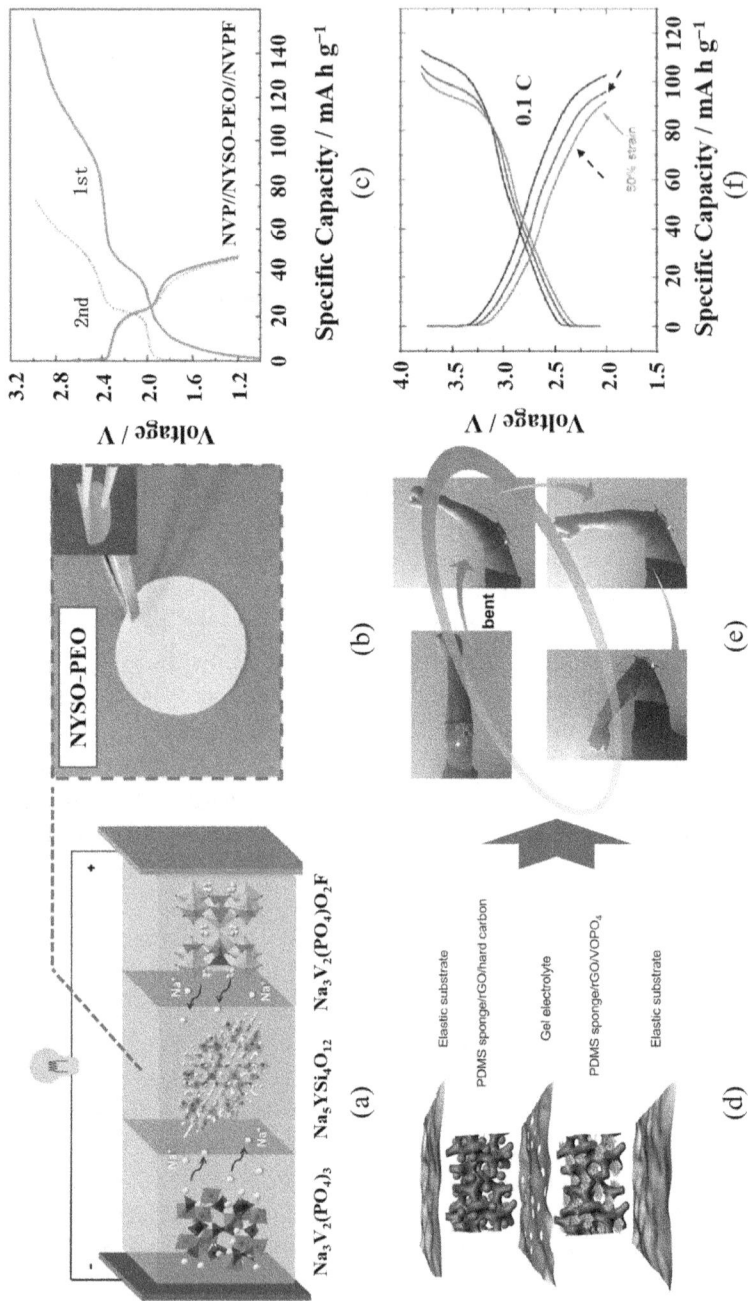

**Figure 6.17:** Sodium-ion full batteries based on NASICON structure material [89]: (a) schematic diagram of structure; (b) solid electrolyte drawings; (c) specific capacity-voltage curve. Flexible full sodium-ion batteries [90]: (d) schematic diagram of structure; (e) bending tests; and (f) cycle properties under different tensile states.

dendrites [90]. The solid-state electrolyte acts as a key component of the all-solid-state sodium battery, acting as an ion conductor and separating the positive and negative electrodes. However, the practical application of solid-state sodium batteries is still hampered by some defects of solid-state electrolytes, including low ionic conductivity at room temperature, narrow electrochemical window, poor chemical/electrochemical stability, poor interfacial compatibility, and deterioration of interfacial contacts due to electrode volume changes during sodium-ion de-embedding. Therefore, two critical issues must be considered in the study of all-solid-state sodium batteries: how to prepare a solid electrolyte with high ionic conductivity at room temperature and how to make a tight interface between the solid electrolyte and the electrode. Goodenough et al. used the sol-gel method to prepare a solid electrolyte with a NASICON structure ($Na_3Zr_2Si_2PO_{12}$) and assembled it with a $Na_2MnFe(CN)_6$ (NMHCF) cathode and a sodium metal cathode to form an all-solid-state sodium battery [91]. During charging and discharging, the solid electrolyte at the negative electrode interface effectively suppressed the interfacial side reactions and sodium dendrite growth. In addition, problems such as chemical erosion of the decomposition products in the liquid electrolyte to the cathode and dissolution of transition metal cations in the NMHCF were largely eliminated in the solid-state sodium battery. As shown in Figure 6.18, the cycling performance and coulombic efficiency of the Na//NMHCF solid-state sodium cell are significantly better than those of the Na//NMHCF cell using a conventional organic liquid electrolyte.

A solid-state polymer electrolyte (SPE) constructed from sodium bis(trifluoromethylsulfonyl)imide (NaTFSI)/polyoxyethylene (PEO) was prepared by Hu et al. by a simple solution-casting method [92]. This electrolyte has relatively high ionic conductivity, oxidation resistance, and high thermal stability, and it was assembled with $NaCu_{1/9}Ni_{2/9}Fe_{1/3}Mn_{1/3}O_2$ cathode and sodium metal cathode to form an all-solid-state sodium cell, which exhibited 121.4 mA h g$^{-1}$ at 0.1 C, 0.2 C, 0.5 C, 1 C and 2 C multiplicities, and 115.8 mA h g$^{-1}$, 105.7 mA h g$^{-1}$, 95.8 mA h g$^{-1}$, and 67.7 mA h g$^{-1}$ discharge-specific capacities, respectively, demonstrating excellent multiplicity performance. Mecerreyes et al. designed a novel ionic gel electrolyte using poly(dimethyltrifluoroethylene)-poly(ionic liquid) as the polymer matrix, C3mpyrFSI as the ionic liquid, and NaFSI as the sodium salt and constructed an all-solid-state sodium battery together with a sodium metal cathode [93]. The cell exhibited good capacity retention and very high coulombic efficiency (>97%), with a specific capacity of 114 mA h g$^{-1}$ at a discharge multiplicity of 0.05 C. This is the first report on an all-solid-state sodium battery based on ionic gel, which demonstrates the potential of developing multifunctional sodium-metal-based batteries for applications.

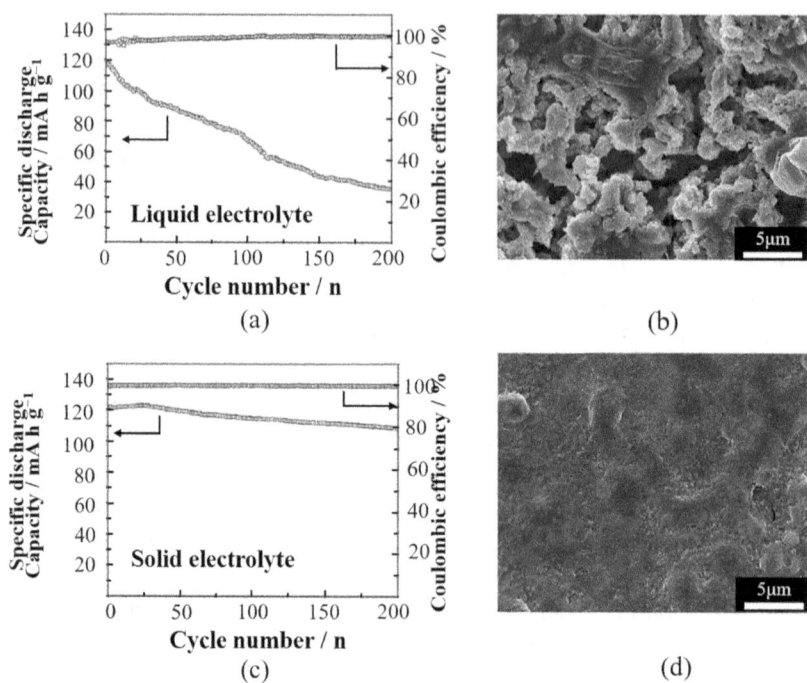

Figure 6.18: Na//Na$_2$MnFe(CN)$_6$ batteries using liquid electrolyte [91]: (a) cycle performance; (b) sodium metal surface after cycling. Na//Na$_2$MnFe(CN)$_6$ battery using solid electrolyte (Na$_3$Zr$_2$Si$_2$PO$_{12}$) [91]: (c) cycle performance; and (d) sodium metal surface after cycling.

## 6.3 Structural design of sodium-ion batteries

The technology for fabricating sodium-ion battery structures is still in the early stages of development, and the basic principles of manufacturing and cell design are similar to those of lithium-ion batteries, although each manufacturer keeps their patents, designs, and specific processes strictly confidential; however, some basic principles and processes can be understood.

Batteries can be divided into three types of forms based on the differences in packaging forms: cylindrical, square, and soft pack, of which cylindrical and square are collectively known as rigid-case batteries. The internal structure of the three types of batteries is not very different; the core difference is that cylindrical and square batteries mainly use metal materials as the shell, while soft pack batteries use aluminum-plastic film as the encapsulation shell. According to China Energy Storage Network statistics, the current penetration rate of soft pack batteries in the power battery is 12–15%, on the one hand, because the current production process is still immature, the degree of standardization is low, and the battery consistency is poor; on the other hand, it is due to the dependence on imports of the main raw material, aluminum plastic film, resulting in relatively high unit cost. However, these development bottlenecks are expected to be eliminated in the future through the scale of production, automation, enhanced battery management system, aluminum plastic film quality improvement, etc., driven by demand. In particular, the lamination process used in soft pack cells has brought forward high requirements for the production process control, highlighting the importance of high-end integrated equipment in the backend.

The sodium-ion full batteries prepared in the laboratory are mainly of soft pack structure, and the sodium-ion battery enterprises represented by HiNa BATTERY and Zhejiang Sodiumtron have produced three types of batteries: cylindrical, square, and soft pack. The future structure of sodium-ion batteries can be appropriately adjusted according to the different market demands to achieve maximum utilization of active materials, while ensuring the battery performance. The following sections will introduce the current soft pack, square, and flexible/stretchable structures, and analyze the advantages and problems faced by each structure in production processes and applications.

## 6.3.1 Soft pack batteries

Soft pack battery refers to the secondary battery with aluminum plastic film as the shell; taking the secondary lithium battery as an example, its structure is shown in Figure 6.19, and its penetration rate in the 3 C field has exceeded 60% and is gradually expanding to the new energy automobile industry. Compared to hard-shell batteries, soft pack batteries have the characteristics of flexible design, light weight, small internal resistance, more cycles, and higher energy density, and are less prone to explosion . However, cost and consistency issues are the limiting factors in the development of soft pack batteries, at present. However, these disadvantages can be eliminated in the future through production scale, automation, enhanced battery management system, and improved quality of aluminum plastic film. The material properties of aluminum plastic film are critical to ensure the long-term stable operation of the battery, and are the main difference between soft pack batteries and cylindrical and square pouch batteries. The design and manufacture of aluminum plastic film is considered to be one of the three major technical challenges in the soft pack battery industry. Aluminum plastic film generally consists of an outer resistance layer (nylon), an aluminum layer, and a high barrier layer (polypropylene), with a glue layer sandwiched between the two layers. According to the functional use, aluminum plastic film can be divided into two categories: power and nonpower. The thickness of the former film is 152 µm, while the thickness of the latter film is 88–123 µm. In comparison, the former film has higher technical requirements.

Soft pack battery cells usually use a laminated sheet process, while nonsquare battery cells usually use a winding method. In the process of die-cutting and stacking

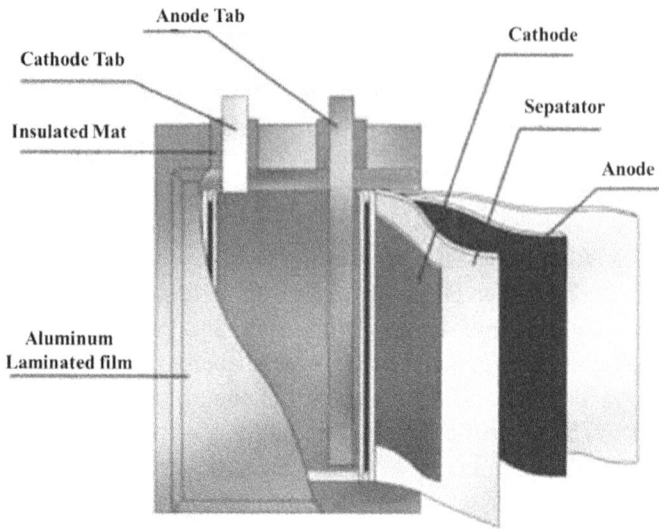

**Figure 6.19:** Schematic diagram of pouch battery structure.

of soft pack batteries, the burrs and dust generated are likely to cause short circuit of the battery, which is a great battery safety hazard, so it is especially important to control the burrs and dust during the production and preparation of soft pack batteries. In the die-cutting and stacking processes, controlling the size of the burr during punching, reducing the dust generated during punching, and avoiding the burr during the transfer process have become the most important problems faced by these two processes. To solve these problems, it is necessary to understand the causes of burr and dust. There are three main reasons for the generation of burr and dust: (i) the punching method; (ii) the structure of the punching die; and (iii) the material and processing accuracy of the punching die. According to the causes of burr and dust, the solution can be improved by the following aspects: ① optimize the structure of existing die; ② improve the precision of die manufacturing and assembly; ③ select laser die-cutting machine; and ④ adopt die-cutting and stacking machine, which can directly enter the stacking platform after punching and cutting the pole piece, avoiding the collision and friction between the pole piece and the material box, completely solving the potential risk of bad pole piece.

### 6.3.2 Cylindrical/square cells

Cylindrical battery is a kind of battery with high capacity, long cycle life, and wide range of application environment temperature. The products are often used in solar lamps, lawn lamps, back-up energy, and many other applications. The International Electrotechnical Association has established a conventional nomenclature for different models and chemical systems of cylindrical batteries, such as the conventional 18,650-type cylindrical battery (Figure 6.20): 18 is the diameter of the battery (unit: mm), 65 is the height of the battery (unit: 0.1 mm). Similarly, for a square battery 366,509, 36 mm refers to the width of the battery, 60 mm to the length, and 9 mm to the thickness. The current housing of square batteries is either steel or aluminum; with the improvement of the production process, the housing is now mainly aluminum, and the cells are wound or stacked with a sheet process. Compared to the traditional cylindrical battery, the energy density of the square structure of the battery is relatively high, and the battery cycle life is long. Domestic square battery technology is very mature, and the material technology (such as gas expansion rate, expansion rate, and other indicators) requirement is not high, and there is no technical barrier in the industry. In terms of manufacturing cost, the material of square battery has been completely localized, its requirements for battery material technology are lower than those of soft pack battery, and the overall material cost is about 10% lower than that of soft pack battery under the same capacity. In addition, the square battery packaging adopts laser sealing technology, which has a stronger seal and can effectively avoid the problem of aging and air leakage at the seal. However, for the square battery manufacturing process and application standards, there is no clear standard division; at the same time, the cylindrical lithium battery lugs are easier to weld than the square lithium battery, and the square battery can easily produce false welding that affects the quality of the battery.

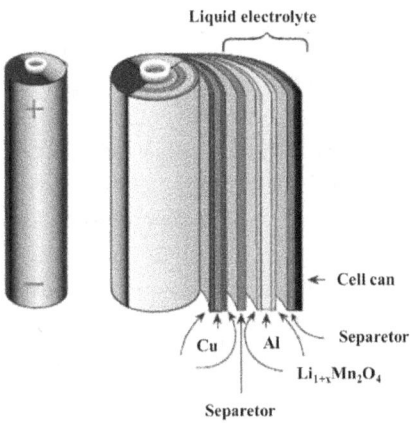

**Figure 6.20:** Schematic diagram of model 18,650 cylindrical battery.

The manufacturing process of typical sodium-ion battery can be similar to the lithium-ion battery process, which is shown below: a metal foil is coated with positive and negative active materials; then, the diaphragm is placed between the positive and negative electrodes and wound, the wound electrode set is inserted into the battery case, the electrolyte is injected, and then sealed, which can be roughly divided into five steps.

(1)  Mixing, stirring, coating, and laminating of positive and negative materials
(2)  Coiling of the same diaphragm
(3)  Injection of electrolyte
(4)  Sealing
(5)  Final sorting

In the low-speed electric field, compared to lead-acid batteries, sodium-ion square batteries have more prominent application advantages in terms of cost, energy density, and safety. For organic sodium-ion batteries, when the energy density of the system increases, the cell capacity also follows; if the energy density is to be 300 W h kg$^{-1}$, the cell capacity will reach about 200 A h, which not only increases the technical difficulty, but also reduces the production efficiency. Importantly, in terms of safety, square batteries use a fully sealed process, which can easily cause explosions when thermal runaway and other safety issues occur. At present, lead-acid batteries are very mature in the field of battery maintenance and recycling, and if sodium-ion batteries want to achieve a breakthrough, it is necessary to establish a perfect recycling system while realizing the commercial application of the product.

### 6.3.3 Flexible/stretchable cells

Flexible batteries, generally referred to as batteries that can be bent and used repeatedly, as shown in Figure 6.21, have properties that include being bendable, stretchable, foldable, and twistable; they can be lithium-ion batteries, zinc-manganese, or sodium-ion batteries, or even supercapacitors. Since each part is subjected to a certain deformation during folding and stretching, the materials and structures of each part of the flexible battery has to maintain its performance after several folds and stretches, which naturally requires very high technology in this field. Current rigid lithium/sodium-ion batteries are subject to severe performance damage and even safety hazards after undergoing deformation. Nevertheless, compared to traditional rigid batteries, flexible batteries have higher environmental adaptability and crash resistance, and better safety. Moreover, flexible batteries enable electronic products to move in a more ergonomic direction, significantly reduce the cost and size of smart hardware, add new capabilities and improve existing ones, and enable an unprecedented depth of integration between smart hardware and the physical world.

Flexible electronic devices have developed very rapidly in recent years, and the ideal flexible battery should have high flexibility, energy density, and power density at the same time; however, these factors are often constrained by each other in flexible batteries. Here, we have done a careful analysis based on the structural design at the cell component and overall device level and compiled the latest advances in

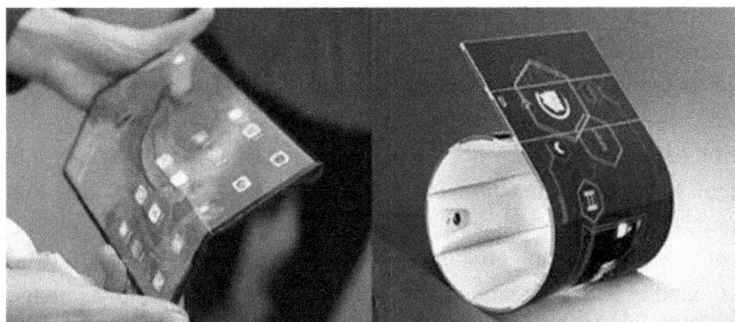

**Figure 6.21:** Flexible electronic device.

flexible batteries to summarize the current academic development ideas into four strategies, as follows.

### 6.3.3.1 Develop porous structure deformable battery components, such as porous collectors, porous electrodes, flexible solid electrolytes, etc

Conventional planar-shaped secondary batteries consist of a metal collector, a negative electrode, a diaphragm, an electrolyte. and a positive electrode with a multilayer structure and the battery is rigid and has a large mass. Although reducing the electrode layer can make the cell flexible, the electrode material made by the conventional slurry coating method can easily separate from the collector during repeated bending, thus affecting the electrochemical performance of the cell. The problem of detachment can be effectively avoided by growing the active material directly on the metal collector, which is one of the most popular and effective methods to obtain flexible planar sodium-ion batteries. The direct growth on the metal collector, usually in the form of nanoarrays (nanowires, nanosheets, nanotubes, etc.), allows close contact between the electrode material and the collector, obtaining optimized mechanical stability and flexibility. In addition, this approach reduces the diffusion distance of ions, and the large surface area of the grown nanoarrays meets the increasing demand for high energy density and high power density. However, the metal foil still has limited flexibility and mechanical strength; furthermore, the planar shape allows limited active electrode material to grow on the collector, which, in turn, affects the energy density of the flexible cell. To address this issue, researchers have tried other three-dimensional or porous nonmetallic collectors, such as carbon nanotube paper, carbon cloth, and three-dimensional graphene foam, instead of flat metal collectors to improve the performance of flexible planar surfaces.

Flexible porous structures are now widely used in battery assemblies to buffer the strain generated when the battery unit is subjected to bending and twisting. Hu et al. reported a graphene oxide conductive porous film that has an electrical conductivity of up to 3,112 S cm$^{-1}$. A flexible battery assembled with this film as a collector, after 100 cycles at a high charge/discharge multiplier (5 C), showed no significant degradation of capacity [94]. A single-walled carbon nanotube-polymer composite cathode material was used to assemble a flexible lithium battery. The battery exhibited a discharge-specific capacity of 182 mA h $g^{-1}$ at low currents (50 mA $g^{-1}$) and still achieved a specific capacity of 75 mA h $g^{-1}$ when discharged at very high currents (5,000 mA $g^{-1}$) [95]. In the field of sodium-ion flexible batteries, it was reported that Prussian blue material was grown on nickel-metal coated carbon cloth, and the poor cycling stability of Prussian blue material was effectively improved by taking advantage of the good flexibility and high conductivity of carbon cloth, which still exhibited a specific capacity of 110 mA h $g^{-1}$ at a 5 C multiplicity of 1,800 cycles [96].

### 6.3.3.2 Ultrathin cell designs, such as single pair (or double pair) positive/diaphragm/negative structures

Compared to battery designs with porous structures, ultrathin battery designs require more thought at the overall device level. As shown in Figure 6.22, Panasonic has released a flexible battery that is only 0.55 mm thick and can be used in all types of wearable devices. This flexible battery maintains 99% capacity even after being bent at a bending radius of 25 mm or twisted to an angle of ±25° for more than 1,000 times. Robert Kun et al. reported a $Li_4Ti_5O_{12}/LiPON/Li$ thin-film solid-state battery prepared by flame spray pyrolysis based on a flexible polyimide as a support substrate. After charging and discharging at a 1 C multiplier and cycling 90 times successively in a flat-bend-flat morphology, the battery still retained more than 98% of its discharge capacity, exhibiting excellent cycling performance [97].

**Figure 6.22:** Schematic diagram of an ultra-thin flexible battery.

### 6.3.3.3 Cell design with geometric topology, such as linear structure, Origami, Kirigami structure, etc

Stretchable electronics, a "soft" electronic device, can not only bend like a flexible electronic device, but can achieve stretching properties that are not possible with conventional flexible devices. There are usually two ways to achieve stretchable secondary cells. The first is to achieve stretchability of traditional rigid components through special structural design, and the second is to replace rigid components (essentially stretchable batteries) with stretchable components. "Structurally" stretchable batteries are not inherently stretchable, but their stretchability is achieved through mechanical structures, such as bending or serpentine interconnections. Although such stretchable sodium-ion cells show good stretchability and stable electrochemical properties under stretching, there are several problems. First, the preparation process requires complex, lithography-related procedures and is difficult to produce on a large scale. Second, the stretchable lithium-ion batteries exhibit capacity degradation after only 20 cycles due to the reaction of the packaging material with water and the poor conductivity of the slurry particles. Therefore, there is a great need to further simplify the manufacturing process and improve the battery components. "Intrinsically" stretchable secondary cells typically exhibit better stretchability. In this context, "intrinsic" has two main meanings. The first is that all rigid parts of the cell are replaced with intrinsically stretchable parts. The active electrode is replaced by an inherently stretchable electrode (e.g., an intrinsically elastic organic material). The second meaning of "intrinsic" is that the special design of the rigid components (made into a wavy spiral coil spring structure) makes the cell itself stretchable, as opposed to the "structurally" stretchable cell described above. Although replacing the rigid component with an intrinsically stretchable material such as a polymer electrode is thought to produce a superior stretchable secondary battery, there is still a performance gap compared to conventional ion batteries. Therefore, special designs of wavy or helical spiral spring-structured sodium-ion batteries are widely used to fabricate intrinsically stretchable secondary batteries. The wave design leads to stretchable characteristics by depositing a rigid material on a pre-stretched elastic substrate through the release of pre-strain. The design of spiral coil spring structure electrodes is another effective strategy for fabricating highly stretchable sodium-ion batteries. For example, stretchable fiber optic lithium ions with helical spiral spring electrodes can be deformed in all sizes and can be woven into a variety of textiles.

In addition to improving the flexibility of the cell module material itself, cell structures designed using the geometric topology principle can reduce the stress changes generated within the cell during deformation. This strategy was originally reported by Kim et al. in their work on linear batteries, which are able to accommodate not only bending deformations but also more complex shape changes such as folding and twisting. Gao et al. designed and assembled a self-healing flexible lithium battery using spring-like $LiCoO_2$/reduced graphene oxide as the anode material in combination with a gel electrolyte. Under complex deformations (bending and twisting), the battery was charged and discharged at a current density of 1 A $g^{-1}$, and the battery still maintained a discharge specific capacity of 82.6 mA h $g^{-1}$; even after cutting off and healing the cell five times, the battery still exhibited a discharge specific capacity of 50.1 mA h $g^{-1}$ capacity [98]. In addition to linear structures, paper folding techniques are also widely used for flexible batteries; Jiang et al. used the Origami paper folding scheme, which enables the creation of compact deformable three-dimensional structures by folding two-dimensional sheet materials along predetermined folds, which can withstand high strength deformation. Soon after, the group combined folding and cutting techniques to develop a Kirigami scheme. The battery was able to achieve more than 85% capacity retention and 8% coulombic efficiency after 100 charge/discharge cycles, and the maximum output power of the battery did not show significant degradation after 3,000 cell deformations [99].

### 6.3.3.4 Decoupling the flexible and energy storage parts of batteries, such as ridged batteries, zigzag batteries, etc

For the flexible cell design described above, misalignment, peeling, and dislodgement between the active material and the collector fluid will still occur during the complex deformation process. The increased overpotential and cell internal resistance due to poor contact will reduce the capacity retention and coulombic efficiency of the full cell, which is detrimental to the cycling performance of the cell. Redesigning the cell architecture to separate the parts that store energy and provide flexibility is a solution to these problems. De Volder et al. demonstrated a layered, tapered carbon nanotube structure, similar to a plant, petunia, with a wide corolla used to carry the positive and negative active material particles and a slender stalk portion below that is tightly bonded to the collector portion. During the cell deformation process, most of the stress is applied to the collector itself, and the conical structure hardly generates any strain during this period, thus exhibiting a high degree of flexibility. $Fe_2O_3/LiNi_8Co_{0.2}O_2$ full cells assembled with this tapered structure still have 88% capacity retention when charged and discharged for 500 cycles at a multiplicity of 1C [100]. Inspired by the fact that animal spines possess good mechanical strength and flexibility, Guoyu Qian et al. reported a method for preparing flexible lithium-ion batteries with high energy density that can be prepared on a large scale: by wrapping a thick, rigid part (corresponding to the spine) along the axis to store energy, while the thin, unwrapped flexible part (corresponding to the bone marrow and intervertebral disc) is used to connect the "spine," thus achieving good flexibility and high energy density of the whole device. Since the rigid electrode part is much larger than the flexible connection part, occupying more than 95% of the cell volume, the overall cell energy density can reach 242 W h $L^{-1}$ [101]. The reasonable bionic design allows it to pass strong dynamic mechanical load tests. Although there have been a large number of papers on flexible secondary batteries, most focus on the preparation of flexible/stretchable electrodes or collectors, and few involve changes in device structure. As shown in Figure 6.23, one of the limiting factors for batteries in portable/wearable electronic devices is the shape of the battery. Therefore, designing flexible secondary batteries with different shapes would be a disruptive technology that would open the way for innovation in flexible device design.

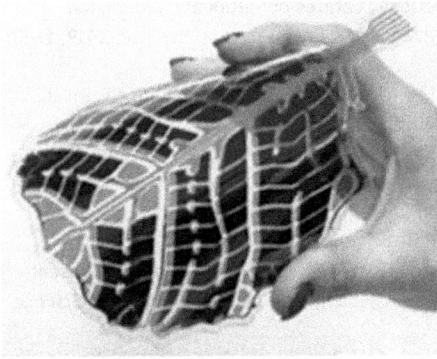

**Figure 6.23:** Schematic diagram of a spinal battery.

At present, flexible batteries still have a long way to go in terms of practicality indicators – battery capacity, energy density, cycle life and other electrochemical performance. The existing laboratory-developed batteries, in general, have high process requirements, low production efficiency, and high costs, and are not suitable for large-scale industrial production. Therefore, in the future, the search for flexible electrode materials and solid-state electrolytes with excellent overall performance, innovative battery structure design, and the development of new solid-state battery preparation processes are the main breakthrough directions. In addition, the biggest pain point of the current battery industry is still the range. In the future, the battery manufacturers that can achieve a dominant position must solve both the range problem and the flexible production problem. The application of new energy (such as solar energy and bioenergy) or new materials (such as graphene) is expected to be able to solve these two problems at the same time. Flexible batteries are becoming the main direction of future consumer electronics, and in the foreseeable future, technological breakthroughs in the entire field of flexible electronics, represented by flexible batteries, are bound to bring about huge changes in the upstream and downstream industries.

# References

[1]    中国能源研究会. 中国能源发展报告2018[J]. 中国电业, 2018, (10): 2.
[2]    肖宏伟. 2018年我国能源形势分析与2019年预测[J]. 发展研究, 2019, (1): 54–58.
[3]    Dunn B, Kamath H, Tarascon J M. Electrical Energy Storage for the Grid: A Battery of Choices[J]. Science, 2011, 334(6058): 928–935.
[4]    Fares R L, Webber M E. The Impacts of Storing Solar Energy in the Home to Reduce Reliance on the Utility[J]. Nature Energy, 2017, 2(2): 17001.

[5]  Liu T, Zhang Y, Jiang Z. et al., Exploring Competitive Features of Stationary Sodium Ion Batteries for Electrochemical Energy Storage[J]. Energy & Environmental Science, 2019, 12(5): 1512–1533.

[6]  Chen H, Cong T N, Yang W. et al., Progress in Electrical Energy Storage System: A Critical Review[J]. Progress in Natural Science, 2009, 19(3): 291–312.

[7]  Etacheri V, Marom R, Elazari R. et al., Challenges in the Development of Advanced Li-Ion Batteries: A Review[J]. Energy & Environmental Science, 2011, 4(9): 3243–3262.

[8]  Li M, Du Z, Khaleel M A. et al., Materials and Engineering Endeavors Towards Practical Sodium-Ion batteries[J]. Energy Storage Materials, 2020, 25(3): 520–536.

[9]  Palomares V, Serras P, Villaluenga I. et al., Na-Ion Batteries, Recent Advances and Present Challenges to Become Low Cost Energy Storage Systems[J]. Energy & Environmental Science, 2012, 5(3): 5884–5901.

[10]  Perrin M, Saintdrenan Y M, Mattera F. et al., Lead–Acid Batteries in Stationary Applications: Competitors and New Markets for Large Penetration of Renewable Energies[J]. Journal of Power Sources, 2005, 144(2): 402–410.

[11]  Yabuuchi N, Kubota K, Dahbi M. et al., Research Development on Sodium-Ion Batteries[J]. Chemical Reviews, 2014, 114(23): 11636–11682.

[12]  Evarts E C. Lithium Batteries: To the Limits of Lithium[J]. Nature, 2015, 526(7575): 93–95.

[13]  Goodenough J B, Park K. The Li-Ion Rechargeable Battery: A Perspective[J]. Journal of the American Chemical Society, 2013, 135(4): 1167–1176.

[14]  Wadia C, Albertus P, Srinivasan V. Resource Constraints on the Battery Energy Storage Potential for Grid and Transportation Applications[J]. Journal of Power Sources, 2011, 196(3): 1593–1598.

[15]  Liu Q, Hu Z, Chen M. et al., Recent Progress of Layered Transition Metal Oxide Cathodes for Sodium-Ion Batteries[J]. Small, 2019, 15(32): 1805381.

[16]  Liu R, Liang Z, Gong Z. et al., Research Progress in Multielectron Reactions in Polyanionic Materials for Sodium-Ion Batteries[J]. Small Methods, 2019, 3(4): 1800221.

[17]  Qian J, Wu C, Cao Y. et al., Prussian Blue Cathode Materials for Sodium-Ion Batteries and Other Ion Batteries[J]. Advanced Energy Materials, 2018, 8(17): 1702619.

[18]  方铮, 曹余良, 胡勇胜, 等. 室温钠离子电池技术经济性分析[J]. 储能科学与技术, 2016, 5(2): 149–158.

[19]  Vaalma C, Buchholz D, Weil M. et al., A Cost and Resource Analysis of Sodium-Ion Batteries[J]. Nature Reviews Materials, 2018, 3(4): 18013.

[20]  Okoshi M, Yamada Y, Yamada A. et al., Theoretical Analysis on De-Solvation of Lithium, Sodium, and Magnesium Cations to Organic Electrolyte Solvents[J]. Journal of the Electrochemical Society, 2013, 160(11): A2160.

[21]  Yamada Y, Iriyama Y, Abe T. et al., Kinetics of Lithium Ion Transfer at the Interface between Graphite and Liquid Electrolytes: Effects of Solvent and Surface Film[J]. Langmuir, 2009, 25 (21): 12766–12770.

[22]  Yamada Y, Koyama Y, Abe T. et al., Correlation between Charge–Discharge Behavior of Graphite and Solvation Structure of the Lithium Ion in Propylene Carbonate-Containing Electrolytes[J]. Journal of Physical Chemistry C, 2009, 113(20): 8948–8953.

[23]  Ong S P, Chevrier V, Hautier G. et al., Voltage, Stability and Diffusion Barrier Differences between Sodium-ion and Lithium-ion Intercalation Materials[J]. Energy & Environmental Science, 2011, 4(9): 3680–3688.

[24]  Kuratani K, Uemura N, Senoh H. et al., Conductivity, Viscosity and Density of $MClO_4$ (M = Li and Na) Dissolved in Propylene Carbonate and γ-Butyrolactone at High Concentrations[J]. Journal of Power Sources, 2013, 223: 175–182.

[25]  Yabuuchi N, Kubota K, Dahbi M. et al., Research Development on Sodium-Ion Batteries[J]. Chemical Reviews, 2014, 114(23): 11636–11682.

[26]  Yu C, Park J, Jung H. et al., NaCrO$_2$ Cathode for High-Rate Sodium-Ion Batteries[J]. Energy & Environmental Science, 2015, 8(7): 2019–2026.

[27]  Ren W, Zheng Z, Xu C. et al., Self-Sacrificed Synthesis of Three-Dimensional Na$_3$V$_2$(PO4)$_3$ Nanofiber Network for High-Rate Sodium-Ion Full Batteries[J]. Nano Energy, 2016, 25: 145–153.

[28]  Zhu Z, Cheng F, Zhe H. et al., Highly Stable and Ultrafast Electrode Reaction of Graphite for Sodium Ion Batteries[J]. Journal of Power Sources, 2015, 293: 626–634.

[29]  Goktas M, Bolli C, Berg E J. et al., Graphite as Cointercalation Electrode for Sodium-Ion Batteries: Electrode Dynamics and the Missing Solid Electrolyte Interphase (SEI)[J]. Advanced Energy Materials, 2018, 8(16): 1702724.

[30]  Jache B, Adelhel P. Use of Graphite as a Highly Reversible Electrode with Superior Cycle Life for Sodium-Ion Batteries by Making Use of Co-Intercalation Phenomena[J]. Angewandte Chemie, 2015, 126(38): 10333–10337.

[31]  Wen Y, He K, Zhu Y. et al., Expanded Graphite as Superior Anode for Sodium-Ion Batteries[J]. Nature Communications, 2014, 5(1): 1–10.

[32]  Chen J, FAN X, JI X. et al., Intercalation of Bi Nanoparticles into Graphite Enables Ultra-Fast and Ultra-Stable Anode Material for Sodium-Ion batteries[J]. Energy & Environmental Science, 2018, 11(5): 1218–1225.

[33]  Tarascon J M, Armand M. Issues and Challenges Facing Rechargeable Lithium Batteries.[J]. Nature, 2001, 414(6861): 359–367.

[34]  Xu J, Thomas H R, Francis R W. et al., A Review of Processes and Technologies for the Recycling of Lithium-Ion Secondary Batteries[J]. Journal of Power Sources, 2008, 177(2): 512–527.

[35]  Gies E. Recycling: Lazarus Batteries[J]. Nature, 2015, 526(7575): S100.

[36]  Liu T, Zhang Y, Chen C. et al., Sustainability-Inspired Cell Design for a Fully Recyclable Sodium Ion Battery[J]. Nature Communications, 2019, 10(1): 1–7.

[37]  陈义鹏, 焦斌亮, 王海明, 等. 光伏发电储能电池放电控制的温度补偿研究[J]. 电源技术, 2014, 38(6): 1078–1080.

[38]  Liu T, Wang B, Gu X. et al., All-Climate Sodium Ion Batteries Based on the NASICON Electrode Materials[J]. Nano Energy, 2016, 30: 756–761.

[39]  Ponrouch A, Palacin M R. On the High and Low Temperature Performances of Na-Ion Battery Materials: Hard Carbon as A Case Study[J]. Electrochemistry Communications, 2015, 54: 51–54.

[40]  You Y, Yao H, Xin S. et al., Subzero-Temperature Cathode for a Sodium-Ion Battery[J]. Advanced Materials, 2016, 28(33): 7243–7248.

[41]  Guo J, Wang P, Wu X. et al., High-Energy/Power and Low-Temperature Cathode for Sodium-Ion Batteries: In Situ XRD Study and Superior Full-Cell Performance[J]. Advanced Materials, 2017, 29(33): 1701968.

[42]  Ponrouch A, Marchante E, Courty M. et al., In Search of an Optimized Electrolyte for Na-Ion Batteries[J]. Energy & Environmental Science, 2012, 5(9): 8572–8583.

[43]  Sun Y, Shi P, Xiang H. et al., High-Safety Nonaqueous Electrolytes and Interphases for Sodium-Ion Batteries[J]. Small, 2019, 15(14): 1805479.

[44]  Ge P, Fouletier M. Electrochemical Intercalation of Sodium in Graphite[J]. Solid State Ionics, 1988, s28–30((part-P2)): 1172–1175.

[45]  Doeff M M, Ma Y, Visco S J. et al., Electrochemical Insertion of Sodium into Carbon[J]. Journal of the Electrochemical Society, 1993, 140(12): L169.

[46] Liu Z, Huang Y, Huang Y. et al., Voltage Issue of Aqueous Rechargeable Metal-Ion Batteries[J]. Chemical Society Reviews, 2020, 49(1): 180–232.

[47] Suo L, Borodin O, Wang Y. et al., "Water-in-salt" Electrolyte Makes Aqueous Sodium-Ion Battery Safe, Green, and Long-Lasting[J]. Advanced Energy Materials, 2017, 7(21): 1701189.

[48] 2018年可再生能源并网运行情况介绍[EB/OL]. 国家能源局. 2019-01-28. http://www.nea.gov. cn/2019-01/28/c_137780519.htm.

[49] Deng J, Luo W, Chou S. et al., Sodium-Ion Batteries: From Academic Research to Practical Commercialization[J]. Advanced Energy Materials, 2017, 8(4): 1701428.

[50] Komaba S, Murata W, Ishikawa T. et al., Electrochemical Na Insertion and Solid Electrolyte Interphase for Hard-Carbon Electrodes and Application to Na-Ion Batteries[J]. Advanced Functional Materials, 2011, 21(20): 3859–3867.

[51] Zhu C, Kopold P, Van Aken P A. et al., Sodium-Ion Batteries: High Power-High Energy Sodium Battery Based on Threefold Interpenetrating Network[J]. Advanced Materials, 2016, 28(12): 2408–2408.

[52] Wang Y, Xiao R, Hu Y. et al., P2-$Na_{0.6}[Cr_{0.6}Ti_{0.4}]O_2$ Cation-Disordered Electrode for High-Rate Symmetric Rechargeable Sodium-Ion Batteries[J]. Nature Communications, 2015, 6(1): 6954–6955.

[53] Guo S, Yu H, Liu P. et al., High-Performance Symmetric Sodium-Ion Batteries Using a New, Bipolar $O_3$-Type Material, $Na_{0.8}Ni_{0.4}Ti_{0.6}O_2$[J]. Energy & Environmental Science, 2015, 8(4): 1237–1244.

[54] Zhang B, Dugas R, Rousse G. et al., Insertion Compounds and Composites Made by Ball Milling for Advanced Sodium-Ion Batteries[J]. Nature Communications, 2016, 7(1): 10308–10308.

[55] Sathiya M, Thomas J, Batuk D. et al., Dual Stabilization and Sacrificial Effect of $Na_2CO_3$ for Increasing Capacities of Na-Ion Cells Based on P2-$Na_xMO_2$ Electrodes[J]. Chemistry of Materials, 2017, 29(14): 5948–5956.

[56] Wan F, Guo J Z, Zhang X H. et al., In-Situ-Binding Sb Nanospheres on Graphene via Oxygen Bonds as Superior Anode for Ultrafast Sodium Ion Batteries[J]. ACS Applied Materials & Interfaces, 2016, 8(12): 7790–7799.

[57] Walter M, Doswald S, Kovalenko M V. Inexpensive Colloidal SnSb Nanoalloys as Efficient Anode Materials for Lithium- and Sodium-Ion Batteries[J]. Journal of Materials Chemistry A, 2016, 4(18): 7053–7059.

[58] Liu J, Yang Z, Wang J. et al., Three-Dimensionally Interconnected Nickel–Antimony Intermetallic Hollow Nanospheres as Anode Material for High-Rate Sodium-Ion Batteries[J]. Nano Energy, 2015, 16: 389–398.

[59] Li H, Peng L, Zhu Y. et al., An Advanced High-Energy Sodium Ion Full Battery Based on Nanostructured $Na_2Ti_3O_7$/$VOPO_4$ Layered Materials[J]. Energy & Environmental Science, 2016, 9(11): 3399–3405.

[60] Kim H, Hong J, Park Y. et al., Sodium Storage Behavior in Natural Graphite Using Ether-based Electrolyte Systems[J]. Advanced Functional Materials, 2015, 25(4): 534–541.

[61] Yuang D, Liang X, Wu L. et al., A Honeycomb-Layered $Na_3Ni_2SbO_6$: A High-Rate and Cycle-Stable Cathode for Sodium-Ion Batteries[J]. Advanced Materials, 2014, 45(46): 6301–6306.

[62] Wang N, Bai Z, Qian Y. et al., Double-Walled $Sb@TiO_{2-x}$ Nanotubes as a Superior High-Rate and Ultralong-Lifespan Anode Material for Na-Ion and Li-Ion Batteries[J]. Advanced Materials, 2016, 28(21): 4126–4133.

[63] Yu D Y, Prikhodchenko P V, Mason C W. et al., High-Capacity Antimony Sulphide Nanoparticle-Decorated Graphene Composite as Anode for Sodium-Ion batteries[J]. Nature Communications, 2013, 4(1): 2922–2922.

[64] Zhang K, Hu Z, Liu X. et al., FeSe$_2$ Microspheres as a High-Performance Anode Material for Na-Ion Batteries[J]. Advanced Materials, 2015, 46(31): 3305–3309.

[65] Komaba S, Murata W, Ishikawa T. et al., Electrochemical Na Insertion and Solid Electrolyte Interphase for Hard-Carbon Electrodes and Application to Na-Ion Batteries[J]. Advanced Functional Materials, 2011, 21(20): 3859–3867.

[66] Li Y, Yang Z, Xu S. et al., Air-Stable Copper-Based P2-Na$_{7/9}$Cu$_{2/9}$Fe$_{1/9}$Mn$_{2/3}$O$_2$ as a New Positive Electrode Material for Sodium-Ion Batteries[J]. Advanced Science, 2015, 2(6): 1500031.

[67] Mu L, Xu S, Li Y. et al., Prototype Sodium-Ion Batteries Using an Air-Stable and Co/Ni-Free O3-Layered Metal Oxide Cathode[J]. Advanced Materials, 2015, 27(43): 6928–6933.

[68] Zhang B, Dugas R, Rousse G. et al., Insertion Compounds and Composites Made by Ball Milling for Advanced Sodium-Ion Batteries.[J]. Nature Communications, 2016, 7(1): 10308–10308.

[69] Wang L, Song J, Qiao R. et al., Rhombohedral Prussian White as Cathode for Rechargeable Sodium-Ion Batteries[J]. Journal of the American Chemical Society, 2015, 137(7): 2548–2254.

[70] Song J, Wang L, Lu Y. et al., Removal of Interstitial H$_2$O in Hexacyanometallates for a Superior Cathode of a Sodium-Ion Battery[J]. Journal of the American Chemical Society, 2015, 137(7): 2658–2664.

[71] Abouimrane A, Weng W, Eltayeb H. et al., Sodium Insertion in Carboxylate Based Materials and Their Application in 3.6 V Full Sodium Cells[J]. Energy & Environmental Science, 2012, 5(11): 9632–9638.

[72] Wang H, Hu P, Yang J. et al., Renewable-Juglone-Based High-Performance Sodium-Ion Batteries[J]. Advanced Materials, 2015, 27(14): 2348–2354.

[73] Hou Z, Li X, Liang J. et al., An Aqueous Rechargeable Sodium Ion Battery Based on a NaMnO$_2$-NaTi$_2$(PO$_4$)$_3$ Hybrid System for Stationary Energy Storage[J]. Journal of Materials Chemistry A, 2015, 3(4): 1400–1404.

[74] Li Z, Young D, Xiang K. et al., Towards High Power High Energy Aqueous Sodium-Ion Batteries: The NaTi$_2$(PO$_4$)$_3$/Na$_{0.44}$MnO$_2$ System[J]. Advanced Energy Materials, 2013, 3(3): 290–294.

[75] Liu Z, Pang G, Dong S. et al., An Aqueous Rechargeable Sodium–Magnesium Mixed Ion Battery Based on NaTi$_2$(PO$_4$)$_3$–MnO$_2$ system[J]. Electrochimica Acta, 2019, 311: 1–7.

[76] Nakamoto K, Kano Y, Kitajou A. et al., Electrolyte Dependence of the Performance of a Na$_2$FeP$_2$O$_7$//NaTi$_2$(PO$_4$)$_3$ Rechargeable Aqueous Sodium-Ion Battery[J]. Journal of Power Sources, 2016, 327: 327–332.

[77] Kumar P R, Kheireddine A, Nisar U. et al., Na$_4$MnV(PO$_4$)$_3$-rGO as Advanced Cathode for Aqueous and Non-Aqueous Sodium Ion Batteries[J]. Journal of Power Sources, 2019, 429: 149–155.

[78] Wu X, Cap Y, Ai X. et al., A Low-Cost and Environmentally Benign Aqueous Rechargeable Sodium-Ion Battery Based on NaTi$_2$(PO$_4$)$_3$-Na$_2$NiFe(CN)$_6$ Intercalation Chemistry[J]. Electrochemistry Communications, 2013, 31: 145–148.

[79] Pasta M, Wessells C D, Liu N. et al., Full Open-Framework Batteries for Stationary Energy Storage[J]. Nature Communications, 2014, 5(1): 3007.

[80] Whitacre J, Tevar A, Sharma S. Na$_4$Mn$_9$O$_{18}$ as a Positive Electrode Material for an Aqueous Electrolyte Sodium-Ion Energy Storage Device[J]. Electrochemistry Communications, 2010, 12(3): 463–466.

[81] Whitacre J, Wiley T, Sharma S. et al, An Aqueous Electrolyte, Sodium Ion Functional, Large Format Energy Storage Device for Stationary Applications[J]. Journal of Power Sources, 2012, 213: 255–264.

[82] Zhang B H, Liu Y, Chang Z. et al., Nanowire Na$_{0.35}$MnO$_2$ from A Hydrothermal Method as A Cathode Material for Aqueous Asymmetric Supercapacitors[J]. Journal of Power Sources, 2014, 253: 98–103.

[83] Feng Y, Zhang Q, Liu S. et al., A Novel Aqueous sodium–Manganese Battery System for Energy Storage[J]. Journal of Materials Chemistry A, 2019, 7(14): 8122–8128.

[84] Zhang S, Liu Y, Han Q. et al., Development and Characterization of Aqueous Sodium-Ion Hybrid Supercapacitor Based on NaTi$_2$(PO$_4$)$_3$//Activated Carbon[J]. Journal of Alloys and Compounds, 2017, 729: 850–857.

[85] Kong Y, Sun J, Gai L. et al., NaTi$_2$(PO$_4$)$_3$/C//LiMn$_2$O$_4$ Rechargeable Battery Operating with Li$^+$/Na$^+$-Mixed Aqueous Electrolyte Exhibits Superior Electrochemical Performance[J]. Electrochimica Acta, 2017, 255: 220–229.

[86] Dong X, Chen L, Liu J. et al., Environmentally-Friendly Aqueous Li (Or Na)-Ion Battery with Fast Electrode Kinetics and Super-Long Life[J]. Science Advances, 2016, 2(1): e1501038.

[87] Huang Y, Zhao L, Li L. et al., Electrolytes and Electrolyte/Electrode Interfaces in Sodium-Ion Batteries: From Scientific Research to Practical Application[J]. Advanced Materials, 2019, 31(21): 1808393.

[88] Sun H, Guo J, Zhang Y. et al., High-Voltage All-Solid-State Na-Ion-Based Full Cells Enabled by All Nasicon-Structured Materials[J]. ACS Applied Materials & Interfaces, 2019, 11(27): 24192–24197.

[89] Li H, Ding Y, Ha H, et al. An All-Stretchable-Component Sodium-Ion Full Battery[J]. Advanced Materials, 2017, 29(23): 1700898.

[90] Zhou W, Li Y, Xin S. et al., Rechargeable Sodium All-Solid-State Battery[J]. ACS Central Science, 2017, 3(1): 52–57.

[91] Gao H, Xin S, Xue L. et al., Stabilizing a High-Energy-Density Rechargeable Sodium Battery with a Solid Electrolyte[J]. Chem, 2018, 4(4): 833–844.

[92] 马强, 胡勇胜, 李泓, 等. 双(三氟甲基磺酰)亚胺钠基聚合物电解质在固态钠电池中的性能[J]. 物理化学学报, 2018, 34(2): 213–218.

[93] De Anastro A F, Lago N, Berlanga C. et al., Poly(ionic Liquid) Ion-Gel Membranes for All Solid-State Rechargeable Sodium Battery[J]. Journal of Membrane Science, 2019, 582: 435–441.

[94] Amin K, Meng Q, Ahmad A. et al., A Carbonyl Compound-Based Flexible Cathode with Superior Rate Performance and Cyclic Stability for Flexible Lithium-Ion Batteries[J]. Advanced Materials, 2018, 30(4): 1703868.

[95] Liu Y, Zhang P, Sun N. et al., Self-Assembly of Transition Metal Oxide Nanostructures on MXene Nanosheets for Fast and Stable Lithium Storage[J]. Advanced Materials, 2018, 30(23): 1707334.

[96] Yu L, Hu L, Anasori B. et al., MXene-Bonded Activated Carbon as A Flexible Electrode for High-Performance Supercapacitors. ACS Energy Letters, 2018, 3(7): 1597–1603.

[97] Chang X, Zhu Q, Sun N. et al., Graphene-Bound Na$_3$V$_2$(PO$_4$)$_3$ Film Electrode with Excellent Cycle and Rate Performance for Na-Ion Batteries[J]. Electrochimica Acta, 2018, 269: 282–290.

[98] Wang Y, Chen C, Xie H. et al., 3D-Printed All-Fiber Li-Ion Battery toward Wearable Energy Storage[J]. Advanced Functional Materials, 2017, 27(43): 1703140.

[99] Kohlmeyer R R, Blake A J, Hardin J O. et al., Composite Batteries: A Simple yet Universal Approach to 3D Printable Lithium-Ion Battery Electrodes[J]. Journal of Materials Chemistry, 2016, 4(43): 16856–16864.

[100] Kim S, Choi K, Cho S. et al., Printable Solid-State Lithium-Ion Batteries: A New Route toward Shape-Conformable Power Sources with Aesthetic Versatility for Flexible Electronics[J]. Nano Letters, 2015, 15(8): 5168–5177.

[101] Hwang C, Song W, Han J. et al., Foldable Electrode Architectures Based on Silver-Nanowire-Wound or Carbon-Nanotube-Webbed Micrometer-Scale Fibers of Polyethylene Terephthalate Mats for Flexible Lithium-Ion Batteries[J]. Advanced Materials, 2018, 30(7): 1705445.

# Index

https://doi.org/10.1515/9783110749069-007

www.ingramcontent.com/pod-product-compliance
Lightning Source LLC
Chambersburg PA
CBHW080712220326

41598CB00033B/5396